Lecture Notes in Statistic

Edited by S. Fienberg, J. Gani, K. Kri(
I. Olkin, and N. Wermuth

P. Cheeseman
R. W. Oldford (Eds.)

Selecting Models from Data

Artificial Intelligence and Statistics IV

Springer-Verlag
New York Berlin Heidelberg London Paris
Tokyo Hong Kong Barcelona Budapest

P. Cheeseman
Mailstop 269-2
NASA
Ames Research Center
Moffett Field, CA 94035
USA

R.W. Oldford
Department of Statistics
and Actuarial Science
University of Waterloo
Waterloo Ontario, N2L 3G1
CANADA

Library of Congress Cataloging-in-Publication Data
Selecting models from data : artificial intelligence and statistics IV/
P. Cheeseman, R.W. Oldford (eds.)
p. cm. - - (Lecture notes in statistics ; 89)
Includes bibliographical references and index.
ISBN 0-387-94281-5 (New York : acid-free paper). - - ISBN
3-540-94281-5 (Berlin : acid-free paper)
1. Artificial intelligence - - Mathematical models. 2. Artificial
intelligence - - Statistical methods. 3. Statistics. I. Cheeseman,
P. II. Oldford, R. W. III. Series: Lecture notes in statistics
(Springer-Verlag) ; v. 89.
Q335.S4119 1994
519.5'4'028563 - - dc20 94-10820

Printed on acid-free paper.

Camera ready copy provided by the author.
Printed and bound by Braun-Brumfield, Inc., Ann Arbor, MI.
Printed in the United States of America.

9 8 7 6 5 4 3 2 1

ISBN 0-387- 94281-5 Springer-Verlag New York Berlin Heidelberg
ISBN 3-540- 94281-5 Springer-Verlag Berlin Heidelberg New York

Preface

This volume is a selection of papers presented at the Fourth International Workshop on Artificial Intelligence and Statistics held in January 1993. These biennial workshops have succeeded in bringing together researchers from Artificial Intelligence and from Statistics to discuss problems of mutual interest. The exchange has broadened research in both fields and has strongly encouraged interdisciplinary work. The theme of the 1993 AI and Statistics workshop was: "Selecting Models from Data".

The papers in this volume attest to the diversity of approaches to model selection and to the ubiquity of the problem. Both statistics and artificial intelligence have independently developed approaches to model selection and the corresponding algorithms to implement them. But as these papers make clear, there is a high degree of overlap between the different approaches. In particular, there is agreement that the fundamental problem is the avoidence of "overfitting"–i.e., where a model fits the given data very closely, but is a poor predictor for new data; in other words, the model has partly fitted the "noise" in the original data.

The excitement of the AI and statistics workshop, reflected in these papers, comes from the realization that computers will increasingly be required to draw robust inferences from data, sometimes very large quantities of data, despite the presence of incomplete and inaccurate information. And, because the scale of the problems arising from large computer databases quickly overwhelms the human analyst, it is desirable to have a computer assume as much of the role of the analyst as possible. This requires us to have methodological and domain expert information encoded in such a way that it can be used to produce useful results automatically.

Pioneering efforts in this regard have been made in both the Statistics and the AI communities. Statisticians have developed computationally intensive methods for building flexible and interpretable models. AI, particularly machine learning, has pioneered computer based methods of inference from databases with many notible sucesses, as well as a few failures. One solid inference that can be drawn from these workshops is that AI could benefit from making statistical methods the heart of future programs.

It seems to us that there is enormous potential for development at the intersection of statistics, artificial intelligence and computer science. The papers in this book constitute an important step in this direction.

Finally, the production of a volume of this nature requires the dedicated effort of many people. We owe considerable thanks to the programme committee for their effort in the development of the scientific programme, to the Society for AI and Statistics who initiate and finance the workshops, to Nandanee Basdeo who handled the electronic submission of papers and who bore much of the burden for the logistics of this volume and the workshop, to Gwen Sharp who ably handled much of the workshop administration, and to the Department of Statistics and Actuarial Science at the University of Waterloo which was generous with its staff and computational resources in the production of this volume. Finally, we would like to thank the authors in this volume and the workshop participants who have made the workshops a success.

P. Cheeseman
Ames, California

R.W. Oldford
Waterloo, Ontario

January, 1994.

Previous workshop volumes:

Gale, W.A.,ed.(1986) *Artificial Intelligence and Statistics*, Reading, MA: Addison-Wesley.
Hand, D.J., ed. (1990) Artificial Intelligence and Statistics II. *Annals of Mathematics and Artificial Intelligence*, **2**.
Hand, D.J., ed. (1993) *Artificial Intelligence Frontiers in Statistics: AI and Statistics III*, London, UK: Chapman & Hall.

1993 Programme Committee:

General Chair:	R.W. Oldford	U. of Waterloo, Canada
Programme Chair:	P. Cheeseman	NASA (Ames), USA
Members:		
	W. Buntine	NASA (Ames), USA
	M. Deutsch-McLeish	U. of Guelph, Canada
	Wm. Dumouchel	Columbia U, USA
	W.A. Gale	AT&T Bell Labs, USA
	D.J. Hand	Open University, UK
	H. Lenz	Free University, Germany
	D. Lubinsky	AT&T Bell Labs, USA
	E. Neufeld	U. of Saskatchewan, Canada
	J. Pearl	UCLA, USA
	D. Pregibon	AT&T Bell Labs, USA
	P. Shenoy	U. of Kansas, USA
	P. Smythe	JPL, USA

Sponsors: Society for AI And Statistics

International Association for Statistical Computing

Contents

Part I

Overviews: Model Selection

1

Statistical strategy: step 1

D.J.Hand

Department of Statistics
Faculty of Mathematics
The Open University
Milton Keynes
MK7 6AA

ABSTRACT Before one can select a model one must formulate the research question that the model is being built to address. This formulation – deciding precisely what it is one wants to know – is the first step in using statistical methods. It is the first step in any statistical strategy. This paper contends that often the research question is poorly formulated, so that inappropriate models are built and inappropriate analyses undertaken. To illustrate this three examples are given: explanatory versus pragmatic comparisons in clinical trials, confused definitions of interaction, and two ways to measure relative change. A plea is made that statistics teaching should focus on higher level strategic issues, rather than mathematical manipulation since the latter is now performed by the computer.

1.1 Introduction

The primary theme of this conference is model selection. Many other papers here present and compare strategies and criteria for such selection. This paper, however, takes a step back, to begin with the question 'Why do we want to select a model in the first place?'

The answer, of course, is that we wish to apply the model in addressing some particular research question. For example, some papers at this meeting try to find the best model to answer the question 'What class does this object belong to?' and others address the question 'What are the causal influences on this variable?'

This means that, before one can even consider model selection methods one has to be quite clear what one wants the model for - - or what research question one is attempting to address with the model.

This process of formulating the research question is thus the first step in a 'statistical strategy'.

A statistical strategy is an explicit statement of the steps, decisions, and actions to be made during the course of a statistical analysis. Such explicit statements are, of course, necessary in order to be able to communicate **how** to do an analysis. One would have thought that this was necessary in order to teach statistics. It is certainly necessary if one intends to build a statistical expert system.

Strangely enough, most of the work on statistical strategy seems to have been motivated by this interest in statistical expert systems, with relatively little coming from other areas such as teaching (one exception is Cox and Snell, 1981). Indeed teaching seems to have concentrated on the formal aspects of statistical techniques. (A step in the right direction is the use of project work, though here no principles of strategy are introduced and it is really an attempt to introduce the 'apprentice' role described below into the teaching curriculum.)

---

[1] *Selecting Models from Data: AI and Statistics IV.* Edited by P. Cheeseman and R.W. Oldford. ©1994 Springer-Verlag.

One reason for this is undoubtedly the mathematical complexity of statistical methods: knowing when a technique should be used is of little value if one is incapable of undertaking the numerical manipulations. Moreover, the mathematical manipulations had to be understood in their entirety in order to apply them. Half a calculation of an estimator is of little value. In contrast, partial understanding of the role and type of questions a technique can address will allow one to apply the methods (often incorrectly, no doubt) so that it might appear that less effort is needed to teach these aspects.

Similarly, the mathematical aspects are easy to formalise (the mathematics itself is a formalisation) while the non-mathematical aspects are difficult to formalise, so that again the former lends itself to teaching. Perhaps we should here acknowledge the fact that statistics has unfortunately often been perceived as being a part of mathematics, so that in the teaching there will be a natural emphasis on the mathematical aspects at the expense of the other, less clearly defined but at least as important, aspects.

This has meant that while conventional statistics teaching has concentrated on the manipulative aspects of deriving estimators, conducting tests, fitting models, and so on, issues of when to apply what tools and what to do when things go wrong have been left to be learnt in an apprentice role, 'on the job'. It is as if one was being given the individual bricks and left to watch others on the building site to learn how the bricks should be put together to make the house.

However, all of this is changing under the impact of the computer. Nowadays statistical software handles the formal arithmetic and mathematical manipulations, so that a researcher undertaking an analysis need not understand the algorithms involved. A trivial example would be the calculation of least squares regression coefficients. The mathematics involves knowledge of differential calculus, and the arithmetic involves knowledge of how to calculate sums of products and squares. But if a computer is to undertake the calculations, what the researcher (or, more generally, the user of least squares regression) needs to know is that a sum of squared deviations criterion is used and that the resulting estimated regression coefficients are those that minimise this criterion. Being able to reproduce what the computer is doing (much faster than a human could do it) is all very well, but is of little practical value, whereas being able to formulate the research question and then being able to use the tool (the computer) in an appropriate way and place is vital.

Of course, we acknowledge here the importance of knowing what is going on inside the computer if one is concerned with developing new methodology. Most users of statistics, however, are not. They want to apply standard techniques (probably using standard software) – and they therefore need to know if their problems fit such techniques.

This, therefore, leads us to the first step of a statistical strategy: the clear formulation of research aims, and their expression in terms of statistical tools.

It is my contention that all too often clients seek statistical advice without a clear formulation of their research objectives and, perhaps worse, statisticians undertake analyses without clarifying these objectives, so that inappropriate or incorrect conclusions are often drawn.

This conference is the fourth in the series. The previous papers I have presented represent steps towards my present position (Hand, 1986, 1990, 1992). The first was on high level statistical strategy, attempting to illustrate with a strategy for multivariate analysis of variance. Then I thought that the difficulties in formulating strategies, while by no means trivial, could be tackled by straightforward attempts to formalise statisticians' actions in tackling problems. The second paper described the construction and experiences with systems for helping researchers apply statistics to problems. I had here stepped back from the notion of an 'expert system' and described statistical 'knowledge enhancement systems', aimed at supporting the user's knowledge rather than providing a whole additional layer of expertise. The third focussed on the use of metadata in statistical expert systems, but concluded: 'it is not clear that the researcher involved can always make sufficiently precise statements about the research objectives, and a statistician is needed to identify subtle differences between questions.' In the present paper I go a step further and suggest that even statisticians often do not think carefully enough about the initial step in a strategy – precisely what it is the researcher wants to know. This paper presents some examples to illustrate this contention.

1.2 Some examples

1.2.1 Pragmatic versus explanatory comparisons

The distinction between pragmatic and explanatory studies is best known in medical areas, but it is in fact ubiquitous. It is discussed at length by Schwartz, Flamant, and Lellouch (1980). In brief, and in medical terms, a pragmatic trial is concerned with the practical effectiveness of a treatment, as administered in the way it would be used in practice, while an explanatory trial aims at discovering real biological differences. Although perhaps superficially identical, these two questions in fact are very different. The differences have consequence for both design and analysis and yet the distinction is all too rarely made in statistical practice.

Consider, for example, a comparison between two propose treatments. In a pragmatic study the overriding concern would be that one should not identify the poorer treatment as the better. (Such errors have been termed 'Type III' errors.) In such a situation the Type I error would not matter (if the two treatments were equally effective – the null hypothesis – then it would not matter which was chosen). In contrast, if the study was explanatory then it would be important to keep both Type I error and Type II error small. Since the two types of question have different error structures, we might expect that different sample sizes would be needed to address the questions.

In many studies, especially in clinical trials, measurements are taken over a period of time so that withdrawals from the programme can cause a problem. In a pragmatic study one might decide to retain the dropouts in the analysis, coding such cases as treatment failures – they certainly are not successes. In contrast, in an explanatory study, one wishes to make statements about patients who have stuck rigorously to the treatment regimen, and hence would exclude such dropouts from the analysis. (This is, of course, something of a simplification, and in practice one should consider carefully precisely what the objectives are.)

Yet another difference arises in the subjects being studied. In a pragmatic study one is hoping to make inferences which apply to a future population to which the treatments may be given. The test sample should thus be sampled from that population and should adequately represent it – including all its heterogeneities. In contrast, in an explanatory study one will aim to minimise variation – and hence use as homogeneous a sample as possible (relying on implicitly non-statistical inferences to subjects which differ from those in the sample?).

Schwartz et al (1980) give the following striking example of the difference between the two types of research question. Consider a study to explore the effectiveness of a sensitising drug prior to radiotherapy in treating cancer. We wish to compare two treatments:

T1: radiotherapy alone.

T2: radiotherapy preceded for 30 days by a sensitising drug.

In an explanatory trial one will wish to see how the sensitising drug influences the outcome, without there being any other differences between the two treatment groups apart from the fact that one has used the drug and the other has not. Now, group T2 have to wait 30 days before they can commence radiotherapy. Therefore, to make the groups comparable in this regard, group T1 should also have to wait 30 days. Presumably they should be given a placebo medication throughout this period.

In contrast, in a pragmatic trial one wants to compare the treatments that would actually be used in practice. And in practice, the radiotherapy in T1 would commence immediately, and not after 30 days treatment with a drug known to have no effect!

The two questions, explanatory or pragmatic, are quite distinct and it is essential for the researchers to know which they wish to address even before the data can be collected.

1.2.2 Interaction

Table 1, from Brown and Harris (1978), shows a cross- classification of a sample according to whether or not they had recently experienced a life event, whether or not they had a close relationship with a husband or boyfriend, and whether or not they developed depression. The hypothesis was that a close relationship was protective to some extent against the effect of life events in inducing depression. To explore this hypothesis, Brown and Harris computed the proportions developing depression in each of the four conditions resulting from the cross-classification of the relationship variable and the life event variable, to give Table 2. From this table it is easy to see that $(0.32 - 0.03) = 0.29$ is not equal to $(0.11 - 0.01) = 0.10$, and Brown and Harris consequently concluded that there was an interaction in the effect of these variables on the probability of suffering from depression.

Analysing the same data, however, Tennant and Bebbington (1978) questioned their conclusion. They reasoned that since $0.32/0.03 = 10.7$ was approximately equal to $0.11/0.01 = 11.0$ there was no evidence for an interaction.

The resolution, of course, lies in the definition of interaction. That is, it lies in making precise the initial research question. Brown and Harris are using an additive model whereas Tennant and Bebbington are using a multiplicative model. As with the pragmatic/explanatory distinction outlined above, attempting to choose a model is futile unless one considers why one wants it – precisely what question one intends to use the model to address.

Another example of the confusion arising from the distinction between additive and multiplicative models arises in investigations of the interaction between medicines. A combination of two drugs may be more effective than if the drugs act independently (in which case the combination is said to exhibit 'synergy') or less effective than if the drugs act independently (and the combination is said to exhibit 'antagonism'). For example, childhood acute lymphatic leukaemia shows a 40-50% remission with individual drugs but a 94-95% remission with a combination of three drugs. However, these sorts of statements (and, indeed, the definitions of 'synergy' and 'antagonism') are only meaningful if there is a clearly defined 'independence' model with which to compare the results of the combination. Unfortunately there is no unique definition and many proposals have been made.

Obvious and common ones are the additive and multiplicative ones. Let $E(da)$ signify the effect of a dose da of drug A, $E(db)$ that of a dose b of drug B, and $E(da, db)$ that of a combination of dose da of drug A with db of drug B. Then the additive definition says that there

Table 1. Depression and life events

	No close relationship		Close relationship	
	Event	No event	Event	No event
Depression	24	2	10	2
No depression	52	60	78	191

Table 2. Proportions developing depression

	Event	No event
No close relationship	0.32	0.03
Close relationship	0.11	0.01

is no interaction between the two drugs if

$$E(da, db) = E(da) + E(db) \tag{M1}$$

and the multiplicative definition says that there is no interaction if

$$S(da, db) = S(da) * S(db) \tag{M2}$$

where $S = 1 - E$, defined when E is a fractional effect.

Suppose, however, that $E(da) = E(db) = 50\%$ and $E(da, db) = 90\%$. Then measure $M1$ (with 100% expected under the no interaction model) suggests that the combination is antagonistic while measure $M2$ (with 75% expected under the no interaction model) suggests that the combination is synergistic.

Measure $M1$ is satisfactory if both drugs have linear dose/response relationships and $M2$ is satisfactory if both have exponential dose-response relationships. However, neither is generally valid: neither works if the dose-response relationship is (for example) sigmoidal and neither works if the relationship differs between the drugs.

1.2.3 Measuring relative change

Consider the exchange rate between the £ and the \$. At the time of writing it is about $1 = \$1.5$. Not so long ago, however, it was $1 = \$2$. This means that the £ has dropped to 75% $(= 1.5/2)$ of its previous value relative to the \$. In contrast, however, the \$ has increased to 133% $(= (1/1.5)/(1/2))$ of its previous value. Thus while one currency has lost only 25% in value the other has gained 33%. Which of these it is appropriate to use clearly depends on the question one is trying to answer (or, perhaps, the point one is trying to make?). Tornqvist, Vartia, and Vartia (1985) have more to say about this.

1.3 Conclusion

The purpose of these examples has been to illustrate the assertion that often too little care is taken in formulating the reasons for constructing a model, with the consequence that sometimes an inappropriate model may be selected. Resolution of the problem will not be achieved through more sophisticated search algorithms or ways of playing off accuracy against numbers of parameters. Before such considerations arise it is necessary to decide what the model is for – what question it is being constructed to answer.

In the introductory section it was remarked that adequate formulation of the research question is the first step in any statistical strategy. Given that, it is legitimate to ask how far back one can push the 'question formulation' aspects without requiring extensive application domain knowledge. That is, what general aspects are there which might be used in the question formulation stage? Hand (1993) has considered this in more detail. Certain basic issues can be identified. These are things which all statisticians will know about but which would benefit from being explicitly stated in the early discussions with researchers.

One example is the issue of control: when deciding on the appropriate model form, what other variable, terms, or contrasts should be controlled for? For example, should one use simple or multiple regression? Both result in legitimate models; they merely address different issues. The same phenomenon is evident in Simpson's paradox, which arises because of a confusion between two questions, one conditional and the other unconditional. Another way of looking at this issue, and another interpretation which has wide applicability, is to ask whether one wishes the results of the analysis to be a statement about populations or about individuals.

Trying to identify basic issues and explicitly consider them during the question formulation stage is one strategy for alleviating the difficulties. Another is to try to formulate a model such that the different questions are equivalent under the model assumptions. A trivial example would be the case when the researcher cannot decide whether the mean or median is an appropriate summary statistic to use in addressing the question 'Which of two groups has the larger average?' If it is legitimate to assume symmetry of the distributions then the questions become logically equivalent (and one can then choose on other grounds, such as power of the test). The model – that the distributions are symmetric – has meant that the issue of precisely which question is being asked is irrelevant. Of course, in using this approach, one must be confident that the chosen model is legitimate.

The computer has given us the freedom to concentrate on higher level strategic issues, instead of being forced to focus on lower level arithmetic and algebraic manipulations. These higher level issues are arguably even more important than the lower level issues. No matter how accurate an answer is, it is useless if it addresses the wrong question.

1.4 References

[1] Brown G.W. and Harris T. (1978) Social origins of depression: a reply. *Psychological Medicine*, *8*, p577-588.

[2] Cox D.R. and Snell E.J. (1981) *Applied statistics: principles and examples*. London: Chapman and Hall.

[3] Hand D.J. (1986) Patterns in statistical strategy. *In Artificial intelligence and statistics*, ed. W.A.Gale. Reading, Massachusetts: Addison-Wesley, p355-387.

[4] Hand D.J. (1990) Practical experience in developing statistical knowledge enhancement systems. *Annals of Mathematics and Artificial Intelligence*, *2*, p197-208.

[5] Hand D.J. (1992) Measurement scales as metadata. In *Artificial Intelligence Frontiers in Statistics*, ed. D.J.Hand. London, Chapman and Hall.

[6] Hand D.J. (1993) Deconstructing statistical questions. In preparation.

[7] Schwartz D., Flamanat R., and Lellouch J. (1980) (Trans. M.J.R.Healy) *Clinical Trials*. London: Academic Press.

[8] Tennant C. and Bebbington P. (1978) The social causation of depression: a critique of the work of Brown and his colleagues. *Psychological Medicine*, *8*, 565-575.

[9] Tornqvist L., Vartia P., and Vartia Y.O. (1985) How should relative changes be measured? *The American Statistician*, *39*, p43-46.

2

Rational Learning: Finding a Balance Between Utility and Efficiency

Jonathan Gratch, Gerald DeJong and Yuhong Yang

Beckman Institute for Advanced Studies
University of Illinois
405 N. Mathews
Urbana, IL 61801

Beckman Institute for Advanced Studies
and
Department of Statistics
Yale University
New Haven, CT 06520

ABSTRACT Learning is an important aspect of intelligent behavior. Unfortunately, learning rarely comes for free. Techniques developed by machine learning can improve the abilities of an agent but they often entail considerable computational expense. Furthermore, there is an inherent tradeoff between the power and efficiency of learning techniques. This poses a dilemma to a learning agent that must act in the world under a variety of resource constraints. This article considers the problem of *rational* learning algorithms that dynamically adjust their behavior based on the larger context of overall performance goals and resource constraints.

2.1 Introduction

The field of machine learning has developed a wide array of techniques for improving the effectiveness of performance elements. For example, learning techniques can discover effective classifiers and enhance the speed of planning systems. Learning techniques can take general performance systems and tailor them to the eccentricities of particular domains. Unfortunately, this capability is not without sometimes considerable cost. This expense arises from three sources: first is the cost of attaining a sufficiently large sample of training problems; second is the cost of properly configuring the learning technique (e.g. the determination of parameter settings) which can substantially influence performance; finally there is the computation complexity of the learning algorithm.

Specific learning algorithms can mitigate some of the sources of learning expense. Learning systems incorporate many compromises, or biases, to insure efficient learning. Typically these compromises appear as implicit design commitments, and are fixed into the implementation of learning techniques. These compromises impose tradeoffs between the efficiency and power of a learning technique - more powerful learning techniques generally require greater expense - and the fixed nature of these commitments limit generality. Ideally, a learning system would possess the flexibility to adapt its bias to the demands of a particular learning situation.

---

[0]This research is supported by the National Science Foundation under grant NSF-IRI-92-09394.

[1]*Selecting Models from Data: AI and Statistics IV.* Edited by P. Cheeseman and R.W. Oldford.

In this article we discuss the use of decision theory to control the behavior of a learning system. Many authors have illustrated the usefulness of decision theory as a framework for controlling inference [Doyle90, Horvitz89]. Decision theory provides a well-understood framework for expressing tradeoffs and reasoning about them under uncertain information. The work described in this article can be seen as an application of decision-theoretic meta-reasoning to the control of learning algorithms. We describe the general principles involved in such an approach. We then present an instantiation of these principles, showing how they can be applied within the COMPOSER learning approach, a statistical approach to improving the efficiency of a planning system [Gratch92a]. The extended system dynamically adjusts its learning behavior based on the resources available for learning.

2.2 Rationality in learning

The goal of learning is to tailor a performance element to be effective in some specific environment. For example, the performance element could be a planner where the input is problem specifications; the output is plans. The environment is simply the tasks faced by a performance element and adaptation to this environment must be judged against some criteria for success. In planning, criteria include accuracy, planning efficiency, and plan quality. We adopt a decision-theoretic view of learning. An environment is characterized by a probability distribution over the set of possible tasks. The user provides a *utility function* that specifies a criterion for success on individual tasks. The effectiveness of a performance element is characterized by its *expected utility* over the task distribution. This is the sum of the utility of each task weighted by the probability of a task's occurrence. For example, in classification problems, the standard utility function assigns a value of one to a correctly classified feature vector, and a value of zero to an incorrectly classified vector. The expected utility of a classifier is equivalent its accuracy over the distribution of feature vectors.

This view of learning facilitates a close analogy between learning and work in rational reasoning. In reasoning there is a decision procedure that must choose, from a set of possible actions, an action with high expected utility. In learning, there is a decision maker that must choose, from amongst a set of possible changes to a performance element, a change that produces a performance element with high expected utility. In fact, one might make the argument that reasoning subsumes learning, by allowing "learn something" to be one of the reasoner's actions. Nonetheless, there are issues unique to learning that justify their individual investigation as a specialization of general reasoning. For example, deliberating about possible actions involves past information on these action's effects. Deliberating about learning involves future information, and its impact on action deliberation.

In reasoning, a reasoner that always chooses the action with maximal expected utility exhibits *substantively rationality* [Simon76] (also called Type 1 rationality [Good71]). Similarly, we can define a *substantively rational learning system* as a system that always identifies the transformed performance element with maximum expected utility. Substantive rationality is seldom attainable in that it may require the use of infinite resources. This has led to a focus on rationality under limited resources. Simon refers to this as *procedural rationality* (also called Type 2 rationality [Good71]) because the focus is on identifying efficient procedures for making adequate decisions. A procedurally rational agent relaxes the strict requirements of substantive rationality in the interest of reasoning efficiency. The analogy between reasoning and learning applies here as well. It is rarely possible to expend sufficient resources for a learning system to find optimal solutions. Instead we demand that our learning techniques quickly identify adequate but possibly sub-optimal solutions.

2.2.1 *Fixed policy*

Since it is unrealistic to produce substantively rational learning systems, the question then becomes how best to limit the behavior of a learning technique. For non-trivial learning problems there is a clear tradeoff between the efficiency of a learning technique and expected improvements it provides. The typical approach is to adopt a fixed policy towards resolving this tradeoff. The learning system implementor commits to some fixed restrictions on the learning approach. For example, many learning systems use restrictive representation languages or make use of local search procedures.

Fixed policies can be effective in restricted situations. Unfortunately, the same policy may not apply equally well in all circumstances. We may demand different behavior from our learning system depending on if we have few or many resources to commit towards learning. The former requires a highly restricted learning technique while the latter would be better served by a more liberal policy.

2.2.2 *Parameterized policy*

An alternative is to provide flexibility in how a learning technique behaves. This can be seen as incorporating some *degrees of freedom* into the learning technique that must be resolved by the agent that utilizes the technique. An example is the user specified confidence parameter provided by PAC-learning techniques. Higher confidence requires more examples and thus higher learning cost. The user is free to resolve this tradeoff based on the demands of his or her particular circumstances. *Anytime algorithms* provide another example [Dean88]. These algorithms can be interrupted at any point with a useful result, and the results become monotonically better over time. With an anytime learning algorithm, the user can dictate how much time to expend on learning.

Parameterized policies can greatly enhance the flexibility of a learning technique, but they also place greater demands on the user. For this flexibility to be useful, the user must understand how this freedom affects the tradeoff between utility and efficiency. Sometimes the relationship is only apparent *after* learning has begun. For example, given an anytime learning algorithm, the decision of when to terminate might depend on the current rate of improvement (how is the expected utility of the performance element changing with learning time), or worse, on the future rate of improvement. If this information is not available to the user, the additional flexibility is only a burden.

2.2.3 *Rational policy*

Incorporating degrees of freedom into a learning system increases the flexibility of an approach, but it also increases the demands on the user. There may be a quite complex mapping between the goals of the user and the setting of the various learning parameters. Ideally, the user should be able to articulate his or her goals and leave it to the learning system to configure its policy to best satisfy these goals. If the learning system can estimate the cost of a learning policy and the expected improvement that results from it, it can use these quantities to dynamically tailor a policy that is suited to the particular learning task. We say a learning system incorporates a *rational policy* if it dynamically balances the trade-off between learning utility and learning efficiency. A *rational learner* is a learning system that uses a rational policy.

A rational policy requires the user to explicitly specify the relationship between the utility of the performance element and the learning cost required to improve that utility. Just as the user of a substantively rational learning approach supplies a utility function that describes the behavior of the performance element (henceforth called a Type 1 utility function), the user of a procedurally rational learning system must supply a utility function that indicates how utility is discounted by the cost of learning (henceforth called a Type 2 utility function).

2.3 A specific rational learning task

The notion of rationally allocating learning resources is quite general and can be realized in many ways. Our primary interest is in area of speed-up learning and therefore we are investigating learning agents specialized to this task. In particular, we consider learning tasks of the following form. The performance element is viewed as a problem solver that must solve a sequence of tasks provided by the environment according to a fixed distribution. The learning system possesses operations that change the average resources required to solve problems. Learning precedes problem solving. This follows the typical characterization of the speed-up learning paradigm: there is an initial learning phase where the learning system consumes training examples and produces an improved problem solver; this is followed by a utilization phase where the problem solver is used repeatedly to solve problems.

We define an obvious Type 2 goal for this task: to maximize *the expected number of problems that can be solved within a given resource limit.* This relates learning cost to improvement (a Type 2 utility function) as learning operations consume resources that could be available for problem solving. A rational learning system for this task must assess how best allocate resources between learning and problem solving given the amount of resources available.

We have developed a rational learning approach for this task. The approach dynamically assesses the relative benefits and costs of learning and flexibly adjusts its tradeoff between these factors depending on the resource constraints placed on it. We call the approach RASL (for RAtional Speed-up Learner) and it builds upon our previous work on the COMPOSER learning system, a general statistical framework for probabilistically identifying transformations that improve Type 1 utility of a problem solver [Gratch92a]. COMPOSER has demonstrated its effectiveness in artificial planning domains [Gratch92a] and in a real-world scheduling domain [Gratch93]. The system was not, however, developed to support rational learning. Using the low-level learning operations of the COMPOSER system, RASL incorporates rational decision making procedures that provide flexible control of learning in response to varying resource constraints. RASL acts as a learning agent that, given a problem solver and a fixed resource constraint, balances learning and problem solving with the goal of maximizing expected Type 2 utility.

2.3.1 Overview

RASL must solve many problems as possible within the resources. RASL can start solving problems immediately. Alternatively, it can devote some resources towards learning in the hope that this expense is more than made up by improvements to the problem solver. Type 2 utility corresponds to the number of problems that can be solved with the available resources. RASL should only invoke learning if it increases the *expected number of problems* (*ENP*) that can be solved. The learner embodies decision making capabilities to perform this evaluation.

The expected number of problems is governed by the following relationship:

$$ENP = \frac{available\,resources\;LESS\;resources\;expended\;during\;learning}{average\;resource\;cost\;to\;solve\;a\;problem\;with\;learned\;problem\;solver}$$

The top term reflects the resources available for problems solving. If no learning is performed, this is simply the total available resources. The bottom term reflects the average cost to solve a problem with the learned problem solver. RASL is reflective in that it guides its behavior by explicitly estimating *ENP*. To evaluate this expression, RASL must develop estimators for the resources that will be spent on learning and the performance of the learned problem solver.

Like COMPOSER, RASL adopts a hill-climbing (myopic) approach to maximizing utility. The system incrementally transforms a problem solver. At each step in the search, a transformation generator proposes a set of possible changes to the current problem solver. The learning system then evaluates these changes with training examples and chooses one transformation that improves expected utility. At each step in the search, RASL has two options: (1) stop learning and start solving problems, or (2) investigate new transformations, choosing one with the largest expected increase in *ENP*. In this way, the behavior of RASL is driven by its own estimate of how learning influences the expected number of problems that may be solved. The system only continues the hill-climbing search as long as the next step is expected to improve the Type 2 utility. RASL proceeds through a series of decisions where each action is chosen to produce the greatest increase in *ENP*. However, as it is a hill-climbing system, the *ENP* exhibited by RASL will be locally, and not necessarily globally, maximal.

2.3.2 *Estimating ENP*

The core of RASL's rational analysis is a statistical estimator for the *ENP* that results from possible learning actions. This section describes the derivation of this estimator. At a given decision point, RASL must choose between terminating learning or investigating a set of transformations to the current problem solver. RASL uses COMPOSER's statistical procedure to identify beneficial transformations. Thus, the cost of learning is based on the cost of this procedure. After reviewing some statistical notation we describe this statistical procedure. We then develop from this statistic a estimator for *ENP*.

Let X be a random variable. An observation of a random variable can yield one of a set of possible numeric outcomes where the likelihood of each outcome is determined by an associated probability distribution. X_i is the ith observation of X. EX denotes the expected value of X, also called the mean of the distribution. $\bar{X}_n$ is the *sample mean* and refers to the average of n observations of X. $\bar{X}_n$ is a good estimator for EX. *Variance* is a measure of the dispersion or spread of a distribution. We use the *sample variance* as an estimator for the variance of a distribution.

The function $\Phi(x)$ is the cumulative distribution function of the standard normal (also called standard gaussian) distribution. $\Phi(x)$ is the probability that a point drawn randomly from a standard normal distribution will be less than or equal to x. This function plays a important role in statistical estimation and inference. The Central Limit Theorem shows that the difference between the sample mean and true mean of an arbitrary distribution can be accurately represented by a standard normal distribution, given sufficiently large sample. In practice we can perform accurate statistical inference using this "normal approximation."

COMPOSER's statistical evaluation

Let T be the set of hypothesized transformations. Let *PE* denote the current problem solver. When RASL processes a training example it determines, for each transformation, the change in resource cost that the transformation would have provided if it were incorporated into *PE*. This is the difference in cost between solving the problem with and without incorporating the transformation into *PE* . This is denoted $\Delta r_i(t|PE)$ for the *i*th training example. Training examples are processed incrementally. After each example the system evaluates a statistic called a *stopping rule*. This rule determines when enough examples have been processed to state with confidence that a transformation will help or hurt the average problem solving speed of PE. The particular rule we use was proposed by Arthur Nádas [Nádas69].

After processing a training example, RASL evaluates the Nádas rule for each transformation:

$$n \geq n_0 \; AND \; \frac{S_n^2(t|PE)}{[\bar{\Delta r}_{n_0}(t|PE)]^2} \leq \frac{n}{a^2}, \; where \; \Phi(a) = \frac{\delta}{2|T|} \tag{2.1}$$

where n_0 is a small finite integer indicating an initial sample size, *n* is the number of examples taken so far, $\bar{\Delta r}_n(t|PE)$ is the transformation's average improvement, $S_n^2(t|PE)$ is the observed variance in the transformation's improvement, and δ is a confidence parameter. If the expression holds, the transformation will speed-up (slow down) *PE* if $\bar{\Delta r}_{n_0}(t|PE)$ is positive (negative) with confidence 1-δ. The number of examples taken at the point when Equation 1 holds is called the *stopping time* associated with the transformation and is denoted $ST(t|PE)$.

Estimating Stopping Times

Given that we are using the stopping rule in Equation 1, the cost of learning a transformation is the stopping time for that transformation times the average cost to process an example. Thus, one element of an estimate for learning cost is an estimate for stopping times. In Equation 1, the stopping time associated with a transformation is a function of the variance, the square of the mean, and the sample size *n*. We can estimate these first two parameters using a small initial sample of n_0 examples: $S_n^2(t|PE) \approx S_{n_0}^2(t|PE)$ and $\bar{\Delta r}_n(t|PE) \approx \bar{\Delta r}_{n_0}(t|PE)$. We can estimate the stopping time by using these estimates, treating the inequality in Equation 1 as an equality, and solving for *n*. The stopping time cannot be less than the n_0 initial examples so the resulting estimator is:

$$\hat{ST}_{n_0}(t|PE) = max\left\{n_0, \lceil a^2 \frac{S_{n_0}^2(t|PE)}{[\bar{\Delta r}_{n_0}(t|PE)]^2} \rceil\right\}, \; where \; \Phi(a) = \frac{\delta}{2|T|} \tag{2.2}$$

Learning Cost

Each transformation can alter the expected utility of the performance element. To accurately evaluate potential changes we must allocate some of the available resources towards learning. Under our statistical formalization of the problem, learning cost is a function of the stopping time for a transformation, and the cost of processing each example problem.

Let $\lambda_j(T, PE)$ denote the learning cost associated with the *j*th problem under the transformation set T and the performance element *PE*. The total learning cost associated with a transformation, t, is the sum of the per problem learning costs over the number of examples needed to apply the transformation: $\lambda(t, T, PE) = \sum_{j=1}^{ST(t|PE)} \lambda_j(T, PE) = \bar{\lambda}_{ST(t|PE)}(T, PE) \times ST(t|PE)$.

ENP Estimator

We can now describe how to estimate ENP for individual transformations based on an initial sample of n_0 examples. Let R be the available resources. The resources expended learning a transformation can be estimated by multiplying the average cost to process an example by the stopping time associated with that transformation: $\bar{\lambda}_{n_0}(t, T, PE) \times \hat{ST}_{n_0}(t|PE)$. The average resource use of the learned performance element can be estimated by combining estimates of the resource use of the current performance element and the change in this use provided by the transformation: $\bar{r}_{n_0}(PE) - \bar{\Delta r}_{n_0}(t|PE)$. Combining these estimators yields the following estimator for the expected number of problems that can be solved after learning transformation t:

$$E\hat{N}P_{n_0}(R, t|PE) = \frac{R - \bar{\lambda}_{n_0}(t, T, PE) \times \hat{ST}_{n_0}(t|PE)}{\bar{r}_{n_0}(PE) - \bar{\Delta r}_{n_0}(t|PE)} \tag{2.3}$$

Given the available resources, RASL estimates, based on a small initial sample, if a transformation should be learned. If the current *ENP* is greater than the *ENP* associated with any transformation (learning is at a local maximum), learning terminates and the remaining resources are used to solve problems. If there are some transformations with greater *ENP*, resources are devoted to investiging these further. A complete description of the algorithm is provide in [Gratch92b].

2.4 Empirical evaluation

RASL's mathematical framework predicts that the system can appropriately control learning under a variety of time pressures. Given an initial problem solver and a fixed amount of resources, the system should allocate resources between learning and problem solving to solve as many problems as possible. The theory that underlies RASL makes several predictions that we can evaluate empirically. (1) Given an arbitrary level of initial resources, RASL should solve a greater than or equal to number of problems than if we used all available resources on problem solving.[3] As we increase the resources, RASL should (2) spend more resources learning, and (3) exhibit a greater increase in the number of problems that can be solved.

We empirically test these claims in the real-world domain of spacecraft communication scheduling. The task is to allocate communication requests between earth-orbiting satellites and the three 26-meter antennas at Goldstone, Canberra, and Madrid. These antennas make up part of NASA's Deep Space Network. Scheduling is currently performed by a human expert, but the Jet Propulsion Laboratory (JPL) is developing a heuristic scheduling technique for this task. In a previous article, we describe an application of COMPOSER to improving this scheduler over a distribution of scheduling problems. COMPOSER acquired search control heuristics that improved the average time to produce a schedule [Gratch93]. We apply RASL to this domain to evaluate our claims.

[3] RASL pays an overhead of taking n_0 initial examples to make its rational deliberations. If learning provides no benefit, this can result in RASL solving less examples than a non-learning agent.

A learning trial consists of fixing an amount of resources, evoking RASL, and then solving problems with the learned problem solver until resources are exhausted. This is compared with a non-learning approach that simply uses all available resources to solve problems. Learning trials are repeated multiple times to achieve statistical significance. The behavior of RASL under different time pressures is evaluated by varying the amount of available resources. Results are averaged over thirty learning trials to show statistical significance. We measured several aspects of RASL's behavior to evaluate the claims. The principle component of behavior is the number of problems that are solved. In addition we track the time spent learning, the Type I utility of the learned scheduler, and the number of hill-climbing steps adopted by the algorithm. RASL has two parameters. The statistical error of the stopping rule is controlled by δ, which we set at 5%. RASL requires an initial sample of size n_0 to estimate *ENP*. For these experiments we chose a value of thirty.

The results are summarized in Figure 1.

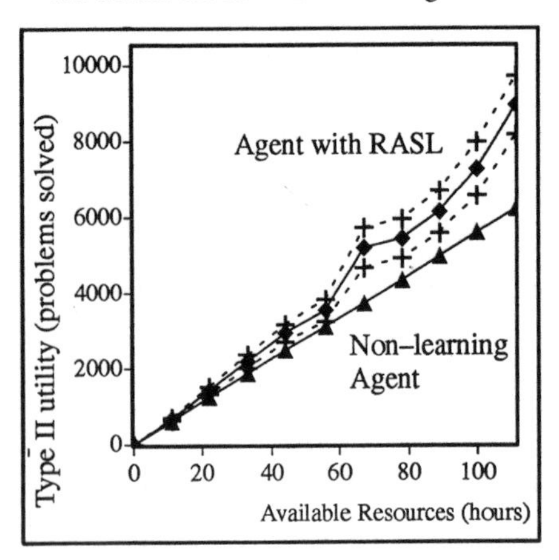

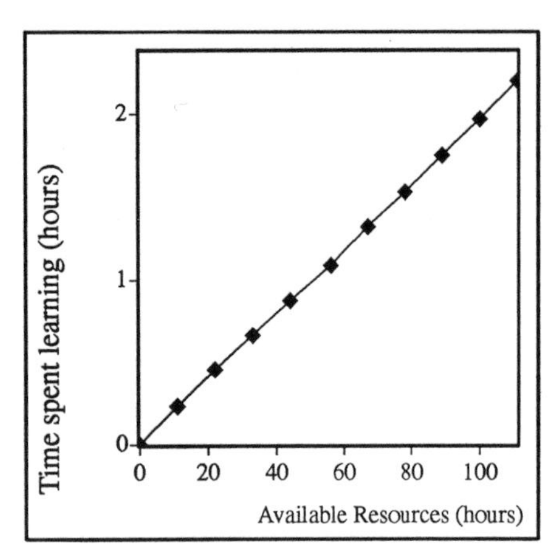

RASL yields an increase in the number of problems that are solved compared to a non-learning agent, and the improvement accelerates as the resources are increased. As the resources are increased, RASL spends more time in the learning phase, taking more steps in the hill-climbing search, and acquiring schedulers with better average problem solving performance.

The improvement in *ENP* is statistically significant. The graph of problems solved shows 95% confidence intervals on the performance of RASL over the learning trials. Intervals are not shown for the non-learning system as the variance was negligible over the thirty learning trials. We are somewhat surprised by the strict linear relationship between the available resources and the learning time. At some point learning time should level off as learning produces diminishing returns.

It is an open issue how best to set the initial sample parameter, n_0. The size of the initial sample influences the accuracy of the estimate of *ENP*, which in turn influences the behavior of the system. Poor estimates can degrade RASL's decision making capabilities. Making n_0 very large will increase the accuracy of the estimates, but increases the overhead of the technique.

It appears that the best setting for n_0 depends on characteristics of the data, such as the variance of the distribution. Further experience on real domains is needed to assess the impact of this sensitivity. There are ways to address the issue in a principled way (e.g. using cross-validation to evaluate different setting) but these would increase the overhead of the technique.

2.5 Conclusion

Learning systems cannot produce maximal increases in performance *and* be maximally efficient. Instead, they must adopt policies that balance these two needs. Most learning techniques adopt a fixed policy to this tradeoff. Unfortunately, fixed policies limit the generality of learning techniques. Furthermore, the usefulness of a given policy is frequently dependent on information that is only available once learning has begun. In this article we have explored some of the issues involved in rational policies. These are learning policies that adapt their behavior based on information obtained during the learning process. We describe an extension to the COMPOSER system that implements a rational policy. While this is only a first attempt at the problem, it raises a number of interesting issues. Furthermore, it highlights an issue that is not typically discussed in the learning community – the trade-off between learning efficiency and utility.

2.6 References

[Dean88] Dean, T. and Boddy, M., "An Analysis of Time-Dependent Planning," *Proceedings of The Seventh National Conference on Artificial Intelligence*, Saint Paul, MN, August 1988, pp. 49-54.

[Doyle90] Doyle, J., "Rationality and its Roles in Reasoning (extended version)," *Proceedings of the National Conference on Artificial Intelligence*, Boston, MA, 1990, pp. 1093-1100.

[Good71] Good, I. J., "The Probabilistic Explication of Information," in *Foundations of Statistical Inference*, V. P. Godambe, D. A. Sprott (ed.), Hold, Rienhart and Winston, Toronto, 1971, pp. 108-127.

[Gratch92a] Gratch, J., and DeJong, G., "COMPOSER: A Probabilistic Solution to the Utility Problem in Speed-up Learning," *Proceedings of the National Conference on Artificial Intelligence*, San Jose, CA, July 1992.

[Gratch92b] Gratch, J., DeJong, G. and Yang, Y., "Rational Learning: (extended version)," Technical Report UIUCDCS-R-92-1756, Department of Computer Science, University of Illinois, Urbana, IL, 1992.

[Gratch93] Gratch, J., Chien, S., and DeJong, G., "Learning Search Control Knowledge for Deep Space Network Scheduling," *Proceedings of the Ninth International Conference on Machine Learning,* Amherst, MA, 1993.

[Horvitz89] Horvitz, E. J., Cooper, G. F., and Heckerman, D. E., "Reflection and action under scarce resources: theoretical principles and empirical study," in *Proceedings of the Eleventh International Joint Conference on Artificial Intelligence,* Detroit, MI, August 1989, pp. 1121-1127.

[Minton88] Minton, S. N., "Learning Effective Search Control Knowledge: An Explanation-Based Approach," Ph.D. Thesis, Department of Computer Science, Carnegie-Mellon University, Pittsburgh, PA, March 1988.

[Nádas69] Nádas, A., "An extension of a theorem of Chow and Robbins on sequential confidence intervals for the mean," *The Annals of Mathematical Statistics*, 40, 2 (1969), pp. 667-671.

[Simon76] Simon, H. A., "From Substantive to Procedural Rationality," in *Method and Appraisal in Economics*, S. J. Latsis (ed.), Cambridge University Press, 1976, pp. 129-148.

3

A new criterion for selecting models from partially observed data

Hidetoshi Shimodaira

Department of Mathematical Engineering and Information Physics
University of Tokyo
Hongo 7-3-1, Bunkyo-ku, Tokyo, JAPAN
shimo@bcl.t.u-tokyo.ac.jp

ABSTRACT A new criterion PDIO (predictive divergence for indirect observation models) is proposed for selecting statistical models from partially observed data. PDIO is devised for "indirect observation models", in which observations are only available indirectly through random variables. That is, some underlying hidden structure is assumed to generate the manifest variables. For example, unsupervised learning recognition systems, clustering, latent structure analysis, mixture distribution models, missing data, noisy observations, etc., or the models whose maximum likelihood estimator is based on the EM (expectation-maximization) algorithm. PDIO is a natural extension of AIC (Akaike's information criterion), and the two criteria are equivalent when direct observations are available. Both criteria are expressed as the sum of two terms: the first term represents the goodness of fit of the model to the observed data, and the second term represents the model complexity. The goodness of fit terms are equivalent in both criteria, but the complexity terms are different. The complexity term is a function of model structure and the number of samples and is added in order to take into account the reliability of the observed data. A mean fluctuation of the estimated true distribution is used as the model complexity in PDIO. The relative relation of the "model manifold" and the "observed manifold" is, therefore, reflected in the complexity term of PDIO from the information geometric point of view, whereas it reduces to the number of parameters in AIC. PDIO is very unique in dealing with the unobservable underlying structure "positively." In this paper the generalized expression of PDIO is shown using two Fisher information matrices. An approximated computation method for PDIO is also presented utilizing EM iterates. Some computer simulations are shown to demonstrate how this criterion works.

Keywords: model selection criterion, model search, EM algorithm, Bayes theorem, incomplete data, indirect observation, AIC, Fisher information, Kullback-Leibler divergence

3.1 Introduction

Selecting models from observed data is very important and necessary in practical applications. In this paper, we use the word "model" to represent a probability distribution (or density) function $p(x|\theta)$, where x is a random variable, and θ is a parameter. Let x and θ be real vectors, and assume x takes only discrete values if necessary. We also use the word "observed data" to represent n observed samples $x_1, x_2, \cdots, x_n$, where each x_t is a sample of independently and identically distributed (i.i.d.) random variables being subject to the unknown true distribution $q(x)$. The assumption of i.i.d. is just for simplicity. When the model $p(x|\theta)$ and the observed training data x_t's are given, you can tune the parameter θ using some methods, such as the maximum likelihood method,that is, $\prod_{t=1}^{n} p(x_t|\theta) \to \max$. Here you may have a simple but important question: Is the estimated parameter $\hat{\theta}(x_t\text{'s})$ really a good one? The distance between

[1]*Selecting Models from Data: AI and Statistics IV.* Edited by P. Cheeseman and R.W. Oldford. ©1994 Springer-Verlag.

the estimated distribution $p(x|\hat{\theta}(x_t\text{'s}))$ and the true distribution $q(x)$ depends on whether the model $p(x|\theta)$ is good or not.

The model $p(x|\theta)$ is specified by the system designer using the a priori knowledge about the object. If the designer can specify only one model with absolute confidence, there is no need for model selection. But it is often the case that he or she has many competing candidate models $p_1(x|\theta_1), p_2(x|\theta_2), \cdots$, and he or she has to choose a good one among them on the basis of the observed data.

One of the conventional methods of model selection is the statistical hypothesis testing. This procedure has been applied for many years, but some problems are pointed out by Akaike [Akaike 74]: ambiguity in choice of the loss function or in choice of the significance level, and a rather limited application of this procedure. This method should be applied only at the final stage of the model selection, where you have very few competing models. This method works well at the confirmatory situation for model validation.

The Bayesian approach is also available for model selection. In this case, you need a prior distribution for each model and for each parameter value. But in general, no one can obtain these priors beforehand on reasonable ground. In practice, the priors are parameterized with hyper-parameters, and these hyper-parameters are estimated using the observed data. The hyper-parameter is, therefore, equivalent to the parameter in the non-Bayesian sense and the problem reduces to a non-Bayesian case.

In this paper, we adopt the criterion method for model selection. We define some criterion to measure the goodness of models, and choose the model which minimizes the criterion. This criterion should be computed from the specification of the model $p(x|\theta)$ and the observed data x_t's. AIC (Akaike's information criterion) [Akaike 74] [MurYosAm 93] is one of these criteria, and the effectiveness of AIC is confirmed widely in applications. This method can be applied at the first stage of the model selection, where you have many competing models and you want to choose some good models among them. This method works well at the exploratory situation for model search.

Though AIC is a very good criterion, it is intended to consider only observed variables. In some applications, it is often the case that you want to take it into account that the observed variables are generated from unobservable variables. A new criterion PDIO (predictive divergence for indirect observation models) [Shimodaira 92a] [Shimodaira 92b] is proposed for this purpose. PDIO is a natural extension of AIC, and the two criteria are equivalent when only observable variables are considered.

3.2 Indirect Observation Model

We introduce the indirect observation model to deal with the situation in which the object can only be observed indirectly through random variables. Let x denote the unobservable "complete" data, and y denote the observable "incomplete" data. Assume the relation $y = g(x)$, where $g(\cdot)$ is a many-to-one mapping, and note that x cannot be determined uniquely even if y is given. Thus, the model structure is specified by both $p(x|\theta)$ and $g(x)$. When $x = (y, z)$, this model is called the incomplete observation model, where y is a manifest variable and z is a latent variable. Under some transformation of variables, the indirect observation model and the incomplete observation model are equivalent.

A considerably wide class of problems can be expressed with this framework [DemLaiRu 77]

[RednWalk 84]. For example, unsupervised learning recognition systems, cluster analysis, missing data, noise contamination, latent structure analysis, and so forth.

Example 1 (speech recognition) An automatic speech recognizer is one of the examples of unsupervised learning system [LeeHonRe 90]. Let y represent the whole time series $y = \{y(t) : t = 0, 1, \cdots\}$, where each $y(t)$ is a short time power spectrum of observed human speech at time n. Though we only observe y, we want to know the sequence of the words spoken. Let z represent the whole time series of the hidden words. If you know the prior distribution $q(x)$, where $x = (y, z)$, then you can compute the posterior distribution $q(z|y)$ given observed speech y using the Bayes theorem and can perform speech recognition.

Usually it is assumed that the series x obeys the Markov chain model and the series y obeys the hidden Markov chain model. Since the model $p(x|\theta)$ cannot be determined uniquely by the prior knowledge of language, the system designer has to choose a good one from possible alternatives by trial-and-error. The estimated parameter $\hat{\theta}(y)$ is computed from the observed training data and the model $p(x|\theta)$. It can be said to be a good model if the estimated distribution $p(x|\hat{\theta})$ is close enough to the true distribution $q(x)$.

Generally speaking, when you use a complex model with many parameters, it can be fitted very well to the training data. In this case, the estimated distribution $p(x|\hat{\theta})$ can explain very well the training data. But this often causes the situation that the estimated distribution explains poorly the data for recognition generated independent of the training data. Thus, it is important to construct a model, whose amount of information corresponds to that contained in the training data. A good result cannot be expected if you use a much more complex model compared with the training data.

Example 2 (time series analysis for economic systems) The time series analysis using state-space models is one of the examples of using the model whose observation is contaminated with noise. Let s represent the time series of some economic index. Assume this series is generated by autoregressive moving average (ARMA) process. Though we want to examine s, observations are only available on $y = F(s) + w$, where $F(\cdot)$ denotes some linear filter and w denotes white noise series. In this case, the complete data is $x = (s, w)$, where x is estimated from y by the Kalman smoother.

3.3 Estimating the Parameter

Let $x_1, \cdots, x_n$ be n i.i.d. random samples from $q(x)$. Observations are only available on incomplete data $y_1, \cdots, y_n$, not on complete data $x_1, \cdots, x_n$. Here we consider the problem of how to estimate the parameter θ when the model $p(x|\theta)$ and observations are given. We adopt the Kullback-Leibler (K-L) divergence as a measure of distance:

$$D(q(x)|p(x)) = \int q(x) \log \frac{q(x)}{p(x)} \, dx.$$

First, we consider the situation that the complete data x can be observed directly. The empirical distribution (or density) function constructed from observed training data $x_1, \cdots, x_n$ is $\hat{q}(x) = (1/n) \sum_{t=1}^{n} \delta(x - x_t)$. There is no problem in the case that x is discrete, but the delta function $\delta(\cdot)$ should be smoothed with some kernel function when x is continuous. It is necessary that the square root of the variance of the kernel σ_n approaches zero as $n \to \infty$, but slower than the order of $O(1/n)$. Since this problem is not important in an asymptotic situation, we ignore it in the following discussion. Note that smoothing the delta function is primarily for the convergence of the entropy $H(\hat{q}) = -\int \hat{q} \log \hat{q}\, dx$, though this term has no effect on model selection.

Though the optimal value of the parameter θ is the value which achieves the minimum of the divergence $D(q(x)|p(x|\theta))$, we cannot obtain its value because we do not know the true distribution $q(x)$. Thus, we adopt the estimator $\hat{\theta}$, which achieves $D(\hat{q}(x)|p(x|\theta)) \to \min$. Since this is equivalent to $(1/n)\sum_{t=1}^{n} \log p(x_t|\theta) \to \max$, this estimator $\hat{\theta}$ is the maximum likelihood estimator (MLE).

Next, we consider the situation that observations are only available on incomplete data y. Now we do not have $\hat{q}(x)$ but only have $\hat{q}(y) = (1/n)\sum_{t=1}^{n} \delta(y - y_t)$. Thus we adopt the estimator $\hat{\theta}$, which achieves $D(\hat{q}(x)|p(x|\theta)) \to \min$, where $\hat{q}(x)$ is also estimated as well as $\hat{\theta}$, under the condition that the marginal distribution $\int \hat{q}(x)\, dz = \hat{q}(y)$ is given.

From the nature of the K-L divergence, the following fact can be proven: for any given $p(x)$ and $\hat{q}(y)$, the minimum of $D(\hat{q}(x)|p(x))$, under the marginal condition $\int \hat{q}(x)\, dz = \hat{q}(y)$, is $D(\hat{q}(y)|p(y))$, and it is achieved by $\hat{q}(x) = p(z|y)\hat{q}(y)$. From this fact, the estimator $\hat{\theta}$ for indirect observation models is equivalent to the MLE of incomplete data, that is, the maximizer of the log-likelihood of incomplete data $L(\theta) = (1/n)\sum_{t=1}^{n} \log p(y_t|\theta)$.

This result can be viewed from the Bayesian point of view, where z is a parameter and y is a random variable. Let $p(z)$ be a prior of z, and $p(y|z)$ be the distribution of y. Since $p(z|y) = p(y|z)p(z)/\int p(y|z)p(z)\, dz$ is the posterior of z given y, $\hat{q}(z) = \int \hat{q}(x)\, dy = \int p(z|y)\hat{q}(y)\, dy$ can be viewed as the average of the posteriors $p(z|y_t)$'s given y_t's.

Usually iterative procedures, such as EM (expectation-maximization) algorithm [DemLaiRu 77] or the scoring method, are used to estimate the MLE $\hat{\theta}$ for indirect observation models. Generally speaking, these procedures cannot find the global maximum but only find local maxima or saddle points. We ignore this problem in this paper.

The estimation problem of indirect observation models can be viewed from the "information geometry" [Amari 85] (See Figure 1). Considering the space of the whole probability distributions of the complete data x, the model $p(x|\theta)$ can be viewed as a manifold in the space. We call it the model manifold. Similarly, all distributions $\hat{q}(x)$ whose marginal is $\hat{q}(y)$ can be viewed as a manifold. We call it the observed manifold. The E-step of EM algorithm can be viewed as projection onto the observed manifold, and the M-step as projection onto the model manifold. The minimum distance of the two manifolds measured by the K-L divergence is achieved by the estimated empirical distribution $\hat{q}(x)$ and the estimated true distribution $p(x|\hat{\theta})$.

3.4 Criterion for Model Selection

Let us consider criteria to choose good models from a set of candidate models $\{p_i(x|\theta_i)\}$ on the basis of the observed data $\hat{q}(y)$. Though there are many criteria proposed previously, they are intended to measure only observed variables. Here we propose a new criterion PDIO (predictive divergence for indirect observation models), which is a natural extension of the well known information criterion AIC.

Let $q(x)$ be the true distribution of the complete data x, and $p(x|\theta)$ be the model under consideration. We adopt the predictive divergence criterion (PDC) to measure the goodness of the model:

$$\mathrm{PDC}(q,p) = \langle D(q(x)|p(x|\hat{\theta}))\rangle,$$

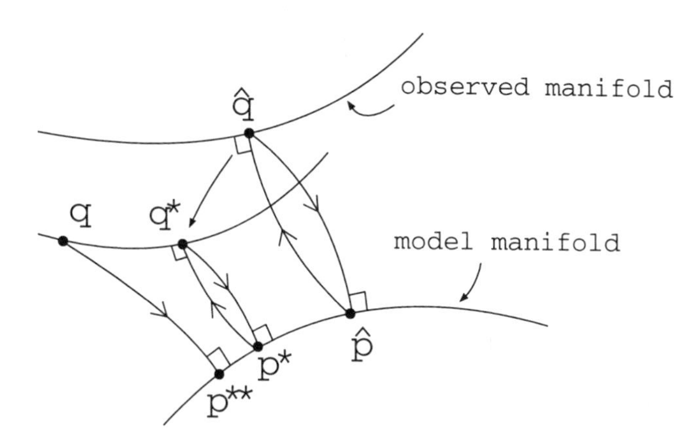

FIGURE 1. geometrical relationship in the space of probability distributions of complete data x. Asymptotically the estimated empirical distribution $\hat{q}$ approaches the optimal empirical distribution q^*. Assume the optimal distribution p^* is close enough to the true distribution q. Note that $p^*(x) = p(x|\theta^*)$, $p^{**}(x) = p(x|\theta^{**})$, $\hat{p}(x) = p(x|\hat{\theta})$.

where $\langle\cdot\rangle$ denotes the expectation concerning $\hat{\theta}$. Note that $\hat{\theta}$ is a random variable because it is a function of n random variables $y_1, \cdots, y_n$. PDC is the expectation of distance between the estimated distribution $p(x|\hat{\theta})$ and the true distribution $q(x)$ measured by the K-L divergence $D(q(x)|p(x|\hat{\theta}))$.

Actually, the value of PDC cannot be obtained because we do not know the true distribution $q(x)$. Thus, we define PDIO as an estimator of PDC:

$$\text{PDIO}(\hat{q}, p) = D(\hat{q}(y)|p(y|\hat{\theta})) + (1/n)\,\text{tr}(I_X(\hat{\theta})I_Y^{-1}(\hat{\theta})),$$

where the Fisher information matrices I_X and I_Y are defined by

$$I_X(\theta) = -\int \frac{\partial^2 \log p(x|\theta)}{\partial\theta^2} p(x|\theta)dx, \quad I_Y(\theta) = -\int \frac{\partial^2 \log p(y|\theta)}{\partial\theta^2} p(y|\theta)\,dy.$$

When x can be observed, I_X coincides with I_Y and the trace term of PDIO reduces to the number of parameters $\dim\theta$. Thus, PDIO coincides with AIC. Since $I_X - I_Y$ is, in general, non-negative definite, then $\text{tr}(I_X I_Y^{-1}) \geq \dim\theta$. Therefore the value of PDIO is not smaller than that of AIC.

Assume n is large enough so that asymptotic discussion is justified. Let θ^* achieve $D(q(y)|p(y|\theta)) \to \min$, θ^{**} achieve $D(q(x)|p(x|\theta)) \to \min$. Note that $\theta^* \neq \theta^{**}$ in general. It can be said that $p(x|\theta^*) = p(x|\theta^{**})$, if the true distribution $q(x)$ is included in the model $p(x|\theta)$, or if the observed manifold and the model manifold are mutually orthogonal, under some conditions. We make an assumption that $q(x)$ is close enough to $p(x|\theta^*)$, which is not always true in practice but is necessary to approximate the value of PDC. The following theorem can be proven [Shimodaira 92b] under this assumption:

Theorem 1

$$\text{PDC}(q, p) \simeq \langle \text{PDIO}(\hat{q}, p) \rangle.$$

Proof We assume that the true distribution $q(x)$ is well approximated by $p^*(x)$ to sketch the proof briefly. This "assumption" is essential to the following proof, though the assumption is not always justified in practice, so we need further consideration on this problem. Note that I_X and I_Y are assumed to be evaluated at θ^* (or θ^{**}) in the following discussion, whereas these are evaluated at $\hat{\theta}$ for computation of PDIO in practice. This difference is a part of higher order terms, so it is negligible.

Two matrices G_Y^* and H_Y^* are introduced by

$$G_Y^* = \int q(y)(\nabla \log p(y|\theta^*))^2 \, dy, \text{ and } H_Y^* = -\int q(y)\nabla\nabla \log p(y|\theta^*) \, dy.$$

Here ∇ represents the vector operator $\partial/\partial\theta$. Under mild regularity conditions, it can be said that $\sqrt{n}(\hat{\theta} - \theta^*)$ is asymptotically normally distributed with mean 0 and covariance matrix $H_Y^{-1} G_Y H_Y^{-1}$. Supposing $p^*(y) = q(y)$, then $G_Y = H_Y = I_Y$. Thus, from the assumption, the covariance matrix can be approximated by $H_Y^{-1} G_Y H_Y^{-1} \approx I_Y^{-1}$.

Expand $D(q(x)|p(x|\hat{\theta}))$ around θ^* ignoring higher order terms to obtain:

$$\begin{aligned} D(q(x)|p(x|\hat{\theta})) \approx\; & D(q(x)|p(x|\theta^*)) & (3.1)\\ & +\nabla D(q(x)|p(x|\theta^*))(\hat{\theta} - \theta^*) & (3.2)\\ & +(1/2)(\hat{\theta} - \theta^*)'\nabla\nabla D(q(x)|p(x|\theta^*))(\hat{\theta} - \theta^*) & (3.3) \end{aligned}$$

Let us evaluate the expectation $\langle\cdot\rangle$ asymptotically for each of the three terms labeled.

First we consider the second term. From the definition, $\nabla D(q(x)|p^{**}(x)) = 0$, so that $\nabla D(q(x)|p^*(x)) \approx 0$ under the assumption. Furthermore, $\langle\hat{\theta} - \theta^*\rangle$ converges zero asymptotically, so that $\langle(3.2)\rangle \approx 0$.

Next we consider the third term. Let $H_X^* = -\int q(x)\nabla\nabla \log p(x|\theta^*) \, dx$. Then we have

$$\begin{aligned} \langle(3.3)\rangle &= (1/2n)\,\mathrm{tr}(H_X H_Y^{-1} G_Y H_Y^{-1}) \\ &\approx (1/2n)\,\mathrm{tr}(I_X I_Y^{-1}). \end{aligned}$$

Note that $H_X \approx I_X$ and $H_Y \approx I_Y$ under the assumption.

Finally we consider the first term. Decompose it into the following three terms:

$$\begin{aligned} D(q(x)|p^*(x)) =\; & D(q(x)|p^*(x)) - D(\hat{q}(x)|p^*(x)) + \\ & D(\hat{q}(x)|p^*(x)) - D(\hat{q}(x)|\hat{p}(x)) + \\ & D(\hat{q}(x)|\hat{p}(x)) \\ \approx\; & (1/\sqrt{n})U + & (3.4)\\ & (1/2)(\hat{\theta} - \theta^*)'\nabla\nabla D(\hat{q}(x)|p(x|\hat{\theta}))(\hat{\theta} - \theta^*) + & (3.5)\\ & D(\hat{q}(y)|\hat{p}(y)) & (3.6) \end{aligned}$$

Asymptotically $\nabla\nabla D(\hat{q}|\hat{p}) \to \nabla\nabla D(q^*|p^*)$ as $n \to \infty$. Thus, $\nabla\nabla D(\hat{q}|\hat{p}) \approx I_X$ from the assumption. Then we have

$$\langle(3.5)\rangle \approx (1/2n)\,\mathrm{tr}(I_X I_Y^{-1}).$$

On the other hand, $\hat{q}(x) \to q^*(x)$ as $n \to \infty$. Then we have $\langle(1/\sqrt{n})U\rangle \approx 0$ considering $q^*(x) \approx q(x)$ from the assumption.

Combining these results, we obtain

$$\langle D(q(x)|p(x|\hat{\theta}))\rangle = \langle D(\hat{q}(y)|p(y|\hat{\theta}))\rangle + (1/n)\,\mathrm{tr}(I_X I_Y^{-1}) + o(n^{-1})$$

asymptotically under the assumption that $q(x) = p^*(x)$. This implies PDIO is an estimator of PDC unbiased up to the terms of order $O(n^{-1})$.

Note that $(1/\sqrt{n})U$ is a random variable of order $O_p(1/\sqrt{n})$, whereas the second term of PDIO $(1/n)\,\mathrm{tr}(I_X I_Y^{-1})$ is intended to correct the bias of order $O(1/n)$. This seems like nonsense, because the disturbance term $(1/\sqrt{n})U$ is asymptotically larger than the correction term. This problem is absolutely the same as in AIC [MurYosAm 93]. When the candidate models to be compared are nested, this disturbance term will be canceled, so that the problem will disappear.

∎

Thus, PDIO, whose value can be computed from the observed data and the model specification, can be used as a model selection criterion instead of PDC, whose value cannot be computed in practice. Since $D(\hat{q}|p(\hat{\theta})) = \int \hat{q}(y)\log\hat{q}(y)\,dy - L(\hat{\theta})$ and the first term does not depend on model specifications, therefore

$$\mathrm{PDIO}'(\hat{q}, p) = -L(\hat{\theta}) + (1/n)\,\mathrm{tr}(I_X I_Y^{-1})$$

can be used for model selection. This modified criterion is better than PDIO in terms of its variance and stability of the result.

The first term of PDIO represents how well the model is fitted to the observed data. Thus, this may take a smaller value as the model becomes more complex. In general, the value of this term can be computed easily. The second term of PDIO represents the model complexity. The first term is an estimate of PDC, but it is biased by the amount of this second correction term, which can be seen as a penalty for using too complex a model compared with the amount of information given by the data. PDIO is expressed as the sum of these two terms. The model which may be selected is both simple and well fitted to the observed data.

3.5 Computation of the Criterion

Let us consider the computation of $\mathrm{tr}(I_X I_Y^{-1})$, the second term of PDIO. Theoretically, the explicit form expression of this term can be obtained from analytical expressions of I_X and I_Y. For example, those for log-linear models are shown by the author [Shimodaira 92b], and the numerical examples shown below are computed using these explicit expressions. It needs, however, hard computation for large scale models. Here, we show that EM iterates can be utilized for computing the trace term.

Let $M(\theta)$ denote the one step operator of EM iterates. The sequence $\theta^0, \theta^1, \theta^2, \cdots$ generated from the starting value θ^0 is written by $\theta^{i+1} = M(\theta^i)$. We define two functions as follows:

$$Q(\theta, \phi) = (1/n)\sum_{t=1}^{n} \int p(z_t|y_t, \phi)\log p(x_t|\theta)\,dz_t,$$

$$H(\theta, \phi) = (1/n)\sum_{t=1}^{n} \int p(z_t|y_t, \phi)\log p(z_t|y_t, \theta)\,dz_t.$$

The missing information matrix is $I_{X|Y} = I_X - I_Y$. The following theorem is easily verified [DemLaiRu 77] [Tanner 91]:

Theorem 2

$$\frac{\partial M(\theta)}{\partial \theta}\bigg|_{\hat{\theta}} \frac{\partial^2 Q(\theta, \hat{\theta})}{\partial \theta^2}\bigg|_{\hat{\theta}} = \frac{\partial^2 H(\theta, \hat{\theta})}{\partial \theta^2}\bigg|_{\hat{\theta}}.$$

TABLE 3.1. Number of selection with PDIO

n	Number of latent classes				
	1	2	3	4	5
200	914	82	0	1	3
500	136	828	32	2	2
1,000	0	561	438	0	1
2,000	0	18	982	0	0
5,000	0	0	1000	0	0
10,000	0	0	1000	0	0

TABLE 3.2. Number of selection with AIC

n	Number of latent classes				
	1	2	3	4	5
200	25	736	238	1	0
500	0	92	881	27	0
1,000	0	2	981	14	3
2,000	0	0	999	1	0
5,000	0	0	996	4	0
10,000	0	0	996	4	0

Considering the observed Fisher information approaches the expected Fisher information as $n \to \infty$,

Corollary 3

$$\operatorname*{plim}_{n\to\infty}\Big\{\frac{\partial M(\theta)}{\partial\theta}\Big|_{\hat\theta} I_X(\hat\theta) - I_{X|Y}(\hat\theta)\Big\} = 0$$

can be proven easily, and hence, when $n \to \infty$, asymptotically,

Corollary 4

$$I_X I_Y^{-1} \simeq \Big(I - \frac{\partial M(\theta)}{\partial\theta}\Big|_{\hat\theta}\Big)^{-1}.$$

This result can be used as an asymptotic approximation of the trace term. Note that the value of $\frac{\partial M(\theta)}{\partial\theta}|_{\hat\theta}$ is easily obtained by numerically differenciating $M(\theta)$ using EM iterates. In this asymptotic situation, the sencond term of PDIO can be seen as the rate of convergence of the EM algorithm.

3.6 Numerical Examples

We made two numerical simulations to illustrate how PDIO works as a model selection criterion, compared with AIC. The result for PDIO is shown in Table 3.1, and that for AIC is in Table 3.2. The complete data is a contingency table of $8 \times 8 \times 3$ levels, and the incomplete data is the 8×8 marginal table. The true model has 3 latent classes, where alternatives have classes ranging from one to five. Each number in the following tables indicates how many times the model is selected in one thousand experiments. Both criteria show a tendency for small models to be selected when the number of samples n is small, and for the true model to be selected as n increases. Though we cannot decide which criterion is better than the other from this simulation, we can say that PDIO is a feasible criterion as well as AIC.

3.7 Conclusion

We proposed a new information criterion for indirect observation models, and obtained the generalized expresssion of it using two Fisher information matrices. We also showed a method of approximated computation of the criterion by utilizing the EM iterates. PDIO is a natural extension of AIC, and is appropriate to examine the unobservable latent variables. Though we showed how this criterion works by the computer simulations, it should be tested in practical applications.

I wish to thank Prof. Kaoru Nakano for giving me the chance to study this theme. I also would like to express my appreciation to Yutaka Sakaguchi, Noboru Murata, and Fumiyasu Komaki for their helpful discussions. I wish to thank Prof. Shun-ichi Amari, Prof. Chihiro Hirotsu, Prof. Genshiro Kitagawa, and Prof. Makio Ishiguro for their helpful advice. I should like to express my grateful thanks to Ms. Julie Winder who corrected my usage of English in this manuscript.

3.8 REFERENCES

[Akaike 74] AKAIKE, H. (1974) "A new look at the statistical model identification," *IEEE Trans. Automat. Contr.* **AC-19**, 716–723.

[Amari 85] AMARI, S. (1985) *Differential-Geometrical Methods in Statistics*, Vol. 28 of *Lecture Notes in Statistics*. Springer-Verlag, Berlin.

[DemLaiRu 77] DEMPSTER, A. P., LAIRD, N. M. and RUBIN, D. B. (1977) "Maximum Likelihood from Incomplete Data via the EM Algorithm," *J. Roy. Statist. Soc. Ser. B* **39** , 1–38.

[LeeHonRe 90] LEE, K., HON, H. and REDDY, R. (1990) "An Overview of the SPHINX Speech Recognition System," *IEEE Trans. Acoust., Speech & Signal Process.* **ASSP-38**, 35–45.

[MurYosAm 93] MURATA, N., YOSHIZAWA, S. and AMARI, S. (1993) "Network Information Criterion — Determining the Number of Hidden Units for an Artificial Neural Network Model," *IEEE Trans. NN.* (to appear).

[RednWalk 84] REDNER, R. A. and WALKER, H. F. (1984) "Mixture densities, maximum likelihood and the EM algorithm," *SIAM Review* **26**, 195–239.

[Shimodaira 92a] SHIMODAIRA, H. (1992) "Learning and Recognition based on Probabilistic Models — An approach using dual coordinate system," Master's thesis, University of Tokyo. (in Japanese).

[Shimodaira 92b] SHIMODAIRA, H. (1992) "A New Information Criterion for Incomplete Observation Log-linear Models," *Proc. 14th Symposium of Japanese Society of Applied Statistics*. (in Japanese).

[Tanner 91] TANNER, M. A. (1991) *Tools for Statistical Inference*, Vol. 67 of *Lecture Notes in Statistics*. Springer-Verlag, Berlin.

4
Small-sample and large-sample statistical model selection criteria

S. L. Sclove

University of Illinois at Chicago

ABSTRACT
Statistical model selection criteria provide answers to the questions, "How much improvement in fit should be achieved to justify the inclusion of an additional parameter in a model, and on what scale should this improvement in fit be measured?" Mathematically, statistical model selection criteria are defined as estimates of suitable functionals of the probability distributions corresponding to alternative models. This paper discusses different approaches to model-selection criteria, with a view toward illuminating their similarities and differences. The approaches discussed range from explicit, small-sample criteria for highly specific problems to general, large-sample criteria such as Akaike's information criterion and variants thereof. Special emphasis is given to criteria derived from a Bayesian approach, as this presents a unified way of viewing a variety of criteria. In particular, the approach to model-selection criteria by asymptotic expansion of the log posterior probabilities of alternative models is reviewed. An information-theoretic approach to model selection, through minimum-bit data representation, is explored. Similarity of the asymptotic form of Rissanen's criterion, obtained from a minimum-bit data representation approach, to criteria derived from a Bayesian approach, is discussed.

KEY WORDS: model selection, model evaluation, Akaike's information criterion, AIC, Schwarz's criterion, Kashyap's criterion, Bayesian inference, posterior probabilities.

4.1 Introduction; sets of alternative models

Usually a statistical analysis involves consideration of a set of alternative models for the data. "Model selection" refers to the choice of "good" models or a "best" model from such a set. A "model-selection criterion" (MSC) is a formula which provides a figure-of-merit for the alternative models. Such a figure-of-merit, enabling "automatic" choice of a model, can be important in artificial-intelligence situations. This sort of situation is typified by an expert system for statistical analysis-and-prediction. For example, such an expert system might deal with repetitive prediction situations in which a training dataset is taken into the system and used to choose a predicting function, and predictions are made accordingly. Then, another training set is presented, these data are processed, a prediction function is chosen, and predictions are made; and so on.

Generally, alternative models will involve different numbers of parameters. The more parameters that are used, the better the fit that can be achieved. Model-selection criteria take account simultaneously of both the goodness-of-fit of a model and the number of parameters used to achieve that fit. They provide answers to the question, "How much improvement in fit should be achieved to justify the inclusion of an additional parameter in a model?"

As an example, consider multiple linear regression. The residual mean square is the residual sum of squares, divided by the residual degrees of freedom, $n - m - 1$, when the sample size is

---

[1] *Selecting Models from Data: AI and Statistics IV.* Edited by P. Cheeseman and R.W. Oldford.

n and m explanatory variables are used. The residual degrees of freedom involves the number of parameters used to achieve the fit. It will be seen below that appropriate criteria for this problem can depend on $n - m$ in more complicated ways.

4.2 Aspects of model selection in multivariate normal regression

We begin with a discussion of an approach to model selection in multivariate normal (multinormal) regression, because it illustrates some general considerations in model selection and because it is interesting in its own right. Some of these ideas are found in Sclove (1969); see also Aitkin (1974) and Bendel and Afifi (1977).

We shall see that residual mean square may be a reasonable MSC for merely *fitting* a set of data, but modification is required to provide a reasonable criterion for choosing a regression equation for *prediction*.

Let the $(p+1)$-dimensional random vector Z have a multinormal distribution. Let Z be partitioned as $Z = (Y, X)$, where Y is a scalar random variable and X is p-dimensional; the variable Y is the response (dependent) variable and the p elements of X are the explanatory (independent) variables.

Suppose we have data $\mathcal{Z}_n$, n independent observations of Z. Now our task is to use $\mathcal{Z}_n$ to choose a regression equation to fit the n Y-values. This regression equation is to be the usual multiple linear regression function involving one, several, or all of the p variables in X. That is, the alternative models correspond to different subsets of explanatory variables. Let these subsets be indexed by c. Let m_c be the number of explanatory variables in subset c; $1 \leq m_c \leq p$.

Let $Y_i'(c)$ be the predicted value of Y_i according to subset c. Then $Y_i - Y_i'(c)$ is the prediction error of subset c for the i-th case, $i = 1, 2, \cdots, n$. Its conditional expected square, given $X_i(c)$, the vector of values of the explanatory variables in subset c for the i-th case, is related to σ_c^2, the error variance in the regression of Y on the explanatory variables in subset c. This error variance is a suitable functional to measure the goodness of subset c for the problem of fitting the Y-values. The "residual sum of squares," $SSRes_c$, is the sum of squares of prediction errors. The residual mean square, i.e., residual sum of squares divided by $n - m_c - 1$, is an estimate of this error variance and as such is a suitable MSC for the problem of *fitting* a regression model to a data set.

Now suppose we shall be given additional observations on X and shall have to *predict* the corresponding values of Y: Our task is to use the data $\mathcal{Z}_n$ to choose a predicting function for Y. What now is an appropriate functional to measure the goodness of a predicting function, and what is a suitable MSC, i.e., a suitable estimate of this functional?

We formalize this prediction problem as follows. In the absence of feedback (i.e., when we are not told the true value of Y before the next prediction must be made, so that no updating of the predictor is possible), each of these predictions is a replicate of the following problem. We are given a new observation X_0 on X, and we have to predict the corresponding value Y_0, using a multiple linear regression equation estimated from $\mathcal{Z}_n$ and involving some subset of the p variables in X.

For any predicting function $Y_0' = f(X_0, \mathcal{Z}_n)$, the mean squared error of prediction (MSEP), $E(Y_0 - Y_0')^2$, where the expectation is over both X_0 and $\mathcal{Z}_n$, is a reasonable measure of goodness of the predicting function Y_0. The expectation is taken over both X_0 and $\mathcal{Z}_n$ because the whole prediction process involves obtaining the initial training set of n observations, choosing a predicting function using this training set, and then being given additional values of the explanatory variables, generated by the same random mechanism, and having to predict the corresponding values of the dependent variable.

Taking the expectation over X_0 and holding $\mathcal{Z}_n$ fixed, one obtains the conditional mean squared error, given $\mathcal{Z}_n$, which is [cf. equation (2.9) of (Stein 1960)]

$$E[(Y_0 - Y_0')^2)|\mathcal{Z}_n] = \sigma_c^2 + [(a-\alpha) + (b-\beta)'\mu_x]^2 + (b-\beta)'\Sigma_{xx}(b-\beta),$$

where the subscript x here refers to the subset c of explanatory variables, α and β are the usual regression parameters and a and b are their respective estimates.

If now the expectation is taken with respect to $\mathcal{Z}_\backslash$, one obtains MSEP(c). From formula (2.23) of Stein (1960) it follows that this gives

$$MSEP(c) = A(n,m)\sigma_c^2 \quad \text{if} \quad n \geq m_c + 3$$

where

$$A(n,m_c) = (n^2 - n - 2)/[n(n - m_c - 2)].$$

The parameter σ_c^2 is the variance of the conditional distribution of Y, given the m_c explanatory variables in subset c. The functional MSEP(c) is infinite if $n \leq m_c+2$. (The maximum likelihood estimator of the regression coefficients is well defined when $n \geq m_c + 1$, but, when $n = m_c + 1$ or $m_c + 2$, the predicting function based on it has infinite mean squared error of prediction.)

Up to this point we have developed a parametric function, MSEP, which is an appropriate functional for measuring the value of any given subset c; i.e., it is a figure-of-merit for scoring alternative subset regressions.

The problem of estimating this figure-of-merit remains. It is estimates of the figure-of-merit that are the "model- selection criteria."

Let $SSRes_c$ denote the residual sum of squares resulting from regression on subset c. The unbiased estimator of σ_c^2 is $SSRes_c/(n - m_c - 1)$ and the maximum likelihood estimator is $SSRes_c/n$. According as we choose the estimate of σ_c^2 to be the unbiased or the maximum likelihood estimator we obtain as an estimator for the MSEP either

$$A(n,m_c)SSRes_c/(n - m_c - 1)$$

or

$$A(n,m_c)SSRes_c/n.$$

Recalling that

$$A(n,m_c) = (n^2 - n - 2)/[n(n - m_c - 2)],$$

we see that these two model-selection criteria are, respectively,

Criterion 1:

$$SSRes_c(n^2 - n - 2)/[n(n - m_c - 1)(n - m_c - 2)]$$

Criterion 2:

$$SSRes_c(n^2 - n - 2)/[n^2(n - m_c - 2)].$$

The statistic $SSRes_c/[(n - m_c - 1)(n - m_c - 2)]$, which is equivalent to Criterion 1, is the criterion of Bendel and Afifi (1977). Tukey (1967), in the discussion of a paper by Anscombe, suggested the criterion $SSRes_c/(n - m_c - 1)^2$, which is similar to Criterion 1, although the approach used in his discussion is different from the present approach. For small values of n the bias of the maximum likelihood estimator can be quite sizeable, so that Criterion 1, based on the unbiased estimator, would seem to have a distinct theoretical advantage. Of course, other estimators would lead to still other criteria.

Further details of the above development are given in Sclove (1993).

4.3 Penalized likelihood model-selection criteria

Next, more general model-selection criteria, taking the form of a penalized likelihood function will be discussed.

We have seen model selection criteria in the regression problem taking the form $SSRes_c C(n, m_c)$ where C is a constant depending upon n and m_c. Note that, for such an MSC,

$$\begin{aligned} \log MSC &= \log[SSRes_c C(n, m_c)] \\ &= \log[SSRes_c] + \log C(n, m_c) \\ &= \begin{bmatrix} \text{lack} \\ \text{of fit} \\ \text{term} \end{bmatrix} + \begin{bmatrix} \text{term involving} \\ \text{the number of} \\ \text{parameters used.} \end{bmatrix} \end{aligned}$$

The criteria to be studied in what follows are similar to this; i.e., they contain a term for lack of fit and a term for the number of parameters used. More precisely, these criteria take the form of a penalized likelihood function, the negative log likelhood plus a penalty term which increases with the number of parameters. In choosing a model from a specified class, that model in the class which minimizes the model-selection criterion is chosen as the best model.

4.3.1 Akaike's, Schwarz's and Kashyap's criteria

These criteria are penalized likelihood criteria. It has become customary to multiply the log maximum likelihood by negative 2 in writing these criteria; they then take the form

$$-2\log(\max L_k) + a(n)m_k + b_k,$$

where here "log" denotes the natural (base e) logarithm, and

k = a subscript indexing the alternative models.

L_k = likelihood under the k-th model.

$\max L_k$ = supremum of L_k with respect to the parameters

m_k = number of parameters in the k-th model

$a(n) = 2,\ \forall\, n,\ b_k = 0,$ in Akaike's criterion AIC

$a(n) = \log n,\ b_k = 0,$ in Schwarz's criterion

$a(n) = \log n,\ b_k = \log[det B_k],$ in Kashyap's criterion

det = determinant

and B_k is the matrix of second partial derivatives of $\log L_k$ with respect to the parameters, evaluated at their maximum likelihood estimates. The mathematical expectation of the matrix of second partials of $\log L_k$ is the Fisher information matrix.

The optimal k is that value which minimizes the criterion.

An additional parameter will be included in the model if the improvement in

$$2\log(\max L_k)$$

is greater than $a(n)$, i.e., if $\max L_k$ improves by a factor greater than $\exp[a(n)/2]$. That is, the function $a(n)$ in a given MSC is the "cost" of fitting a single parameter according to that MSC.

Akaike's criterion generally chooses a model with more parameters than do the others. Since for n greater than 8, $\log n$ is greater than 2, Schwarz's criterion will choose a model having no more parameters than that chosen by Akaike's, when n is greater than 8. Hannan (1980) characterized monotonically nondecreasing functions $a(n)$ such that the resulting rule is consistent, finding $a(n) = \log n$ and $a(n) = \log\log n$. This shows the order required of $a(n)$ for consistency.

In homoscedastic Gaussian models,$-2\log(\max L_k)$ is, except for constants, n times the log of the maximum likelihood estimate of the variance.

4.3.2 *Remarks on the derivations of the criteria*

Khinchin (1957) gives a set of reasonable axioms which leads to *entropy* as the measure of information. Akaike's information criterion AIC (Akaike 1973, 1983, 1985, 1987) is based on a heuristic estimate of the *cross–entropy* true model and any candidate model k (see Parzen 1982). Let us review the background of this.

Let X be an observable random variable, H_f the hypothesis that X is distributed with probability density function (p.d.f.) f, and H_g the hypothesis that X is distributed with p.d.f. g. The quantity $f(x)/g(x)$ is the likelihood ratio. Kullback (1959, p. 5) defines the log of the likelihood ratio, $\log[f(x)/g(x)]$, as "the information in $X = x$ for discrimination" in favor of H_f against H_g, referring the reader to Good (1950, p. 63), who describes it also as the "weight of evidence" for H_f given x. Kullback (1959) then defines the information per observation from f for discriminating between f and g, $I(f:g)$, as the expected log likelihood ratio,

$$I(f:g) = E_f\left[\log\left(\frac{f(X)}{g(X)}\right)\right],$$

where E_f denotes expectation with respect to f. Note that

$$I(f:g) = E_f[\log f(X)] - E_f[\log g(X)].$$

The Kullback-Leibler discrepancy $-\ E_f[\log g(X)]$ (Kullback and Leibler 1951) is thus the essential part of the expected log-likelihood ratio; it is related to the entropy.

Suppose g varies over a specified family (often, a parametric family). Then minimizing $I(f:g)$ with respect to g is a way of finding that member of the family which most closely approximates (in this sense) the true p.d.f., f.

Akaike's criterion AIC is an asymptotic approximation to a sample estimate of the cross-entropy

$$-E_f[\log g(X)].$$

So, AIC is related to Kullback-Leibler information.

In turn, Kullback-Leibler information can be related to the Bayesian approach. Let x denote the sample value of X. (The variable X may be, e.g., a vector of n observations; then x denotes the value of the sample of n.) If H_f is the hypothesis that X is distributed according to f and H_g is the hypothesis that X is distributed according to g, and $P(H_f)$ and $P(H_g)$ are the prior probabilities of H_f and H_g, then we see that the the log-likelihood ratio and the log-posterior odds differ only by constant, for

$$\frac{P(H_f|x)}{P(H_g|x)} = \frac{P(H_f)f(x)}{P(H_g)g(x)},$$

or

$$\log \frac{f(x)}{g(x)} = \log \frac{P(H_f|x)}{P(H_g|x)} - \log \frac{P(H_f)}{P(H_g)}.$$

The approach of Schwarz (1978) and Kashyap (1982) is Bayesian, though these criteria should have large-sample appeal outside the Bayesian framework as they are asymptotically independent of particular prior specifications.

The Bayesian approach to model selection centers upon the posterior probabilities of the alternative models, given the observations. Schwarz's (1978) and Kashyap's (1982) criteria come from an expansion of the log posterior probabilities of the alternative models. Schwarz's criterion comes from the first term. Kashyap (1982), generalizing Schwarz's work beyond exponential families, takes the expansion a term further. Thus Kashyap's criterion comes from the first two terms. Leonard (1982) also discusses the posterior probability expansion approach to model-selection criteria. [Earlier, DeGroot (1970) had given the relevant asymptotic expansion of posterior probabilities.] A somewhat more general expansion is one of the basic theorems of the book by Linhart and Zucchini (1986); they do not, however, reference the work of DeGroot, Kashyap, Leonard or Schwarz, perhaps because the Bayesian approach does not fall within the framework they have set out for their book.

The idea behind the approach of Schwarz-Kashyap-Leonard is to expand the integrand giving the log posterior probability in a Taylor's series around the maximum likelihood estimator and evaluate the results using Gaussian integration. One then notes that there is a term containing m_k and that the multipler of m_k is $\log n$.

Stated in somewhat more detail, the result giving Schwarz's criterion is (cf. Kashyap 1982)

$$\begin{aligned}\log P(H_k|x) \;\approx\; & \log C + \log P(H_k) + \log(\max L_k) + (m_k/2)\log(2\pi) \\ & -(m_k/2)\log n + \log f_k(\theta_k*)\end{aligned}$$

where C is a constant; θ_k is the m_k dimensional parameter of the k-th model, $\theta_k{*}$ is its maximum likelihood estimate, and f_k is the prior over θ_k; $f_k(\theta_k{*})$ is this prior, evaluated at the maximum likelihood estimate. That is,

$$\log P(H_k|x) \approx \log C + \log P(H_k) + (m_k/2)\log(2\pi) - SIC(k)/2 + \log f_k(\theta_k*),$$

where

$$SIC(k) = -2\log(maxL_k) + m_k \log n$$

is Schwarz's criterion.

4.4 Finite mixture models

The finite mixture density is

$$f(x) = p_1 f_1(x) + p_2 f_2(x) + \cdots + p_K f_K(x),$$

$p_c > 0$, $c = 1, 2, \cdots, K$, and $p_1 + p_2 + \cdots + p_K = 1$, where the f_c's are densities, usually from a specified parametric family.

The choice of the number K of components in the mixture is a model-selection problem. It would be helpful if the above model-selection criteria could be applied. However, regularity conditions are required for the validity of the expansions leading to these criteria. Unfortunately, these conditions are not met for the finite mixture model. The problem is that if p_c is set equal to zero, the parameters of the corresponding f_c become meaningless. [See, e.g., Hartigan (1985).] Thus the likelihoods for different values of K may not be comparable.

Nonetheless, use of model-selection criteria for choice of K has met with some success. Examples and further details are given in Sclove (1993).

4.5 Minimum-bit data representation

Next, an information-theoretic approach to model selection, through minimum-bit data representation, is explored, with special reference to cluster analysis.

Consider the following (very small) problem. The dataset 2, 3, 4, 6, 7, 8 is to be described, represented and stored. One could store the six numbers as they are. This would require a certain number of bits. Alternatively, one could store the mean (5), and the deviations from the mean. This would require a certain number of bits (e.g., number of bits needed to represent a number $y > 1$ is approximately $\log_2 y$), and a certain amount of effort to store the mean, recall it, and add it to the deviations if the raw data were to be re-constructed.

Alternatively, since these data fall into two "clusters," 2, 3, 4 and 6, 7, 8, one might consider storing the two cluster means (3 and 7) and deviations from the cluster means. The point is that the deviations from the cluster means will be smaller than the deviations from the overall mean and may require fewer bits. On the other hand, there is the problem of coding, storing and recalling two means to re-construct the data. One could attach costs to the various elements of this data storage problem. The number of clusters used should be the number which minimizes the corresponding overall cost criterion. Further details of this example are given in Sclove (1993).

Such considerations of minimum-bit data representation are evidently related to Rissanen's approach to model-selection criteria. The similarity of the asymptotic form of Rissanen's criterion and Schwarz's criterion is notable: In both cases, the multiplier $a(n)$ of the number of parameters m_k is $a(n) = \log n$. Thus, the choice $a(n) = \log n$ is seen to have arisen from two somewhat different approaches. Rissanen's criterion is of the following form.

$$\begin{bmatrix} \text{total} \\ \text{number of bits} \\ \text{required} \end{bmatrix} = \begin{bmatrix} \text{number of bits} \\ \text{to encode} \\ \text{the parameters} \end{bmatrix} + \begin{bmatrix} \text{number of bits} \\ \text{to encode} \\ \text{the data} \end{bmatrix}$$

If one identifies the first term with log prior probability and the second with log likelihood, then this becomes

$$\text{log prior probability} + \text{log likelihood}$$

$$= \log(\text{prior probability} \times \text{likelihood})$$
$$= \text{Const.} + \log(\text{posterior probability}),$$

giving a link between the minimum-bit and Bayesian approaches (Cheeseman 1993).

4.6 References

[Aitkin 1974] Aitkin, M.A. (1974) "Simultaneous Inference and the Choice of Variable Subsets," *Technometrics* **16**, 221-227.

[Akaike 1969] Akaike, H. (1969) "Fitting Autoregressive Models for Prediction," *Ann. Inst. Statist. Math.* **21**, 243-247.

[Akaike 1973] Akaike, H. (1973) "Information Theory and an Extension of the Maximum Likelihood Principle," *Proc. 2nd International Symposium on Information Theory*, B.N. Petrov and F. Csaki, eds., 267-281. Akademia Kiado, Budapest.

[Akaike 1983] Akaike, H. (1983) "Statistical Inference and Measurement of Entropy," *Scientific Inference, Data Analysis and Robustness*, H. Akaike and C.-F. Wu, eds., 165-189. Academic Press, New York.

[Akaike 1985] Akaike, H. (1985) "Prediction and Entropy," *A Celebration of Statistics: The ISI Centenary Volume*, A.C. Atkinson and S.E. Fienberg, eds., 1-24,. Springer-Verlag, New York.

[Akaike 1987] Akaike, H. (1987) "Factor analysis and AIC," *Psychometrika* **52**, 317-332.

[Bendel and Afifi 1977] Bendel, R.B. and Afifi, A.A. (1977) "Comparison of Stopping Rules in Forward Stepwise Regression," *Jnl. Amer. Statist. Assoc.* **72**, 46-53.

[Cheeseman 1993] Cheeseman, P. (1993) "Overview of Model Selection." *4th Internat'l Workshop on Artificial Intelligence and Statistics*, Society for Artificial Intelligence and Statistics, Jan. 3-6, 1993, Ft. Lauderdale, Florida.

[DeGroot 1970] DeGroot, M. H. (1970) *Optimal Statistical Decisions*. McGraw-Hill, New York.

[Good 1950] Good, I. J. (1950) *Probability and the Weighing of Evidence*. Charles Griffin, London.

[Hannan 1980] Hannan, E. J. (1980) "The Estimation of the Order of an ARMA Process," *Ann. Statist.* **8**, 1071-1081.

[Hartigan 1985] Hartigan, J. (1985) "A Failure of Likelihood Asymptotics for Normal Mixtures," *Proc. Berkeley Conf. in Honor of Jerzy Neyman and Jack Kiefer)*, L.M. LeCam and R.A. Olshen, eds., **II**, 807-811, Wadsworth, Inc., Belmont, Calif.

[Kashyap 1982] Kashyap, R. L. (1982) "Optimal Choice of AR and MA Parts in Autoregressive Moving Average Models," *IEEE Trans. PAMI* **4**, 99-104.

[Khinchin 1957] Khinchin, A. I. (1957) *Mathematical Foundations of Information Theory*. Dover, New York. (Translation of two papers by Khinchin in *Uspekhi Matematicheskii Nauk* **VII, no. 3**, 1953, pp. 3-20 and **XI**, no. 1, 1956, pp. 17-75.)

[Kullback 1959] Kullback, S. (1959) *Information Theory and Statistics*. John Wiley & Sons, New York. Reprinted (1968), Dover Publications, New York.

[Kullback and Leibler 1951] Kullback, S. and Leibler, R.A. (1951) "On Information and Sufficiency," *Ann. Math. Statist.* **22**, 79-86.

[Leonard 1982] Leonard, T. (1982) Comment on "A Simple Predictive Density Function" by M. LeJeune and G. D. Faulkenberry, *J. Amer. Statist. Assoc.* **77**, 657-658.

[Linhart and Zucchini 1986] Linhart, H. and Zucchini, W. (1986) *Model Selection.* John Wiley & Sons, New York.

[Parzen 1982] Parzen, E. (1982) "Maximum Entropy Interpretation of Autoregressive Spectral Densities," *Statist. and Prob. Lttrs.* **1**, 7-11.

[Rissanen 1978] Rissanen, J. (1978) "Modeling by Shortest Data Description," *Automatica* **14**, 465-471.

[Rissanen 1985] Rissanen, J. (1985) "Minimum-Description-Length Principle," *Ency. Statist. Sci.* **5**, 523-527. John Wiley & Sons, New York.

[Rissanen 1986] Rissanen, J. (1986) "Stochastic Complexity and Modeling," *Ann. Statist.* **14**, 1080-1100.

[Rissanen 1987] Rissanen, J. (1987) "Stochastic Complexity," *J. Roy. Statist. Soc.* **B49**, 223-239.

[Schwarz 1978] Schwarz, G. (1978) "Estimating the Dimension of a Model," *Ann. Statist.* **6**, 461-464.

[Sclove 1969] Sclove, S. L. (1969) "On Criteria for Choosing a Regression Equation for Prediction," *Technical Report No. 28*, Dept. of Statistics, Carnegie-Mellon University.

[Sclove 1993] Sclove, S. L. (1993) "Some Aspects of Model-Selection Criteria." To appear in *Multivariate Statistical Modeling, Proc. 1st U.S./Japan Conference on the Frontiers of Statistical Modeling: An Informational Approach*, H. Bozdogan, ed., Kluwer Academic Publishers, Dordrecht, the Netherlands.

[Stein 1960] Stein, C. (1960) "Multiple Regression," *Contributions to Probability and Statistics: Essays in Honor of Harold Hotelling*, I Olkin, ed., 424-443. Stanford Univ. Press, Palo Alto.

[Tukey 1967] Tukey, J.W. (1967) Discussion of "Topics in the Investigation of Linear Relations fitted by the Method of Least Squares" by F.J. Anscombe, *J. Roy. Statist. Soc.* **B29**, 2-52.

5
On the choice of penalty term in generalized FPE criterion

Ping Zhang

Department of Statistics, University of Pennsylvania

ABSTRACT In variable selection, many existing selection criteria are closely related to the generalized final prediction error (FPE) criterion. In the linear regression context, the FPE criterion amounts to minimizing $C(k, \lambda) = \text{RSS}(k) + \lambda k \hat{\sigma}^2$ over k, where k is the number of parameters in the model, RSS(k) is the residual sum of squares and $\hat{\sigma}^2$ is some estimate of the error variance. Different values of λ give different selection criteria. This article presents some useful results on the choice of λ. Some insights are obtained. Application to the study of the multifold cross validation criterion is also discussed.

Key words and phrases: Minimax and admissible criterion; Multifold cross validation; Parsimony principle; Random walk; Variable selection.

5.1 Introduction

One of the goals of statistical modelling is to make intelligent predictions. In regression models, this is achieved by relating the so called response variable Y with a group of covariates $X_1, \ldots, X_K$. In the horse race example considered by Salzberg (1986), Y represents the win/lose status of a particular horse and the X variables are observable characteristics of that horse. Model selection in this context amounts to selecting a subset of these X variables that best predict the outcome of a future race. Another example is medical diagnosis. Here Y indicates whether a patient has one of a few possible diseases. The X variables are health related attributes of that patient. When the number of X variables is large, it takes more than an expert to make a reasonable prediction. A large part of what an expert does is to locate a few key X variables that he feels are most relevant to the Y variable. Such a task can often be accomplished by some intelligent computer programs. The HANDICAPPER program described in Salzberg (1986) provides a good example. A similar, but more sophisticated, program (CART) has been developed by Breiman et. al. (1984). Other less automatic model selection criteria abound. See Akaike (1973), Mallows (1973), Schwarz (1978), Hannan & Quinn (1979), Linhart & Zucchini (1986), Rissanen (1986) and Wei (1992).

Computing intensive methods such as HANDICAPPER and CART, which are application oriented, lack rigorous theoretical justifications. The criteria developed by theorists, on the other hand, rely too heavily on oversimplified assumptions that are hard to verify. One thing in common, however, for all model selection methods is that they all utilize the parsimony principle, either implicitly or explicitly. In fact, the whole area of model selection has close ties with the parsimony principle. Other things being equal, simple models are always preferred. The

---

[0]The research is partially supported by the Research Foundation of the University of Pennsylvania and NSA Grant MDA904-93-H-3014.

focus of this paper is to describe some results that provide insights into the quantitative aspect of the parsimony principle.

Formally speaking, let us consider the linear regression model $Y = X\beta + \epsilon$, where $Y = (y_1, \ldots, y_n)^t$ is the observation vector, $X = \{(x_{ij}) : 1 \leq i \leq n, 1 \leq j \leq K\}$ is the design matrix, and $\epsilon = (\epsilon_1, \ldots, \epsilon_n)^t$ is a vector of independent identically distributed random variables with mean zero and variance σ^2. Suppose that the model is correct except that some of the coefficients might be zero. Let k_0 denote the number of non-zero coefficients. It is conventionally assumed that $\beta = (\beta_1, \ldots, \beta_{k_0}, 0, \ldots, 0)$. In other words, the covariates are often assumed to be preordered so that only the number of variables needs to be determined. The problem of identifying the set of nonzero parameters is called variable selection.

In this article, we consider the generalized final prediction error (FPE) criterion

$$C(k, \lambda) = \text{RSS}(k) + \lambda k \hat{\sigma}^2, \tag{5.1}$$

where $\text{RSS}(k)$ is the residual sum of squares for a model where only k out of K covariates are used, $\hat{\sigma}^2$ is an appropriate estimate of σ^2 and λ is a non-negative penalty for model complexity. The model to be selected is the one that minimizes criterion (1).

Model selection criteria of the form (1) abound in the literature. A popular justification of (1) is that it reflects the spirit of parsimony principle. The traditional AIC and BIC criteria correspond to $\lambda = 2$ and $\lambda = \log n$ respectively. The ϕ-criterion of Hannan & Quinn (1979) suggests the use of $\lambda = c \log \log n$, $c > 0$. A key question in variable selection is the choice of penalty term λ. Considerable efforts have been made in the literature. One knows now, for example, that $\lambda = \log n$ has a nice Bayesian connection (Schwarz, 1978) and $\lambda = c \log \log n$ distinguishes strong consistency from weak consistency (Hannan & Quinn, 1979). Akaike (1973, 1985) argues that $\lambda = 2$ is a magic number that emerges naturally if the problem is set up properly. George & Foster (1992) show that $\lambda = 2 \log k$ is the right choice based on a minimax consideration. Hurvich & Tsai (1989) suggest that $\lambda = 2n/(n - k - 2)$ be used to compensate for the bias incurred when k/n is large. Each one of these criteria seems to have a perfect justification if one looks at the problem in a particular way. However, for a data analyst who is less concerned with theoretical optimality, it is more important to have available a rule of thumb that is simple but flexible enough to be useful under a variety of practical situations. Based on empirical evidence, some authors have suggested that criterion (1) with a λ between 2 and 6 would do well in most situations. See Bhansali & Downham (1977), Atkinson (1980, 1981) and Friedman (1991). Critics of such an argument are concerned that such ad hoc choice of λ lacks theoretical justification. They argue that any finite λ would lead to an inconsistent criterion. Hence a penalty term that depends on the sample size should be used. Specifically, it has been suggested that $\lambda = \lambda_n$ should satisfy $\lambda_n \to \infty$ and $\lambda_n / n \to 0$. See Bozdogan (1987), Nishii (1988) and Zhao et al. (1986a,b).

The purpose of this article is to provide a unified approach to the assessment of criterion (1). The results obtained will in turn allow us to compare the performance of FPE criteria with different penalty terms. Section 2 presents some distributional properties of criterion (1). We argue that a choice of $\lambda \in [3, 4]$ would be adequate for most practical purposes. In section 3, we consider decision theoretic properties of (1) and show that the incorrect models are sometimes preferable to the true model. In section 4, we present an elegant connection between the multifold cross validation criterion and the FPE criterion. The former method is widely applied in classification studies. Our results shed some light on the way MCV should be implemented. Finally in section 5, some remarks are made regarding model selection in general.

5.2 The random walk approach

Let $\hat{k}_\lambda = \arg\min_k C(k, \lambda)$. A selection criterion is consistent if $\hat{k}_\lambda$ converges to k_0 as the sample size n increases. Unfortunately, the FPE criterion is not consistent for any finite λ. Instead, $\hat{k}_\lambda$ converges to a random variable. This section gives the asymptotic distribution of $\hat{k}_\lambda$ from which useful insight can be obtained regarding the choice of λ.

For simplicity, we assume that σ^2 is known so that criterion (1) can be written as $C(k, \lambda) = \text{RSS}(k) + \lambda k \sigma^2$. It is easy to show (see Shibata, 1976) that $P(\hat{k}_\lambda < k_0) \to 0$. On one hand, this means that the FPE criterion will never choose a model that is underfit. On the other hand, the same result implies that one can essentially define $\hat{k}_\lambda$ as the minimizer of $C(k, \lambda)$ over the range $k \geq k_0$.

Let P_k be the projection matrix of the kth model, i.e. $P_k = X_k(X_k' X_k)^{-1} X_k'$ where X_k is the corresponding covariate matrix of the kth model. Define $W_k = P_k - P_{k-1}$. Then for $k \geq k_0$, we can write

$$C(k, \lambda) = \epsilon^t \epsilon - (\epsilon^t P_k \epsilon - \lambda k \sigma^2) = \epsilon^t \epsilon - \sum_{i=1}^{k} Z_i^{(n)},$$

where $Z_i^{(n)} = \epsilon^t W_i \epsilon - \lambda \sigma^2$. Minimizing $C(k, \lambda)$ is thus equivalent to maximizing $S_k^{(n)} = \sum_{i=1}^k Z_i^{(n)}$, $k = k_0, k_0 + 1, \ldots, K$. Observe that the random vector $(Z_1^{(n)}, \ldots, Z_K^{(n)})$ converges in distribution to $(Z_1, \ldots, Z_K)$, where the Z_i's are independent identically distributed random variables with distribution $\sigma^2(\chi_1^2 - \lambda)$. Hence asymptotically, the problem of model selection is equivalent to the problem of finding the global maximizer of the random walk sequence $S_k = \sum_{i=1}^k Z_i$, $k = k_0, k_0 + 1, \ldots, K$. The following result is due to Zhang (1992).

Proposition 1 *For $\lambda > 1$, there exists a random variable $T_\lambda < \infty$ such that $\lim_{K - k_0 \to \infty} \hat{k}_\lambda = T_\lambda$ with probability one. The distribution of T_λ satisfies*
(a) $P(T_\lambda = k_0) = \exp[-\sum_1^\infty k^{-1} P(\chi_k^2 > \lambda k)]$,
(b) $E(T_\lambda) = k_0 + \sum_1^\infty P(\chi_k^2 > \lambda k)$ and
(c) $var(T_\lambda) = \sum_1^\infty k P(\chi_k^2 > \lambda k)$.
If $\lambda \leq 1$, then $\hat{k}_\lambda$ tends to ∞ with probability one.

These results are useful in the evaluation of criterion (1). For instance, it is well know that $\lambda = 2$ is not necessarily the best choice. But how bad is it? From part (a) of the Proposition 1, we have that for the C_p or AIC criteria

$$P(\hat{k}_{AIC} = k_0) \geq P(T_2 = k_0) = 0.7117.$$

In other words, $\lambda = 2$ is not bad at all because the corresponding criterion chooses the correct model with a probability higher than 70% even in the worst case. Clearly, increasing λ will improve the chance of choosing the correct model. There is a point, however, where further increases of λ will not produce significant improvement. From Figure 1, we can see that the interval of λ values that yields reasonable performance is roughly between $\lambda = 3$ and $\lambda = 4$.

5.3 The decision theoretic approach

In this section, we consider the choice of λ from a decision theoretic point of view. It is obvious that different λ values cause criterion (1) to choose different models. Our goal is to find the

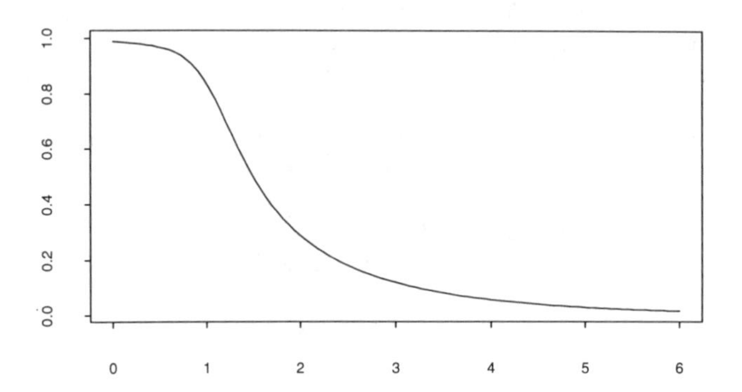

FIGURE 1. The probability of mis-identification as a function of the penalty term. The horizontal axis is λ and the vertical axis is λ v.s. $1 - P(T_\lambda = k_0)$.

optimal λ using decision theory. This issue has been dealt with by many authors. Stone (1981) and Kempthorne (1984) considered the problem of admissibility. Shibata (1986) and George & Foster (1992) studied the minimaxity of criterion (1). Schwarz (1978) and Smith & Speigelhalter (1980) gave the FPE criterion a Bayesian interpretation.

We consider only the simplest case where the selection is between two nested Gaussian linear regression models. Assume that the following are the feasible models:

$$\begin{aligned} M_1 :&\ \ Y = X_1\alpha + \epsilon, \\ M_2 :&\ \ Y = X_1\alpha + X_2\beta + \epsilon, \end{aligned}$$

where X_1 is $n \times k_1$ and X_2 is $n \times m$. Suppose that $\epsilon = (\epsilon_1, \ldots, \epsilon_n)^t$ and the ϵ_i's are iid $N(0, \sigma^2)$ with σ^2 known. In such case, the $\hat{\sigma}^2(K)$ term in (1) should be replaced by σ^2 and criterion (1) will choose the model with dimension $\hat{k}_\lambda = \arg\min_{k=k_1,k_1+m} C(k, \lambda)$. The loss function is chosen to be the squared prediction error. Hence the risk function is $R(\lambda, \theta, k_0) = E\|\tilde{Y} - \hat{Y}(\hat{k}_\lambda)\|^2$, where $\tilde{Y}$ is a future observation vector on the same design points and $\theta = (\alpha, \beta)$.

Let $Y = f + \epsilon$ be the true model, where f is either $X_1\alpha$ or $X_1\alpha + X_2\beta$. Usually, the minimax penalty term λ^* is defined by

$$\lambda^* = \arg\inf_{\lambda} \sup_{(\theta, k_0)} R(\lambda, \theta, k_0). \tag{5.2}$$

It is well known, however, that $\lambda^* = 0$ under the current set up. In other words, the minimax criterion will always include all covariates (Shibata, 1986 and Stone, 1981). To avoid such trivial conclusions, we put constraints on the parameter space. Let $b = f^t(P_2 - P_1)f/\sigma^2$ where P_1 and

P_2 are the projection matrices corresponding to the two linear regression models M_1 and M_2 respectively. Define the constrained minimax penalty term by

$$\lambda^*(M) = \arg\inf_{\lambda} \sup_{b \leq M, k_0} R(\lambda, \theta, k_0).$$

Then the following result is due to Zhang & Krieger (1993).

Proposition 2 *(a) If $M < 1$, then $\lambda^*(M) = \infty$. (b) If the parameter space is restricted to $b \leq 1$, then all $\lambda \geq 0$ are inadmissible.*

Since $b/m = \{f^t(P_2 - P_1)f/m\}/\sigma^2$ resembles the usual F-statistic for testing between two nested models, Proposition 2 suggests that when this population version of F-statistic is less than m^{-1}, criterion (1) is not appropriate because it is not admissible no matter what the value of λ is. Furthermore, the criterion that always chooses the smaller model dominates (1) for all $\lambda \geq 0$. The latter criterion is also minimax. In some sense, this is parsimony principle at work. Notice that the parameter b is a measure of the contribution of the extra terms that are in M_2 but not in M_1. Presumably, an honest selection criterion should select M_2 whenever the extra parameters are not zero. The above result shows, however, that a good criterion should choose M_2 only when the extra terms are important enough.

The conclusion of Proposition 2 can also be extended to more general situations. It is folklore that all nice decision rules are Bayes rules or at least close to Bayes rules. Adopting this point of view, we can obtain an intuitive understanding of why criterion (1) sometimes fails when the parameter space is constrained. For simplicity, we assume that the matrices X_1 and X_2 are orthogonal to each other. Suppose that (α, β) has some prior distribution. Then according to Kempthorne (1984), the Bayes rule chooses M_1 over M_2 if and only if

$$E(\beta^t|Y)X_2^tX_2E(\beta|Y) < (\hat{\beta} - E(\beta^t|Y))^t X_2^tX_2(\hat{\beta} - E(\beta|Y)),$$

where $\hat{\beta} = (X_2^tX_2)^{-1}X_2^tY$. This is equivalent to

$$\hat{\beta}^tX_2^tX_2\hat{\beta} > 2\hat{\beta}^tX_2^tX_2E(\beta|Y). \tag{5.3}$$

Moreover, it is straightforward to show that criterion (1) chooses M_1 over M_2 if and only if

$$\hat{\beta}^tX_2^tX_2\hat{\beta} > \lambda m\sigma^2. \tag{5.4}$$

When there is no constraint on the parameter space, Kempthorne (1984) argues that one can choose a prior distribution such that equality holds in (3). This suggests that the Bayes risk is the same under both model M_1 and M_2. Hence all criteria are admissible. However, when the prior distribution of β is supported on a bounded region, it is obvious that (3) and (4) can not be the same. Hence criterion (1) can not be Bayesian. It is in this sense that criterion (1) can not be admissible whenever β is bounded. A rigorous proof of such a statement, however, is beyond the scope of this article.

5.4 Multifold cross validation

In this section, we apply the results of section 2 to the study of the multifold cross validation (MCV) criterion. Multifold cross validation gives a criterion of the form

$$\text{MCV}_k = \frac{1}{\binom{n}{d}} \sum_s \|Y_s - X_{s,k}\hat{\beta}_{(-s)k}\|^2, \tag{5.5}$$

where s represents a size-d subset of $\{1, 2, \ldots, n\}$, and $Y_s = \{y_i : i \in s\}$, $X_{s,k} = \{(x_{jl}) : j \in s, l \leq k\}$. $\hat{\beta}_{(-s)k}$ is the least squares estimate of β obtained without using observations in s. When $d = 1$, (5) reduces to the simple leave-one-out cross validation (CV).

The idea of deleting more than one observation is actually more natural than the more popular delete one CV. The first theoretical treatment of MCV apparently appeared in Geisser (1975). For applications of MCV in the context of classification and regression trees (CART), see Breiman et al. (1984). Some recent accounts are given by Burman (1989), Breiman & Spector (1989) and Shao (1993). A well known fact is that the simple delete one CV is asymptotically equivalent to (1) with $\lambda = 2$ (Stone, 1977). The parallel result for MCV has been obtained by Zhang (1993). We have

Proposition 3 *Let $P_k = X_k(X_k^t X_k)^{-1} X_k^t$ be the projection matrix of the k-dimensional submodel. Suppose that $p = \lim_{n\to\infty}(d/n)$ is positive. Then under some mild conditions,*

$$n\,MCV_k = \begin{cases} RSS(k) + \lambda k\sigma^2 + o_p(1), & k \geq k_0; \\ \epsilon^t\epsilon + nb_k + o_p(n), & k < k_0, \end{cases}$$

where $\lambda = (2-p)/(1-p)$ and $b_k = \liminf_{n\to\infty} n^{-1}(\beta^t X^t P_k^{\perp} X\beta) > 0$ for $k < k_0$.

An important question concerning the use of MCV criterion is how to choose p, the proportion of data that are deleted. Proposition 3 shows that the MCV criterion is asymptotically equivalent to the FPE criterion (1) with $\lambda = (2-p)/(1-p)$. Some interesting observations can be made here. From Proposition 1, the larger the penalty term λ the better the criterion is as far as choosing the correct model is concerned. Since $(2-p)/(1-p) > 2$, MCV is therefore a better criterion than CV. Moreover, since $(2-p)/(1-p)$ is increasing in p, the larger p is, the better the criterion. This sounds a little pathological because the best MCV criterion would then correspond to $p = 1$, where almost all the observations are deleted (Shao, 1993). Using an argument similar to the discussion at the end of section 2, it is safe to say that $p = 75\%$ is a reasonable choice. This means that the majority of the data should be used for validation.

It is obvious that (5) involves huge amount of computation. Many cost effective variations of MCV have been suggested in the literature. The most convenient approach is the following bootstrap method. Suppose that we draw random subsets of size-d from $\{1, \ldots, n\}$. Let $s_i^*, i = 1, \ldots, N$, be the sampled subsets. Burman (1989) proposes a so called repeated learning-testing (RLT) criterion defined as

$$\mathrm{RLT}_k = N^{-1} \sum_{i=1}^{N} \left\| Y_{s_i^*} - X_{s_i^*,k} \hat{\beta}_{(-s_i^*),k} \right\|^2.$$

If $N/n^2 \to \infty$, we can show that

$$\mathrm{RLT}_k = \mathrm{MCV}_k + o_p(n^{-1}).$$

In other words, the performance of RLT is asymptotically comparable to that of MCV. In simulation studies conducted by Zhang (1993), satisfactory results are obtained with relatively small bootstrap sample size N. See Shao (1993) for other variations of MCV criterion.

5.5 Concluding remarks

The scope of model selection extends beyond linear regression. In general, suppose that there are a number of possible models that can be used to describe the data. Let the candidate parametric models be denoted by $f_k(x, \theta_k), \theta_k \in \Theta_k$, $k = 1, \ldots, K$. The FPE criterion can be defined as

$$C^*(k, \lambda) = -2f_k(x, \hat{\theta}_k) + \lambda \text{dim}(\Theta_k),$$

where $\hat{\theta}_k$ is the maximum likelihood estimate of θ_k under the kth model and $\text{dim}(\Theta_k)$ is the dimension of the parameter space Θ_k. Under regularity conditions, we expect that results similar to the linear regression case would hold for the general $C^*(k, \lambda)$ criterion. In particular, the rule of thumb $\lambda \in [3, 4]$ would still be valid.

A key assumption in the previous sections is that the order of importance of the X variables are given. This has received much criticism in the literature. The so called all subset selection method, which avoids such an awkward assumption, goes to the other extreme by ignoring any prior information on the ordering of the X variables. Obviously, such all subset search can be computationally very expensive. The best solution, in our view, is to employ some kind of a sequential approach. At first, all covariates receive equal weights. As more data become available, knowledge about the importance of the covariates accumulates and the weights assigned to them are updated. At the end, the variables with high weights are selected. It is clear that such methods have strong Bayesian flavor.

One way to assign weights is through a hierachical Bayesian approach. To illustrate, let $\beta_1, \ldots, \beta_K$ be the coefficients corresponding to the K covariates. Suppose that a prior distribution is assumed for these parameters. Let $I = (I_1, \ldots, I_K)$, where the I_k's are independent Bernoulli trials with success probabilities equal to $p_1, \ldots, p_K$ respectively. Conditional on I, $(\beta_1, \ldots, \beta_K)$ follows the marginal distribution of $(\beta_1 I_1, \ldots, \beta_K I_K)$. The probabilities $p_1, \ldots, p_K$ can be viewed as the weights assigned for each covariates. At the beginning, one can assume a non-informative prior for I by letting $p_k = 0.5$, $k = 1, \ldots, K$. Every time a new observation comes, the weight probabilities are updated by the corresponding posterior distribution of I. It is not clear yet how this approach works. We plan to report progress in this direction at a different time.

5.6 REFERENCES

[1] Akaike, H. (1973). Information theory and an extension of the maximum likelihood principle. *In: Second International Symposium on Information Theory. (N.N. Petrov and F. Czái eds), 267-281. Budapest: Akademiai Kiadó.*

[2] Akaike, H. (1985). Prediction and Entropy. *In: A Celebration of Statistics, The ISI Centenary Volume.* Eds. A. Atkinson and S. Fienberg, New York: Springer-Verlag, 1-24.

[3] Atkinson, A.C. (1980). A note on the generalized information criterion for choice of a model. *Biometrika*, **67**, 413-418.

[4] Atkinson, A.C. (1981). Likelihood ratios, posterior odds and information criteria. *Journal of Econometrics*, **16**, 15-20.

[5] Bhansali, R.J. & Downham, D.Y. (1977). Some properties of the order of an autoregressive model selection by a generalization of Akaike's FPE criterion. *Biometrika*, **64**, 547-551.

[6] Bozdogan, H. (1987). Model selection and Akaike's information criterion (AIC): The general theory and its analytical extensions. *Psychometrika*, **52**, 345-370.

[7] Breiman, L., Friedman, J.H., Olshen, R.A. & Stone, C. (1984). *Classification and Regression Trees*. Wadsworth, Belmont, California.

[8] Breiman, L. & Spector, P. (1990). Submodel selection and evaluation in regression: The X-random case. *Internat.Statist.Rev., forthcoming.*

[9] Burman, P. (1989). A comparative study of ordinary cross-validation, v-fold cross-validation and the repeated learning-testing methods. *Biometrika*, **76** 503-514.

[10] Friedman, J. (1991). Multivariate adaptive regression splines (with discussions). *Ann.Statist.*, **19** 1-141.

[11] Geisser, S. (1975). The predictive sample reuse method with applications. *J.Amer.Statist.Assoc.*, **70** 320-328.

[12] George, E. & Foster, D. (1992). The risk inflation of variable selection in regression. Ann.Statist., forthcoming.

[13] Hannan, E. & Quinn, B. (1979). The determination of the order of an autoregression. *J.Roy.Statist.Soc., Series B*, **41**, 191-195.

[14] Hurvich, C. & Tsai, C.L. (1989). Regression and time series model selection in small samples. *Biometrika*, **76**, 297-307.

[15] Kempthorne, P. (1984). Admissible variable selection procedures when fitting regression models by least squares for prediction. *Biometrika*, **71**, 593-474.

[16] Linhart, H. & Zucchini, W. (1986). *Model Selection*. New York: John Wiley & Sons.

[17] Mallows, C.L. (1973). Some comments on C_p. *Technometrics*, **15**, 661-675.

[18] Nishii, R. (1988). Maximum likelihood principle and model selection when the true model is unspecified. *Journal of Multivariate Analysis*, **27**, 392-403.

[19] Rissanen, J. (1986). Stochastic complexity and modelling. *Ann.Statist.*, **14**, 1080-1100.

[20] Salzberg, S. (1986). Pinpointing good hypotheses with heuristics. *In: Artificial Intelligence and Statistics. Eds. W. Gale*, pp. 133-158. Addison-Wesley Publishing Company, Reading, MA.

[21] Schwarz, G. (1978). Estimating the dimension of a model. *Ann.Statist.*, **6**, 461-464.

[22] Shao, J. (1993). Linear model selection by cross-validation. *J.Amer.Statist.Assoc.*, **88**, 486-494.

[23] Shibata, R. (1976). Selection of the order of an autoregressive model by Akaike's information criterion. *Biometrika*, **63**, 117-126.

[24] Shibata, R. (1986). Selection of the number of regression variables: A minimax choice of generalized FPE. *Ann.Inst.Stat.Math.*, **38**, 459-474.

[25] Smith, A.F.M. & Speigelhalter, D.J. (1980). Bayes factors and choice criteria for linear models, *J.Roy.Statist.Soc., Series B*, **42**, 213-220.

[26] Stone, C. (1981). Admissible selection of an accurate and parsimonious normal linear regression model, *Ann.Statist.*, **9**, 475-485.

[27] Stone, M. (1977). An asymptotic equivalence of choice of model by cross-validation and Akaike's criterion. *J.Roy.Statist.Soc., Series B*, **39**, 44-47.

[28] Wei, C.Z. (1992). On predictive least squares principles. *Ann.Statist.*, **20**, 1-42.

[29] Zhang, P. (1992). On the Distributional Properties of Model Selection Criteria. *J.Amer.Statist.Assoc.*, **87**, 732-737.

[30] Zhang, P. (1993). Model selection via multifold cross validation. *Ann.Statist.*, **21**, 299-313.

[31] Zhang, P. & Krieger, A.M. (1993). Appropriate penalties in the final prediction error criterion: A decision theoretic approach. *Statistics and Probability Letters*, forthcoming.

[32] Zhao, L.C., Krishnaiah, P.R. & Bai, Z.D. (1986a). On detection of the number of signals in presence of white noise. *Journal of Multivariate Analysis*, **20**, 1-25.

[33] Zhao, L.C., Krishnaiah, P.R. & Bai, Z.D. (1986b). On detection of the number of signals when the noise covariance matrix is arbitrary. *Journal of Multivariate Analysis*, **20**, 26-49.

6
Cross-Validation, Stacking and Bi-Level Stacking: Meta-Methods for Classification Learning

Cullen Schaffer

Department of Computer Science
CUNY/Hunter College
695 Park Avenue, New York, NY · 10021
schaffer@marna.hunter.cuny.edu

ABSTRACT So many methods have been invented for the problem of classification learning that practitioners now face a meta-level problem of choosing among alternatives or arbitrating between them. A natural idea is to use *cross-validation* to select one of several learning algorithms. A more general approach is Wolpert's *stacking*, which uses the predictions of several methods together; and this may be further generalized to *bi-level stacking*, in which ordinary attributes are allowed to play a role in arbitrating between methods. In this paper, we examine cross-validation, stacking and bi-level stacking and present empirical results to illustrate key points.

6.1 Introduction

Classification learning problems are amenable to rule set, decision tree and neural network induction methods as well as to statistical approaches including Bayesian classifiers, k-nearest-neighbor techniques and linear discriminants. With this variety of sophisticated approaches at their disposal, practitioners are faced with a meta-level problem of choosing or arbitrating between the various alternatives.

A simple, but appealing solution to this problem is to run a cross-validation study to estimate the predictive accuracy that would be attained by each of several representative methods and to select the apparent winner. The benefits—and limitations—of this cross-validation approach are illustrated with an empirical example in Schaffer (1993b).

As Wolpert (1992) has pointed out, however, selecting the apparent winner is only the simplest way of using the information provided by a cross-validation study. A more general idea—Wolpert's *stacking*—is to use cross-validation to tell us what various classification methods would have predicted about the class of each training case and then to induce a relationship between these predictions and the true class. If we think of each classification method as a kind of expert, ordinary cross-validation selects a single expert and ignores the others. By contrast, stacking is a form of arbitration that potentially takes all opinions into account. It might, for example, allow a consensus between several less reliable experts to overrule the opinion of the leading one.

Stacking creates a set of meta-attributes for each training case—one for the prediction of each ordinary classification method—and then conducts induction on these meta-attributes. A still more general idea is to use attributes and meta-attributes together in inducing a class prediction model. This *bi-level stacking* approach makes it possible to rely more on one expert for some

[1] *Selecting Models from Data: AI and Statistics IV.* Edited by P. Cheeseman and R.W. Oldford.

kinds of cases and more on another expert for others. If neural networks are better at identifying a disease in men and decision trees are better at identifying it in women, a bi-level stacked model could adjust the arbitration accordingly.

In this paper, we examine cross-validation, stacking and bi-level stacking approaches to the meta-classification problem in the context of the illustrative example of Schaffer (1993b). Results and conclusions of the original study of cross-validation in this case are reviewed in the following section. Then we report results of an extended study including stacking and bi-level stacking approaches.

6.2 Cross-Validation in the Selection of Classification Methods

Every classification method embodies a bias that makes it more appropriate for certain applications and less appropriate for others [Mitchell 80, Schaffer 93a]. Ideally, given a classification problem, we would like to select the method with the most appropriate bias on the basis of our knowledge of the domain from which the problem is drawn. In practice, however, we often know too little about the application domain and about the conditions under which each method works best to proceed in this ideal manner. Instead, we may be forced to take an empirical tack in selecting a classification method.

In this case, a natural idea is to conduct a cross-validation study [Geisser 75, Stone 74], partitioning the training data into a number of groups, using each in turn as a test set for models produced on the basis of the remaining data and choosing the method that achieves the highest average accuracy. What do we gain by applying this selection technique rather than consistently using a single classification method?

One answer is that we may gain in predictive accuracy if performance is averaged over a suitable mix of problems. Naturally, cross-validation can do no better on an individual problem than the best of its constituent methods. But if a series of problems is not uniformly favorable to a single method and cross-validation consistently identifies the best method for each problem, then cross-validation may turn in a better average performance than any individual.

Note, though, that for an *un*suitable mix of problems the average performance of cross-validation may be worse than one of its constituent methods. If all or nearly all problems are favorable to one method, cross-validation may perform worse than that method, since chance effects will sometimes cause it erroneously to choose another one.

In short, although it is conceptually natural to distinguish base-level methods like decision tree or neural network induction from meta-methods like cross-validation that choose between them, this distinction is arbitrary. Cross-validation is an algorithmic means of inducing prediction models from data—a classification learning method. As a consequence, it embodies a bias that makes it more appropriate than alternatives in some domains and less so in others. If predictive accuracy is our sole objective, we cannot prefer cross-validation of several methods to application of just one—any more than we can prefer neural networks to decision trees—unless we know something about the mix of problems likely to be encountered.

Cross-validation has a more definite advantage, however, if we take risk into account. Although it may not always choose the best of the methods it considers, it is unlikely to select a method *far* worse than the best, since there is only a small chance that such a method will appear best in a cross-validation study. Cross-validation may be viewed as a form of bet-hedging or insurance; we accept slightly suboptimal performance in the eventuality that all problems are favorable to

	Accuracies				CV's choices		
Problem	Tree	Rules	Net	CV	Tree	Rules	Net
Annealing	92.3	93.9	99.0	99.0	0	0	10
Glass	66.8	66.8	45.7	65.9	6	4	0
Image	96.9	97.1	89.8	96.9	10	0	0
Sonar	69.2	71.6	82.6	80.7	0	1	9
Vowels	76.5	72.4	58.4	76.5	10	0	0
Average	80.3	80.0	75.1	83.8			

TABLE 6.1. Cross-validation outperforms constituent strategies.

a single method in order to avoid a catastrophic loss of predictive accuracy in the eventuality that the mix of problems includes some for which this method is highly inappropriate.

6.2.1 *A Cross-Validation Experiment*

These points are illustrated by the results of an experiment first reported in Schaffer [Schaffer 93b]. Four classification methods are compared on five problems. The classification methods include one each for decision tree, rule set and neural network induction and a cross-validation method that selects between the first three on the basis of a ten fold study.[3]

Problems unfamiliar to the author were chosen from the UCI machine learning repository [MurAha 92]. To ensure the heterogeneous mix of problems favorable to cross-validation, however, three problems (Annealing, Glass and Image) were selected from a main directory presumably dominated by examples favorable to symbolic induction and two (Sonar and Vowels) were selected from a subdirectory presumably dominated by examples favorable to connectionist processing.[4]

In line with the analysis above, two predictions regarding the results of this experiment were set out before data was collected: first, that cross-validation would turn in the best average performance and, second, that its accuracy would never be far from the best of the three constituent methods.

6.2.2 *Results of the Cross-Validation Experiment*

Table 6.1 summarizes the results of the experiment. Accuracies for each problem are averaged over ten trials, in each of which a different segment of the data is reserved for testing. The last three columns indicate in how many of these trials the cross-validation method CV chose the decision tree, rule set or neural network model for prediction.

The last row confirms the prediction that CV would turn in the best average performance. It outperforms the best of the constituent methods by an average of 3.5 percent, a result significant at above the .999 level, using a paired t test. This illustrates the superiority of cross-validation

[3]The constituent systems are C4.5 and C4.5rules [Quinlan 93], using pessimistic pruning and default parameter settings, and BP [McClRum 88]. BP was run with one hidden layer of five units and trained for 1000 epochs. The learning rate was set at 1.1 and the momentum at .5; although these parameters are nonstandard, BP was received with scripts using them and no attempt was made to tune them to the test problems.

[4]Basic information about the data sets and how they were transformed or collated for use in this experiment is given in an appendix to [Schaffer 93b].

	Average Accuracy			
Problems	Tree	Rules	Net	CV
Glass-Image-Annealing	85.3	85.6	78.2	87.3
Sonar-Vowels	72.8	72.0	70.5	78.6

TABLE 6.2. Cross-validation performs best even for specialized test suite subsets.

for heterogeneous problem mixes.

In fact, the data turns out to support this point more strongly than expected. Although efforts were made to ensure a heterogeneous problem suite, this attempted balancing turned out to be ineffectual and, hence, irrelevant. As Table 6.2 shows, CV turned in the best average predictive performance both for the three problems presumed favorable to symbolic processing and for the two presumed favorable to connectionist processing. Even where domain knowledge might have suggested application of a single method, cross-validation proved superior.

On the other hand, the same results also illustrate how a single method might outperform cross-validation. If most problems in an environment were like Sonar, for example, the predictive accuracy of induced neural networks would be better than that of CV by about two percentage points. Cross-validation provides no inherent benefit in predictive accuracy—its effect depends on the mix of problems.

In risk protection, however, cross-validation does confer a definite advantage and the data illustrates this as well. Someone using C4.5 uniformly on the problems of the test suite would miss models that accurately classify an additional 13.4 percent of fresh cases on average in the Sonar problem. Someone using the tested version of neural network induction would do even worse, missing models that accurately classify an additional 21.1 percent of fresh cases on average for the Glass data. By comparison, CV is extremely safe; in these trials, its average performance is always within 2 percent of the best of the tested strategies.

6.3 Stacking and Bi-Level Stacking

Cross-validation is a meta-method for arbitrating between base-level classification methods. In this section, we consider two alternative approaches to arbitration. These are most easily understood after an examination of some details of ordinary cross-validation.

Suppose we have three base-level methods, A, B and C and want to use cross-validation to choose between them. Assuming we have n cases for use in training, our real question is: Which of these methods will yield the best predictive accuracy when applied to a set of n training cases?

The key to cross-validation is that we can get data pertinent to this question from the training set. Suppose we use Method A with 9/10 of the training set to build a prediction model. On the assumption that the performance of Method A with $.9n$ training cases is not too different from its performance with n training cases, this yields a prediction model of the sort that we would expect from application of A to the full training set. In cross-validation, we apply this model to the remaining 1/10 of the training cases to see how accurate such a model will be. Note, though, that the test actually yields more detailed information.

It tells us what a model resulting from the use of Method A with approximately n

training cases would have predicted for each holdout case.

In the course of a complete cross-validation study, each segment of the training set serves in turn as a holdout for testing and all three base-level methods are applied. Hence, at the end of the study, we know what a model resulting from each of the three methods on a training set of roughly the relevant size would have predicted for each case on hand.

This information might suggest clearly that the models produced by, say, Method B yield more accurate predictions than those produced by Methods A and C, and this is precisely the kind of pattern ordinary cross-validation is intended to identify. On the other hand, information about what models produced by the three methods would have predicted for the training cases might show a strong pattern of another kind. For example, perhaps the clearest indication of the true class is when two of the three methods agree. Or perhaps Method B is highly accurate when it predicts class 3, but otherwise Method C is most reliable. Patterns of this kind may be quite clear, but the simple winner-take-all approach of cross-validation is not adequate for identifying them.

In essence, a cross-validation study produces a new set of *prediction attributes* for the training cases, each attribute corresponding to the prediction one of the base-level methods might have made for a case when applied to a training set of approximately the available size.[5] Ordinary cross-validation restricts attention to the simplest kind of relationship between the prediction attributes and the true class—one in which a single attribute is used directly. Wolpert (1992) proposes instead to undertake a broader search, conducting a full-scale induction to characterize the relationship between prediction attributes and class. If the best plan is simply to use the predictions associated with Method B, this technique of *stacking* a new level of induction above the original one can discover the fact and reproduce the effect of an ordinary cross-validation. But if it detects a more complex relationship between the various predictions and the true class—one of those suggested above, for example—stacking will result in an arbitration scheme of commensurate complexity. Stacking is thus a generalization of the use of prediction attributes in ordinary cross-validation.

Because of the machinations necessary to obtain them for training cases, prediction attributes may seem somewhat artificial. But for fresh cases—the ones for which predictions are actually desired—the idea is quite natural. In addition to the original attributes associated with each new case, we can easily find out what various models produced from training data have to say about its likely class. The intuition behind stacking is that each such prediction distills some important information out of the original attributes and that, if this is not always the same information, we may do best to use the predictions *together* somehow in making our own best guess.

It is easy to imagine, however, that not all relevant information contained in the original attributes makes its way into the prediction attributes. Or, to make the same point in another way, in arbitrating between methods, we might still like to pay attention to the original features of each case. For example, perhaps Method A is more accurate for older patients and Method C is more accurate for younger ones. Stacking, as proposed by Wolpert, takes only prediction attributes into account in the second tier of induction. We have just shown the potential advantage of what we will call *bi-level stacking*—conducting induction on the original and prediction attributes together to predict the class.

[5]Wolpert uses the qualifiers *level-0* and *level-1* to distinguish between the original and prediction attributes, because he wants to emphasize that we can continue to higher levels if we like. In the present context, the term *prediction attributes* seems clearer than *level-1 attributes*.

6.4 Preliminary Comments on Stacking and Bi-Level Stacking

The following section reports the results of an experiment replicating the one described above and extending it to include stacking and bi-level stacking as well as ordinary cross-validation. Before turning to empirical work, however, it is worth asking what we may reasonably expect. Here are some preliminary answers.

First, again, every meta-method for classification learning is also a method. As such it is no better or worse than its bias is appropriate to the domain of application. Stacking and bi-level stacking confer no inherent benefit in terms of predictive accuracy; performance relative to cross-validation or to the uniform application of a single base-level method depends entirely on the mix of problems.

Second, by contrast with ordinary cross-validation, stacking and bi-level stacking may perform better than any constituent base-level method on an individual problem.

Third, like cross-validation, stacking and bi-level stacking provide bet-hedging insurance against the consequences of choosing an inappropriate learning method. Of the two, stacking is more conservative—since it considers a relatively small space of arbitration schemes—and may be expected to entail less risk.

Fourth, stacking and bi-level stacking offer no advantage unless different base-level induction methods capture different aspects of the relationship between attributes and class. Diversity at the base level is essential.

Fifth, and perhaps most important, after prediction attributes are produced using base-level induction methods, stacking and bi-level stacking require us to carry out a second level of induction using the new attributes. As at the base level, a host of induction methods are applicable and the bias of the one selected may critically influence results. For this reason, Wolpert called stacking a kind of "black art"; its performance depends on the use of a second-level induction method that favors the arbitration schemes most likely to improve predictive accuracy.[6]

6.5 An Extended Meta-Method Experiment

To compare stacking and bi-level stacking with ordinary cross-validation, we consider the results of an experiment extending the one described above. The problems are the same and we again use base-level methods for decision tree, rule set and neural network induction, although the algorithms employed are different for the latter two.[7]

In all, ten induction methods are compared: the three base-level methods, ordinary cross-validation applied to select between these, and stacking and bi-level stacking with each of the base-level methods employed at the second level of induction.[8] Cross-validation and production

[6] See Breiman [Breiman 92]—a study of stacking in multivariate regression—for a good example of both the difficulty and the value of finding an appropriate second-level induction scheme.

[7] For rule induction, we employ CN2 [ClarNib 89, ClarBos 91] in place of C4.5rules. This change is intended to increase the diversity of the base-level methods, since the closely related algorithm C4.5 is used for decision tree induction. For neural network induction, we employ OPT [vdSmagt 92], a conjugate-gradient-based method. This is efficient, and hence convenient, but it also eliminates the dependence of performance on reasonable choices of learning rate and momentum parameters. At the base level, OPT is applied with five hidden units.

Data is coded as described in the appendix to [Schaffer 93b] except that, where BP used a single output variable for the two-class Sonar problem in the original study, OPT uses two output variables.

[8] When neural network induction is used at the second level, the number of hidden units is the maximum of five and the number of classes. The idea is to ensure that it will be easy for a network to base class predictions on a single prediction attribute—as in ordinary cross-validation—if this is optimal.

	Problem					
Method	Annealing	Glass	Image	Sonar	Vowels	Average
Tree	93.1	67.7	96.5	74.9	78.7	82.2
Rules	88.9	68.6	86.8	68.3	60.6	74.6
Net	99.2	63.0	95.0	**80.7**	70.6	81.7
CV	99.2	66.2	96.5	**80.7**	78.7	84.3
ST-Tree	99.2	66.1	96.4	79.2	79.3	84.0
ST-Rules	99.2	68.5	96.4	78.7	79.3	**84.4**
ST-Net	**99.4**	67.1	**96.8**	78.7	**80.2**	**84.4**
BL-Tree	99.2	68.6	96.4	73.5	78.1	83.2
BL-Rules	99.0	**70.9**	96.1	73.5	78.9	83.7
BL-Net	**99.4**	66.6	96.4	80.3	79.0	84.3

TABLE 6.3. Results of an extended comparison experiment.

of prediction attributes are based on 10-fold partitions of the training data.

Table 6.3 shows results averaged over ten runs, in each of which a different segment of the data is reserved for testing. The abbreviation "ST-Tree" designates stacking with decision tree induction at the second level; "BL-Rules" designates bi-level stacking with rule set induction at the second level and so on. For each problem, the highest accuracy is shown in bold.

We have said that stacking and bi-level stacking can outperform their constituent methods on an individual problem and, in fact, the best performance on four of the five sample problems is turned in by one of these meta-methods. In the case of the Glass problem, bi-level stacking with rule induction at the second level nets an additional 2.3 percentage points in accuracy over the best base-level method. Moreover, stacking with neural networks or rule induction gives the best overall performance for this suite of problems. These points illustrate the potential benefits of the new meta-methods.

Caveats are in order, though. With the results before us, it is easy to see that stacking with neural networks is better for these problems than, say, bi-level stacking with decision tree induction. This is clearly not true in general, however, and it is hard to see how we could decide beforehand which meta-method to prefer. Also, as the last column of the table makes clear, the difference between even the best of the stacking-based methods and ordinary cross-validation is negligible here.

Note, though, that *all* the meta-methods, cross-validation included, outperform the base-level methods on average. For the more conservative meta-methods—cross-validation and stacking—this is true, again, even if we consider only the presumed-symbolic-oriented or presumed-connectionist-oriented problems. Unless we have strong reasons to expect a homogeneous distribution of problems, the results illustrate why it may be prudent to adopt a meta-method of some kind.

In addition to the potential for increasing average predictive accuracy, conservative meta-methods also cut risk, as we have seen. Table 6.4 recasts the results of the experiment, showing the difference between the accuracy attained by each method and the accuracy attained by the best base-level method for each problem. In general, the largest deviations are associated with the base-level methods, the next largest with methods for bi-level stacking and the smallest with

Method	Problem				
	Annealing	Glass	Image	Sonar	Vowels
Tree	−6.1	−.9		−5.8	
Rules	−10.3		−9.7	−12.4	−18.1
Net		−5.6	−1.5		−8.1
CV		−2.4			
ST-Tree		−2.5	−.1	−1.5	.6
ST-Rules		−.1	−.1	−2.0	.6
ST-Net	.2	−1.5	.3	−2.0	1.5
BL-Tree			−.1	−7.2	−.6
BL-Rules	−.2	2.3	−.4	−7.2	.2
BL-Net	.2	−2.0	−.1	−.4	.3

TABLE 6.4. Deviations from best base-level performance.

cross-validation and ordinary stacking.[9]

6.6 Remarks

Meta-methods for classification learning provide two potential benefits—an increase in expected accuracy and a decrease in the risk of very poor accuracy. As we have stressed, the first depends on the distribution of problems. In principle, cross-validation helps when the relative performance of constituent methods is expected to vary widely over this distribution, stacking is preferable when different methods are expected to capture different parts of the underlying relationships and bi-level stacking provides an additional advantage if base-level attributes are useful in identifying the cases for which each method is particularly effective. In practice, however, this qualitative characterization will rarely suffice to help us choose between meta-methods. If predictive accuracy is the goal, empirical evidence based on experience with the problem distribution of interest is probably of more use than analysis in meta-method selection.

Regarding risk protection, analysis can be expected to speak more definitively, but it has not done so as yet. Cross-validation has attracted considerable attention from statisticians (e.g. [Efron 82, Stone 77], but its value as a form of insurance through diversification has not been studied. Likewise, Wolpert's work on stacking has emphasized the potential for increasing predictive accuracy. At present we can only observe qualitatively (1) that cross-validation and stacking are less risky than bi-level stacking and (2) that all meta-methods may become more risky if we increase the set of constituent methods indiscriminantly.[10]

Whether our goal is security or performance, the empirical work of this paper shows that meta-methods for classification learning can help. Just as at the base level, however, knowledge or experience is essential in applying techniques appropriately. Successful induction *always* results from appropriate selection of bias and meta-induction is no exception to this rule.

[9] Bi-level stacking with neural network induction is an exception to this pattern, perhaps because the limited number of hidden units employed in our trials effectively narrows the second-level search for patterns.

[10] The more poor methods we add, the higher the probability that one of them will appear to be of value by sheer chance and hence mislead the higher-level induction.

6.7 REFERENCES

[Breiman 92] Breiman, L. (1992) Stacked regressions. Technical Report 367, Department of Statistics, University of California at Berkeley.

[ClarBos 91] Clark, P. and Boswell, R. (1991) Rule induction with CN2: Some recent improvements. In *Machine Learning, EWSL-91*. Springer-Verlag.

[ClarNib 89] Clark, P. and Niblett, T. (1989) The CN2 induction algorithm. *Machine Learning* **3**, 261–283.

[Efron 82] Efron, Bradley (1982) *The Jackknife, the Bootstrap and Other Resampling Plans*. SIAM.

[Geisser 75] Geisser, S. (1975) The predictive sample reuse method with applications. *Journal of the American Statistical Association* **70**, 320–328.

[McClRum 88] McClelland, J. L. and Rumelhart, D. E. (1988) *Explorations in Parallel Distributed Processing*. MIT Press.

[Mitchell 80] Mitchell, T. M. (1980) The need for bias in learning generalizations. Technical Report CBM-TR-117, Rutgers University.

[MurAha 92] Murphy, P. M. and Aha, D. W. (1992) UCI repository of machine learning databases, a machine-readable data repository. Maintained at the Department of Information and Computer Science, University of California, Irvine, CA. Data sets are available by anonymous ftp at ics.uci.edu in the directory pub/machine-learning-databases.

[Quinlan 87] Quinlan, J. R. (1987) Generating productions rules from decision trees. In *Proceedings of the Tenth International Joint Conference on Artificial Intelligence*, 304–307.

[Quinlan 93] Quinlan, J. R. (1993) *C4.5: Programs for Machine Learning*. Morgan Kaufmann, San Mateo, CA.

[Schaffer 93a] Schaffer, C. (1993a) A conservation law for generalization performance. In review.

[Schaffer 93b] Schaffer, C. (1993b) Selecting a classification method by cross-validation. *Machine Learning* **13**, forthcoming.

[Stone 74] Stone, M. (1974) Cross-validatory choice and assessment of statistical predictions. *Journal of the Royal Statistical Society (Series B)* **36**, 111–147.

[Stone 77] Stone, M. (1977) Asymptotics for and against cross-validation. *Biometrika* **64**, 29–35.

[vdSmagt 92] Smagt, P. P. van der (1992) Minimisation methods for training feed-forward networks. Technical Report UIUC-BI-TB-92-17, Beckman Institute, Theoretical Biophysics, University of Illinois at Urbana/Champaign. The OPT program is the conjugate-gradient method with restarts described here.

[Wolpert 92] Wolpert, D. H. (1992) Stacked generalization. *Neural Networks* **5**, 241–259.

7
Probabilistic approach to model selection: comparison with unstructured data set

Victor L. Brailovsky

Departments of Computer Sciences and Mathematical Statistics
Tel-Aviv University
Ramat-Aviv, 69978, Israel

ABSTRACT
The problem of model selection by data is discussed. This is the problem of finding the internal structure of a given data set, such as signal or image segmentation, cluster analysis, curve and surface fitting. In many cases, there is limited background information, and the given set of data is almost the only source of analysis and decision about the model. A brief review of different approaches to the problem is presented and the probabilistic approach, based on the comparison of the results obtained for a given data set with the one obtained for a set of unstructured data, is introduced. This approach is presented in greater detail for two problems of regression analysis: selecting best subset of regressors and finding best piece–wise regression. Some theoretical results are presented and relations with other approaches to model selection are discussed.

7.1 Introduction

1–1. The problem of model selection from data is one of the most difficult problems of data analysis and, at the same time, one of the most important in many modern applications. It suffices to mention such varied problems as selecting the best subset of regressors in Regression Analysis, selecting the optimal number of clusters in Cluster Analysis, and many problems in Signal and Image Processing (such as image and signal segmentation, texture analysis and so on). The difficulty is that in many cases selecting a model that delivers the best agreement with a given set of data (e.g. minimum average square deviation or another measure of agreement) leads to selecting a very complex model, such as regression with a polynomial of a high order, clustering with a great number of clusters, segmentation of an image with a great number of regions. This phenomenon, called 'overfitting', reflects the fact that the set of data represents not only the properties of the general population that the data are taken from, but also some irregularities and fluctuations that are characteristic only for the given data set. Fitting the model to these irregularities and fluctuations leads to the fact that a more complex model fits the data better without regard to real complexity of the underlying model.

As a result, not only does one obtain incorrect information concerning the real complexity of the underlying phenomena, but the predicting quality of the model deteriorates while the complexity of the model becomes excessive.

One mechanism of the overfitting effect is tuning the considerable number of parameters of a complex model to data. Another one is connected with the process of searching the best model, which in fact also leaves room for such tuning (so-called competition bias). Thus, when discussing the problem of optimal model selection, one means a method of separating two

---

[1] *Selecting Models from Data: AI and Statistics IV.* Edited by P. Cheeseman and R.W. Oldford.

processes: fitting properties of a general population and fitting the irregularities of a given data set.

1–2. The problem of model selection received widespread attention in the last years and a number of approaches to its solving were suggested.

1) Regarding the case where complete information about the problem is available and it is possible to select the optimal model with the help of an optimal algorithm, there is not much to add besides the fact that in many applications such a situation is an exception, not a rule. Thus, henceforth, we will consider the case where information about the problem is very limited and a decision should be made under uncertain conditions.

2) The Bayessian approach to the model selecting problem has gained considerable acceptance. As it is well known, the approach is based on ascribing prior probabilities to the models of different complexity, using the Bayes formula for obtaining posterior probabilities, given data, and selecting the model according to the maximum of the posterior probability. Usually the set of priors represents a formalization of the Ockham's razor principle: the simple model explaining the data is the best.

The exact meaning of the word 'explaining' is expressed in the definition of the set of priors, for in terms of these quantities one decides what increase of the quality of fit justifies selection of a more complex model. Often the specification of the set of priors is to a considerable extent arbitrary.

In addition to this the use of Bayessian approach implies the possibility to calculate the probabilities to obtain a given set of data, provided a certain model is correct. Such knowledge of the form of probability distributions is not always available. We will continue the discussion in Section 6.

3) The approach based on regularization theory that deals with ill– posed problems, presents a trade–off between the quality of fit and some other properties of the model (e.g. smoothness, complexity). This trade–off includes specification of some parameters that in many cases is not a simple problem. In many other cases it is not at all clear if it is possible to find a good solution using this approach.

4) Model selection, based on cross–validation estimate [1], [2] often presents a plausible solution. The problem is that in some cases the application of this method is very problematic if not impossible. Firstly, to apply this criterion the problem should be formulated in terms of function approximation (e.g. regression), which is not always the case (e.g. cluster analysis). Secondly, even if a problem is formulated as function approximation the difficulties with the application of the cross–validation criterion are connected with the widespread use of different methods of smoothing and filtering which, in fact, extend the influence of a given sample to a number of neighbors (e.g. in signal and image processing). After such a transformation the transformed data set does not consist of independent elements. In this situation both operations, excluding a single sample from the sample set and using the excluded sample for prediction and estimation, become problematical. Similar difficulties are connected with the use of robust statistics with high breakdown point [3]. Finally, the cross–validation procedure can be very time consuming.

1–3. Hence, the problem of model selection continues to be at the center of interest of a wide section of researchers and it is worthwhile to continue the search for new productive approaches to its solution.

In recent years the author developed a probabilistic approach to the model selection problem, based on the comparison of results obtained for a real problem, described by a set of experimental data, with those obtained for unstructured sets of data (e.g. pure noise). We use the fact that with the help of Monte–Carlo sampling from the source of unstructured data and subsequent analysis of the samples one can obtain information concerning the possible scale of fluctuations in the sample sets of a given size for the problem under consideration and thereby obtain a basis for the separation between fitting a real property of the underlying model and fitting the irregularities of a given set of experimental data (see p.1–1).

The approach was successfully applied to the problem of finding the best subset of regressors [4]–[6], the best clustering [7], and to problems of restoration of structure of signals [8], [9] and 2D images [12]. This approach is not Bayessian and requires only general and limited information about the problem.

7.2 Principles of the probabilistic approach.

2–1. The main idea of the probabilistic approach is close to one used in statistical hypothesis testing. Given a sample set of experimental data of size n from a given population one needs to check a given property of the population (e.g. clustering, fitting by a curve) with the help of an analysis of the sample set. On the first step, a test–statistic is formulated, a high value indicating the presence of the mentioned property. The value of the test–statistic for the given experimental data is calculated and one should decide if this value is high enough to indicate the presence of the property of interest.

To answer this question one should determine the probability distribution of the test–statistic for the population in which the property of interest is absent. We will call it *unstructured population*. In this way one obtains a reference probability distribution of the test–statistic for the case when the null–hypothesis asserting absence of the property of interest, is valid. Now, having the value of the test–statistic for the experimental data, one can determine with what probability one can obtain the same or greater value of the test–statistic for a sample set of the size n drawn from the unstructured population. If the probability is greater than the preestablished significance level, then the null–hypothesis about absence of the property is accepted; if not, it is rejected.

2–2. In the probabilistic approach for each problem under consideration one defines the notion of *unstructured population* that will be the source of *unstructured sets of data*, i.e. the property of interest may be present in the sets of data only as a result of fluctuations. One also should determine a test–statistic. For a given data set of the size n one can find the best solution (i.e. the best fit, the best clustering) within the framework of each of the competing models and calculate the values of the test–statistics for these best solutions.

One can now generate an unstructured set of data of the same format and size as the given data set and perform on it exactly the same operations that were performed on the given data set. As a result one finds the best solutions for each of the competing models, this time for the unstructured data set. One calculates the values of the test–statistic for these best solutions. The last procedure (i.e. generating a new set of unstructured data, finding the best solution for each of the competing models, calculating test–statistic for these best solutions) can be performed many times. As a result for each of the competing models one can obtain probability distribution of the test– statistic for unstructured data sets of given size n.

Next, for each of the models one can compare the value of the test–statistic obtained for the best solution for the given data set with the corresponding probability distribution obtained for the sets of unstructured data. As a result one obtains the probability to find, for unstructured data, the value of the test–statistic equal to or greater than that obtained for the given data set. The model with the minimal probability is considered best since the significance of the results obtained for this model is maximal.

2–3. It is a general approach that may not always be performed literally. The main reason is that for any reasonable data structure the probabilities mentioned in the previous paragraph, are very small (e.g. $\sim 10^{-9}$, see [5]) and Monte–Carlo estimation of such small probabilities is a very computer–costly task. Thus, the problem now is how to find a modification of the approach which is more computer efficient, and, at the same time, retains the basic idea.

We consider two such approaches. One of them is based on the idea to substitute the unstructured data set with a heavily perturbed original data set (e.g. by adding to it a large portion of noise). As a result one defines a source of perturbed data sets and, with the help of Monte–Carlo experiments, one can estimate the same probability distributions of the test–statistics for the competing models as one defined for unstructured data sets in the previous paragraph. After this the whole procedure is performed exactly as it was described there. Meanwhile, due to the fact that one works with a perturbed original data set instead of with an absolutelry unstructured one, the above mentioned probabilities are much greater and the problem of their estimation is much simpler. We used this modification in a number of applications [6]–[9] with fairly good results.

At the same time there is a problem connected with this approach: what proportion of noise should be added to the original data set? This problem is discussed in [8].

2–4. Another method to avoid estimating very small probabilities is as follows. In many cases one has a nestled system of models of growing complexity; more complex models provide better fit, i.e. higher value of the test–statistic. By way of example one can mention all kinds of one–term regression formulas (model 1), two–term formulas (model 2) and so on, with the terms taken from a priori fixed set of possible regressors.

For a given data set one can find the best solutions within the framework of each of the competing models. In our example– the best one–term formula, the best two–term formula and so on. One calculates the values of the test–statistic for these best formulas as well as relative increase of these test–statistic when one makes a transition from a simpler model (e.g. one–term regression) to a more complex model (two–term regression), successively (i.e. transition from one–term to two–term regression, from two–term to three–term and so on).

Now one defines the unstructured population data exactly as it was done in p. 2–2. For each unstructured data set one finds the best solution for each of the competing models, the values of the test–statistic for these solutions, as well as the relative increase of these values when one makes transition from a simpler model to a more complex one. As a result of many Monte–Carlo experiments with different unstructured data sets one obtains probability distributions for relative increase of the test–statistic for the transitions from the most simple model to the next one (e.g. from one–term to two–term regression) and so on, up to transition to the most complex model (from $(k-1)$–term regression to k–term regression).

Now one should fix a significance level α and for each of the above mentioned probability distributions find the percentage points $\theta_{m,m+1}$, such that the probability for an unstructured data set to obtain relative increase of the test–statistic of more than $\theta_{m,m+1}$, while making the transition from the model with complexity m to those with $m+1$, is equal to α.

Next, one should come back to the given data set and compare the relative increase of the test–statistics for it while making transition from the most simple model to the next one (e.g. from one–term regression to two– term regression) with the corresponding percentage point $\theta_{1,2}$. If the relative increase is greater than the threshold, it means that the increase is significant and we can continue and compare the next increase (from two–term regression to three–term one) with the corresponding threshold $\theta_{2,3}$ and so on. If at some stage it appears that the relative increase of the test–statistics for the original data set is smaller than the corresponding percentage point, it means that one should stop the process and the last model achieved is the solution.

Let us note some properties of the procedure. Due to the fact that one uses the values of the significance level accepted in statistics (such as 0.01 or 0.05), one should not find percentage points for very small probabilities, and, thus, there is no problem of their estimating in the considered approach. Secondly, the only fitting parameter here is the significance level α, which is usual for many formulations of hypothesis testing and may easily be chosen for any problem.

Finally, there are many problems in image and signal analysis in which one deals with different sets of data given in a standard format, e.g. piece–wise regression for a standard mesh of values of argument (or arguments). In this case the calculation of α–percentage points $\theta_{m,m+1}$ may be performed regardless of the form of a signal (response function); thus the whole computer–intensive part of the work may be performed as *preprocessing*, once for the set of problems.

7.3 Selecting best subset of regressors

3–1. The problem of selecting best subset of regressors with the help of the suggested approach was considered in [4]–[6]. Here one presents the outline of the approach.

Let X_d be a d–dimensional space of explanatory variables; let y be a scalar response function. The data set $(x_1, y_1), (x_2, y_2), \ldots, (x_n, y_n)$ represents result of n independent experiments (or measurements): $x_i \in X_d$ and y_i are the values of the response function at these points. Let a set of prospective regressors Φ be fixed *a priori*.

As test–statistic that measures the quality of fit, one can use either Residual Sum of Squares

$$RSS = \sum_{i=1}^{n} (y_i - \hat{y}_i)^2 \tag{1}$$

or the square of the multiple correlation coefficient

$$\Delta = 1 - \frac{RSS}{\sum_{i=1}^{n} y_i^2}. \tag{2}$$

Here $\hat{y}_i$ is the estimate of the response function, obtained with the help of a regression formula. In (2) one means that the values of the response function are centered.

Now define the unstructured population for the considered case. Define on the same n points $x_1, x_2, \ldots, x_n$ the values of artificial response function $\xi_1, \xi_2, \ldots, \xi_n$ as i.i.d. random variables from $N(0,1)$.

3–2. The basic procedure described in 2–2, theoretically may be applied to the considered problem directly. This approach to the problem of selecting best subset of regressors was presented in [4], [5]. At the same time, the problem of estimating very small probabilities forces us to consider the method described in 2–3.

3–3. As a first step one should define the meaning of the perturbed original set of data. We do this by defining at each point x_i of the original data set the value of artificial response function $\xi_i = y_i + \sqrt{\kappa} \cdot \eta_i$. Here η_i are i.i.d. random variables from $N(0,1)$, κ stands for the constant that determines the proportion of noise when perturbing the original data set. After this the whole procedure is performed according to 2–2, 2–3. As was mentioned above, if the parameter κ is chosen properly, the probabilities under consideration are not so small and the procedure of model selection is quite feasible. This approach was presented in [6], some properties of the probabilistic estimate were discussed and some applications were considered.

7.4 Piece–wise regression

We will consider the application of the approach described in p.2–4 to finding optimal piece–wise regression in greater detail, to give a clear idea of how it works.

4–1. Consider the array of n sites $X = [1, 2, \ldots, n]$. Let $y_1, y_2, \ldots, y_n$ be the set of values of a response function. One assumes the response function has the form $y_i = f(x_i) + \varepsilon_i;\ i = 1, 2, \ldots, n$. Here $f(x)$ stands for a function with the following piece-wise-smooth structure. The array X is divided into a number (k) of regions. Inside each of the regions the function has a simple form (constant, linear, quadratic), one calls them the regions of smoothness. Meanwhile inside different regions of smoothness, the dependence may be different and there is no requirement of continuity between the neighboring regions. The number of the regions k and the location of the change-points (knots) are unknown. ε_i refers to a random error (noise that corrupts the underlying signal $f(x)$).

The objective is to discover the piece-wise-smooth structure of the underlying signal $f(x)$ with the help of analysis of the corrupted signal $y(x)$. One should find the number of the regions of smoothness, the location of knots and obtain a good approximation of the signal inside each of the regions.

4–2. Assume for a moment that the number of regions of smoothness k and the location of the knots $x^1, x^2, \ldots, x^{k-1}$ are known. So the k regions are

$$D_1 = [1, x^1);\ D_2 = [x^1, x^2);\ \ldots;\ D_k = [x^{k-1}, n]. \tag{3}$$

The round bracket means that the corresponding knot does not belong to the region, the square bracket means the opposite.

Define a linear regression for a region D_p $(p = 1, 2, \ldots, k)$ as a polynomial of the order r:

$$\hat{y}_{D_p} = a_r x^r + a_{r-1} x^{r-1} + \cdots + a_1 x + a_0. \tag{4}$$

The regression is linear in the coefficients a_i and the estimates of the coefficients may be obtained from the least square principle:

$$\sum_{x_i \in D_p} [y(x_i) - \hat{y}_{D_p}(x_i)]^2 = \min \tag{5}$$

4–3. If the location of the knots $x^1, x^2, \ldots, x^{k-1}$ is not known one can obtain their estimates from the least square principle as well. Let some values of the knots be fixed. For $p = 1, 2, \ldots, k$ one can calculate the estimates of the regression coefficients for the polynomials (4) and define the error of the piece-wise regression as

$$E(x^1, x^2, \ldots, x^{k-1}) = \sum_{p=1}^{k} \sum_{x_i \in D_p} [y(x_i) - \hat{y}_{D_p}(x_i)]^2. \tag{6}$$

The location of the knots may be found as

$$\min_{x^1, x^2, \ldots, x^{k-1}} E(x^1, x^2, \ldots, x^{k-1}) = RSS(k). \tag{7}$$

Technically, the minimum (7) may be obtained with the help of dynamic programming. Such an approach was applied to solving the piece-wise regression problem in [10].

4–4. With the help of this approach one can obtain the best solution according to the criterion (6), (7) for a given k. In fact one often does not know the value of k. However in many cases one can assume that $k < k_m$, where k_m is a known quantity. One can now obtain the best solutions for $k = 1, 2, \ldots, k_m$ and the problem is how to select from these k_m solutions the one that corresponds to the underlying model. This is a typical example of the problem of model selection with the nestled system of models: the most simple when $k = 1$, the most complex when $k = k_m$. So, it is natural to use the probabilistic approach presented in 2–4. As unstructured population for this case one defines on the same array of n sites X the values $\xi_1, \xi_2, \ldots, \xi_n$ which are i.i.d. from $N(0, 1)$.

4–5. In the first stage one sets a significance level α and prepares the table of percentage points $\theta_{k,k+1}$ according to 2–4, for different problems that may be defined in a given array of n sites X. So, for unstructured data sets $Pr[((RSS_a(k) - RSS_a(k+1))/RSS_a(k)) > \theta_{k,k+1}] = \alpha$.

4–6. Now, when one has a specific piece–wise regression problem on the array X one can calculate the values $t_{k,k+1} = (RSS(k) - RSS(k+1))/RSS(k)$ for the problem and to compare them with the respective percentage points obtained in p.4–5. Begining with $k = 1$, if $t_{k,k+1} > \theta_{k,k+1}$, set $k = k + 1$ and repeat the comparison. Otherwise fix the final complexity of the model k.

In this procedure we considered transitions from the model with one region of smoothness to the model with two regions, from two to three and so on up to the moment when the transition is not significant, i.e. $t_{k,k+1} \leq \theta_{k,k+1}$. In this way, one selects the model with k regions of smoothness.

4–7. Due to the fact that the preparation of the table of percentage points (p.4–5) may be performed as a preprocessing, the algorithm dealing with the experimental data is rather fast (compare with cross–validation, p.1–2). So, one can apply the considered in this Section procedure to 2D image segmentation in line by line (column by column) mode. Some experiments [12] demonstrate that this approach is effective, especially when the level of noise is high and the patterns are "roof–shaped" (i.e. there are no steps on the edges) and known methods of image segmentation do not work.

7.5 Remarks about properties of the probabilistic approach

5–1. Here we mention in a loosy and qualitative form some properties of the suggested probabilistic approach, the exact formulations and proofs are presented in [4]–[8].

It was proved, in particular, that if for two models one obtains the same value of a test–statistic, less complex model obtains better probabilistic estimate. The similar property is correct for two models with the same value of test–statistics, each of them is selected as the best one in the course of a search. Probabilistic estimate is better for the model obtained from the smaller search.

To demonstrate how the probabilistic estimate corrects the overfitting effect of a sample estimate of fit consider an extreme case of complexity.

Example. Given n experiments (3–1) and a subset of regressors S consisting of n linearly independent regressors, for any data set one obtains the perfect fit, i.e. $\Delta = 1$. The same situation with any artificial response function, i.e. $\Delta_a \equiv 1$. As a result $Pr[\Delta_a \geq \Delta] = 1$. Thus the quality of fit that seems to be the best according to the usual sample estimate, appears to be the worst according to the probabilistic estimate. The example demonstrates how the probabilistic estimate corrects the overfitting effect of the sample estimate of fit.

Now for the same situation consider probabilistic estimate of significance of the transition from a formula with k linearly independent regressors to those with n $(k < n)$, such that $RSS(n) = 0$ (see Sections 3,4). The test–statistic for this case $t_{k,n} = [(RSS(k) - RSS(n))/RSS(k)] \equiv 1$ for both experimental and artificial responce functions ($RSS(k) \neq 0$). As a result the procedure described in p.2–4 does not make transition from the model with k regressors to the model with n regressors. So, the same kind of correction of the overfitting effect holds for this version of the probabilistic approach as well.

5–2. As for the asymptotic properties of the probabilistic estimate, they depend on the value of the Vapnik–Chervonenkis dimension (VCD, see [6], [11]) of a parametric class of regression formulas, possible in the framework of a given class of models. It is possible to prove that if VCD is finite and an additional condition of the boundeness is fulfilled, then for unstructured data sets for any $\kappa > 0$

$$Pr[\Delta_a > \kappa] \rightarrow 0, \text{ when } n \rightarrow \infty.$$

It means that under considered conditions asymptotically any positive fit $\Delta > 0$ obtained for an original data set, becomes significant.

Another result is that under the same conditions for any $\alpha > 0$, α-percetage point $\theta_{k,k+1} \rightarrow 0$, when $n \rightarrow \infty$. So, asymptotically, under considered conditions any improvement of fit obtained for an original data set justifies the transition to a more complex model.

7.6 Discussion and conclusion

6–1. In Section 1 we began a discussion about the Bayessian approach to model selection and we now compare it with the probabilistic approach presented in this paper.

In Section 5 it was demonstrated how Ockham's razor principle works in the framework of the probabilistic approach. In particular, it works through estimating the significance of an obtained improvement of fit for increasing model complexity. For example, the probabilistic approach ensures against overfitting models.

Similar preference of simple models may be obtained in the framework of the Bayessian approach through special distribution of priors among the models of different complexity. But in this case there is no built-in insurance against overfitting.

6–2. In the framework of the approach suggested here, if sample size is growing we, generally speaking, arrive at the situation, where any improvement of fit becomes significant and the corresponding transition to a more complex model becomes justifiable. (However, it is easy to present an example where if VCD of the models under consideration is growing rapidly with the growth of the sample size, this is not the case.) Under conditions, when this is the case, the exact point where this transition becomes justifiable, depends on the relation between the size and structure of the given data set and the volume of the search and complexity of the regression formulas, specified by the models under consideration.

Regarding the Bayessian approach, if more complex models are not forbidden by ascribing to them zero priors, when sample size grows the data evidences (e.g. in the form of improvement of estimate of fit for a complex model) overweigh the influence of priors and the choice of the complex model becomes justifiable from the maximum posterior probability standpoint. The exact point where this transition to the complex model becomes justifiable depends on the detailed distribution of priors among different models, which is often to a considerable extent arbitrary.

6–3. When one formulates the concept of unstructured data one should remember that while it is unstructured in one way, it may be quite structured in another way. For the problem of finding the best subset of regressors one defines the artificial response function $\xi_1, \xi_2, \ldots, \xi_n$ as i.i.d. random variables from $N(0,1)$ on *the same* n points $x_1, x_2, \ldots, x_n$, that the original sample set was defined on. It means that in Monte–Carlo experiments with unstructured data sets one preserves the covariance structure of the sample points in the space of arguments X_d. It is easy to see that the thresholds $\theta_{k,k+1}$, that define the decision concerning transition to a more complex model, depend on the structure. For example, if these n points in the space X_d are located along a straight line, one has, in fact, a one–dimensional problem. As for regressors, their behavior along this line only, is essential. The thresholds are automatically adjusted to both these conditions. Their values for the essentially d-dimensional problem would generally speaking be different.

6–4. In conclusion it is possible to say that the suggested probabilistic approach to model selection is based on a "built-in" mechanism of estimating significance. This mechanism takes into account both the size and structure of the given data set and power and degrees of freedom of the models under consideration. It selects the model that gives the most significant result under given circumstances. It is worth mentioning that if the probabilistic estimate of the best solution calculated according to 2–2, 3–3, appeared to be not so small, it indicates that the significance of the obtained result may be questionable.

7.7 References

[1] A.L. Lunts, V.L. Brailovsky, Evaluation of attributes obtained in statistical decision rules, *Engineering Cybernetics*, 3, 98-109, 1967.

[2] M. Stone, Cross–validatery choice and assessment of statistical predictions, *Journal of the Royal Statistical Society (Series B)*, 36, 111-147, 1974.

[3] P.J. Rousseeuw, A.M. Leroy, *Robust regression and outlier detection*, Wiley, N.Y., 1987.

[4] V.L. Brailovsky, A predictive probabilistic estimate for selecting subsets of regressor variables, *Ann. N.Y. Acad. Sci.*, 491, 233-244, 1987.

[5] V.L. Brailovsky, On the use of a predictive probabilistic estimate for selecting best decision rules in the course of search, *Proc. IEEE Comp. Society Conf. on Computer Vision and Pattern Recognition*, Ann Arbor, MI, 469-477, 1988.

[6] V.L. Brailovsky, Search for the best decision rules with the help of a probabilistic estimate, *Ann. of Math. and AI*, 4, 249-268, 1991.

[7] V.L. Brailovsky, A probabilistic approach to clustering, *Pattern Recognition Letters*, 12, No 4, 193-198, 1991.

[8] V.L. Brailovsky, Yu. Kempner, Application of piece-wise regression to detecting internal structure of signal, *Pattern Recognition*, 25, No 11, 1361-1370, 1992.

[9] V.L. Brailovsky, Vector piece–wise regression versus clustering (definition and comparative analysis), *Pattern Recognition Letters*, 13, No 4, 227-235, 1992.

[10] R.Bellman, R.Roth, Curve fitting by segmented straight lines, *J. Am. Stat. Assoc.*, Vol 64, 1074-1079, 1969.

[11] V.N. Vapnik, *Estimation of dependencies, based on empirical data*, Springer, New York, 1982.

[12] V.L. Brailovsky, Yu. Kempner, Restoring the original range image structure using probabilistic estimate, *Accepted for 10-th Israeli Symposium on Artificial Intelligence and Computer Vision*, 1993.

8
Detecting and Explaining Dependencies in Execution Traces

Adele E. Howe and Paul R. Cohen

Computer Science Department
Colorado State University
Fort Collins, CO 80523
howe@cs.colostate.edu and

Experimental Knowledge Systems Laboratory
Department of Computer Science
University of Massachusetts
Amherst, MA 01003
cohen@cs.umass.edu

ABSTRACT AI systems in complex environments can be hard to understand. We present a simple method for finding dependencies between actions and later failures in execution traces of the Phoenix planner. We also discuss failure recovery analysis, a method for explaining dependencies discovered in the execution traces of Phoenix's failure recovery behavior.

Dependencies are disproportionately high co-occurrences of particular precursors and later events. For the execution traces described in this paper, the precursors are failures and failure recovery actions; the later events are later failures. In complicated environments, it can be difficult to know whether actions produce long-term effects, in particular, whether certain actions cause or contribute to later plan failures. Statistical techniques such as those discussed in this paper can help designers determine how recovery actions affect the long-term function of a plan and whether recovery actions are helping or hindering the progress of plans.

8.1 Identifying Contributors to Failure in Phoenix

Phoenix is a simulated environment populated by autonomous agents. It is a simulation of forest fires in Yellowstone National Park and the agents that fight the fires. Agents include watchtowers, fuel trucks, helicopters, bulldozers and, coordinating (but not controlling) the efforts of all, a fireboss. Fires burn in unpredictable ways due to wind speed and direction, terrain and elevation, fuel type and moisture content, and natural boundaries such as rivers, roads and lakes. Agents behave unpredictably, too, because they instantiate plans as they proceed, and they react to immediate, local situations such as encroaching fires.

Lacking a perfect world model, neither a Phoenix planner nor its designers can be absolutely sure of the long term effects of actions: Does an action interact detrimentally with a later action in the plan? Will an action provide short term gain with long term loss? Are failures caused by

[0]This research was supported by DARPA-AFOSR contract F49620-89-C-00113, the National Science Foundation under an Issues in Real-Time Computing grant, CDA-8922572, and a grant from the Texas Instruments Corporation.

[1]*Selecting Models from Data: AI and Statistics IV.* Edited by P. Cheeseman and R.W. Oldford.

	F_{ip}	$F_{\overline{ip}}$
R_{sp}	52	33
$R_{\overline{sp}}$	240	643

TABLE 8.1. Contingency table for testing the pattern $R_{sp} \rightarrow F_{ip}$.

a mismatch between the planning system and its environment? These questions are extremely difficult to answer for large, complex systems, and yet, these are precisely the systems in which detrimental interactions and failures are most likely [Corbato 91]. To identify the sources of failure and expedite debugging, we have developed a technique called *Failure Recovery Analysis* (FRA) [Howe 92]. FRA detects dependencies between failure recovery actions – those taken to recover from plan failures – and later failures. FRA also explains how some failure recovery actions might have caused later failures.

FRA involves four steps. First, execution traces are analyzed for statistically significant dependencies between failure recovery actions and subsequent failures; we call this step dependency detection. The remaining three steps explain failures by using the dependencies to focus the search for flaws in the planner that may have caused the observed failures.

8.1.1 Detecting Dependencies

Dependency detection is syntactic and requires little knowledge of the planner or its environment; thus, it can be applied to any planner in any environment. To begin, we gather execution traces of the planner. To determine how the planner's actions might lead to failure, the execution traces include failures and the recovery actions that repaired them, as in the following short trace:

$$F_{ner} \rightarrow R_{sp} \rightarrow F_{ip} \rightarrow R_{rp} \rightarrow F_{prj} \rightarrow R_{sp} \rightarrow F_{ip}$$

F's are failures (e.g., F_{ip} is the Insufficient Progress failure in Phoenix) and R's are recovery actions (e.g., R_{sp} is the Substitute Projection action in the Phoenix Planner's recovery action set). It appears from this short trace that the failure *ip* is always preceded by the recovery action *sp*. We call disproportionately high co-occurrences between failures and particular precursors *dependencies*. In this example, the precursor of *ip* is the action *sp*, but conceptually, it could be any combination of predecessors in the trace: the preceding failure, the combination of the failure and the recovery action that repaired it, or even longer combinations of previous actions and failures. Currently, the analysis looks at only singles (i.e., failures or recovery actions) and pairs (i.e., failures and recovery actions) as precursors.

With more data, we can test whether the observed relationship between *sp* and *ip* is statistically significant. We build a contingency table of four cells: one for each combination of precursor and failure and their negations. For example, the contingency table in Table 8.1 tests whether F_{ip} depends on the precursor R_{sp}.

We test the significance of the observed relationship, $R_{sp} \rightarrow F_{ip}$, with a G-test on the contingency table. A G-test on this table is highly significant, $G = 42.86, p < .001$, meaning that it is highly unlikely that the observed dependence is due to chance or noise.

We construct contingency tables for three types of immediate precursors: failures, recovery actions, and a combination of a failure and the recovery action that repaired it. We denote these cases $F \rightarrow F$, $R \rightarrow F$, and $FR \rightarrow F$, respectively. The three types overlap. In particular,

$FR \rightarrow F$ is a special case of both $F \rightarrow F$ and $R \rightarrow F$ (because they subsume all possible values of the missing member), so if the former dependency is significant, we do not know whether it is truly a dependency between a F_1R and the subsequent F_2, or between F_1 irrespective of the intervening action, or between R with the initial failure playing no role. In practice, all three dependencies might be present to varying degrees.

We sort out the strengths of the dependencies by running a variant on the G-test called the Heterogeneity G-test [Sokal and Rohlf 81]. The intuition is that we compare the contributions of subsets to that of the superset; one can imagine looking at a Venn diagram (as in Figure 1) to gauge whether the failure *ner*, the recovery action *sp* or the combination seems to account for most of the area in the intersection with the subsequent failure *ip*. Both F_{ner} and R_{sp} overlap with part of the subsequent F_{ip}, but it is easy to see that a larger proportion of R_{sp} than F_{ner} overlaps with F_{ip}. Thus, R_{sp} is a more reliable precursor for F_{ip}.

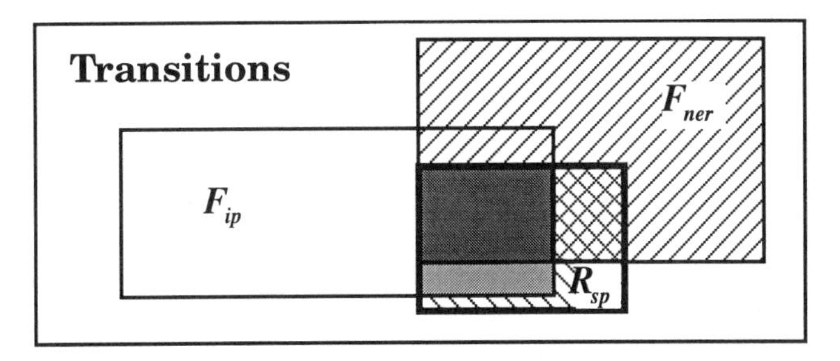

FIGURE 1. Venn diagram representing the data for the precursors and the following failure.

Similarly, but somewhat harder to see, the proportion of $F_{ner}R_{ip}$ (the darker shaded area plus the cross hatched area) that overlaps with F_{ip} (just the darker shaded area) is about the same as the proportion of R_{sp} (the area in the heavy box) that overlaps with F_{ip} (the two shaded areas), suggesting that we can generalize the relationship without loss of information. To compute the overlap, we add the G values for each of the subsets together (e.g., for $F \rightarrow F$, we add G values for all possible R's) and compare the result to the G value for the superset; if the difference is significant, then the subsets account for more of the variance in the dependency and so cannot be generalized. For this example, the G values for the subsets add to 45.89; the difference between that and G for the superset R_{sp} is 3.035, which is not a significant difference at the .05 level.

8.1.2 *Explaining Dependencies*

The first step in failure recovery analysis identifies potential problems in the planner's interaction with its environment; the remaining three steps explain how those problems may have been produced by the planner's actions and suggest redesigns to avoid the resulting failures. The statistical dependencies are mapped to structural dependencies in the planning knowledge bases suspected to be vulnerable to failure. Then, the interactions and vulnerable plan structures are used to generate explanations of how the observed failures might occur. Finally, the explanations serve to recommend redesigns of the planner and recovery component. These steps do not rely on statistical techniques or arguments, so we will not describe them in detail here. Interested readers should consult [Howe 92].

8.1.3 Sensitivity of Dependency Detection to the Size of Execution Traces

Execution traces are often expensive to collect. Consequently, much of the effort required to execute dependency detection is expended collecting execution traces. We expect that the results of dependency detection will vary based on how many execution traces we collect; the total number of patterns (i.e., possible combinations of different types of precursors and failures) and the ratios of the patterns (i.e., the ratio of the counts in the first column to the counts in the second column) in a contingency table influences the results of the G-test. To determine how the size and number of execution traces collected influences the results of the G-test, we need to answer two questions: How does the value of G change as the number of patterns in the execution traces increases? How does the value of G change as the precursor to failure co-occurrence (i.e., the ratio of the upper right to upper left cells in the contingency table) varies from the rest of the execution traces (i.e., the ratio of the lower left to the lower right cells in the contingency table)? The first question addresses the sensitivity of the test to the size of the execution traces; the second addresses the sensitivity to noise: how much of a difference is required to detect a dependency?

G-Test Sensitivity to Execution Trace Size.

We selected the G-test over the more common Chi-square test because the G-test is *additive*. Additivity means that G values for subsets of the sample can be added together to get a G value for the superset. If the ratios remain the same but the total number of counts in the contingency table double, then the G value for the contingency table doubles as well. For example, the G value for the contingency table in Table 1 is 42.86; the G value for the contingency table with 10 times fewer counts (i.e., a contingency table with 5, 3, 24 and 64 in its cells) is 4.319 or roughly (as close as one gets when rounding the counts to the nearest integer) one tenth of 42.86. Additivity means that the value of G increases linearly with the amount of data (or in this case, the number of patterns in the execution traces).

A linear relationship between the number of patterns in the execution traces and the value of G is convenient for several reasons. First, the additivity property is exploited for the second step in dependency detection: pruning overlapping dependencies. We can divide the patterns into their subparts (e.g., a precursor with both a failure and recovery method in it) and add the resulting G values to get the same value as if we had calculated a G for all the subsets together. Second, a linear relationship is predictable. We know that the more patterns in the execution traces, the more likely we are to detect dependencies. Linearity is convenient because we are unlikely to be surprised by new dependencies suddenly showing up if we gather a few more execution traces (meaning the new dependencies were not even close to being dependencies before the additions). The bottom line is that given execution traces with few patterns, the G-test can find strong dependencies, but given more patterns, it will also find rare dependencies. If a user of FRA is interested in detecting *any* dependencies, then a few execution traces will be adequate to do so; if the user wishes to find rare or obscure dependencies, then it will be necessary to gather more execution traces. The level of effort expended in gathering execution traces depends on what kinds of dependencies one wishes to find.

Empirically Testing for Data Sensitivity.

We know that the value of G increases linearly with increases in the number of patterns in the execution traces, but only if the ratios in the contingency table remain the same, as the number of

	Exec. Traces 1	Exec. Traces 2	Exec. Traces 3	Exec. Traces 4
R-F	0/4	4/8	10/15	7/12
F-F	9/13	15/19	7/15	10/12
FR-F	5/7	3/4	11/16	0/0
Total	14/24	22/31	28/46	17/24

TABLE 8.2. Dependencies remaining after tweaking contingency tables. The table includes the number of dependencies remaining after tweaking over the total number of dependencies found in the execution traces from the four experiments.

patterns increases. In trying to decide how many execution traces to gather, we also need to know whether the results will be vulnerable to noise, which is more apparent with few patterns. Unlike the sensitivity to total number of patterns, the sensitivity of the G-test to noise is complicated.

We can evaluate empirically whether, in practice, getting slightly more or fewer execution traces would have significantly changed which dependencies were detected in execution traces gathered from Phoenix. We do so by seeing how many of the dependencies would not have been detected if the counts in row one in the contingency table varied by a small amount. For example, if the contingency table in Table 8.1 had [52,35,240,643] instead of [52,33,240,643], then $G = 40.42$, which is not much different than the value for Table 8.1 of $G = 42.86$. To determine whether the dependencies detected in the execution traces are vulnerable to noise, we can do the following test: 1) construct the contingency table for dependencies detected in execution traces, 2) vary, one at a time, the counts of row one, column one and row one, column two by ± 2, and 3) run a G-test on the resulting contingency tables. By tweaking the contingency table cell values in this manner, we check the sensitivity of G to noise in the data. Both columns of the first row were varied because some of the first column counts were 1, which makes it impossible to test whether a lower ratio of first column to second column might not have influenced the value of G more than a higher ratio. We tweaked the counts by ± 2 because many contingency tables contained cell counts of less than 5, varying by ± 2 spans that range.

Table 8.2 shows how many of the dependencies found in each of four sets of execution traces for Phoenix would remain if their contingency tables are so tweaked. About 65% of the dependencies remain after tweaking their contingency table values, meaning that the counts in the first row of the contingency table can be changed by ± 2 without dropping the significance of G below the level of α. So, 35% of the dependencies detected would disappear if a few patterns more or less were included in the execution traces. Based on this testing of execution traces from Phoenix, dependency detection is sensitive to small differences in the content of the execution traces. Most of the dependencies that were vulnerable to the tweaking were based on execution traces that included few instances of the precursor/failure pattern, 23 out of 44 or 52% of the dependencies that disappeared were based on contingency tables in which one of the counts in the first row was less than five.

The implication of the sensitivity of dependency detection to noise in the execution traces is that rare patterns are especially sensitive to noise and so should be viewed skeptically. One must interpret the results of dependency detection with care: if "sensitive" dependencies are discarded, then rare events may remain undetected; at the same time, one does not wish to chase chimeras. Interpreting dependencies requires weighing false positives against misses. If we are trying to identify dependencies between precursors that occur rarely or failures that occur rarely, then additional effort should be expended to get enough execution traces to ensure that the dependency is not due to noise.

8.2 Applications and Extensions of Dependency Detection

We discovered by chance that dependencies can be used to track modifications to planners and their environments. We ran Phoenix in one configuration, call it A, and collected execution traces from which we derived a set of significant dependencies, D_A. Then we modified Phoenix slightly–we changed the strategy it used to select failure recovery actions–and ran the modified system and collected execution traces, and, thus, another set of dependencies, D_B. Finally, we added two new failure recovery actions to the original set, ran another experiment, and derived another set of dependencies D_C. To our surprise, the intersections of the sets of dependencies were small. However, both $D_A \cap D_B$ and $D_B \cap D_C$ contained more dependencies than $D_A \cap D_C$, suggesting that the size of an intersection mimics the magnitude of modifications to a system.

These results are only suggestive, but they raise the possibility that particular planner-environment pairs can be characterized by sets of significant failure-action dependencies. If true, this technique might enable us to identify classes of environments and planners. Today, we assert on purely intuitive grounds that some environments are similar; in the future we might be able to measure similarity in terms of the overlap between sets of dependencies derived from a single planner running in each environment. Conversely, we might measure the similarity of *planners* in terms of dependencies common to several planners in a single environment.

8.2.1 Further Work

More Complex Dependencies.

The dependencies examined so far have been limited to temporally adjacent failures, actions or the combination of each. The combinatorial nature of dependency detection precludes arbitrarily long sequences of precursors. More complex dependencies can be discovered either by controlling the collection of data to selectively test for particular dependencies (through experiment design) or by heuristically controlling the construction and comparison of dependencies (through enhancements to the Heterogeneity G-Test). A new experiment design would selectively eliminate actions from the available set to test whether each precipitates or avoids particular failures (i.e., an ablation or lesion study). Rather than examining all possible chains of which some action is a member, the new analysis removes the action from consideration, which results in execution traces free from the interaction of the missing action. Dependency sets from the different execution traces would be compared to assess the influence of the missing action.

Alternatively, dependency sets can be built iteratively from subsets; the Heterogeneity G-Test suggests a method of doing so for singletons and pairs, but cannot be applied in a straightforward fashion to longer combinations. We need to enhance the technique to compare longer combinations and use the results of comparing sets of shorter precursors to motivate the search for longer ones. For example, if some singleton subsumes a set of pairs, it seems unlikely to be necessary to look at longer combinations beyond the pairs. In effect, Heterogeneity testing becomes a means of controlling heuristic search through the potentially combinatorial space of possible dependencies.

Marker Dependencies.

Some dependencies might function as markers for particular characteristics of environments. For example, severely resource constrained environments might be characterized by resource contention failures repeating over and over, leading to the observed dependency that one resource contention failure leads to another. We would expect this dependency to appear in any type of resource constrained environment, however superficially different, whether it is a transportation planner, an air traffic control system, or a forest fire fighter dispatcher. To look for such markers, we will need to describe a hierarchy of failures and actions such that dependency sets for different task environments can be compared.

8.3 REFERENCES

[Corbato 91] Fernando J. Corbato. On building systems that will fail. *Communications of the ACM*, 34(9):72–81, September 1991.

[Howe 92] Adele E. Howe. Analyzing failure recovery to improve planner design. In *Proceedings of the Tenth National Conference on Artificial Intelligence*, pages 387–393, July 1992.

[Sokal and Rohlf 81] Robert R. Sokal and F. James Rohlf. *Biometry: The Principles and Practice of Statistics in Biological Research.* W.H. Freeman and Co., New York, second edition, 1981.

9

A method for the dynamic selection of models under time constraints

Geoffrey Rutledge and Ross Shachter

Section on Medical Informatics
Knowledge Systems Laboratory
and
Department of Engineering-Economic Systems
Stanford University, Stanford, CA 94305-5479
rutledge@camis.stanford.edu

ABSTRACT Finding a model of a complex system that is at the right level of detail for a specific purpose is a difficult task. Under a time constraint for decision-making, we may prefer less complex models that are less accurate over more accurate models that require longer computation times. We can define the optimal model to select under a time constraint, but we cannot compute the optimal model in time to be useful. We present a heuristic method to select a model under a time constraint; our method is based on searching a set of alternative models that are organized as a graph of models (GoM). We define the application-specific level of prediction accuracy that is required for a model to be *adequate*, then use the probability of model adequacy as a metric during the search for a minimally complex, adequate model. We compute an approximate posterior probability of adequacy by applying a belief network to compute the prior probability of adequacy for models in the GoM, then by fitting the models under consideration to the quantitative observations. We select the first adequate model that we find, then refine the model selection by searching for the minimally complex, adequate model. We describe work in progress to implement this method to solve a model-selection problem in the domain of physiologic models of the heart and lungs.

9.1 Manual construction of models

The most important task of a model builder is to develop models that fulfill the complexity and tractability requirements of an application [Neelankavil 87]. In certain applications, a high degree of model-prediction accuracy may be essential; in others, accuracy may be sacrificed to increase simplicity, or to decrease the time taken to evaluate the model. For example, a detailed model of the motion of objects in space would include relativistic effects. Under typical assumptions for the motion of spacecraft, relativistic effects can be ignored, since they have a negligible effect on prediction accuracy. A model of orbital motion that includes only Newtonian interactions is tractable, accurate, and conceptually simple. The Navy teaches the diagnosis and maintenance of boiler plants by allowing trainees to interact with a numerical simulation of such plants. The simulation model must reproduce behavior that is generally consistent with that of a boiler plant, but the model does not need to predict accurately the behavior of any specific plant [Hollan 84]. Doctors frequently make use of conceptual physiologic models to interpret the significance of abnormal serum electrolyte measurements of their patients. Such conceptual models should be only as complex as needed to explain the causal pathways of the abnormality [Kuipers 84].

---

[1] *Selecting Models from Data: AI and Statistics IV.* Edited by P. Cheeseman and R.W. Oldford. ©1994 Springer-Verlag.

Traditionally, model-building experts have hand-crafted simulation models of complex domains to meet the required accuracy of a specified task with a minimum of complexity. Finding the appropriate assumptions and simplifications that will lead to tractable, yet accurate, models requires knowledge of statistical and numerical methods, experience in model building, and expertise in the domain of the application. Modeling experts must place constraints on the modeling task for real-world systems, because the number of possible models grows exponentially as the size of a domain increases. Weld has suggested that human model builders cope with the size of the space of possible models by applying simplifying assumptions, and by choosing a perspective that simplifies the modeling task. He points out that these techniques are analogous to the abstraction methods implemented in the fields of planning and search [Weld 92].

9.2 Automated construction of models

There is a growing interest in methods to assist the model builder with the tasks of creating and applying models that are at an appropriate level of detail. Prior investigators developed programs that found a system-specific model by making a selection from a set of alternative models [Addanki 91, Weld 89, Weld 92]. They organized the set of alternative models as a graph of models (GoM), in which the nodes represented models, and the arcs represented the simplifying assumptions that distinguished adjacent models. In the GoM formulation, the search algorithms explored paths from an unsuitable model by evaluating the consequences of asserting or retracting the corresponding simplifying assumptions. Another approach to model selection involves composing models from individually selected components, or submodels [Falkenhainer 90, Nayak 91]. Because the rules constraining the selection of model components are a form of simplifying assumptions, compositional modeling is analogous to search in the GoM. We can think of compositional modeling as a search in a conceptual GoM that consists of all valid combinations of model components.

9.3 The tradeoff of accuracy and complexity

Both the GoM and the compositional modeling techniques applied rules that defined constraints on the use of alternative models; these constraints represented conditions in which the assumptions of the models were likely to be true. The programs evaluated each model's simplifying assumptions to determine whether the assumptions were violated. When multiple models satisfied all constraints, the programs selected the simplest model. In situations where every model violated at least one assumption, the programs were unable to reason about the assumption violations to find the model with the least significant constraint violation. These rule-based constraint-satisfaction techniques did not reason directly about the complexity or tractability of the selected models. In situations where timely model predictions are required, we may prefer a simpler, tractable model that has reduced accuracy, over a more detailed, intractable model that has greater accuracy. We illustrate this tradeoff of accuracy and complexity with a modeling problem in the domain of physiology of critically ill patients.

9.4 A real-world problem

We previously described VentPlan, a program that monitored patients in the intensive-care unit (ICU), and made recommendations for the settings of the ventilator (mechanical breathing device) [Rutledge 89, Rutledge 92]. This program implemented a simplified model of the physiology of the heart and lungs to interpret patient observations and to predict the effect of alternative settings of the ventilator. The model was a series of linked first-order differential equations that described the effects of ventilator treatment on the flows of oxygen and carbon dioxide in the body. Each time that observations for a patient were recorded, VentPlan estimated the patient-specific model parameters, then evaluated the model repeatedly during a search for the optimal settings for the ventilator.

9.5 A time constraint on model evaluation

In this time-critical domain, a monitoring program must make a recommendation for action as soon as possible, and at least before the physician caring for the patient makes the next treatment decision. We determined that a response time of 1 minute was acceptable, and selected a model that would allow VentPlan to meet that constraint. The number of model evaluations required during each cycle of data interpretation and treatment recommendation varies with the number of patient-specific parameters and with the number of observations to be fitted; VentPlan typically evaluated its model more than 1000 times during each cycle. To meet the 1 minute response time, VentPlan's model had to take less than 50 milliseconds to simulate the effects of a control setting. We achieved this evaluation time by selecting a model that had an equilibrium solution, and then by solving for the equilibrium solution using a root-finding method. We built a model, called VentSim, that expands the VentPlan model in three areas. VentSim has no closed-form solution at equilibrium, and requires orders of magnitude more computation time to be solved by numeric integration. Although VentSim represents a wider range of physiologic abnormalities than the VentPlan model does, VentSim is too complex to be used in the VentPlan application (see Table 1).

9.6 The tradeoff of model complexity and accuracy in the ICU domain

The problem of finding a model that is at the ideal level of detail is difficult to solve. For some ICU patients, the simplified physiologic model in VentPlan predicted accurately the effects of changes to the ventilator settings [Rutledge 92]. For other patients who have specific physiologic abnormalities, only models with detailed representation of the areas of abnormality give accurate predictions [Baisingthwaite 85]. Although the VentSim model is intractable for use in a program such as VentPlan, we have simplified the model under various assumptions, to create a set of models that vary in their complexity and in the time they require to compute a solution. For varying physiologic states, the different models represent different tradeoffs of computation complexity and prediction accuracy. We show the range of computation complexity of these models in Table 1. We are developing a method for dynamic selection of a model that is at the right level of detail to balance the tradeoff of model complexity and prediction accuracy. Our model-selection method reasons within a graph of hierarchically organized structural models to find a minimally complex model suitable for use in a time-critical control application. We are implementing this method to select physiologic models that should improve the performance of an ICU-patient monitoring application.

9.7 The graph of models

We organize the set of alternative models as a GoM, in which the arcs are labeled with the assumption that separates the adjacent models (see Figure 1) [Murthy 87, Penberthy 87]. Finding a model to select involves searching within the GoM for the model that optimizes the tradeoff of model complexity and prediction accuracy.

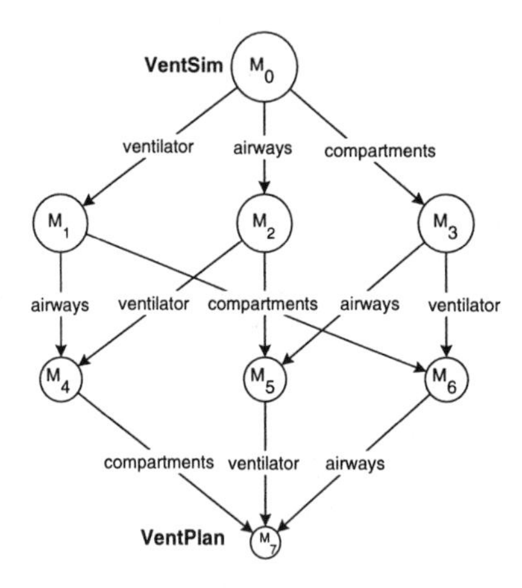

FIGURE 1. A graph of models for a ventilator-management advisor. Nodes represent models, and arcs are labeled with the simplifying assumption that separates the adjacent models. The models are arranged from the most complex model ($\mathbf{M_0}$, the base model) that makes no additional assumptions, to the least complex model ($\mathbf{M_7}$) that makes three additional assumptions. Simplifying assumptions: ventilator = the total lung compliance and total airway resistance are within a near-normal range; airways = the distributions of airway resistance and of lung compliance are homogeneous; compartments = the distribution of ventilation to perfusion is nearly uniform—in which case, a single ventilation and perfusion compartment is adequate.

9.8 Definition of the optimal model

We can define the optimal model in a time-critical application, according to the value of the model's predictions. In a control application, a model's predictions for alternative, proposed control actions are compared during a search for the model's recommended control action. To decide which model is optimal, we compare each model's recommended control actions; we assume that no change in the control settings is made until the model has computed a new control action, and we evaluate the expected effect of a recommended control action by simulating that action with the base model (the most detailed and accurate model available) [Zeigler 76]. Each control action has an instantaneous value that we calculate with a decision-theoretic value model. We define a loss function to be the integral of the value of the system state resulting from control actions over a specified time interval. The loss function calculates the loss of a model's recommended actions with respect to "ideal" control actions—the base-model's recommended actions implemented with no delay due to model computation. We calculate this loss function for a set of alternative models to find the model that would have been the optimal model to select if we had known to select it initially. Since the loss function requires that we evaluate the most complex models, it does not help us to find the optimal model dynamically, in a real-time application. We presented details of the loss function elsewhere [Rutledge 91].

9.9 An approximate method to select the optimal model dynamically

We reason that we can set a minimum prediction accuracy such that, if a model meets that prediction accuracy, then the model is adequate. We search for the least complex model, from the set of adequate models, to find a model that is likely to make a good tradeoff of computation time and prediction accuracy. We define the event that a model is adequate, then we compute a probability that each model within the GoM is adequate. Finally, we search the GoM for the least complex, adequate model; we assume that this model is likely to represent the optimal model. We begin our search for the optimal model with the model that has the highest prior probability of being adequate. We compute a prior probability of adequacy for each model with a belief network. Belief networks, or causal probabilistic networks, are directed, acyclic graphs, in which the nodes are variables, and the arcs represent the conditional distributions of the child nodes given the parent-node values [Pearl 86].

In our belief network, the top-level nodes correspond to diagnoses, such as asthma and pulmonary edema, and the bottom-level nodes correspond to the models in the GoM (see Figure 2). The middle nodes in this network correspond to physiologic parameters, such as the mean static lung compliance (C_L). The arcs from the diagnosis nodes to the parameter nodes define the conditional distributions for each parameter, given the diagnoses. The arcs from the parameter nodes to the model nodes define the conditional probabilities that the models are adequate given the the parameter values. Arcs between models express the relationships that, if a model is not adequate, then the models that are simplifications of it are also not adequate. We apply the Lauritzen-Spiegelhalter algorithm to evaluate the network and to compute the probability distributions on all nodes [Lauritzen 87].

If we have observations of the system that we are modeling, we update the prior probability (computed by the belief network) to compute an approximate posterior probability of model adequacy. We apply Bayes' theorem to compute the posterior probability that a model is adequate, given the system observations ($\mathbf{y}$). We use the goodness-of-fit of the model to the observations to compute the probability that,if the model were *correct* (M_i^C), the observations might have occurred ($\Pr(\mathbf{y}|M_i^C)$). Since $\Pr(\mathbf{y}|M_i^A) \geq \Pr(\mathbf{y}|M_i^C)$, we can compute $\Pr(\mathbf{y}|M_i^C)$, as a lower bound on $\Pr(\mathbf{y}|M_i^A)$, and the calculation of the posterior probability that a model is adequate may be an underestimate. This approximation leads to conservative model-selection behavior, since we may reject a model that is adequate according to our definition, but we are confident that a model that we do select is adequate. We perform a local search within the GoM, starting with the least complex model that has a prior probability of adequacy exceeding a threshold value. We first compute the posterior probability of adequacy for this model. If the posterior probability also exceeds our threshold, then we make a control recommendation with this model before attempting to refine the model selection. If the posterior probability is less than the threshold—that is, the initial model does not fit the observations—then we compute the posterior probability

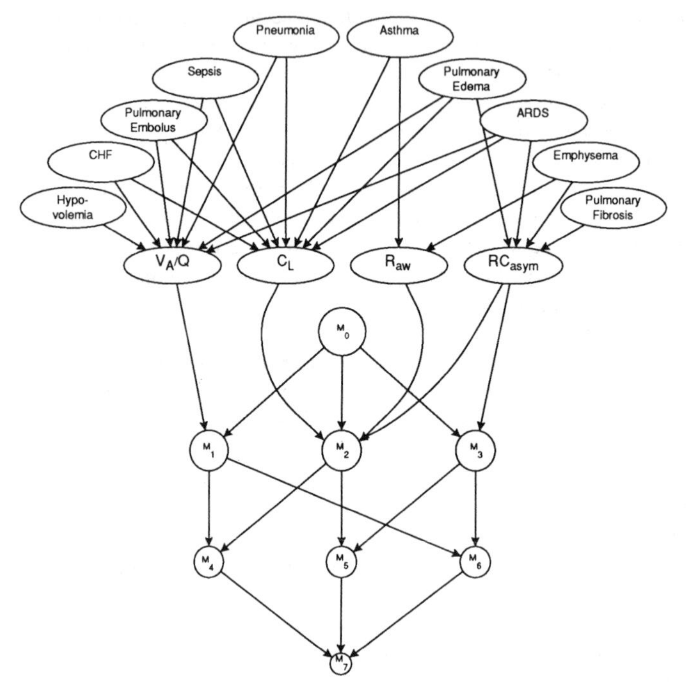

FIGURE 2. Belief network for computing prior probability of model adequacy. The lower nodes, labeled $\mathbf{M_0}$ to $\mathbf{M_7}$, are true-false nodes that correspond to the eight models in Figure 1. The probability that a model node is true is the probability that the corresponding model in the graph of models is adequate. The top nodes are true-false nodes that correspond to diagnoses; these nodes are used as inputs to the network when one or more diagnoses are known. Middle nodes are multivalued nodes that correspond to parameters of the physiologic state. CHF: congestive heart failure; ARDS: adult respiratory distress syndrome; V_A/Q: ratio of lung ventilation to perfusion; $\mathbf{R_{aw}}$: mean airway resistance; $\mathbf{C_L}$: mean static lung compliance; $\mathbf{RC_{asym}}$: degree of asymmetry of the resistance-time constants of the lung.

for each of the adjacent models, until we find a model with a posterior probability that exceeds the threshold. We move up to the next level of complexity within the GoM when observations demonstrate that our model is in error. Conversely, once we have selected a model and made our recommendation for control action, we refine our model selection by evaluating the adequacy of the less complex models.

Table 1 Model complexity for a set of physiologic models

Model [a]	Additional assumptions [b]	Solution method	Number of model variables	Equilibrium-solution time [c] (seconds)
M_0 (VentSim)	none	integration	143	52.7
M_3	airways assumption	integration, separating the short and long time constants	129 [c]	5 [c]
M_5	airways and ventilator assumptions	analytic equilibrium solution	107 [c]	0.8 [c]
M_7 (VentPlan)	airways, compartment, and ventilator assumptions	analytic equilibrium solution	71	0.02

[a] Model labels correspond to the labels in Figures 1 and 2.
[b] This number is an estimate based on a preliminary implementation or on an analysis of model structure and the expected solution method.
[c] Computation time to find the equilibrium solutions (M_5 and M_7), or to integrate the model for 30 minutes (M_0 and M_3). During the parameter-estimation procedure, each model must be evaluated more than 1000 times. All times are computed on a 25-MHz, Motorola 68040 processor (NeXTstation).

9.10 Implementation details

We are implementing the model-selection program in the Mathematica programming language [Wolfram 91]. The interactive features of the Mathematica environment assist us to develop incrementally the tools for comparing alternative models. The symbol-manipulation features allow us to combine a symbolic representation of the GoM with a numerical analysis of the quantitative evaluation of each model. We are thus able to achieve the benefits of an interactive development environment without loss of computation power for evaluating the more complex models. We solve the models in the GoM by calling external routines that are written in C and compiled separately. We solve the most complex model (M_0) by integrating the differential equations numerically, and the simplest model (M_7) by searching for the roots of the corresponding equilibrium equations [Press 89]. We solve each intermediate model ($M_1 - M_6$) by separating the model into components and solving each component either by numerical integration, or, when possible, by searching for the roots of the component's equilibrium equations.

The range of computation times to find the equilibrium solutions for the models varies from 20 milliseconds for the simplest model (M_7) to more than 50 seconds for the most complex model (M_0).

The parameters of our models are underdetermined in almost all cases. We estimate the patient-specific model parameters by the technique of empirical Bayesian estimation. This method adjusts the prior distribution of each model parameter to obtain an approximate posterior distribution in light of the observations. We implement this method as a modified Levenberg-Marquardt fitting procedure, in which prior distributions on the fitted parameters are treated as if they were observations of the parameters [Press 89, Sheiner 82].

The implementation and testing of the GoM for a VMA is in progress. We have built the GoM-symbolic manipulation routines, the parameter-estimation routines and the most complex and the least complex models. We plan to test our model-selection method initially with data generated by the most complex model, and later with observations from ICU patients who are being treated with a ventilator.

9.11 Discussion

The approach we have taken to dynamic model selection builds on prior efforts to automate reasoning about model accuracy [Addanki 89, Weld 92]. Our representation of modeling constraints in a probabilistic framework allows us to compare alternative models that have some degree of constraint violation. This capability is essential when no model satisfying all constraints (including constraints on computation time) is available. In addition, our loss function defines the optimal model to select in a decision-theoretic sense, and allows us to evaluate the performance of our dynamic selection method. In the high-stakes decision-making environment of the ICU, the tradeoff of computation time and prediction accuracy is acute. Our model-selection method finds models at the ideal level of detail for a computation-resource limited task. However, the underlying problem of searching for an appropriate model is similar to other model-construction problems for which minimizing model complexity is a goal. Our approach to the selection of a minimally complex, adequate model does not require that we enumerate all alternative models, since each model could be constructed dynamically, if it is needed, during the search in the GoM. However, the time constraint for model selection imposes a penalty for the additional computation time that would be required to compose models dynamically. In addition, for models of heart and lung physiology, the interactions among modeling components would make a compositional modeling approach difficult, if not impossible.

9.12 Acknowledgments

We thank Lewis Sheiner and Lawrence Fagan for helpful discussions. We thank Adam Galper for providing us with the routines for belief-network evaluation. We especially thank Edward Shortliffe for providing the research environment in which we work. We are grateful to Lynda Dupre for her review of an earlier version of this manuscript. This research was supported in part by Grant IRI-9108359 from the National Science Foundation and Grants LM-07033 and LM-04136 from the National Library of Medicine. Computing facilities were provided by the SUMEX-AIM Resource LM-05208.

9.13 REFERENCES

[Addanki 91] Addanki, S., Cremonini, R. and Penberthy, J.S. (1991) Graphs of models. *Artificial Intelligence,* **51**:145-177.

[Baisingthwaite 85] Baisingthwaite, J. B. (1985) Using computer models to understand complex systems. *The Physiologist,* **28**:439-442.

[Fagan 80] Fagan, L.M. (1980) *VM: Representing Time-Dependent Relations in a Medical Setting,* (1980) Ph.D. thesis, Department of Computer Science, Stanford University, Stanford, CA, June, 1980.

[Falkenhainer 91] Falkenhainer, B. and Forbus, K.D. (1991) Compositional modeling of physical systems. *Artificial Intelligence,* **51**:95-143.

[Farr 91] Farr, B.R. (1991) *Assessment of Preferences Through Simulated Decision Scenarios,* Ph.D. thesis, Section on Medical Informatics, Stanford University, Stanford, CA, June, 1991.

[Hollan 84] Hollan, J.D., Hutchins, E.L. and Weitzman, L. (1984) STEAMER: An interactive inspectable simulation-based training system. *AI Magazine*, Summer issue, pp. 15-27.

[Kuipers 84] Kuipers, B. and Kassirer, J.P. (1984) Causal reasoning in medicine: Analysis of a protocol. *Cognitive Science,* **8**:363-385.

[Lauritzen 88] Lauritzen, S.L. and Spiegelhalter, D.J. (1988) Local computations with probabilities on graphical structures and their application to expert systems. *Journal of the Royal Statistical Society,* **50**:157-224.

[Murthy 87] Murthy, S. and Addanki, S. (1987) PROMPT: An innovative design tool. *Proceedings of the Sixth National Conference on Artificial Intelligence (AAAI-87),* Seattle, WA, pp. 637-642, Palo Alto, CA: Morgan Kaufmann, August, 1987.

[Neelankavil 87] Neelankavil, F. (1987) *Computer Simulation and Modeling.* Chichester: John Wiley and Sons.

[Nayak 92] Nayak, P., Addanki, S. and Joskowicz, L. (1992) Automated model selection using context-dependent behaviors. *Proceedings of the Tenth National Conference on Artificial Intelligence (AAAI-92),* San Jose, CA, pp. 710-716, Menlo Park, CA: AAAI Press, July, 1992.

[Pearl 86] Pearl, J. (1986) Fusion, propagation and structuring in belief networks. *Artificial Intelligence,* **29**:241-248.

[Penberthy 87] Penberthy, S. (1987) *Incremental Analysis in the Graph of Models.* S.M. thesis, Department of Computer Science, Massachusetts Institute of Technology, Cambridge, MA.

[Press 89] Press, W.H., Flannery, B.P., Teulokolsky, S.A. and Vetterling, W.T. (1989) *Numerical Recipes in C: The Art of Scientific Programming.* Cambridge: Cambridge University Press.

[Rutledge 89] Rutledge, G., Thomsen, G., Beinlich, I., Farr, B., Sheiner, L., Fagan, L. (1989) Combining qualitative and quantitative computation in a ventilator therapy planner. *Proceedings of the Thirteenth Annual Symposium on Computer Applications in Medical Care (SCAMC-89),* Washington, D.C., pp. 315-319, Washington, D.C.: IEEE Press, November, 1989.

[Rutledge 91] Rutledge, G.W. (1991) Dynamic selection of models under time constraints. *Proceedings of the Second Annual Conference on AI, Simulation and Planning in High Autonomy Systems,* Cocoa Beach, FL, pp. 60-67, Los Alamitos, CA: IEEE Press, April, 1991.

[Rutledge 92] Rutledge, G.W., Thomsen, G.E., Farr, B.R., Tovar, M.A., Polaschek, J.X., Beinlich, I.A.; Sheiner, L.B., Fagan, L.M. (1993) The design and implementation of a ventilator-management advisor. *Artificial Intelligence in Medicine Journal,* **5**:67-82.

[Sheiner 82] Sheiner, L.B. and Beal, S.L. (1982) Bayesian individualization of pharmacokinetics: simple implementation and comparison with non-Bayesian methods. *Journal of Pharmaceutical Sciences,* **71**:1344-1348.

[Sittig 88] Sittig, D.F. (1988) *COMPAS: A Computerized Patient Advice System to Direct Ventilatory Care,* Ph.D. thesis, Department of Medical Informatics, University of Utah, Salt Lake City, UT, June, 1988.

[Weld 89] Weld, D. (1989) Automated model switching: Discrepancy-driven selection of approximation reformulations. Technical Report 89-08-01, Department of Computer Science, University of Washington, Seattle, WA, October, 1989.

[Weld 92] Weld, D.S. (1992) Reasoning About Model Accuracy, *Artificial Intelligence,* **56**:255-300.

[Wolfram 91] Wolfram, S. (1991) *Mathematica, A System for Doing Mathematics by Computer, 2nd ed.,* Redwood City, CA: Addison Wesley.

[Zeigler 76] Zeigler, B.P. (1976) *Theory of Modelling and Simulation.* New York: John Wiley and Sons.

Part II

Graphical Models

10
Strategies for Graphical Model Selection

David Madigan, Adrian E. Raftery, Jeremy C. York, Jeffrey M. Bradshaw, and Russell G. Almond

Dept. of Statistics
University of Washington

University of Washington

Carnegie-Mellon University

Fred Hutchinson Cancer Research Center and EURISCO

and

Statistical Sciences Inc., Seattle Washington.

ABSTRACT
We consider the problem of model selection for Bayesian graphical models, and embed it in the larger context of accounting for model uncertainty. Data analysts typically select a single model from some class of models, and then condition all subsequent inference on this model. However, this approach ignores model uncertainty, leading to poorly calibrated predictions: it will often be seen in retrospect that one's uncertainty bands were not wide enough.
The Bayesian analyst solves this problem by averaging over all plausible models when making inferences about quantities of interest. In many applications, however, because of the size of the model space and awkward integrals, this averaging will not be a practical proposition, and approximations are required. Here we examine the predictive performance of two recently proposed model averaging schemes. In the examples considered, both schemes outperform any single model that might reasonably have been selected.

10.1 Introduction

A typical approach to data analysis and prediction is to initially carry out a model selection exercise leading to a single "best" model and to then make inference as if the selected model were the true model. However, as a number of authors have pointed out, this paradigm ignores a major component of uncertainty, namely uncertainty about the model itself (Raftery, 1988, Breslow, 1990, Draper, 1994, Hodges, 1987, Self and Cheeseman, 1987). As a consequence uncertainty about quantities of interest can be underestimated. For striking examples of this see York and Madigan (1992), Regal and Hook (1991), Raftery (1993), Kass and Raftery (1993), and Draper (1994).

There is a standard Bayesian way around this problem. If Δ is the quantity of interest, such as a structural characteristic of the system being studied, a future observation, or the utility of a course of action, then its posterior distribution given data D is

$$\mathrm{pr}(\Delta \mid D) = \sum_{k=1}^{K} \mathrm{pr}(\Delta \mid M_k, D)\mathrm{pr}(M_k \mid D). \tag{10.1}$$

[0]This work is supported in part by a NSF grant to the University of Washington and by a NIH Phase I SBIR Award "Computing environments for graphical belief modeling" to Statistical Sciences.

This is an average of the posterior distributions under each of the models, weighted by their posterior model probabilities. In equation (10.1), $M_1, \ldots, M_K$ are the models considered, the posterior probability for model M_k is given by

$$\mathrm{pr}(M_k \mid D) = \frac{\mathrm{pr}(D \mid M_k)\mathrm{pr}(M_k)}{\sum_{l=1}^{K} \mathrm{pr}(D \mid M_l)\mathrm{pr}(M_l)}, \tag{10.2}$$

where

$$\mathrm{pr}(D \mid M_k) = \int \mathrm{pr}(D \mid \theta, M_k)\mathrm{pr}(\theta \mid M_k)d\theta, \tag{10.3}$$

θ is a vector of parameters, $\mathrm{pr}(\theta \mid M_k)$ is the prior for θ under model M_k, $\mathrm{pr}(D \mid \theta, M_k)$ is the likelihood, and $\mathrm{pr}(M_k)$ is the prior probability that M_k is the true model, all conditional on the class of models being considered. Notice that the "conventional" approach is a special case of (10.1), where one particular model has been assigned a prior probability of one.

Averaging over *all* the models in this fashion provides better predictive ability, as measured by a logarithmic scoring rule, than using any single model M_j (Madigan and Raftery, 1991, hereafter referred to as MR).

However, implementation of the above strategy is difficult. There are two primary reasons for this: first, the integrals in (10.3) can be hard to compute, and second, the number of terms in (10.1) can be enormous.

For graphical models for discrete data (and other important model classes), efficient solutions to the former problem have been developed. Two approaches to the latter problem, i.e. the enormous number of terms in (10.1), have recently been proposed. MR do not attempt to approximate (10.1) but instead, appealing to standard norms of scientific investigation, adopt a model selection procedure. This involves averaging over a much smaller set of models than in (10.1) and delivers a parsimonious set of models to the data analyst, thereby facilitating effective communication of model uncertainty. Madigan and York (1993) on the other hand suggest directly approximating (10.1) with a Markov chain Monte Carlo method.

MR examined the predictive performance of their method. Our purpose in this paper is to examine the predictive performance of the Markov chain Monte Carlo method and compare the predictive performance of both approaches. This work is of direct relevance to probabilistic knowledge-based systems systems where model uncertainty abounds (Bradshaw *et al.*, 1993).

10.2 Model selection and Occam's window

Two basic principles underly the approach presented in MR. First, they argue that if a model predicts the data far less well than the model which provides the best predictions, then it has effectively been discredited and should no longer be considered. Thus models not belonging to:

$$\mathcal{A}' = \left\{ M_k : \frac{\max_l\{\mathrm{pr}(M_l \mid D)\}}{\mathrm{pr}(M_k \mid D)} \leq C \right\}, \tag{10.4}$$

should be excluded from equation (10.1) where C is chosen by the data analyst. Second, appealing to Occam's razor, they exclude complex models which receive less support from the data than their simpler counterparts. More formally they also exclude from (10.1) models belonging to:

$$\mathcal{B} = \left\{ M_k : \exists M_l \in \mathcal{A}, M_l \subset M_k, \frac{\mathrm{pr}(M_l \mid D)}{\mathrm{pr}(M_k \mid D)} > 1 \right\} \tag{10.5}$$

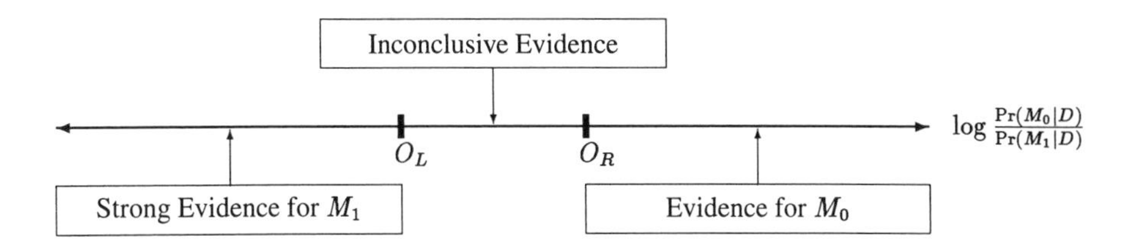

FIGURE 1. Occam's Window: Interpreting the log posterior odds, $\log \frac{\Pr(M_0|D)}{\Pr(M_1|D)}$, where M_0 is a submodel of M_1

and equation (10.1) is replaced by

$$\operatorname{pr}(\Delta \mid D) = \frac{\sum_{M_k \in \mathcal{A}} \operatorname{pr}(\Delta \mid M_k, D)\operatorname{pr}(D \mid M_k)\operatorname{pr}(M_k)}{\sum_{M_k \in \mathcal{A}} \operatorname{pr}(D \mid M_k)\operatorname{pr}(M_k)} \tag{10.6}$$

where

$$\mathcal{A} = \mathcal{A}' \backslash \mathcal{B}. \tag{10.7}$$

This greatly reduces the number of models in the sum in equation (10.1) and now all that is required is a search strategy to identify the models in $\mathcal{A}$. Two further principles underly the search strategy. First, if a model is rejected then all its submodels are rejected. This is justified by appealing to the independence properties of the models. The second principle — "Occam's Window" — concerns the interpretation of the log of the ratio of posterior model probabilities $\operatorname{pr}(M_0 \mid D)/\operatorname{pr}(M_1 \mid D)$. Here M_0 is one link "smaller" than M_1. We show the essential idea in Figure 1. If there is evidence for M_0 then M_1 is rejected but to reject M_0 we require strong evidence *for* the larger model, M_1. If the evidence is inconclusive (falling in Occam's Window) neither model is rejected. MR set the edges of the window at -3 and 0.

These principles fully define the strategy. Typically the number of terms in (10.1) is reduced to fewer than 20 models and often to as few as two. MR provide a detailed description of the algorithm.

10.3 Markov chain Monte Carlo model composition

Our second approach is to approximate (10.1) using Markov chain Monte Carlo methods, such as in Hastings (1970) and Tierney (1991), generating a process which moves through model space. Specifically, let $\mathcal{M}$ denote the space of models under consideration. We can construct an irreducible, aperiodic Markov chain $\{M(t), t = 1, 2, \ldots\}$ with state space $\mathcal{M}$ and equilibrium distribution $\operatorname{pr}(M_i \mid D)$. Then, under mild regularity conditions, for any function $g(M_i)$ defined on $\mathcal{M}$, if we simulate this Markov chain for $t = 1, \ldots, N$, the average:

$$\hat{G} = \frac{1}{N} \sum_{t=1}^{N} g(M(t)) \tag{10.8}$$

converges with probability one to $E(g(M))$, as N goes to infinity. To compute (10.1) in this fashion, we set $g(M) = \operatorname{pr}(\Delta \mid M, D)$.

To construct the Markov chain we define a neighborhood nbd(M) for each $M \in \mathcal{M}$ which is the set of models with either one link more or one link fewer than M and the model M itself. Define a transition matrix q by setting $q(M \to M') = 0$ for all $M' \notin \mathrm{nbd}(M)$ and $q(M \to M')$ constant for all $M' \in \mathrm{nbd}(M)$. If the chain is currently in state M, we proceed by drawing M' from $q(M \to M')$. If the model is decomposable it is then accepted with probability:

$$\min\left\{1, \frac{\mathrm{pr}(M' \mid D)}{\mathrm{pr}(M \mid D)}\right\}.$$

Otherwise the chain stays in state M. It has been our experience that this process is highly mobile and runs of 10,000 or less are typically adequate.

10.4 Analysis

The efficacy of a modeling strategy can be judged by how well the resulting "models" predict future observations (Self and Cheeseman, 1987). We have assessed the predictive performance of Markov chain Monte Carlo model composition (MC^3) method for the three examples considered by MR. The MR results are reproduced for comparison purposes. In each case we started the Markov chain at a randomly chosen model and ran the chain for 100,000 iterations, discarding the first 10,000. The data sets each have between six and eight binary variables. Performance, measured by the logarithmic scoring rule, is assessed by randomly splitting the complete data sets into two subsets. One subset, containing 25% of the data, is used to select models with the other subset being used as set of test cases. Repeating the random split, varying the subset proportions, or starting the Markov chain from a different location produces very similar results.

The first example concerns data on 1,841 men cross-classified according to risk factors for Coronary Heart Disease. This data set was previously analysed by Edwards and Havránek (1985) and others. The risk factors are as follows: A, smoking; B, strenuous mental work; C, strenuous physical work; D, systolic blood pressure; E, ratio of β and α proteins; F, family anamnesis of coronary heart disease.

The second example concerns a survey which was reported in Fowlkes *et al.* (1988) concerning the attitudes of New Jersey high-school students towards mathematics. A total of 1190 students in eight schools took part in the survey. The variables collected were: A, lecture attendance; B, Sex; C, School Type (suburban or urban); D, "I'll need mathematics in my future work"(agree or disagree); E, Subject Preference (maths/science or liberal arts); F, Future Plans (college or job). In what follows we refer to this as the "Women and Mathematics" example.

The final example concerns the diagnosis of scrotal swellings. Data on 299 patients were presented in MR, cross-classified according to one disease class, Hernia (H), and 7 binary indicants as follows: A, possible to get above the swelling; B, swelling transilluminates; C, swelling separate from testes; D, positive valsalva/stand test; E, tender; F, pain; G, evidence of other urinary tract infections.

Results are presented in Tables 1, 2 and 3 for each of the examples. Given in each case are the models selected by MR and the logarithmic score summed over the test cases for each individual model. Next the score resulting from averaging over these models is given. For the Coronary Heart Disease example, the score is also included for the model selected by Whittaker (1990) on the basis of the full data set. This represents the score that would result from using a typical model selection procedure. Finally the score for MC^3 is given.

TABLE 10.1. Coronary Heart Disease: Predictive Performance

Model	*Posterior probability %*	Logarithmic Score
$[AE][BC][BE][DE][F]$	26	4986.7
$[AC][BC][BE][DE][F]$	16	4980.9
$[AC][AE][BC][DE][F]$	13	4981.0
$[A][BC][BE][DE][F]$	9	4989.4
$[AE][BC][BE][D][F]$	8	4987.4
$[AE][BC][DE][F]$	7	4989.5
$[AC][BC][BE][D][F]$	5	4981.6
$[AC][BC][DE][F]$	4	4983.7
$[AC][AE][BC][D][F]$	4	4981.7
$[A][BC][BE][D][F]$	3	4990.1
$[A][BC][DE][F]$	2	4992.2
$[AE][BC][D][F]$	2	4990.2
$[AC][BC][D][E][F]$	1	4984.4
$[ABCE][ADE][BF]$	Whittaker	4990.2
Model Averaging		4953.6
Markov Chain Monte Carlo Model Composition		4933.7

TABLE 10.2. Women and Mathematics: Predictive Performance

Model	*Posterior probability %*	Logarithmic Score
$[A][B][CDF][DE]$	75	3318.9
$[A][B][CF][DE][DF]$	21	3317.3
$[A][B][CF][DE]$	4	3320.4
Model Averaging		3313.9
Markov Chain Monte Carlo Model Composition		3271.5

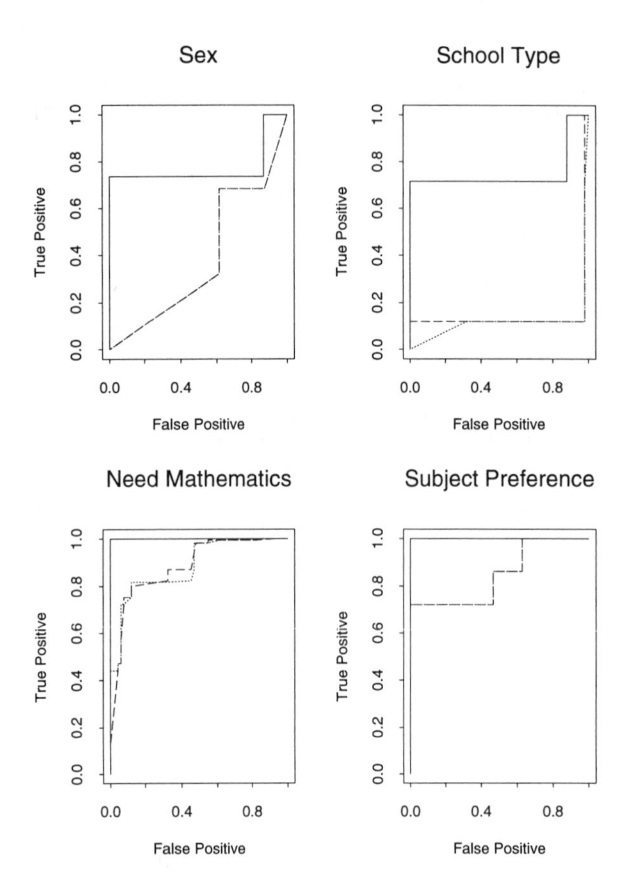

FIGURE 2. Women and Mathematics: ROC Curves. The dashed ROC curves show how well the model with the highest posterior probability performs, the dotted curves show the performance averaging over the models in Occam's window, while the solid curves are for MC^3.

TABLE 10.3. Scrotal Swellings: Predictive Performance

Model	*Posterior probability %*	Logarithmic Score
$[AH][AD][BDE][CD][EF][FG]$	3	605.3
$[AH][DH][BDE][CD][EF][FG]$	3	599.6
$[AH][DH][BDE][CDE][EF][FG]$	5	600.6
$[AH][AD][BDE][CDE][EF][FG]$	5	606.3
$[AH][AD][BDE][CD][EF][EG]$	15	603.4
$[AH][DH][BDE][CD][EF][EG]$	15	597.7
$[AH][DH][BDE][CDE][EF][EG]$	27	598.7
$[AH][AD][BDE][CDE][EF][EG]$	27	604.4
Model Averaging		594.2
Markov Chain Monte Carlo Model Composition		590.1

In each case, methods that average over models, provide predictive performance which is superior to the performance resulting from basing the inference on any single model which might reasonably have been selected. In the coronary heart disease data for example, the Occam's window models outperform the "best" model (i.e. that with the highest posterior probability) by 33 points of log predictive probability, or 66 points on the scale of twice the log probability on which deviances are measured. MC^3 provides a further performance improvement of 20 points (or 40 points on the deviance scale).

A ROC analysis was also carried out for each of the examples and in Figure 2 we show the ROC curves for four of the variables in the women and mathematics data set. Here 25% of the data was used for testing. The dashed ROC curves show how well the model with the highest posterior probability performs, the dotted curves show the performance averaging over the models in Occam's window, while the solid curves are for MC^3. For each of the variables, MC^3 provides substantially improved performance. Such clear differences do not occur in each of the examples, although typically, methods which average over models provide superior ROC curves.

10.5 Discussion

In the cases where the evidence overwhelmingly points to one particular model then model averaging will typically provide only minor improvements in predictive performance. However, we believe that this happy situation is the exception rather than the rule. For example, consider again the Women and Mathematics survey:

Using the complete dataset, the Occam's Window procedure selects $[A][BDE][CDF]$ and $[A][BDE][DF][CF]$. These two models are shown in Figure 3. Upton (1991) reports that a model selection procedure based on the AIC criterion selects $[ABCE][CDF][BCD][DEF]$, while a procedure based on the BIC criterion selects the much simpler (non-graphical) $[A][BE][CE][CF]$ $[BD][DE][DF]$. We show the interaction graphs of these models in Figure 4. Such results are typical—AIC selects overly complex models while BIC provides a remarkably good approximation to the Bayesian procedure.

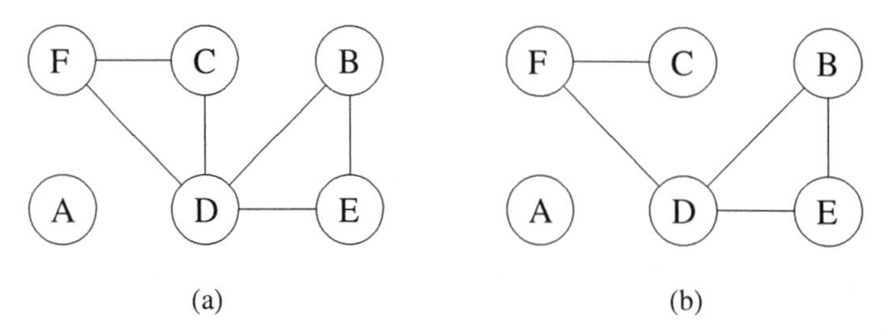

FIGURE 3. Women and Mathematics: Decomposable Models Selected by the Occam's Window Procedure

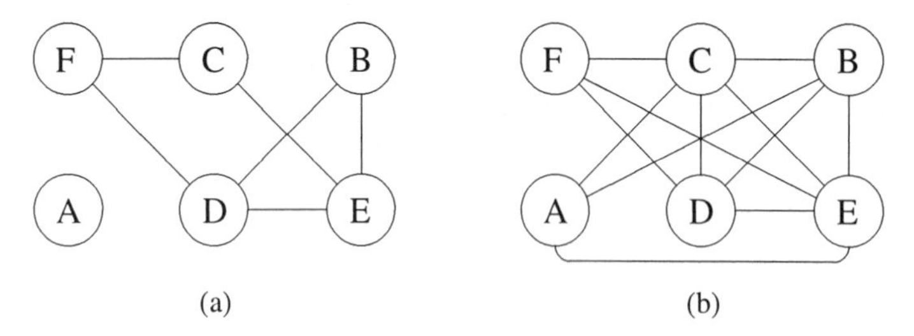

FIGURE 4. Women and Mathematics: Interaction Graphs of the Models Selected by Upton (1991) using (a) the BIC Criterion, and (b) the AIC Criterion.

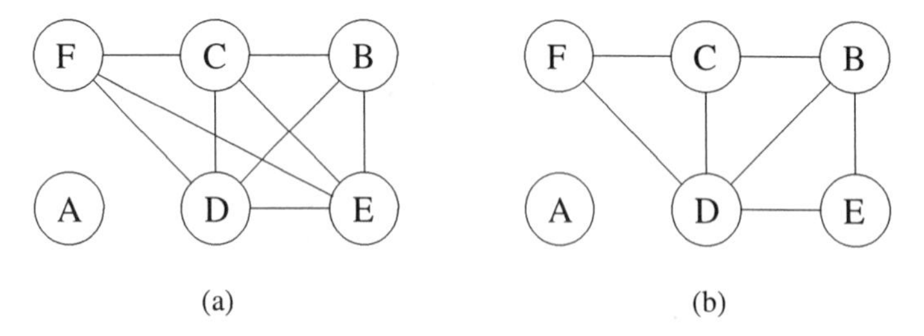

FIGURE 5. Women and Mathematics: (a) Interaction Graph of one of the Models Selected by MIM's headlong procedure; (b) Model Selected by MIM using the "Fast" Procedure.

MIM's backward stepwise procedure produces the remarkably complex model,
$[CDEF][BCDE][ABCE]$.
The backward stepwise procedure with the "headlong" option selected a variety of models on different runs, including
$[CDF][CDE][BDE][A]$,
$[CDEF][BCDE][ABCE]$,
$[CDEF][BCDE][A]$
(shown in Figure 5 (a)), and
$[CDF][BCDE][ABCE]$.
The forward selection procedure, on the other hand, selects
$[DEF][CDF][BDE][A]$.

The "fast" procedure of Edwards and Havránek (1985), as implemented in David Edward's graphical modeling software, MIM, results in a single model being selected, namely $[A][BCD][BDE][CDF]$. This is model (b) of Figure 5 and is similar to model (a) of Figure 3, with the addition of a link from B to C (in fact, the p-value associated with this extra link is 0.04). We used a critical value of 0.05 for this analysis.

In the face of such model uncertainty, basing predictions on any single model is unsatisfactory. Furthermore, these assorted models have widely varying out-of-sample predictive abilities. Model averaging represents a practical way out of these difficulties. Indeed, Hodges (1987) argues that "what is clear is that when the time comes for betting on what the future holds, one's uncertainty about that future should be fully represented and model [averaging] is the only tool around". We expect that proper accounting for model uncertainty will be of importance in belief networks, which are now routinely used in knowledge-based system applications (Bradshaw *et al.*, 1993).

Concerning the two model averaging methods discussed here, MC^3 generally provides better performance than Occam's window. However, communication of model uncertainty is often important, and the insight into model uncertainty provided by the Occam's window method will be important in many applications. Furthermore, the execution times for the Occam's window method are typically much shorter than those for MC^3.

10.6 References

[1] Bradshaw, J.M., Chapman, C.R., Sullivan, K.M., Boose, J.H., Almond, R.G., Madigan, D., Zarley, D., Gavrin, J., Nims, J. and Bush, N. (1993) "KS-3000: An application of DDUCKS to bone-marrow transplant patient support," *Proc. 7th European Knowledge Acquisition for Knowledge-Based Systems Workshop (EKAW-93), Toulouse and Caylus, France*, 57–74.

[2] Breslow, N. (1991) "Biostatistics and Bayes," *Stat. Sci.* **5**, 269–298.

[3] Draper, D., (1994) "Assessment and propagation of model uncertainty," *JRSS (B)*, to appear.

[4] Edwards, D. and Havránek, T. (1985) "A fast procedure for model search in multidimensional contingency tables," *Biometrika* **72** , 339–351.

[5] Fowlkes, E.B., Freeny, A.E. and Landwehr, J.M. (1988) "Evaluating logistic models for large contingency tables," *JASA* **83**, 611–622.

[6] Hastings, W.K. (1970) "Monte Carlo sampling methods using Markov chains and their applications," *Biometrika* **57**, 97–109.

[7] Hodges, J.S. (1987) "Uncertainty, policy analysis and statistics," *Stat. Sci.* **2**, 259–291.

[8] Kass, R.E. and Raftery, A.E. (1993) "Bayes factors and model uncertainty." *Technical Report 254*, Department of Statistics, University of Washington.

[9] Madigan, D. and Raftery, A.E. (1991) "Model selection and accounting for model uncertainty in graphical models using Occam's window." *Technical Report 213*, Department of Statistics, University of Washington.

[10] Madigan, D. and York, J. (1993) "Bayesian graphical models for discrete data." *Technical Report 259*, Department of Statistics, University of Washington.

[11] Raftery, A.E. (1988) "Approximate Bayes factors for generalised linear models." *Technical Report 121*, Department of Statistics, University of Washington.

[12] Raftery, A.E. (1993) "Approximate Bayes factors and accounting for model uncertainty in generalised linear models." *Technical Report 255*, Department of Statistics, University of Washington.

[13] Regal, R. and Hook, E. (1991) "The effects of model selection on confidence intervals for the size of a closed population," *Stat. Med.* **10**, 717–721.

[14] Self, M. and Cheeseman, P. (1987) "Bayesian prediction for artificial intelligence," *Proc. 3rd Workshop on Uncertainty in Artificial Intelligence, Seattle,* 61–69.

[15] Tierney, L. (1991) "Markov chains for exploring posterior distributions." *Technical Report 560*, School of Statistics, University of Minnesota.

[16] Upton, G.J.G. (1991) "The exploratory analysis of survey data using log-linear models," *The Statistician* **40**, 169–182.

[17] Whittaker, J. (1990) *Graphical models in Applied Mathematical Multivariate Statistics.* John Wiley & Sons, Chichester, England.

[18] York, J.C. and Madigan, D. (1992) "Bayesian methods for estimating the size of a closed population," *Technical Report 234*, Department of Statistics, University of Washington.

11
Conditional dependence in probabilistic networks

Remco R. Bouckaert

Utrecht University
Department of Computer Science
P.O.Box 80.089, 3508 TB Utrecht
The Netherlands
remco@cs.ruu.nl

ABSTRACT In general, a probabilistic network is considered a representation of a set of conditional independency statements. However, probabilistic networks also represent dependencies. In this paper an axiomatic characterization of conditional dependence is given. Furthermore, a criterion is given to read conditional dependencies from a probabilistic network.
Keywords: probabilistic network, conditional independence, causal input list, minimal I-map.

11.1 Introduction

The graphical representation of probabilistic relationships between variables gets a lot of attention in different areas of research the last years. A probabilistic networks, also known by the name belief network [Pea88], causal network [LS88], and influence diagram [Sha86], is such a representation. In the field of artificial intelligence, systems have been developed for efficient computation of inferences [LS88, Pea88, Sha86] using probabilistic networks.

Independencies between variables are practically indispensable when making inferences in large knowledge-based systems. Probabilistic networks are a powerful means for representing conditional independency statements on variables. A probabilistic network has associated a semantics that allows for reading independencies between variables [LDLL90, Pea88]. Independencies read from the network can be used to decide on relevance of variables to inference problems.

However, in a probabilistic network it is not generally true that if variables are not shown to be independent they are actually dependent. In other words, there may exist independencies that cannot be read from the network. Therefore, it is interesting to find out where these hidden independencies reside in the network. In order to do so, we consider variables that are definitely dependent in a given network.

In Section 2, we give an overview of conditional *in*dependence and its relation with probabilistic networks. In Section 3, we study properties of conditional dependence and in Section 4, we present a criterion for reading dependencies from a probabilistic network.

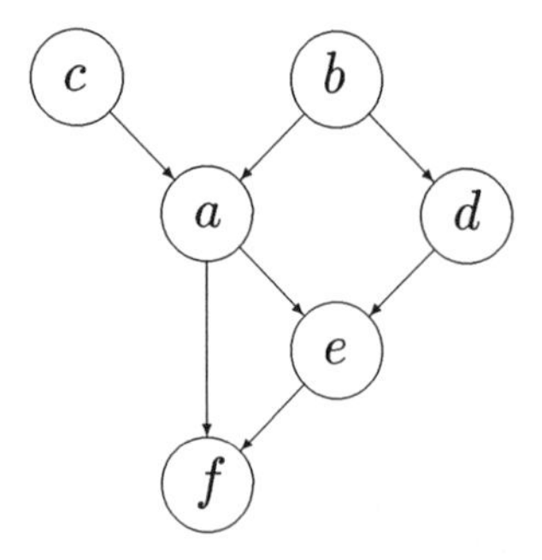

FIGURE 1. DAG in which $<c, \emptyset, d>$, $<c, b, d>$ and $<c, ab, d>$, but not $<c, a, d>$ or $<c, abf, d>$.

11.2 Conditional independence

We consider a joint probability distribution P over a set of variables U. In this paper we use capital letters to denote sets of variables and lower case letters to denote single variables. All variables or sets of variables mentioned are elements or subsets of U unless stated otherwise. We call X and Y *conditionally independent* given Z, written $I(X, Z, Y)$, if $P(XY|Z) = P(X|Z)P(Y|Z)$ for all values of the variables in XYZ (for sets we write XY to denote the union of X and Y); $I(X, Z, Y)$ is called an *independency statement*. By definition $I(X, Z, \emptyset)$ for any X and Z. An *independency model* over U is a set of independency statements. A *complete independency model* M_I of a distribution P over U is the set of all valid independency statements in P. For positive definite distributions, the following axioms called *independency axioms* apply [Daw79, PP86, PV87].

symmetry	$I(X, Z, Y)$	$\Leftrightarrow I(Y, Z, X)$
decomposition	$I(X, Z, WY)$	$\Rightarrow I(X, Z, Y)$
weak union	$I(X, Z, WY)$	$\Rightarrow I(X, ZW, Y)$
contraction	$I(X, ZW, Y) \wedge I(X, Z, W)$	$\Rightarrow I(X, Z, WY)$
intersection	$I(X, ZW, Y) \wedge I(X, ZY, W)$	$\Rightarrow I(X, Z, WY)$

With these axioms independency statements can be derived from other independency statements. For instance, let M_I be an independency model for a given distribution P such that $I(a, b, c) \in M_I$. Then, by symmetry we have that $I(c, b, a)$ must also be in M_I. We sometimes omit braces to prevent an overflow of them. So, we write $I(a, b, c)$ for $I(\{a\}, \{b\}, \{c\})$.

A directed acyclic graph (DAG) is a directed graph that does not contain paths starting and ending at the same node. A *trail* in a DAG is a path that does not consider the direction of the arcs. We denote a trail by the ordered sequence of nodes that are in the trail. For example, in the DAG in Figure 1 *cabd* is a trail. A *head-to-head node* in a trail is a triple of consecutive nodes x, y, z in the trail such that $x \rightarrow y \leftarrow z$ in the DAG. A *probabilistic network* is a pair (G, Γ) where G is a DAG and Γ is a set for every variable $u \in U$ of conditional probability tables $P(u|\pi_u)$ that enumerate the probabilities of all values of u given the values of its parents π_u in the DAG. The distribution represented by this network is $\prod_{u \in U} P(u|\pi_u)$ [Pea88]. Independency statements that hold in the distribution represented by a probabilistic network can be read from the structure of the DAG using the notions of blocked trail and d-separation [Gei90, Pea88].

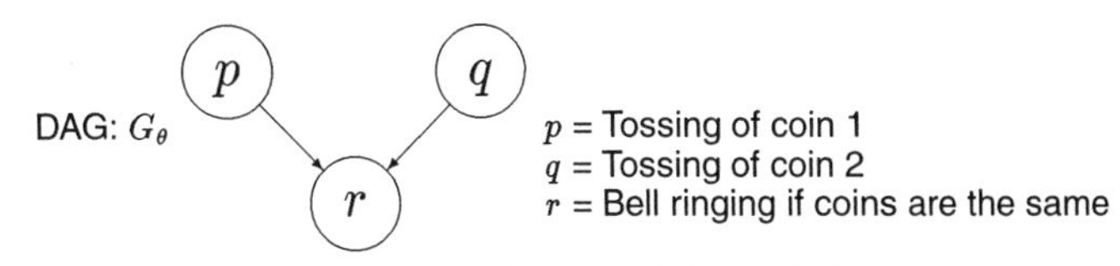

Independency model: $M_I = \{I(p,\emptyset,q), I(p,\emptyset,r), I(q,\emptyset,r) + \text{sym.}\}$
Causal ordering: $\theta = \{p,q,r\}$
Causal input list: $L_\theta = \{I(p,\emptyset,\emptyset), I(p,\emptyset,q), I(r,pq,\emptyset)\}$
Associated model: $M_{I,\theta} = \{I(p,\emptyset,q), I(q,\emptyset,p)\}$

FIGURE 2. A DAG that is a minimal I-map of the coins and bell-example.

Definition 11.2.1 *Let G be a DAG. A trail in G between two nodes x and y is* **blocked** *by a set of nodes Z if at least one of the following two conditions hold:*
- *the trail contains a head-to-head node e and e $\notin$ Z and every descendant of e in G is not in Z.*
- *there is a node e in the trail with e $\in$ Z and e is not a head-to-head node in the trail.*

Definition 11.2.2 *In a DAG G, let X, Y and Z be sets of nodes. We say that X is* **d-separated** *from Y given Z, written $<X, Z, Y>$, if every trail between any node $x \in X$ and any node $y \in Y$ is blocked by Z.*

In Figure 1, an example DAG is depicted. $<c, \emptyset, d>$ and $<c, b, d>$ are valid separation statements since all trails between c and d, i.e., *cabd*, *caed* and *cafed*, are blocked. $<c, ab, d>$ is valid since the trail *cabd* is blocked by $\{a, b\}$ because it contains node b that is not a head-to-head node. The trail *caed* is blocked by $\{ab\}$ since it contains node a that is not a head-to-head node in the trail. However, $<c, a, d>$ is not valid since the trail *cabd* is not blocked by a because it contains node a which is a head-to-head node

A DAG G is an *I-map* of an independency model M_I if $<X, Z, Y>$ in G implies $I(X,Z,Y) \in M_I$; G is a *minimal I-map* of M_I if no arc can be removed from G without destroying its I-mappedness. G is a *D-map* of M_I if $I(X,Z,Y) \in M_I$ implies $<X, Z, Y>$ in G; G is a *perfect map* of M_I if it is both an I-map and a D-map of M_I.

It is not always possible to find a DAG that is a perfect map of a distribution. For example, consider the situation that a bell rings if the outcomes of two tossed coin is the same [Pea88]. Let p and q represent the outcome of the coins and r the ringing of the bell. Just to keep the distribution positive definite, put the bell in a noisy factory such that it is not always clear if it rings or not. Then, all variables are pairwise independent: $M_I = \{I(p,\emptyset,q), I(p,\emptyset,r), I(q,\emptyset,r) + \text{sym.}\}$. But, given the third variable they are dependent. This non-monotonic behavior cannot perfectly be represented by a DAG because composition ($<X,Z,Y> \wedge <X,Z,W> \Rightarrow <X,Z,WY>$) holds for d-separation [Pea88]. So, if the DAG represents both $I(p,\emptyset,r)$ and $I(p,\emptyset,q)$ then it also represents $I(p,\emptyset,rq)$. The best we can do is represent the model by a minimal I-map as in Figure 2. However, this DAG does not represent the independence between p and r since $<p,\emptyset,r>$ does not hold in the DAG.

A minimal I-map can be constructed from an independency model M_I using the notion of causal input lists [PGV90].

Definition 11.2.3 *Let θ be a total ordering on U. Let M_I be a complete independency model of a distribution P over U. A* **causal input list** *L_θ over M_I is a set of independency statements*

such that for every $x \in U$, L_θ contains exactly one independency statement of the form:

$$T = I(x, \pi_x, U_x \backslash \pi_x)$$

in which $U_x = \{y | y \in U,\ \theta(y) < \theta(x)\}$ and π_x is the smallest subset of U_x such that T holds in M_I. π_x is called the **parent set** *of x.*

For positive definite distributions, a causal input list can be constructed in $O(|U|^2)$ consultations of M_I: for $x \in U$, we have that a node $y \in U_x$ is in π_x if and only if $I(x, U_x \backslash y, y)$ is not in M_I.

Let $\theta = \{p, q, r\}$ be a causal ordering for the coins and bell example. Then the causal input list would be $L_\theta = \{I(p, \emptyset, \emptyset), I(p, \emptyset, q), I(r, pq, \emptyset)\}$. A DAG G_θ is associated with the ordering θ by letting G_θ be the DAG constructed in the following way: start with an arc-less graph and place an arc for each node u from every node in its parent set to node u. For the coins and bell example, the nodes p and q don't have incoming arcs since their parent sets are empty. The parent set of r contains both p and q so node r will get incoming arcs from both these nodes. The result is the graph in Figure 2. Note that the parent set depend on the ordering θ. Let $\theta' = \{p, r, q\}$ be a causal ordering, then the causal input list $L_{\theta'}$ is $\{I(p, \emptyset, \emptyset), I(r, \emptyset, p), I(q, rp, \emptyset)\}$.

An independency model $M_{I,\theta}$ can be associated with a DAG G_θ constructed from a causal input list L_θ by letting $I(X, Z, Y) \in M_{I,\theta}$ if and only if $<X, Z, Y>$ holds in G_θ. Now $M_{I,\theta}$ is the closure of L_θ under the independency axioms (follows from [VP88]). Furthermore, it is known $M_{I,\theta} \subseteq M_I$ for any θ [PGV90].

11.3 Conditional dependencies

As we have seen in the previous section, a probabilistic network can be used to represent independencies. However, it is not always possible to find a perfect map for a given distribution. So, $I(X, Z, Y)$ does not always imply $<X, Z, Y>$.

We call X and Y *conditionally dependent* given Z, written $D(X, Z, Y)$, if not $I(X, Z, Y)$; $D(X, Z, Y)$ is called a *dependency statement*. A *dependency model* over U is a set of dependency statements $D(X, Z, Y)$ with X, Z, Y disjoint subsets of U. The *complete dependency model* M_D of a distribution over U is a dependency model containing all dependency statements that hold in the distribution. We define the following *dependency axioms*:

symmetry	$D(X, Z, Y)$	$\Leftrightarrow D(Y, Z, X)$
composition	$D(X, Z, Y)$	$\Rightarrow D(X, Z, WY)$
weak reunion	$D(X, ZW, Y)$	$\Rightarrow D(X, Z, WY)$
extraction	$D(X, Z, WY) \wedge I(X, Z, W)$	$\Rightarrow D(X, ZW, Y)$
extraction+	$D(X, Z, WY) \wedge I(X, ZY, W)$	$\Rightarrow D(X, Z, Y)$
intersection	$D(X, Z, WY) \wedge I(X, ZY, W)$	$\Rightarrow D(X, ZW, Y)$

Theorem 11.3.1 *For any probability distribution all dependency axioms but intersection hold. Furthermore, for positive definite distributions also the intersection axiom holds.*

Proof: We will only proof that symmetry holds for any probability distribution P. Let $D(X, Z, Y)$ be a valid dependency statement for P. By definition we have $D(X, Z, Y) \Leftrightarrow not\ I(X, Z, Y)$. Now, assume that $I(Y, Z, X)$ holds. Then, by symmetry for independency statements it follows that $I(X, Z, Y)$. This last independency statement is false, however. So, we have $not\ I(Y, Z, X)$ from which it follows that $D(Y, Z, X)$. The proofs for the other axioms are analogous. □

These dependency axioms can be used to deduce new dependency statements from given statements. For example, let M_D be a dependency model for a given distribution P and $D(a, b, c)$ is in M_D. Then, by symmetry we have that $D(c, b, a)$ must also be in M_D.

Lemma 11.3.1 *For any positive definite distribution P $D(X, Z, Y) \in M_D$ if and only if two nodes $x \in X, y \in Y$ exist such that $D(x, XYZ \backslash \{x, y\}, y) \in M_D$.*

A proof can be found in the appendix. >From the lemma it follows that for $D(X, Z, Y)$ for a positive definite distributions, it is sufficient to show that two variables $x \in X$ and $y \in Y$ exist such that $D(x, XYZ \backslash \{x, y\}, y)$ is in the complete dependency model.

Let L_θ be a causal input list over an independency model of a positive definite distribution over U. Then, the *dependency base* associated with L_θ is the set of dependency statements $\Sigma_\theta = \{D(x, \pi_x \backslash \{y\}, y) | x \in U, y \in \pi_x\}$. We define the dependency model associated with θ, denoted as $M_{D,\theta}$, as the closure of the dependency base Σ_θ under the dependency axioms. It can be shown that the closure of Σ_θ under symmetry, weak reunion, composition, extraction and intersection is equal to the closure of Σ_θ under all dependency axioms [Bou92].

Theorem 11.3.2 *Let M_I be a complete independency model of a positive definite distribution over U and M_D be its complete dependency model. Let L_θ be a causal input list over M_I. Let $M_{D,\theta}$ be the dependency model associated with L_θ. Then, $M_{D,\theta} \subseteq M_D$.*

Proof: The property stated in the theorem will be proved by contradiction. Assume for a statement $D(x, \pi_x \backslash y, y) \in \Sigma_\theta$ that it is not in M_D. Then, by definition $I(x, \pi_x \backslash y, y) \in M_I$. $I(x, \pi_x \backslash y, y)$ and $I(x, \pi_x, U_x \backslash \pi_x)$ imply $I(x, \pi_x \backslash y, U_x \backslash (\pi_x \backslash y))$ using contraction. But this is not a valid statement since it implies that π_x was not the smallest subset of U_x such that $I(x, \pi_x, U_x \backslash \pi_x)$. So, all statements in Σ_θ are in M_D. Since the dependency axioms are sound for positive definite distributions, the theorem follows from the definition of $M_{D,\theta}$. □

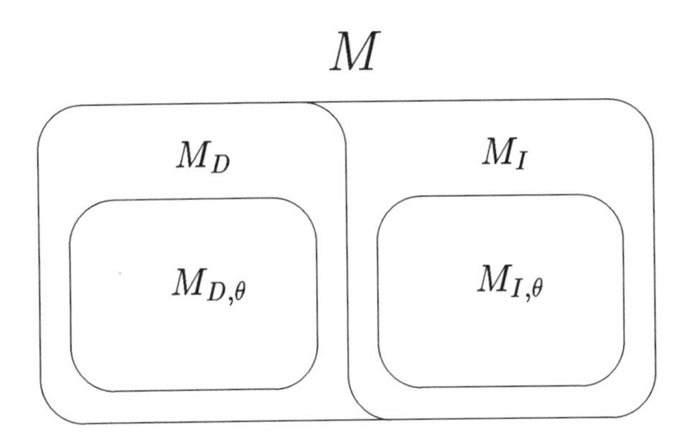

FIGURE 3. Division of the dependency pool

The *dependency pool* M over U is the set of all triples (X, Z, Y) with X, Y and Z disjoint subsets of U. For a given causal input list L_θ over U, we can divide the pool M into three disjoint sets: $M_{I,\theta}$, $M_{D,\theta}$ and $M \backslash (M_{I,\theta} \cup M_{D,\theta})$ as depicted in Figure 3. Note that from a given causal input list of an independency model, not all independency and dependency statements may be known. So, there are statements for which it cannot be decided from the structure of the graph alone whether it is a independency statement or a dependency statement.

For the coins and bell example, as shown in Figure 2, $M_{I,\theta} = \{I(p,\emptyset,q) + \text{sym.}\}$. The associated dependency model $M_{D,\theta}$ is

$$\{I(r,q,p), I(r,p,q), I(r,\emptyset,pq), I(p,\emptyset,rq), I(q,\emptyset,rp), + \text{sym.}\}$$

which leaves the statements $\{I(p,\emptyset,r), I(q,\emptyset,r), + \text{sym.}\}$ to be unknown.

11.4 Graphical criterion for conditional dependencies

In Section 2 we have argued that all independency statements in the independency model $M_{I,\theta}$ associated with the causal input list L_θ can be read from G_θ constructed from L_θ using the d-separation criterion. It would be useful to have a similar graphical criterion for reading dependency statements from the DAG G_θ. In this section we investigate the properties of such a criterion.

Consider a DAG G_θ constructed from a causal input list L_θ over an independency model M_I. Let $D(X,Z,Y)$ be a dependency statement in $M_{D,\theta}$. Then, a derivation exists starting with $D(x,\pi_x\backslash y,y)$ and ending with $D(X,Z,Y)$. By structural induction over the steps in the derivation (that use symmetry, composition, weak reunion, extraction and intersection and even extraction+) it can be shown [Bou92] that in every step the following properties are preserved: Let $D(X',Z',Y')$ be the result of a step in the derivation then two nodes $x \in X'$ and $y \in Y'$ or $x \in Y'$ and $y \in X'$ exist such that $y \to x$ and $\pi_x \subset XYZ$. So, any graphical criterion for reading dependency statements $D(X,Z,Y)$ from G_θ must satisfy these conditions.
Some conditional dependency statements can be read from the graph using the following criterion:

Definition 11.4.1 *In a DAG G, we say that X and Y are* **coupled** *given Z, written* $>X,Z,Y<$, *if nodes* $x \in X$ *and* $y \in Y$ *or* $x \in Y$ *and* $y \in X$ *exist such that all following conditions hold:*
• $y \to x$ *is an arc in G.*
• $\pi_x \subset XYZ$.
• *a set Q exists such that* $Z \subseteq Q \subseteq XYZ\backslash\{x,y\}$ *and* $<x,Q,y>$ *in G when the arc* $y \to x$ *is removed.*

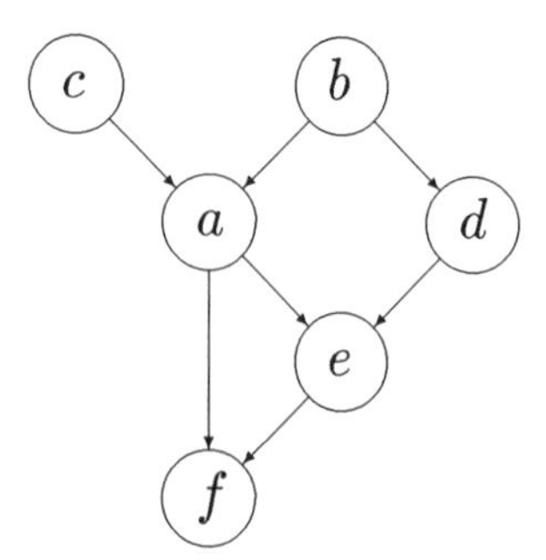

FIGURE 4. Minimal I-map for which $>a,c,b<$, $>a,cd,b<$ and $>ac,e,db<$, but not $>a,ce,bf<$ or $>a,de,b<$.

In Figure 4 some examples of coupling statements are given. We have that $>a, c, b<$ since $b \rightarrow a$ is an arc in the graph, $\pi_a = \{b, c\} \subset \{a, b, c\}$ and all trails between a and b not containing $b \rightarrow a$, i.e., $aedb$ and $afedb$, contain a head-to-head node and therefore are blocked. Also we have that $>ac, e, bd<$ since still $b \rightarrow a$ is an arc in G, $\pi_a \subset abcde$ and a set $Q = \{d, e\}$ exists such that all trails are blocked. The statement $>a, ce, bf<$ does not hold since the trail $aedb$ is not blocked and no set Q such that $ce \subseteq Q \subseteq cef$ can be found that does so. The statement $>a, de, b<$ does not hold since π_a is not subset of $adeb$.

Theorem 11.4.1 *In a DAG G that is a minimal I-map of an independency model M_I, the following property holds:*

$$>X, Z, Y< \Rightarrow D(X, Z, Y)$$

A proof can be found in the appendix. As a consequence we have that if the first two conditions for coupled hold for a triple of sets (X, Z, Y) and the set Z does not contain descendants of x where x corresponds to the node in the definition then $D(X, Z, Y)$.

The theorem implies that the definition of coupled gives sufficient conditions for reading dependency statements from a minimal I-map. However, it does not give necessary conditions to do so. For example, if the DAG in Figure 4 is a minimal I-map of a model then we know from the dependency base that $D(a, c, b)$, so by symmetry $D(b, c, a)$ and using composition $D(b, c, aef)$. From the DAG we have $<b, ace, f>$ implying $I(b, ace, f)$. And, $D(b, c, aef)$ and $I(b, ace, f)$ imply $D(b, cf, ae)$ using intersection. However, $>b, cf, ae<$ does not hold in the DAG: the trail $aedb$ will never be blocked by a set containing f and not d.

The question that arises is: does a mathematical esthetic and not too complex criterion exists such that this criterion can be used to read dependency statements from the graph. Such a criterion must contain the following property. As we saw in the previous example, $D(b, cf, ae)$ in the DAG of Figure 4. However, $D(be, cf, a)$ can not be derived using the dependency axioms and the independency statements in the DAG. This example shows that the derivation depends on the place of the variable e in the statement. A graphical criterion must reckon with this possibility.

For dependency statements $D(X, Z, Y)$ where X and Y are single variables this problem does not arise. We have the following result.

Theorem 11.4.2 *In a DAG G_θ constructed from a causal input list L_θ of a complete independency model M_I, the following property holds:*

$$>x, Z, y< \Leftrightarrow D(x, Z, y) \in M_{D,\theta}$$

For a proof we refer to the appendix. This theorem says that at least for all dependency statements in $M_{D,\theta}$ concerning single nodes can be read from the graph.

11.5 Conclusions

Not all independency statements can always be represented by a probabilistic network. Therefore, if the d-separation criterion does not hold for three sets of variables, it does not necessarily mean the sets are dependent. In this paper we gave a sound axiomatic characterization of conditional dependencies. Furthermore, a graphical criterion is presented to read most of the dependency statements from a DAG that is a minimal I-map. It is shown that for some statements it cannot be deduced from the structure of the probabilistic network whether it is an independency or a dependency statement.

Acknowledgement

I thank Linda van der Gaag for helpful comments on earlier drafts resulting in a much better presentation of the paper.

11.6 Appendix

Lemma 11.3.1 *For any positive definite distribution P* $D(X,Z,Y) \in M_D$ *if and only if two nodes* $x \in X, y \in Y$ *exist such that* $D(x, XYZ\backslash\{x,y\}, y) \in M_D$.

Proof: First we show the $\Leftarrow$ part. Assume that two nodes $x \in X$ and $y \in Y$ exist such that $D(x, XYZ\backslash\{x,y\}, y)$. Then,

$D(x, XYZ\backslash\{x,y\}, y)$
$\Rightarrow$ {Weak reunion }
$D(x, XZ\backslash\{x\}, Y)$
$\Rightarrow$ {Symmetry }
$D(Y, ZX\backslash\{x\}, x)$
$\Rightarrow$ {Weak reunion }
$D(Y, Z, X)$
$\Rightarrow$ {Symmetry }
$D(X, Z, Y)$

Now the $\Rightarrow$ part. We assume $D(X,Z,Y)$ that holds, Let $X = \{x_1, ..., x_n\}, Y = \{y_1, ..., y_m\}$. Now suppose that for all $x \in X, y \in Y$, the statement $I(x, XYZ\backslash\{x,y\}, y)$ is valid. Then,

$\forall_{x\in X} I(x, XYZ\backslash\{x, y_1\}, y_1)$
$\Rightarrow$ {Intersection with $I(x, XYZ\backslash\{x, y_2\}, y_2)$ }
$\forall_{x\in X} I(x, XYZ\backslash\{x, y_1, y_2\}, y_1 y_2)$
$\Rightarrow$ {Intersection with $I(x, XYZ\backslash\{x, y_3\}, y_3)$ }
$\forall_{x\in X} I(x, XYZ\backslash\{x, y_1, y_2, y_3\}, y_1 y_2 y_3)$

...

$\forall_{x\in X} I(x, XYZ\backslash\{x, y_1, y_2, .., y_{m-1}\}, y_1 y_2 .. y_{m-1})$
$\Rightarrow$ {Intersection with $I(x, XYZ\backslash\{x, y_m\}, y_m)$ }
$\forall_{x\in X} I(x, XZ\backslash\{x\}, Y)$
$\Rightarrow$ {Symmetry }
$\forall_{x\in X} I(Y, XZ\backslash\{x\}, x)$

A similar observation holds for $x_1, \ldots, x_n$. So, we can derive $I(X,Z,Y)$ which contradicts the assumption $D(X,Z,Y)$. Therefore, the assumption that for all $x \in X, y \in Y$ the statement $I(x, XYZ\backslash\{x,y\}, y)$ holds is false. □

For the proof of Theorem 11.4.1 we use the following lemma.

Lemma A1 *Let* $G = (U, A(G))$ *be a DAG. Let* $x, y \in U$ *be two nodes such that* $y \rightarrow x$ *in G. Let* $Z \subseteq U\backslash\{x,y\}$ *be a set of nodes such that* $\pi_x \subset Z \cup \{y\}$. *Let* $c \in U\backslash Z$ *be a node that is a*

descendant of node x and has no descendants in Z. Furthermore, let any trail between x and y not containing $y \rightarrow x$ be blocked by Zc. Then, either $<c, Zx, y>$ or $<c, Zy, x>$.

Proof: Assume two trails $t(x, c)$ and $t(c, y)$ exist such that both $t(x, c)$ is not blocked given Zy and $t(c, y)$ is not blocked given Zx. >From the conditions in the lemma then also $t(x, c)$ is not blocked given Z and $t(c, y)$ is not blocked given Z. Since c does not contain descendants in Z both trails must have an incoming arrow into c.

So, it follows that the trail $t(x, y)$ that arises when $t(x, c)$ and $t(c, y)$ are concatenated is not blocked by Zc (the part between x and c is not blocked by Z, c is a head-to-head node and the part between y and c is also not blocked by Z) and the condition would not be fulfilled. Therefore, either all trails $t(x, c)$ are blocked by Zy or all trails $t(y, c)$ are blocked by Zx. □

Theorem 11.4.1 *In a DAG G that is a minimal I-map of an independency model M_I, the following property holds:*

$$>X, Z, Y< \Rightarrow D(X, Z, Y)$$

Proof: We assume that $>X, Z, Y<$ holds in G. Without loss of generality we take $x \in X$ and $y \in Y$ to be the nodes such $y \rightarrow x$ is an arc in G, $\pi_x \subset XYZ$ and let Q be the set such that $Z \subseteq Q \subseteq XYZ \backslash \{x, y\}$ and $<x, Q, y>$ in G when the arc $y \rightarrow x$ is removed.

Let $Q_u \subseteq Q$ be the set of nodes that are not descendants of x and let $Q_d = Q \backslash Q_u$. Since G is a minimal I-map we have that,

$$\begin{aligned}
& D(x, \pi_x \backslash y, y) \\
\Rightarrow\ & \{\text{Composition}\ \} \\
& D(x, \pi_x \backslash y, yQ_u) \\
\Rightarrow\ & \{\text{Intersection with } I(x, \pi_x, Q_u)\ \} \\
& D(x, Q_u \pi_x \backslash y, y)
\end{aligned}$$

Now we will proof that $D(x, Q\pi_x \backslash y, y)$ also holds by adding nodes in Q that are descendants of x one by one using Lemma A1. If all trails between x and y are blocked by $Q_u \backslash y$ then they are also blocked by $Q_u \pi_x \backslash y$. Let $q \in Q_d$ be the node that is lowest in the ordering θ then Lemma A1 implies either $I(q, xQ_u\pi_x \backslash y, y)$ or $I(q, Q_u\pi_x, x)$. So,

$$\begin{aligned}
& D(x, Q_u \pi_x \backslash y, y) \\
\Rightarrow\ & \{\text{Intersection with } I(q, xQ_u\pi_x \backslash y, y) \text{ or } I(q, Q_u\pi_x, x)\ \} \\
& D(x, qQ_u \pi_x \backslash y, y)
\end{aligned}$$

Repeat this derivation step with $Q_u := Q_u \cup q$ and $Q_d = Q_d \backslash q$ until $Q_d = \emptyset$. Observe that if all trails between a and b not containing $y \rightarrow x$ are blocked by a set D then they are also blocked by a set $D \backslash p$ where p is a descendant of a that has no descendants in D. Therefore, Lemma A1 applies every time the derivation step is repeated. This derivation results in $D(x, Q\pi_x \backslash y, y)$. So,

$$\begin{aligned}
& D(x, Q\pi_x \backslash y, y) \\
\Rightarrow\ & \{\text{Composition and symmetry}\ \} \\
& D(x \cup (X \backslash Q\pi_x), Q\pi_x \backslash y, y \cup (Y \backslash Q\pi_x)) \\
\Rightarrow\ & \{\text{Weak reunion and symmetry}\ \} \\
& D(X, Z, Y)
\end{aligned}$$

So, $D(X, Z, Y) \in M_{D,\theta}$ and therefore by Theorem 11.3.2 $D(X, Z, Y) \in M_D$. □

Theorem 11.4.2 *In a DAG G_θ constructed from a causal input list L_θ of a complete independency model M_I, the following property holds:*

$$>x, Z, y< \Leftrightarrow D(x, Z, y) \in M_{D,\theta}$$

Proof: The $\Rightarrow$ part follows from Theorem 11.4.1. For the $\Leftarrow$ part we have to show that for all derivations resulting in a dependency statement of the form $D(x, Z, y)$ $>x, Z, y<$ holds.

Assume $D(x, Z, y)$ but not $>x, Z, y<$. By structural induction on the steps of the derivation of $D(x, Z, y)$ we know that two nodes $a \in \{x\}, b \in \{y\}$ or $a \in \{x\}, b \in \{y\}$, must exist such that $b \to a$ and $\pi_a \subset XYZ$. Without loss of generality we assume that $a = x$ and $b = y$. Since no $Q \subset XYZ \backslash \{x, y\} = Z$ exists such that all trails between x and y are blocked (otherwise we have $>x, Z, y<$), a trail not containing $y \to x$ must exist between x and y that is unblocked. Since, $\pi_x \subset Zy$ the trail cannot contain an arc from a parent of x to x. So the trail contains a descendant of x (thus also of y) that forms a head-to-head node. For the trail to be blocked a node $w \in Z \cap desc(x)$ must exists such that w is in a trail between x and y.

If $D(x, Z, y)$ can be derived from $D(a, \pi_a \backslash b, b)$ using the dependency axioms then $I(a, \pi_a \backslash b, b)$ can be derived from $I(x, Z, y)$ using the independency axioms. Since we know $\pi_a \backslash b \subseteq Z$ we have to remove the nodes $Z \backslash (\pi_a \backslash b)$ from Z. How can w be removed from Z in the derivation? This can only be done when nodes Q are introduced such that the trail between x and y via w is blocked using contraction. However, introducing such a node cannot be done since the set Q must fulfill $<Q, Zx, y>$ or $<Q, Zy, x>$. But, these will hold for no such Q since the trail between x and y via w is not blocked. So, $D(x, Z, y)$ cannot be derived. □

11.7 References

[Bou92] R.R. Bouckaert, *Conditional dependence in probabilistic networks*, Tech. Report RUU-CS-92-34, Utrecht University, The Netherlands, 1992.

[Daw79] A.P. Dawid. *Conditional independence in statistical theory*, J.R. Stat. Soc. (Series B), 1979, pp. 1–31.

[Gei90] D. Geiger, *Graphoids: a qualitative framework for probabilistic inference*, Ph.D. thesis, UCLA, Cognitive Systems Laboratory, Computer Science Department, 1990.

[LDLL90] S.L. Lauritzen, A.P. Dawid, A.P. Larsen, and H.G. Leimer, *Independence properties of directed markov fields*, Networks, Vol 20 (1990), 491–505.

[LS88] S.L. Lauritzen and D.J. Spiegelhalter, *Local computations with probabilities on graphical structures and their applications to expert systems (with discussion)*, J.R. Stat. Soc. (Series B), Vol. 50 (1988), pp. 157–224.

[Pea88] J. Pearl, *Probabilistic reasoning in intelligent systems: Networks of plausible inference*, Morgan Kaufman, inc., San Mateo, CA, 1988.

[PGV90] J. Pearl, D. Geiger, and T. Verma, *The logic of influence diagrams*, Influence Diagrams, Belief Nets and Decision Analysis (R.M. Oliver and J.Q. Smith, eds.), John Wiley & Sons Ltd., 1990, pp. 67–87.

[PP86] J. Pearl and A. Paz, *Graphoids: a graph based logic for reasoning about relevance relations*, Proceedings ECAI, 1986.

[PV87] J. Pearl and T. Verma, *The logic of representing dependencies by directed acyclic graphs*, Proceedings AAAI, 1987, pp. 374–79.

[Sha86] Ross D. Shachter. *Evaluating influence diagrams.*, Operations Research, Vol. 34 (1986), pp. 871–882.

[VP88] T. Verma and J. Pearl, *Causal networks: Semantics and expressiveness*, Proceedings Uncertainty in Artificial Intelligence, 1988, pp. 352–359.

12

Reuse and sharing of graphical belief network components

Russell Almond, Jeffrey Bradshaw, and David Madigan

StatSci division of MathSoft, Inc., Seattle, WA

Research and Technology, Boeing Computer Services, Seattle, WA
and
Dept. of Statistics, University of Washington, Seattle, WA.

ABSTRACT A team of experts assemble a graphical belief network from many small pieces. This paper catalogs the types of knowledge that comprise a graphical belief network and proposes a way in which they can be stored in *libraries*. This promotes reuse of model components both within the team and between projects.

12.1 Introduction

Graphical belief networks (Bayesian networks, influence diagrams, graphical belief models) have become a popular method for representing uncertain knowledge (Almond, 1990; Heckerman, 1991; Henrion, Breese, and Horvitz, 1991; Howard and Matheson, 1984; Pearl, 1988; Shachter, 1986; Shafer and Shenoy, 1988). Their attractiveness stems from the fact that they combine an easy to understand graphical notation with a rigorous computational model.

In our own work, we often encounter situations where modeling involves several people. Imagine, for example, that we are trying to model the system reliability of a complex machine. One engineer, the overall designer, might put together the overall structure of the model. A second engineer, a reliability expert, might determine what the failure states of the various components are and how they propagate. A third engineer, an expert in purchasing, might develop the models for individual component reliability, and so forth. The effectiveness of such a team will hinge on the quality of their communication. Of particular importance is the degree to which the components they are developing can be clearly described and easily shared among team members.

If the results of modeling efforts could be catalogued, others could build upon previous work, rather than starting from scratch. For example, the same component might appear many times in the completed system, and the model for its reliability can be reused. Or else the same combination of components (say a valve/actuator pair) may be used repeatedly, in each case replicating the same model fragment. Later, a different team may want to use the same components for a slightly different problem. To make this possible, they need a means to discover and exploit similarity among components in the graphical model.

---

[0]This research was supported in part by the GRAPHICAL-BELIEF project, NASA SBIR Phase I Grant NAS 9-18669 and NIH SBIR Phase I Grant 1 R43 RR07749-01.

The importance of effective sharing and reuse of model components has become even more apparent in light of recent developments in automated probabilistic and decision model construction systems (Bradshaw*et al.*, 1992a; Bradshaw *et al.*1991; Holtzman, 1989, Wellman, Breese and Goldman, 1991; Edgar, Puerta and Musen, 1992). In such systems, knowledge-based systems guide the configuration of situation-specific belief and decision models with components selected from an electronic library.

Although this paper concentrates on the problem of sharing and classifying knowledge *within* groups, there is an equally large an important problem of sharing knowledge between groups. Here the lack of standardization in terminology (just look at the number of synonyms for "graphical model") hinders the efforts of research groups from various schools to share examples and methods. Although this paper was derived from the design of a single system (GRAPHICAL-BELIEF, Almond, 1992b), a useful ontology for graphical belief networks will only be achieved by collaboration among many scientists involved in similar efforts.

To build and use libraries of resuable model components, we need at least three things: (1) a rigorous specification of the kinds of components such a library must contain—an *ontology* for graphical belief networks,—(2) a rich description of model components, and (3) a formalism to describe how components may be combined. Sections 2 and 3 briefly describe the first two elements of our approach.

12.2 An ontology for graphical belief networks

A number of authors have argued the benefits for making conceptual commitments explicit in the form of ontologies (Bradshaw, *et al.*, 1992b; Gruber, 1991; Gruber, 1992b; Neches *et al.*, 1991; Skuce and Monarch, 1990). The term *ontology* is borrowed from the philosophical literature where it describes a theory of what exists. Such an account would typically include terms and definitions only for the very basic necessary categories of existence. However, the common usage of ontology in the knowledge sharing and reuse community is as a vocabulary of representational terms and their definitions at any level of generality. A knowledge-based system's *ontology* defines what exists for the program: in other words, what can be represented by it.

In this section, we discuss preliminary results in our efforts to define an ontology for graphical belief models. In Section 4 we describe some of the mechanisms we are exploring for exchanging these ontologies in computer interpretable form with other groups.

12.2.1 The central role of valuations

The central theme of all graphical models is that the variables are represented by nodes connected by edges, which in some sense represent relationships between them. In Bayesian networks (Pearl, 1988), the edges are directed, and the relationships are conditional probability functions. In influence diagrams (Howard and Matheson, 1984), a value variable with utility to the decision maker is designated and compute so as to maximize is expected value. In graphical belief models (Almond, 1990, Dempster and Kong, 1988), the graph is a hypergraph and the "hyperedges" correspond directly to component belief functions.

Note that the graphical model provides a visual description of the structure of a mathematical model. Separation in the graph implies statistical independence (Pearl, 1988). The graph also implies a factorization of the model into distinct relationships (usually associated with edges).

There is a close association between the independence conditions and the factorization, and under certain circumstances they are equivalent (Kong, 1988). As for the purposes of model construction the factorization is more important that the independence (indeed, for belief function models independence does not always imply factorization), we will leave aside issues of independence here.

To capture these diverse relationships in a single notation, Shenoy and Shafer(1990) introduce the term *valuation*. A valuation is defined over a set of variables called a *frame*. It maps sets of outcomes in that frame to values. We will define valuations and frames more formally below, but for now, we can consider them to be the generalization of the familiar probability relationship to include variations on the theme, such as utilities and belief functions.

Shenoy and Shafer define a set of operations, in particular, *combination* and *projections*, on valuations. Combination is the ability to combine two valuations defined over the same frame, for example by multiplying probability potentials together. Projection is the ability to change the frame (set of variables over which the valuation is defined) to a larger or smaller set; the most familiar example of this operation is the marginalization of probability distributions. Together with a theorem which allows limited commutativity of projection and combination, these operations can be used as the basis of local computation techniques.

Furthermore, the valuations define the graphical structure of the problem. Because for all graphical models there is one-to-one correspondence between the graphical structure and the factorization of the problem into valuations of appropriate types, the set of valuations must define the graphical model. Each valuation has an associated small fragment of graphical structure—the connection between the variables over which the valuation is defined. For example a valuation representing "If X and Y then (with probability ϕ) Z" would be represented by a directed edges from X and Y to Z or by undirected edges linking X, Y and Z. These fragments are assembled to form the full graphical model, although it is often more useful to run the association the other way, first defining the graphical structure and then defining the corresponding relationships (valuations).

Although it is easy to see how to assemble many valuations into graphs, one can also subdivide the valuations which describe the relationship structure into components. The frames of variables over which valuations are defined is worthy of further explanation, as are the variables themselves. Also, it is useful to partition the set of outcomes into groupings reflecting natural symmetry in the relationship. Finally, uncertainty about values can be represented by parameters with their own distributions. These substructures within a valuation are described below.

12.2.2 Variables and frames

The place to start in defining a complex problem is with the *variables* or variables of the problem domain. Each variable has a set of *outcomes* which define the values it can take on. For example, a binary variable might be associated with the outcome set $\{0, 1\}$ or $\{\texttt{True}, \texttt{False}\}$.

There is a one-to-one correspondence between the variables of a problem domain and the nodes in the graphical model, although the node may have some additional information attached (such as the location, or the list of neighbors).

The domain of a valuation is a set of tuples over an ordered set of variables. We will refer to that domain as the *frame of discernment* or *frame*. There three different representations of the frame: (1) the *frame of variables* or list of variables, (2) the *frame vector* or tuple of outcome spaces associated with the variables, and (3) the *frame set* or set of possible tuples of outcomes, the cross product of the outcomes sets in the frame vector. For example, for three binary variables, the frame of variables might be (X, Y, Z), the frame vector would be $(\{0,1\},\{0,1\},\{0,1\})$ and the frame set would be $\{(0,0,0),(0,0,1),\ldots,(1,1,1)\}$. The term *frame* is used when the distinction between the three views of the frame is unimportant.

If a valuation is defined over a given frame, one can think of marginalizing it to a smaller frame, or extending it to a larger frame. Extension can be defined very naturally for belief functions, but also can be done for probability potentials by replicating over the appropriate variables. Projecting a valuation onto a new frame is achieved by a combination of marginalization and extension.

12.2.3 Groups, groupings and partitions

It is often useful to think of partitioning the frame set (the set of outcomes) into a number of groups. For example, one could partition a frame defined over two variables X and Y into those in which $X = Y$ and those in which that relationship does not hold. As another example, consider a system S with n identical components $C_1, \ldots, C_n$ in parallel. One might be interested in the probability of system failure, given that $0, 1, 2, \ldots, n$ of the components have failed. Thus there is a natural partition of the conditional part of the frame into sets representing k-out-of-n failures for $k = 0, \ldots, n$.

The *noisy-or* model (Pearl, 1988) is a simple form of this partitioning idea. Let $X_1, \ldots, X_n$ be a collection of binary input variables and Y be an output variable. Rather than specify a complete probability distribution for Y for each configuration of the n input variables, Pearl advocates assessing the conditional probability when two groups of input configurations: one in which at least one of the inputs has occurred and one in which none have occurred. Note that we do not need to restrict ourselves to "ors" and "ands"; any logical grouping of the attributes which we believe to be equivalent can be used.

We will formally define a *group* of outcomes as a set of outcomes about which we share a common pool of information. For example, the outcomes corresponding to the system with exactly k-out-of-n failures and where $X = Y$ both form a *group* of outcomes. Note that there is an implied frame associated with each group.

A set of groups over a particular frame is a *grouping*. Note that groups in a grouping need not be disjoint, nor need they span the entire frame set. Groupings are meant to reflect logical divisions of the valuation domain, and need not be true partitions.

An important subset of groupings is the *partition*. The groups in a *partition* must be pairwise disjoint and must span the entire outcome space. Partitions are particularly important, because probability valuations correspond to value assignments over partitions, as do simple utility valuations.

12.2.4 Formal definition of valuations

We define a *valuation* as a mapping from a *grouping* over a particular *frame* to numeric *values*. In analogy with belief functions, the groups in the grouping are called *focal elements*. Perhaps the most obvious example of this is the mass function of a belief function. However, if the grouping over which the valuation is defined is a partition, then we can define a probability distribution as well. Recall the probability functions and belief functions are subclasses of valuations as they imply certain normalization constraints among the values.

An important subclass of valuations are those for which the grouping over which they are defined forms a partition. These valuations can be represented by an array of values, one element of that array corresponding to each tuple in the frame set. Such an array is called a *potential* which can be used to represent probabilities and utilities. The class of valuations which are defined over partitions and hence can be represented with potentials are called *simple valuations*. Valuations which are not simple must be represented by a more complex scheme, such as an association list between groups and values or an array indexed by subsets of the frame set (superpotentials).

Users of simple valuations, that is people who restrict their modelling effort to probabilities or probabilities and utilities, may not see the necessity of first defining a grouping, but may rather prefer to go directly to the array of values corresponding to the primitive tuples. This strategy, however, can quickly get out of hand for large problems, such a the system with many components. Pearl(1988) introduces the *noisy-or* model (see previous section) to address these situations. Identifying a grouping which reduces the effective domain of the valuation from the frame set to the set of groups could drastically reduce the number of values which must be specified. For example, in representing the knowledge "If X and Y hold then Z usually holds." we may only be concerned about assigning values to the two groupings which correspond to whether or not the rule holds. Furthermore, there is often uncertainty about the values (see Section 2.5), but usually not about the structure of the grouping.

Another important subclass of valuations are the conditional valuations. These valuations divide the frame variables into two groups, the conditions and the consequences. The value is thought to be a conditional value associated with the consequence group given the condition group, for example, a conditional probability of the consequence set given the condition set. Conditional valuations are usually represented by directed edges, where unconditional valuations are usually represented by undirected edges. Conditional valuations can come in both simple (maps to array) and complex varieties.

12.2.5 Parameters and laws

Often there will be uncertainty about the numeric values of a valuation, that is an uncertainty about the strength of the relationship, but not the structure. Furthermore, the same numeric value, representing the same fragment of knowledge may appear in many valuations. In order to be able to trace and revise that knowledge, as well as express uncertainty about it, we must define an indirect pointer to the numeric values.

A *parameter* is just such a pointer to a numeric value. It is used as an alternative to the actual number in order to express uncertainty about the numeric value and promote re-use. As an example of both, the failure probability of a particular valve may be expressed as a parameter. Any place the valve is placed in the model, the same value for its failure probability should be used. If the valve is a new component, about which very little information is known, information about its failure rate may be uncertain or imprecise (or both). As test of the valve and other experience about it become available, the information will become more certain and precise, and the value of all parameters for that valve should be adjusted accordingly.

In order to express uncertainty about the value of parameters, parameters are allowed to have *laws*. These are probability distributions over the space of possible values for the parameters. Because some parameters are functionally linked (for example, the probability of A and not A), generally speaking the parameters will be dependent. In certain cases, it may be possible to make reasonable independence assumptions about some of the parameters.

Note that the term *law* is reserved for probability distributions over parameters, the term *valuation* is used for probability functions over variables. Parameters (in the statistical sense) of laws over parameters are called *hyperparameters*; this usage is consistent with the standard usage in Bayesian statistics. Distributions for hyperparameters are conceivable but hopefully unnecessary.

Spiegelhalter and Lauritzen(1990) use parameters to define a layered graphical model. The upper *quantitative* layer contains the distribution over the parameters, and the lower *qualitative* layer contains the graphical structure of the problem and the structure (groupings) within valuations. To answer questions in the *qualitative* layer, the best (average) values of the parameters are disseminated into the lower layer (in other words, each parameter is assigned a numeric value, the mean of its distribution). The now parameter free valuations are propagated through the graphical structure to answer questions. Finally, data from the consultation, can be used to update the distributions of the parameters in a Bayesian fashion. Almond(1990) uses a similar device, sampling from the distributions of the parameters to capture uncertainty about the parameters in the final estimation.

12.3 Model component libraries

We now turn from the structure within valuations to the structure of many valuations. This is the *graph* of the graphical model. The strength of graphical modelling lies in the independence assumptions represented by separation in the graph, which in turn imply a factorization of the problem into component valuations. This in turn implies that an entire graphical model could be constructed by "dragging and dropping" a collection of valuations from a library into the model. This approach suggest how a design engineer might build a model from the work of a reliability engineer (the library designer); here the selection and placement is accomplished by a drag and drop interface. It is also a good model for how knowledge based model construction might work; here the selection and placement are accomplished by meta-rules which determine which knowledge is applicable when.

It is also possible to group the graphical structure into larger fragments. For example a subsystem might be a graph fragment which is repeated several times in the system. Modellers can obviously take advantage of such parallelism to reduce the modelling effort. Similarly, an intelligent program can take advantage of these symmetries to reduce computational cost.

Such a graph fragment, because of its portable nature, must be slightly different from a graph object. In particular, it will be necessary to duplicate the nodes (variables) in the graph fragment before adding it into the graph. Furthermore, there may be stub-nodes in the fragment which are meant to determine where the fragment will attach to other fragments already placed in the graph. Such stub-nodes will be resolved at model construction time.

Almond(1992a) has implemented a prototype library system that assists users in finding and reusing model components. Reusable model components in the library are packaged as *books*. A book consists of its *contents*—the associated model fragment;—it is labeled with a *title*—a brief description of the contents—and a set of *authors*—a list of contributors, allowing one to trace the sources of knowledge used in its construction—and wrapped in a *jacket*—a more thorough and detailed description of its function.

Users explore the contents of a library by means of two graphical interfaces: the *book editor* (Figure 1) and the *bookcase browser* (Figure 2). Figure 1 shows a *book editor* for a fragment of a graphical model. The display allows the user to examine the title, authors and jacket. The

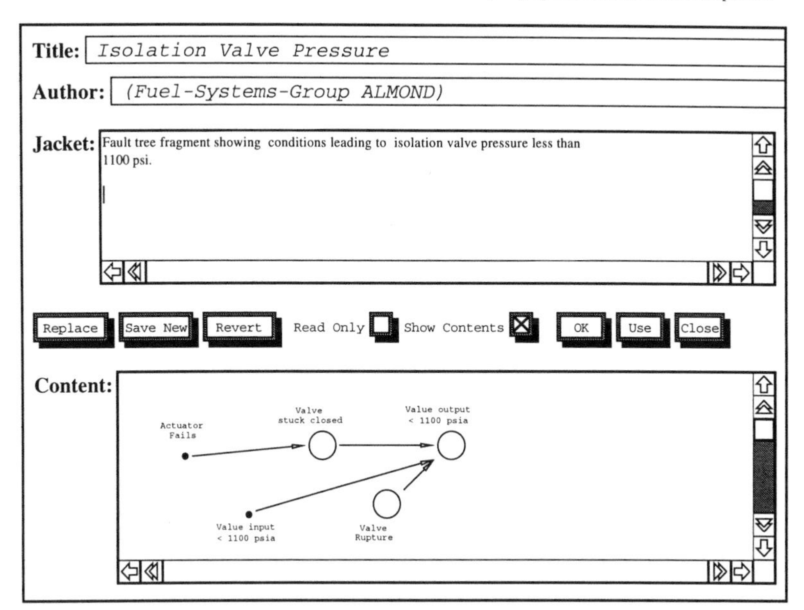

FIGURE 1. Book Editor for Graph Fragment
The book editor allows the user to inspect the title, author, jacket—detailed description of model fragment—and optionally the contents.

actual contents can be optionally examined for a more precise and detailed picture.

The *bookcase browser* presents a list of books by titles and allows the user to select or to open a *book editor* for any of them. If no appropriate book is found, a new one can be created from scratch or by editing an existing book. Ideally, the bookcase browser should be augmented by tools which would the list of books to be filtered via selection criteria (like electronic searching systems in libraries).

12.4 Conclusions and future directions

The rate of progress in ontological issues will be largely determined by how well knowledge can be shared among those in the graphical belief networks community. Results of such analyses are currently shared very little, and where sharing takes place, it is usually either a) in the form of paper reports that take time to distribute and get outdated rapidly, or b) among scientists using some specific piece of not-widely-distributed or supported software. To facilitate development of ontologies it will be necessary for determine how diverse software tools can exchange model fragments in computer-interpretable form.

Gruber's work on Ontolingua (Gruber, 1992a; Gruber, 1992b) currently provides the most promising mechanism for sharing ontologies between different tools and formalisms. Ontolingua extends the knowledge interchange format (KIF; Genesereth and Fikes, 1992) defined by the DARPA knowledge sharing effort with standard primitives for defining classes and relationships,

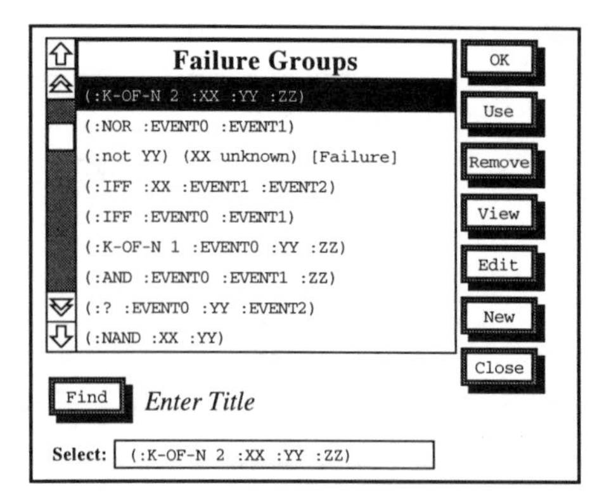

FIGURE 2. Bookcase Browser for Logical Groups

The bookcase browser allows the user to inspect a list of titles. Viewing the book shows an expanded description of the the contends and using the book employs the group in building a valuation.

and organizing knowledge in object-centered hierarchies with inheritance. Ontolingua facilitates the translation of KIF-level sentences to and from forms that can be used by various knowledge representation systems. Bradshaw *et al.*(1992b) and Lethbridge and Skuce (1992) describe the effort to integrate Ontolingua with other knowledge engineering tools.

12.5 References

[Almond 90] Almond, Russell G. (1990) *Fusion and Propagation in Graphical Belief Models: An Implementation and an Example.* Ph.D. dissertation and Harvard University, Department of Statistics, Technical Report S-130.

[Almond 92a] Almond, Russell G. (1992a) "Libraries, Books and Bookcases: A System for Organizing Re-Usable Knowledge." *Statistical Science Research Report 5,* StatSci, 1700 Westlake Ave, N, Seattle, WA.

[Almond 92b] Almond, Russell G.[1992b] "GRAPHICAL-BELIEF: Project Overview and Review." *Statistical Science Research Report 14* StatSci.

[Bradshaw 91] Bradshaw, Jeffery M., Covington, Stanley P., Russo, Peter J. and Boose, John H. (1991) "Knowledge acquisition techniques for decision analysis using AXOTL and AQUINAS." *Knowledge Acquisition* **3**, 49–77.

[Bradshaw 92a] Bradshaw, J.M., Chapman, C.R., Sullivan, K.M., Almond, R.G., Madigan, D., Zarley, D., Gavrin, J., Nims, J., and Bush, N. (1992a) "KS-3000: an application of DDUCKS to bone-marrow transplant patient support." Presented at the *Sixth Annual Florida AI Research Symposium (FLAIRS '93),* Ft. Lauderdale, FL.

[Bradshaw 92b] Bradshaw, J. M., Holm, P. D., Boose, J. H., Skuce, D., and Lethbridge, T. C. (1992b) "Sharable ontologies as a basis for communication and collaboration in conceptual modeling." *Proceedings of the Seventh Knowledge Acquisition for Knowledge-Based Systems Workshop.* Banff, Alberta, Canada.

[DempKong 88] Dempster, Arthur P. and Kong, Augustine (1988) "Uncertain Evidence and Artificial Analysis." *Journal of Statistical Planning and Inference,* **20**, 355-368.

[Edgar 92] Edgar, J.W., Puerta, A.R., Musen, M.A. (1992) "Graph-grammar assistance for modelling of decision." *Proceedings of the Seventh Banff Knowledge Acquisition for Knowledge-Based Systems Workshop,* pp 7:1–19. Banff, Alberta, Canada.

[Genesereth 92] Genesereth, M. R., and Fikes, R. (1992) "Knowledge Interchange Format Version 3.0 Reference Manual" *No. Logic Group Report, Logic-92-1.* Stanford University Department of Computer Science.

[Gruber 91] Gruber, T. R. (1991) "The role of common ontology in achieving sharable, reusable knowledge bases." In J. A. Allen, R. Fikes, and E. Sandewall (Eds.), *Principles of Knowledge Representation and Reasoning: Proceedings of the Second International Conference* (pp 601-602). Morgan Kaufmann, San Mateo, CA.

[Gruber 92a] Gruber, T. R. (1992a) "Ontolingua: A mechanism to support portable ontologies, Version 3.0." *No. Stanford Knowledge Systems Laboratory Technical Report KSL 91-66.* Stanford University Department of Computer Science.

[Gruber 92b] Gruber, T. R. (1992b) "A translation approach to portable ontology specifications." *Proceedings of the Seventh Knowledge Acquisition for Knowledge-Based Systems Workshop.* Banff, Alberta, Canada.

[Heckerman 91] Heckerman, D. (1991) *Probabilistic Similarity Networks,* ACM Press, New York.

[Henrion 91] Henrion, M., Breese, J.S., and Horvitz, E.J. (1991). "Decision analysis and expert systems." *AI Magazine,* 1991, 64-91.

[Holtzman 89] Holtzman, S. H. (1989) *Intelligent Decision Systems.* Addison-Wesley, Reading, Massachusetts.

[HowMath 81] Howard, R. A. and Matheson, J. E.[1981] "Influence diagrams." *Principles and Applications of Decision Analysis.* Strategic Decisions Group, Menlo Park, California.

[Kong 88] Kong, Augustine(1988) "A Belief Function Generalization of Gibbs Ensembles," *Joint Technical Report S-122* Harvard University and *No. 239* University of Chicago, Departments of Statistics.

[LethSkuce 92] Lethbridge, T. C., and Skuce, D. (1992) "Informality in knowledge exchange." Working Notes of the *AAAI-92 Knowledge Representation Aspects of Knowledge Acquisition Workshop*, 92.

[Neches 91] Neches, R., Fikes, R., Finin, T., Gruber, T., Patil, R., Senator, T., and Swartout, W. R. (1991) "Enabling technology for knowledge sharing." *AI Magazine,* pp 36-55.

[Pearl 88] Pearl, Judea [1988] *Probabilistic Reasoning in Intelligent Systems: Networks of Plausible Inference.* Morgan Kaufmann, San Mateo, California.

[Shacter 86] Shachter, R. D.(1986) "Evaluating Influence Diagrams." *Operations Research,* **34**, 871–82.

[ShafShen 88] Shafer, Glenn and Shenoy, Prakash P.(1988) "Bayesian and Belief-Function Propagation." *School of Business Working Paper No. 192.* University of Kansas.

[ShenShaf 90] Shenoy, Prakash P. and Shafer, Glenn (1990) "Axioms for Probability and Belief-Function Propagation." in *Uncertainty in Artificial Intelligence,* **4**, 169-198.

[SkuceMon 90] Skuce, D., and Monarch, I. (1990) "Ontological issues in knowledge base design: Some problems and suggestions." *Proceedings of the Fifth Knowledge Acquisition for Knowledge-Based Systems Workshop,* Banff, Alberta, Canada.

[SpiegLaur 90] Spiegelhalter, David J. and Lauritzen, Steffen L. (1990) "Sequential Updating of Conditional Probabilities on Directed Graphical Structures." *Networks,* **20**, 579–605.

[Wellman 91] Wellman,M. Breese, J.S. and Goldman, R.P. (1991) "Working notes from the AAAI-91 Knowledge-Based Construction of Probabilistic and Decision Models." Anahiem, CA.

13
Bayesian Graphical Models for Predicting Errors in Databases

David Madigan, Jeremy C. York, Jeffrey M. Bradshaw, and Russell G. Almond

Dept. of Statistics
University of Washington,
Seattle, WA,
Carnegie-Mellon University,
Fred Hutchinson Cancer Research Center and EURISCO,
and
Statistical Sciences Inc.,
Seattle WA.

ABSTRACT
In recent years, much attention has been directed at various graphical "conditional independence" models and at the application of such graphical models to probabilistic expert systems. However, there exists a broad range of *statistical* problems to which *Bayesian* graphical models, in particular, can be applied.
Here we demonstrate the simplicity and flexibility of Bayesian graphical models for one important class of statistical problems, namely, predicting the number of errors in a database. We consider three approaches and show how additional approaches can easily be developed using the framework described here.

13.1 Introduction

In recent years, Major developments have taken place in graphical models and in their application to probabilistic knowledge-based systems (Bradshaw *et al.*, 1993). Motivated by these applications, breakthroughs have also taken place in the development of a Bayesian framework for such models. In effect the local probabilities which define the joint distribution of the nodes in the graphical model, themselves become random variables and can be added to the graph. Within this framework Spiegelhalter and Lauritzen (1990) and Dawid and Lauritzen (1993) show how independent beta/dirichlet distributions placed on these probabilities can be updated locally to form the posterior as data becomes available.

It is evident that this work can be useful in a much broader range of applications. For example, recent work by Madigan and Raftery (1991), Madigan and York (1993), and York (1992) shows how this Bayesian graphical framework unifies and greatly simplifies many standard problems such as Bayesian log linear modeling, model selection and accounting for model uncertainty, closed population estimation, multinomial estimation with misclassification and double sampling. Markov chain Monte Carlo methods have facilitated many of these developments.

---

[0]This work is supported in part by a NSF grant to the University of Washington and by a NIH Phase I SBIR Award "Computing environments for graphical belief modeling" to Statistical Sciences.

There are many advantages to using Bayesian graphical models for analyzing discrete data:

1. They provide a unified and conceptually simple mathematical framework for handling a diverse range of problems,
2. The graph as a model representation medium is a great step forward—model assumptions become entirely transparent,
3. Model uncertainty can be accounted for in a straightforward fashion (Madigan and Raftery, 1991),
4. Missing data and latent variables are easily catered for (Madigan and York, 1993), and crucially,
5. Informative subjective opinion can realistically be elicited and incorporated.

The purpose of this article is to demonstrate the effectiveness of Bayesian graphical models in dealing with the important problem of predicting errors in databases.

13.2 Database error prediction

13.2.1 Introduction

A recent article by Strayhorn (1990) introduced an important class of problems in data quality management. The techniques developed potentially have wide application in quality control or indeed in any environment where flawed items must be detected and counted. Strayhorn was motivated specifically by the quality control of research data. He points out that while large numbers of journal pages are devoted to the quantification and control of measurement error, possible errors in data are rarely mentioned (see Fiegl *et al.*, 1982, for a notable exception).

Strayhorn (1990) presented two methods for estimating error rates in research data: the duplicate performance method and the known errors method. However, his analysis was heavily criticized by West and Winkler (1991), hereafter referred to as WW, who present Bayesian analyses of the two methods. Here we show how Bayesian Graphical models provide for a simpler analyses of both methods. We also describe a third method, the duplicate checking method.

13.2.2 Duplicate performance method

Suppose that a large number, N, of paper-based medical records must be entered into a computer database, and further suppose that two data entry personnel, α and ω are available to carry out this task. The idea is that both independently key in the data and then the resulting computer files are compared item by item by a method assumed to be error free. Where there is disagreement, the original paper record is consulted and the disagreement settled. Let d be the total number of disagreements found in this way, $d = x_\alpha + x_\omega$, where x_j is the number of errors attributable to $j, j = \alpha, \omega$. The only errors remaining are the subset of the $N - d$ records where *both* α and ω were in error. The intuition is that if the ratio of disagreements to total items, $\frac{d}{N}$, is low, the individual error rates of α and ω are low, and the probability of joint errors is lower still.

Because α and ω carry out their tasks independently a trivial Bayesian graphical model for this situation has two unconnected nodes A_α and A_ω, where A_j is a binary random variable

	A_ω	$\overline{A}_\omega$	
A_α	z	x_ω	
$\overline{A}_\alpha$	x_α	$N - x_\alpha - x_\omega - z$	
			N

TABLE 13.1. Duplicate Performance Method Table

indicating whether j entered a particular record correctly or not, $j = \alpha, \omega$. As per Spiegelhalter and Lauritzen (1990), a prior distribution must be elicited for pr(A_α) and pr(A_ω). WW suggest that in practice d/N will typically be small so that agreement between α and ω will occur for most records. For binary records, this sort of data will often be equally consistent with both typists being almost always correct or both being almost always incorrect. Consequently, uniform $[0, 1]$ priors on pr(A_α) and pr(A_ω) will result in heavily bimodal posterior distributions. To counteract this problem, WW put prior distributions on pr(A_α) and pr(A_ω) which only include values larger than 0.5 in their support. This takes them outside the class of conjugate priors however. This bimodality problem can also be avoided by using informative priors which are centered on a value greater than 0.5, thereby assuming *a priori* that the typists are more likely to enter data correctly than not. This approach retains conjugacy which proves very useful when performing the calculations. Furthermore prior distributions which are truncated at 0.5, especially the uniform prior on $[0.5, 1]$, will typically provide a poor model for prior expert knowledge.

The framework for this method may be represented as a 2×2 table—see Table 13.1. Here z represents the number of records where α and ω are both correct. Then we have:

$$\begin{aligned} \text{pr}(z \mid x_\alpha, x_\omega, N) &\propto \text{pr}(z, x_\alpha, x_\omega \mid N) \\ &= \int_\theta \text{pr}(D \mid N, \theta)\text{pr}(\theta)d\theta \end{aligned}$$

where D represents complete data and θ is the vector parameter for the cell probabilities.

Strayhorn's analysis assumes a common error rate for α and ω while WW also present an analysis assuming individual error rates. Since the former will typically be more realistic, we only pursue this approach. In Table 13.2 we present results for the hypothetical datasets considered by Strayhorn and WW assuming "informative" prior beta(1,3) distributions for pr(A_α) and pr(A_ω)[3]. This assigns both quantities a prior mean of 0.75 and standard deviation of 0.19. For each dataset we show the probabilities of various undetected error counts. Also provided is the probability assigned to the event that all the events on which there is agreement are in error–this is to demonstrate that the bimodality problem is adequately addressed through the use of reasonable informative priors. The probability of zero undetected errors is included from the WW analysis for comparison purposes. For the more realistic datasets in the lower part of the tables, the results are very similar.

A reviewer pointed out that the assumption that α and ω detect errors independently may be unrealistic. For example, many of the errors may be due to the poor quality of the original record. An analysis incorporating this dependence could proceed by including a link between A_α and A_ω although we do pursue this approach here.

[3]Programs to do the calculations for the datasets in this paper and the other datasets in WW are available from the first author.

				WW	$\text{pr}(z \mid x_\alpha, x_\omega)$							
n	x_α	x_ω	$\text{pr}(z > 0 \mid x_\alpha, x_\omega)$	z=0	z=0	z=1	z=2	z=3	z=4	z=5	z=6	z=max
20	2	3	0.46	0.27	0.54	0.26	0.10	0.04	0.02	0.01	0.01	0.00
20	1	1	0.17	0.59	0.83	0.14	0.03	0.01	0.00	0.00	0.00	0.00
20	1	0	0.09	0.71	0.91	0.08	0.01	0.00	0.00	0.00	0.00	0.00
50	5	5	0.53	0.32	0.47	0.30	0.14	0.06	0.02	0.01	0.00	0.00
50	2	3	0.21	0.65	0.79	0.17	0.03	0.01	0.00	0.00	0.00	0.00
50	2	2	0.16	0.71	0.84	0.14	0.02	0.00	0.00	0.00	0.00	0.00
50	1	0	0.04	0.88	0.96	0.04	0.00	0.00	0.00	0.00	0.00	0.00
100	10	10	0.73	0.20	0.27	0.31	0.21	0.11	0.05	0.02	0.01	0.00
100	5	5	0.31	0.60	0.69	0.24	0.06	0.01	0.00	0.00	0.00	0.00
100	2	3	0.11	0.81	0.89	0.10	0.01	0.00	0.00	0.00	0.00	0.00
100	1	0	0.02	0.94	0.98	0.02	0.00	0.00	0.00	0.00	0.00	0.00
500	25	25	0.76	0.21	0.24	0.32	0.23	0.12	0.05	0.02	0.01	0.00
500	5	5	0.07	0.91	0.93	0.07	0.00	0.00	0.00	0.00	0.00	0.00
500	2	3	0.02	0.96	0.98	0.02	0.00	0.00	0.00	0.00	0.00	0.00
1000	50	50	0.93	0.06	0.07	0.17	0.23	0.21	0.15	0.09	0.05	0.00
1000	25	25	0.50	0.47	0.50	0.34	0.12	0.03	0.01	0.00	0.00	0.00
1000	5	5	0.04	0.95	0.96	0.03	0.00	0.00	0.00	0.00	0.00	0.00
1000	2	3	0.01	0.98	0.99	0.01	0.00	0.00	0.00	0.00	0.00	0.00
5000	50	50	0.41	0.58	0.59	0.31	0.08	0.02	0.00	0.00	0.00	0.00
5000	25	25	0.13	0.86	0.87	0.12	0.01	0.00	0.00	0.00	0.00	0.00
5000	5	5	0.01	0.99	0.99	0.01	0.00	0.00	0.00	0.00	0.00	0.00
5000	2	3	0.00	1.00	1.00	0.00	0.00	0.00	0.00	0.00	0.00	0.00

TABLE 13.2. Duplicate Performance Method: Hypothetical Data and Predictive Probabilities for Undetected Errors, Independent Be(1,3) Priors for $\text{pr}(A_\alpha)$ and $\text{pr}(A_\omega)$. n is the total number of records, x_α and x_ω are the number of errors attributable to each of the two checkers and z is the number of undetected errors.

		X	$\overline{X}$	
D_α	D_ω	0	x_0	
	$\overline{D}_\omega$	0	x_α	
$\overline{D}_\alpha$	D_ω	0	x_ω	
	$\overline{D}_\omega$	$N - x_0 - x_\alpha - x_\omega - z$	z	
				N

TABLE 13.3. Duplicate Checking Method Table

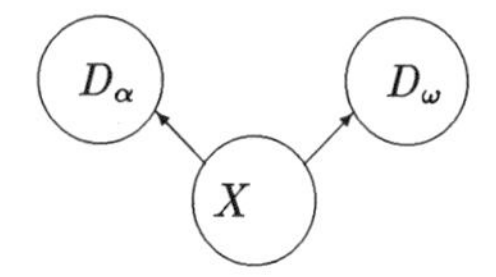

FIGURE 1. Duplicate Checking Method

13.2.3 *Duplicate checking method*

A somewhat different approach was alluded to but not analyzed by WW, and we refer to this as the duplicate checking method. Here we assume that the database *already exists* and the task of our two friends α and ω is to independently check each record in the database. WW make an important assumption that error free records are classified correctly although our analysis does not require this assumption. Thus the method may be represented as in Table 13.3 where D_i now indicates whether or not i detected an error and X is a binary random variable indicating whether or not the record in the database is correct. The key point is that we have an extra piece of information here, namely x_0, the number of records for which both α and ω detect errors. A reasonable Bayesian graphical model for the duplicate checking method is given in Figure 1. Here prior probability distributions are required for pr$(D_\alpha \mid X)$, pr$(D_\omega \mid X)$, pr$(D_\alpha \mid \overline{X})$, pr$(D_\omega \mid \overline{X})$ and pr(X) although the assumption of no false positives renders the first two redundant. The results shown in Table 13.4 assume Beta (1,5) distributions for the other three parameters implying prior means of 0.83 and standard deviations of 0.14.

13.2.4 *Known errors method*

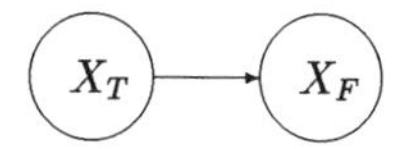

FIGURE 2. Known Errors Method

				$\mathrm{pr}(z \mid x_0, x_\alpha, x_\omega)$								
n	x_0	x_α	x_ω	$z > 0$	z=0	z=1	z=2	z=3	z=4	z=5	z=6	z=max
20	4	2	3	0.36	0.64	0.25	0.08	0.02	0.01	0.00	0.00	0.00
20	0	2	3	0.43	0.57	0.27	0.10	0.04	0.01	0.01	0.00	0.00
20	3	1	1	0.16	0.84	0.13	0.02	0.00	0.00	0.00	0.00	0.00
20	0	1	0	0.07	0.93	0.06	0.01	0.00	0.00	0.00	0.00	0.00
50	5	5	5	0.78	0.22	0.26	0.20	0.13	0.08	0.05	0.03	0.00
50	0	5	5	0.86	0.14	0.20	0.19	0.15	0.11	0.07	0.05	0.00
50	5	1	1	0.18	0.82	0.14	0.03	0.01	0.00	0.00	0.00	0.00
50	0	1	0	0.08	0.92	0.07	0.01	0.00	0.00	0.00	0.00	0.00
100	10	10	10	0.98	0.02	0.06	0.10	0.12	0.13	0.12	0.10	0.00
100	0	10	10	1.00	0.00	0.01	0.02	0.04	0.05	0.06	0.07	0.00
100	5	5	5	0.81	0.19	0.24	0.20	0.14	0.09	0.06	0.03	0.00
100	0	5	5	0.88	0.12	0.18	0.18	0.15	0.11	0.08	0.06	0.00
100	0	1	0	0.08	0.92	0.07	0.01	0.00	0.00	0.00	0.00	0.00
500	25	25	25	1.00	0.00	0.00	0.00	0.00	0.00	0.00	0.01	0.00
500	0	25	25	1.00	0.00	0.00	0.00	0.00	0.00	0.00	0.00	0.00
500	5	5	5	0.82	0.18	0.23	0.20	0.14	0.09	0.06	0.04	0.00
500	0	5	5	0.89	0.11	0.17	0.17	0.14	0.11	0.08	0.06	0.00
500	0	1	0	0.09	0.91	0.07	0.01	0.00	0.00	0.00	0.00	0.00
5000	50	50	50	1.00	0.00	0.00	0.00	0.00	0.00	0.00	0.00	0.00
5000	0	50	50	1.00	0.00	0.00	0.00	0.00	0.00	0.00	0.00	0.00
5000	5	5	5	0.83	0.17	0.23	0.20	0.14	0.10	0.06	0.04	0.00
5000	0	5	5	0.89	0.11	0.16	0.17	0.14	0.11	0.08	0.06	0.00
5000	1	2	3	0.52	0.48	0.28	0.13	0.06	0.03	0.01	0.01	0.00

TABLE 13.4. Duplicate Checking Method: Hypothetical Data and Predictive Probabilities for Undetected Errors, Independent Be(1,5) Priors for $\mathrm{pr}(D_\alpha \mid \overline{X})$, $\mathrm{pr}(D_\omega \mid \overline{X})$ and $\mathrm{pr}(X)$. n is the total number of records, x_0 is the number of errors detected by both checkers, x_α and x_ω are the number of errors detected exclusively by each of the checkers and z is the number of undetected errors.

The known errors method is described by Strayhorn (1990) as follows: "In this method, a member of the research staff completes the data operation in question. The data are then presented to a second person, for example, the supervisor of the staff member, who introduces a certain number of 'known errors' into the data set. The locations and forms of these errors are recorded elsewhere. Then the data set together with known and unknown errors, is given to another staff member, who checks the data set."

WW provide two elegant analyses of this method. Our purpose here is to point out that the known errors method is a special case of the double sampling scheme introduced by Tenenbein (1970) and further analyzed by Madigan and York (1993). We present the results in Table 13.5. Here we have a simple Bayesian graphical model with two nodes, X_T, representing the true state of the record, and X_F, a "fallible" version representing what the checker has recorded. For the original data we only have observations on X_F while for the known errors, both nodes are recorded. To be consistent with the analysis of WW, uniform prior distributions were used. A Markov chain Monte Carlo technique was used to compute the reported probabilities - for details see Madigan and York (1993). Note that the *known* values of X_T are not used when updating the distribution of $\mathrm{pr}(X_T)$.

The results of Table 13.5 are very close to those of WW.

13.3 Discussion

We have shown how directed Bayesian graphical models provide for a straightforward analysis of three database error checking methodologies. In each case informative prior distributions can easily be specified in terms of readily understood quantities and modeling assumptions are transparent. The calculations in each case are straightforward, providing outputs which are much easier to interpret than Strayhorn's confidence intervals.

The real strength of the Bayesian graphical modeling approach for these problems is that it can be generalised in a simple fashion. In particular, the generalizations suggested by WW can easily be incorporated. These include relaxation of the no-false-positive assumptions, varying error rate probabilities according to some characteristic of the data records, adding additional checkers and sampling only a portion of the database.

Because of their simplicity and flexibility, we expect that Bayesian graphical models can and will be increasingly applied to a broad range of applications such as the one considered in this paper.

13.4 References

[1] Bradshaw, J.M., Chapman, C.R., Sullivan, K.M., Boose, J.H., Almond, R.G., Madigan, D., Zarley, D., Gavrin, J., Nims, J. and Bush, N. (1993) "KS-3000: An application of DDUCKS to bone-marrow transplant patient support," *Proc. 7th European Knowledge Acquisition for Knowledge-Based Systems Workshop (EKAW-93), Toulouse and Caylus, France*, 57–74.

[2] Dawid, A.P. and Lauritzen, S.L. (1993) "Hyper-Markov laws in the statistical analysis of decomposable graphical models," *Ann. Stat.* to appear.

[3] Fiegl, P., Polissar, L., Lane, W.E. and Guinee, V. (1982) "Reliability of basic cancer patient data," *Stat. Med.* **1**, 191–204.

[4] Madigan, D. and Raftery, A.E. (1991) "Model selection and accounting for model uncertainty in graphical models using Occam's window." *Technical Report 213*, Department of Statistics, University of Washington.

[5] Madigan, D. and York, J. (1993) "Bayesian graphical models for discrete data." *Technical Report 259*, Department of Statistics, University of Washington.

[6] Spiegelhalter, D.J. and Lauritzen, S.L. (1990) "Sequential updating of conditional probabilities on directed graphical structures," *Networks* **20**, 579–605.

[7] Strayhorn, J.M. (1990) "Errors remaining in a data set: techniques for quality control," *American Statistician* **44**, 14–18.

[8] Tenenbein, A. (1970) "A double sampling scheme for estimating from binomial data with misclassification"' *JASA* **65**, 1350–1361.

[9] West, M. and Winkler, R.L. (1991) "Data base error trapping and prediction," *JASA* **86**, 987–996.

[10] York, J. (1992) "Bayesian methods for the analysis of misclassified or incomplete multivariate discrete data." *PhD Thesis*, Department of Statistics, University of Washington.

Number	k_t	k_f	u_f	$E[z \mid u_f, k_f]$	$pr(z > 0 \mid u_f, k_f)$	$95\%z$
1	10	10	0	0.12	0.10	1
2	10	10	1	0.23	0.17	1
3	10	10	5	0.68	0.37	3
4	10	10	10	1.19	0.52	5
5	10	10	20	2.44	0.68	9
6	10	9	0	0.25	0.18	1
7	10	9	1	0.50	0.32	2
8	10	9	5	1.53	0.62	5
9	10	9	10	2.80	0.79	9
10	10	9	20	5.12	0.90	14
11	10	7	0	0.69	0.37	3
12	10	7	1	1.33	0.57	5
13	10	7	5	3.90	0.88	11
14	10	7	10	7.12	0.96	18
15	10	7	20	13.62	0.99	33
16	15	15	0	0.07	0.06	1
17	15	15	1	0.14	0.12	1
18	15	15	5	0.41	0.27	2
19	15	15	10	0.77	0.42	3
20	15	15	20	1.45	0.58	6
21	15	14	0	0.16	0.12	1
22	15	14	1	0.31	0.23	2
23	15	14	5	0.94	0.50	3
24	15	12	0	0.38	0.26	2
25	15	12	1	0.74	0.42	3
26	15	12	5	2.19	0.77	6
27	20	20	0	0.05	0.05	0
28	20	20	1	0.11	0.09	1
29	20	20	5	0.33	0.23	2
30	20	20	10	0.60	0.36	3
31	20	20	20	1.08	0.50	4
32	20	18	0	0.18	0.14	1
33	20	18	1	0.34	0.26	2
34	20	18	5	1.06	0.56	4
35	20	18	10	1.91	0.74	6

TABLE 13.5. Known Errors Method: Hypothetical Data and Predictive Means and Probabilities for Undetected Errors. 'number' indexes WW datasets, k_f is the number of known errors found out of the total k_t, u_f is the unknown errors found in 100 records, and z is the number of unknown errors remaining. Also, $95\%z$ denotes the upper 95% point of $pr(z \mid u_f, k_f)$ to the nearest integer.

14

Model Selection for Diagnosis and Treatment Using Temporal Influence Diagrams

Gregory M. Provan

Computer and Information Science Department
University of Pennsylvania
Philadelphia PA 19104-6389

ABSTRACT This paper describes the model-selection process for a temporal influence diagram formalization of the diagnosis and treatment of acute abdominal pain. The temporal influence diagram explicitly models the temporal aspects of the diagnosis/treatment process as the findings (signs and symptoms) change over time. Given the complexity of this temporal modeling process, there is uncertainty in selecting (1) the model for the initial time interval (from which the process evolves over time); and (2) the stochastic process describing the temporal evolution of the system. Uncertainty in selecting the initial Bayesian network model is addressed by sensitivity analysis. The choice of most appropriate stochastic process to model the temporal evolution of the system is also discussed briefly.

14.1 Introduction

This paper examines the model-selection process for the diagnosis and treatment over time of a dynamic system $\mathcal{S}$ which is malfunctioning. The main focus of this paper is how to select the set of model parameters Θ_τ which best describes $\mathcal{S}$ over a series of time intervals $\tau = 1, 2, ...$, thereby facilitating the computation of the best diagnosis/repair strategy. The parameters must be selected given uncertainty in (1) the model for the initial time interval (from which the process evolves over time); and (2) the stochastic process describing the temporal evolution of the system.

A Temporal Influence Diagram (TID) is used to model the system $\mathcal{S}$. The TID consists of a sequence of influence diagrams ID_1, $ID_2, ...$ [7,9] (with ID_i modeling the system at time interval i) which evolve according to a time-series process. A parameter set θ_τ defines ID_τ at time τ, and an additional set of parameters defines the time-series process for the temporal evolution of the system over time intervals $\tau = 1, 2, ...$

The particular diagnostic application studied here is acute abdominal pain (AAP). The model used for diagnosis and treatment of AAP is selected from a knowledge base (KB) which encodes the physiological inter-relationships between findings of AAP and the underlying diseases. As discussed in Section 14.2, this KB defines a complete probabilistic model, which can be tailored to specific input data, i.e. a set of findings for a particular patient. Since reasoning over time is crucial for this task, model selection requires searching over models which approximate the *temporal evolution* of the data.

The two parts of the model selection process are examined as follows: (a) The uncertainty in choosing the model for the initial time interval is addressed via sensitivity analysis over models in "local" time intervals. In addition, the use of future time intervals to confirm/disconfirm the

[1] *Selecting Models from Data: AI and Statistics IV.* Edited by P. Cheeseman and R.W. Oldford.

correctness of the initial choice is used. (b) The temporal evolution of system variables is modeled using a semi-Markov process. Other stochastic processes which might also be employed, such as higher-order Markov processes, i.e. processes in which conditional dependencies exist for more than one past time-interval, are also discussed briefly.

14.2 Static Model Structure

A TID consists of the union of a sequence of IDs. Each ID models the system during a time interval, i.e. the system is assumed to have a static state during that interval. This section summarizes the ID used for the static structure.

An ID consists of a Bayesian Network (BN) [9] augmented with decision and value nodes. The BN consists of a directed acyclic graph (called the qualitative level of the BN) augmented with conditional probabilities (called the quantitative level of the BN). The BN's main advantage is that the graph specifies the dependency relations of the problem, thus necessitating only the probability distributions as determined by the graph, and not the $O(2^n)$ potential conditional probability distributions. This reduces the data requirements and computational expense of evaluating the BNs.

A BN is specified using two levels, qualitative and quantitative. On the *qualitative level*, the BN consists of a directed acyclic graph (DAG). In the causal graph $\mathcal{G}(N, A)$ consisting of nodes N and arcs A, the nodes represent state variables, and arcs exist between pairs of nodes related causally. The nodes in the graph, chance nodes, correspond to random variables Arcs in the BN represent the dependence relationships among the variables. Arcs into chance nodes represent probabilistic dependence and are called conditioning arcs. We define the parents of X_i, $pa(X_i)$, as the direct predecessors of X_i in $\mathcal{G}$. The absence of an arc from node i to j indicates that the associated variable X_j is conditionally independent of variable X_i given $pa(X_j)$.

On the *quantitative level*, the BN specifies a frame of data with each node. For each chance node X in the DAG, the domain Ω_X of possible outcomes for X is specified. The conditional probability distributions $P(X_i|pa(X_i))$ necessary to define the joint distribution $P(X_1, X_2, ..., X_n)$ are also specified.

FIGURE 1. Influence diagram for patient X

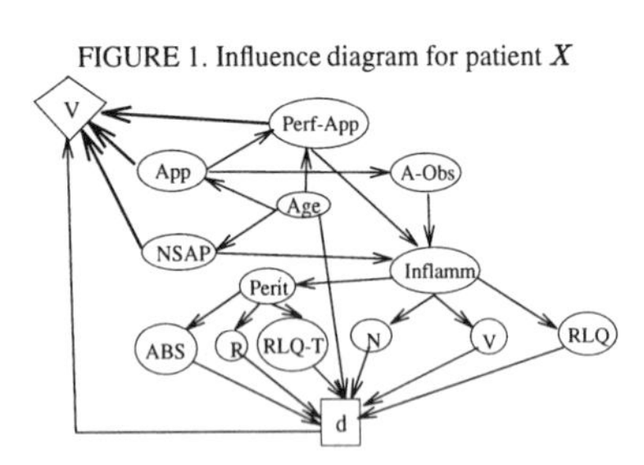

Figure 1 presents an example of the type of network created for this AAP domain. In this figure, chance, decision and value nodes are represented diagrammatically by ovals, rectangles and diamonds respectively. For example, the chance node for Inflammation (*Inflamm*) is conditionally dependent on the chance nodes for Perforated-Appendix (*Perf-App*), Non-Specific Abdominal Pain (*NSAP*) and Appendicial-Obstruction (*A-Obs*). In this ID there is one decision node d and one value node V.

An ID is obtained by adding decision and value nodes to a BN. Given the BN, the decision and value nodes of the ID model the management decisions taken and values obtained by these decisions, respectively. Decision nodes correspond to variables whose value is determined by the decision-maker. A value node corresponds to the utility (or objective function) which the decision-maker attempts to maximize (in expectation).

An arc from a chance node corresponding to variable X to a decision node corresponding to decision d, called an information arc, specifies that information for X is available to the decision-maker when making decision d. In Figure 1, the information arcs go to the decision node from nodes representing findings known to the decision-maker, e.g. nausea (N), vomiting (*V*), right-lower-quadrant pain (*RLQ*), patient age (*Age*), etc. Arcs into a value node indicate that the value $\mathcal{V}$ is based on the value of the connecting nodes, whether chance or decision nodes. For each decision node d in the DAG, the domain of possible outcomes Ω_d for d is specified. For each value node v in the DAG, the domain Ω_v of possible utilities (real numbers) for v is specified. A utility is specified for each of the conditioning variables of v.

14.3 Application Domain: The Diagnosis of Acute Abdominal Pain

14.3.1 Diagnostic Task: Medical Background

Appendicitis, a common cause of acute abdominal pain, is a relatively important disease, in that one in every 16 people can expect to get it [8]. The diagnosis is probabilistic in that two patients with the same findings may not both have appendicitis and two patients with appendicitis may not have the same findings. In addition, no findings are pathognomonic, i.e. are distinctively characteristic of the disease, so a set of findings might equally well indicate any of the several diseases which can cause acute abdominal pain. Because the disease progresses over a course of hours to days, one might be tempted to wait until the complex of signs and symptoms is highly characteristic of appendicitis before removing the appendix. However, the inflamed appendix may perforate during the observation period, causing a more generalized infection and five-fold increased risk of death. Thus, the tradeoff is between the possibility of an unnecessary operation on someone whose findings are similar to early appendicitis and a perforation in someone whose appendicitis is allowed to progress [3].

Constructing a model to diagnose acute abdominal pain for a single time interval may lead to inaccuracy, since many findings take on different meanings as diseases evolve over time, both in terms of their inter-relationships and the diseases indicated by the particular findings [13]. Possible sources of inaccuracy include: (1) static models can be wrong if they model an inappropriate time instant or interval; (2) static models do not capture the evolution of the system (e.g. the location of acute abdominal pain may change over time, providing significant diagnostic information)[3]; or (3) a single stage diagnosis is inadequate for systems requiring multiple tests and/or treatments, both of which can provide significant diagnostic evidence. This is the first model to explicitly incorporate temporal information into the diagnosis/treatment process, in contrast to existing static models, e.g. [3,5,13]. It is expected that the model will have higher accuracy than existing models, although clinical tests have yet to be done to confirm this.

[3] For example, if a woman has abdominal pain, noting that this pain is followed by gastrointestinal distress could help identify a possible case of appendicitis, whereas the presence of gastrointestinal distress prior to the pain would make the presence of gastroenteritis more likely.

14.3.2 State Space for Acute Abdominal Pain

The AAP domain is formalized as follows. The state variables consist of disease, observable finding and intermediate finding variables. The **disease variables** consist of the diseases Appendicitis, Non-Specific Abdominal Pain, Salpingitis and Ruptured Ovary. The **observable finding variables** consist of findings which the physician can observe directly, such as Anorexia (A), Nausea (N), Vomiting (V), Right-Lower-Quadrant pain (RLQ), abdominal Tenderness (T), etc. The **intermediate variables** consist of all other variables, e.g. Appendicial Obstruction (*A-Obs*), Inflammation (*Inflamm*), small bowel Obstruction *Obs*), etc.

More formally, we define the *qualitative* physiological model using a set $\mathcal{X} = \{X_1, X_2, ..., X_n\}$ of variables. Each X_i corresponds to a finding variable, an intermediate disease variable, or a disease variable. Let $\{\theta_\tau, \tau = 0, 1, ...\}$ denote the finite sequence of physiological model states over time. The state space θ_τ for a particular time interval τ has outcome space defined by the cross-product of $\mathcal{X}$, i.e. $X_1 \times X_2 \times \cdots \times X_n$. For example, a particular state vector may be $[X_1 = App(YES), X_2 = NSAP(NO), X_3 = RLQ(YES), X_4 = N(YES), X_5 = (YES), ...]$. Figure 1 shows an example of the causal relations defined over $\mathcal{X}$. This causal structure has been defined by a domain expert, J.R. Clarke, MD (cf. [12]).

The *quantitative information* necessary for this application consists of probability values for the BN, and a utility function over Ω_d. The KB for the acute abdominal pain domain covers over 50 findings, 20 intermediate disease states and 4 diseases, namely *App, SALP, NSAP* and *Rupt-Ov.* Statistical data has been obtained for a portion of the KB, such as disease nodes conditioned on observable finding nodes, e.g. P(appendicitis | anorexia, nausea, vomiting). However, the causal model proposed in this paper also requires the physician's estimate of conditional probability distributions for many intermediate nodes. This data was collected using several thousand cases of acute abdominal pain, and is discussed in further detail in [12].

For this application, the decision domain Ω_d={WAIT, TEST, OPERATE, SEND-HOME, ADMINISTER-ANTIBIOTICS), SYMPTOMATIC-TREATMENT}. In addition to a standard set of utility values valid for any case, patient-specific utility functions must be elicited.

14.4 Temporal Influence Diagrams

This section describes how Markov models are embedded into the ID to create a TID. First we introduce the structure of TIDs, and then the temporal and stochastic process assumptions made in the modeling process.

14.4.1 TID Structure

A TID consists of the union of a sequence of IDs, with "temporal" arcs joining ID_τ to $ID_{\tau+1}$, $\tau \geq 1$. The causal structure of these temporal arcs defines the dynamic process of the model. There is a tradeoff involved in defining time-series processes for TIDs: the more complex the time-series process, the more difficult the TID is to evaluate. Thus, the higher the order the Markov processes, the more temporal arcs that need to be added. In general, one needs to evaluate the tradeoff of adding "temporal" arcs to the TID (to better model the temporal evolution of the system) versus the complexity of evaluating and obtaining data for these larger networks. In this paper we examine the adequacy of a first-order Markov model; in [11] we compare and contrast this model with higher-order models.

14.4.2 Stochastic Process Assumptions

A stochastic process is a family of random variables $X(\tau)$ defined on the same probability space and indexed by a time parameter τ. A wide range of stochastic processes of varying complexity have been studied in the literature, and we examine the potential application to this domain of but a few.

Define the stochastic process $\{X_\tau | \tau = 0, 1, 2, ...\}$ to denote the state of a physiological model variable x at time τ. If X_τ has outcome space $\{a, b, c\}$, X_τ tracks which value a, b or c occurs at time τ over the course of time. A *Markov process* is a widely-used model which captures a simple notion of temporal dependence. In this paper, the physiological model variables are assumed to obey the Markov property, i.e. $P(X_\tau | X_{\tau-1}, ..., X_1, X_0\} = P\{X_\tau | X_{\tau-1}\}$. In addition, because the set of physiological model variables can be represented by the "macro" state variable θ_τ, the stochastic process $\{\theta_\tau | \tau = 0, 1, 2, ...\}$ denoting the state of the patient at time τ, θ_τ, is also Markov, i.e. $P(\theta_\tau | \theta_{\tau-1}, ..., \theta_1, \theta_0\} = P\{\theta_\tau | \theta_{\tau-1}\}$. The patient spends time ψ_{ij} with finding (state) i before exhibiting state j. These times ψ_{ij}, called *holding times*, are positive random variables, each governed by a holding-time probability distribution $F_{ij}(\cdot)$.

The Markov process has an associated state transition matrix P. Given that a patient is in state i, the probability that the patient will be in state j in the next time interval is given by p_{ij}. Hence the transition probabilities p_{ij} form a Markov chain with a transition probability matrix P denoting $\{p_{ij}\}$.

A *semi-Markov process* introduces a second source of randomness into the model: the time spent in any state is now defined as a random variable. A discrete-time semi-Markov process uses an indexed set of state transition times, and allows the time spent in any state to obey an arbitrary probability distribution. A continuous-time semi-Markov process generalizes the discrete-time semi-Markov process further by allowing state transitions at any time instant, and the time spent in any state to obey an arbitrary probability distribution.

A semi-Markov process is characterized by the state space θ, an initial set of probabilities π_0, and the matrix of transition distributions $Q(\cdot)$, where each $Q_{ij}(\cdot)$ can be written $Q_{ij}(\cdot) = p_{ij} F_{ij}(\cdot)$. Section 14.4.3 presents some examples of transition matrices. This semi-Markov process is used because this domain needs transition distributions with holding-time probability distributions $F_{ij}(\cdot)$ which are not necessarily constant, as in a Markov chain, or exponentially-distributed, as in a continuous-time Markov process.

Even more general models are possible by using higher-order Markov assumptions, e.g. allowing X_τ to depend on variables X_{t-2} as well as $X_{\tau-1}$. We compare such models with the discrete-time semi-Markov process described here in [11].

14.4.3 Transition Functions

The TID is completely specified by the state space description θ and by the transition functions $Q_{ij}(\tau)$. This section describes the transition functions used for this domain. The transition matrices for the process, $Q_{ij}(\tau)$, are defined over intervals of 3 hours since onset of symptoms. The times of state transitions are defined in this way primarily because of data availability, and to simplify the model. Thus transitions which occur at 5 hours, for example, are modeled as if they occurred at 6 hours.

Transition functions govern the change of findings over time as follows. For example, Figure 2 shows the temporal variation of the true positive rate for the occurrence of various findings given a diagnosis of appendicitis. These graphs summarize the transition function data, which are

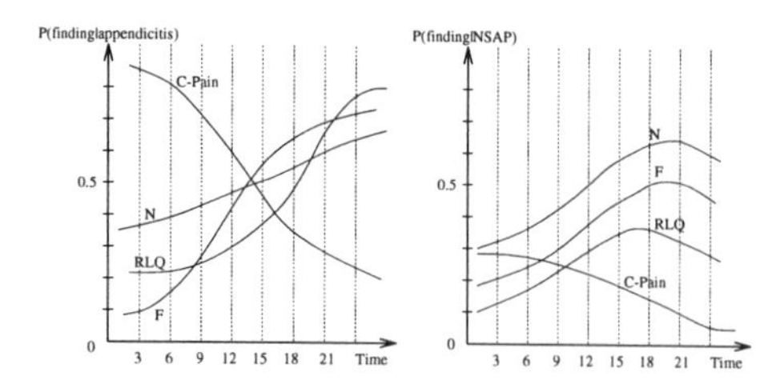

FIGURE 2. Change over time of true positive rate (i.e. P(finding|disease)) for the occurrence of various findings given a diagnosis of appendicitis and non-specific abdominal pain (NSAP). Time is the number of hours since onset of symptoms.

tabulated in the system database. The transition functions attempt to capture temporal changes such as those depicted in these graphs.

A transition matrix p_{ij} denotes the transition probabilities from interval τ_i to τ_{i+1}. For example the transition matrix for RLQ given Appendicitis, $p_{34}(RLQ|App)$, might denote the probability that a patient with App would exhibit RLQ during interval τ_4 given RLQ during τ_3 is 1; and of those patients without RLQ for τ_3, $1/6$ will exhibit RLQ during τ_4 and $15/16$ will not.

At present, some data is available over time, but much of the data is based on physicians' estimates.[4] Estimates are used when precise data is unavailable to assign distributions to the F_{ij}. For simplicity, holding-time distributions are estimated with normal and log-normal distributions. When more temporal data becomes available, more accurate distributions can be calculated using standard statistical parametric estimation methods. As an example, the F_{ij} distribution for C-Pain given appendicitis is approximated by a Gaussian with mean 6 hours and standard deviation 7 hours, i.e. $N(6, 7)$.

14.4.4 The Dynasty Network Construction System

Existing methods of temporal ID construction typically replicate a static Bayesian network over multiple time intervals to construct a temporal BN (e.g. [4,6]). The approach taken here is to compute the most parsimonious model for each time interval, and to parsimoniously connect models between any two time intervals based on a simple time-series process.[5]

The DYNASTY network construction system is used for constructing TIDs, and is implemented on top of the Decision Network manipulation system IDEAL [14]. DYNASTY stores domain knowledge in a Knowledge Base (KB) containing (1) a set Σ of causal rules, and (2) a set Π of conditional probability distributions over Σ, and (3) a set Ω_d of possible decisions, and utility functions $\mathcal{V}$ associated with those decisions.

DYNASTY constructs TIDs using input data over time. The process can be summarized as follows:

1. Construct initial ID for time interval τ, given observations O; test sensitivity to choice of τ.
2. Construct ID for next time interval, $ID_{\tau+1}$, given new data.
    - Update previous network ID_τ, if possible, to create $ID_{\tau+1}$.

[4] A large data collection program is planned, so that the database can be extended with data over a larger number of time intervals.

[5] See [12] for more details of the construction process.

- Add temporal arcs between ID_τ and $ID_{\tau+1}$.

3. Repeat Step 2 until the diagnostic/treatment process is complete.
4. Evaluate TID.[6]

The first step represents the decision-theoretic information for the observations O and the domain knowledge relevant to O; a sensitivity analysis determines if network updating is necessary to better model the data. Then, DYNASTY constructs for the system a TID which is used to compute the sequence of diagnoses and actions taken given these diagnoses [12].

14.5 Model Selection for Initial Time Interval

The first stage of model construction requires selection of an appropriate time interval (e.g. τ_4, the interval for $\tau = 12 - 15$ hours as shown in figure 2) to add the set of probability distributions for the conditional probabilities (indexed by time τ_4) to the causal network. Figure 1 presents an example of an ID created for an actual case. This model is created for a woman who experiences right lower quadrant pain (RLQ), followed by nausea and vomiting. Other information relevant to determining a treatment is also encoded in the ID.

Sensitivity analysis is used primarily to check the sensitivity of diagnoses to network parameters, in order to determine if the correct time interval has been modeled. Sensitivity analysis of diagnoses reveals the threshold probabilities of network variables, among other things, that would change the diagnosis, just given data from that single time interval. Analysing sensitivity of the initial model to choice of time intervals is done in a "local" manner for computational efficiency: if interval τ_i is chosen for modeling, only models from the "local" intervals τ_{i-1} and τ_{i+1} are compared to the model from τ_i. Global analysis, e.g. constructing models from all potential time intervals and averaging over all such models, is computationally expensive, and the models from different time intervals often contain different variables and topologies, so are difficult to average over. Hence, this approach compromises completeness for efficiency. Note that the evolution of the model over time provides important cues to determine the accuracy of the original model. Hence incorrect choice of the initial model may not be a complete disaster.

Sensitivity analysis uses an equivalence class approach to diagnosis [10,12]: decision-equivalent diagnoses, i.e. diagnoses for which the same decision is taken (e.g. administration of particular drugs to a patient), are considered as the same diagnosis. The aim of diagnostic reasoning is to provide a decision (e.g. a treatment of antibiotics) for a set of observations. >From an equivalence-class point of view, this reduces to refining the set of decision-equivalent possibilities; i.e. one does not care about distinct diagnoses, but *distinct decisions* (and their associated distinct equivalence classes). Thus, decision-equivalence induces a partition on the set of diagnoses, where each partition corresponds to a possible distinct decision. This approach increases the decision-making efficiency, compared to approaches that try to distinguish between decision-equivalent diagnoses and/or update the model even if the decision did not change.

Diagnostic Management Example 1 *A sensitivity analysis* was conducted on the network of Figure 1, given the uncertainty of pain onset time, to determine the influence on the decision taken of the uncertainty in time slice modeled. In this case, the data of $\tau = 14$ hours is assumed to be the interval $\tau = 12 - 15$. Given data over time such as that used to construct the graphs

[6] Note the TID evaluation can be done at any point in the process, and not just at the end.

in Figure 2, the effect on the network of using time from the two intervals around the chosen interval, i.e. $\tau = 9 - 12$ and $\tau = 15 - 18$, was computed. In this case, this sensitivity analysis showed that both the interval $\tau = 9 - 12$ and $\tau = 15 - 18$ produce a decision-equivalent network, so no network updating was necessary.

Sensitivity analysis of the decision tradeoff reveals the threshold probability of appendicitis supporting appendectomy to be 0.97. Because the calculated probability at the time of the first examination was near, but below, the threshold, close observation was warranted.

The notion behind the sensitivity analysis is as follows: consider a model constructed at time interval τ, such that decision ω_d^i is the optimal decision. Call β the expected utility for decision ω_d^i. If the probabilities of certain variables are time-dependent, then these new probabilities need to be substituted into the model to check if the equivalence class of the decision would change. Note that different diagnoses may be computed, but if the equivalence class of the decision is unchanged, then, under this decision-equivalent approach, no network updating is necessary. For network updating to be necessary, the threshold β must be exceeded by the expected utility of another decision ω_d^j given probabilities for time interval $\tau' \neq \tau$. In brief, if (1) the equivalence class of the decision indicated by any alternative data set does not change (i.e. the network is decision-equivalent), and (2) data mis-matches between the chosen time interval and the KB model are not better explained by the alternative intervals,[7] then no network updating is done.

14.6 Selection of the Time-Series Process

A Markov process is employed since is has been successfully used in a variety of medical applications [1], and it has well-understood formal and computational properties. The intent is to try one of the simplest possible models, check the adequacy of the model, and increase the complexity of the model only if necessary. The benefits of this assumption, relative to the assumption of more complex models, include the ability to use well-known theoretical properties, and to run complete forward simulations given some input data, using a set of Markov transition matrices. In addition, smaller models may be constructed, entailing less data and allowing more efficient model evaluation procedures.

However, parameter estimation and validation of the selected stochastic process are necessary. The Markov assumption, in noting that the state of the patient is independent of the history of the disease given the previous state, is a simplification. This leads to a relatively weak model in terms of capturing longer-term temporal dependencies, trends which may be important in computing correct diagnoses, even though the theoretical properties of the model are well-understood. The alternative, an attempt to characterize the precise character of the stochastic process (such as that done in [4]), is a focus of current research [11]. Provan [11] compares higher-order Markov models, among others, with the discrete-time semi-Markov process described here. In addition, the parameters for the semi-Markov process need to be estimated with greater accuracy, so that the forward simulation can provide more accurate results.

Note that, for this domain, computational feasibility and data-availability have dictated the use of discrete intervals. Since the continuous-time evolution of diseases necessitates a continouous-time model, the accuracy of our use of a *discrete-time* semi-Markov process with 3-hour time intervals needs to be determined.

---

[7] A data mis-match may be that patient X had no guarding, but the KB model *required* a positive finding of guarding.

14.7 Related Literature

We now describe the relationship to a selection of other work. This paper is related to recent work on Temporal Influence Diagrams [2,4,6] and work on model selection (e.g. the articles by Sclove and Lauritzen *et al.* in this volume).

The system proposed in [4] is primarily interested in the statistical process underlying the temporal Bayesian networks. To this end, the paper focuses on computing the conditional dependence relations; in other words, if $BN_1, BN_2, ...BN_k$ are a temporal sequence of Bayesian networks, Dagum et al. address a method of defining the interconnections among these temporally-indexed BNs. In contrast, this paper assumes a Markov property, in which the dependence relations are fixed; hence, the interconnections among the Bayes networks at different times are pre-specified, once the Bayesian network topology is specified. Thus, for example, arcs are allowed between networks differing by one time interval, e.g. τ and $\tau + 1$ or $\tau - 1$, but *not* τ and $\tau + 2$. Berzuini et al. [2] propose the use of TIDs for reasoning. This paper uses a semi-Markov process for the temporal statistical model, the process which is also employed in our research (and was partially inspired by [2]). Berzuini et al. focus on simulation schemes for solving TIDs. In particular, they propose combining Gibbs sampling and forward sampling as an efficient means of avoiding functional dependencies in the ID design. Dean and Kanazawa [6] use TIDs for robotics applications, and the approach presented has influenced our work. The application to the medical domain of AAP has necessitated our investigation of alternative means of constructing and evaluating TIDs. In particular, DYNASTY constructs a TID in which each networks for different time intervals can be different; Dean and Kanazawa replicate the same network for each time interval.

Research on model selection (as in this volume) typically focuses on static models. Here we are interested in time-series process selection, given that the static model for this dynamic process (AAP) has been defined by an expert. We plan to compare to other model selection approaches the sensitivity analysis approach used to define the initial model.

14.8 References

[1] J. Beck and S. Pauker. The Markov Process in Medical Prognosis. *Medical Decision Making*, 3:419–458, 1983.

[2] C. Berzuini, R. Bellazi, and S. Quaglini. Temporal Reasoning with Probabilities. In *Proc. Conf. Uncertainty in Artificial Intelligence*, pages 14–21, 1989.

[3] J.R. Clarke. A Concise Model for the Management of Possible Appendicitis. *Medical Decision Making*, 4:331–338, 1984.

[4] P. Dagum, R. Shachter, and L. Fagan. Modeling Time in Belief Networks. Technical Report STAN-KSL-91-49, Stanford University, Knowledge Systems Laboratory, November 1991.

[5] F.T. de Dombal, D.J. Leaper, J.R. Staniland, A.P. McCann, and J.C. Horrocks. Computer-aided Diagnosis of Acute Abdominal Pain. *British Medical Journal*, 2:9–13, 1972.

[6] T. Dean and K. Kanazawa. A Model for Reasoning about Persistence and Causation. *Computational Intelligence*, 5(3):142–150, 1989.

[7] R.A. Howard and J.E. Matheson. Influence diagrams. In R. Howard and J. Matheson, editors, *The Principles and Applications of Decision Analysis*, pages 720–762. Strategic Decisions Group, CA, 1981.

[8] H.I. Pass and J.D. Hardy. The appendix. In *Hardy's Textbook of Surgery*, pages 574–581. J.B. Lippincott Co., 2nd edition, Philadelphia, 1988.

[9] J. Pearl. *Probabilistic Reasoning in Intelligent Systems*. Morgan Kaufmann, 1988.

[10] G. M. Provan and D. Poole. The Utility of Consistency-Based Diagnostic Techniques. In *Proc. Conf. on Principles of Know. Representation and Reasoning*, pages 461–472, 1991.

[11] G.M. Provan. Tradeoffs in Knowledge-Based Construction of Probabilistic Models. 1993, submitted.

[12] G.M. Provan and J.R. Clarke. Dynamic Network Construction and Updating Techniques for the Diagnosis of Acute Abdominal Pain. *IEEE Transactions on Pattern Analysis and Machine Intelligence*, 15:299–306, March 1993.

[13] S. Schwartz, J. Baron, and J. Clarke. A Causal Bayesian Model for the Diagnosis of Appendicitis. In L. Kanal and Lemmer J., editors, *Proc. Conf. Uncertainty in Artificial Intelligence*, pages 229–236, 1986.

[14] S. Srinivas and J. Breese. IDEAL: A Software Package for Analysis of Influence Diagrams. In *Proc. Conf. Uncertainty in Artificial Intelligence*, pages 212–219, 1990.

15
Diagnostic systems by model selection: a case study

S. L. Lauritzen, B. Thiesson and D. J. Spiegelhalter

Aalborg University, Aalborg University and MRC, Cambridge

ABSTRACT Probabilistic systems for diagnosing blue babies are constructed by model selection methods applied to a database of cases. Their performance are compared with a system built primarily by use of expert knowledge. Results indicate that purely automatic methods do not quite perform at the level of expert based systems, but when expert knowledge is incorporated properly, the methods look very promising.

Key words: Contingency tables; expert systems; graphical models; probability forecasting;

15.1 Introduction and summary

In recent years, decision support systems based upon exact probabilistic reasoning have gained increasing interest, see [SDLC 93] for a recent survey. Such systems are constructed by specifying a graph, possibly directed, involving important variables together with a probability specification that satisfies conditional independence restrictions encoded in the graph. Exploiting these conditional independences, the specification can be made through local characteristics, i.e. characteristics that depend on only few variables at a time. Similarly, the calculations needed for use of the system in diagnostic support can be performed by local computations [LS 88, JLO 90, SS 90, Dawid 92].

The network as well as the probabilities can be specified entirely by subjective expert knowledge, or by empirical investigations involving data in a more or less prominent role.

The purpose of this paper is — through a case study — to investigate model selection techniques based on classical significance testing such as described in [Wermuth 76] as well as techniques based upon criteria of information type [Akaike 74], and to compare the diagnostic accuracy with systems constructed from more extensive use of subjective expert knowledge. The paper thus follows up on work similar to [AKA 91].

15.2 The case study

We are considering a system for diagnosing congenital heart disease for newborn babies. This forms part of a study with the Great Ormond Street Hospital for Sick Children in London. The part we consider here is involved with diagnosing babies that present with symptoms of cyanosis (blue babies). Details concerning the problem and alternative systems for decision support can be found in [FSMB 89, FSMB 91, SDLC 93].

---

[0]The authors are strongly indebted to Dr. Kate Bull and Dr. Rodney Franklin, Great Ormond Street Hospital for Sick Children, for giving us access to the data, providing the vital expert knowledge and engaging in helpful discussions and tutorials. The research was supported in part by the Danish Research Councils through the PIFT programme and the SCIENCE programme of the EEC.

A total of 238 relevant cases have been collected in three phases: 1979, 1988 and 1989. The 87 cases from 1989 were reserved for a prospective test set, whereas the remaining 151 cases were used as training set i.e., model selection and estimation of probabilities were based upon the training set, whereas the evaluation of diagnostic performance was carried out on the test set. The data contained information on the following variables:

Birth Asphyxia. States: Yes and No. Label B. In 53 cases of the training set, when information was missing (m. i.), the value was set to 'No'.

Age. States: 0-3, 4-10, and 11-30 days. Label A. Set as '0-3' in 1 case with m. i.

Disease. (As finally diagnosed by expert pediatricians at Great Ormond Street Hospital). States: PFC (persistent foetal circulation), TGA (transposition of great arteries), FALLOT (tetralogy of Fallot), PAIVS (pulmonary atresia with intact ventricular septum), TAPVD (obstructed total anomalous pulmonary venous connection), and LUNG (lung disease). Label D.

Sick. States: Yes and No. Label S. Set as 'Yes' in 1 case with m. i.

LVH report. (Left ventricular hypertrophy, as reported by the referring pediatrician). States: Yes and No. Label L. Set as 'No' in 18 cases of m. i.

RUQ O_2. (Oxygen concentration in right upper quadrant). States: <5, 5-12, >12, and Missing. Label R. In this case subject matter knowledge indicated that m. i. on this variable was considered informative *per se*.

Lower Body O_2: (Oxygen concentration in lower body). States <5, 5-12, >12, and Missing. Label O. M. i. here has same status as for RUQ O_2.

CO_2 **report.** States: <7.5 and ≥ 7.5. Label C. Set to '<7.5' in 41 cases of m. i.

X-ray report: States: Normal, Oligaemic, Plethoric, Ground Glass, and Asymmetric/Patchy. Label X.

Grunting report. States: Yes and No. Label G. Set to 'No' in 4 cases of m. i.

Note that the data thus represent a contingency table with a total of 46080 cells (possible configurations) with only 151 cases. Hence the data are here (and in many similar cases) extremely sparse.

15.3 Model Selection

Graphical Chain Models

The models investigated are all so-called graphical chain models or block recursive graphical models [LW 89b, Kreiner 89, Lauritzen 89, WL 90]. Such models are specified through

A recursive structure. This is an ordered partitioning of the variables into blocks as, for example, $BAD < S < LXROCG$. The interpretation is that the complex of variables *BAD* are explanatory to the variables *SLXROCG* — which are then responses to *BAD* — and further that *BADS* are considered explanatory to the variables *LXROCG* and so on. Each complex of variables is denoted a *block*. Thus the recursive structure above has three blocks.

An edge list. Edges are allowed between any pair of variables. The edges must be undirected if they connect variables within a block such as, for example, an edge *XR*, and directed from explanatory variable to response if the edge involves variables in different blocks. Hence an edge *AR* is directed from A to R.

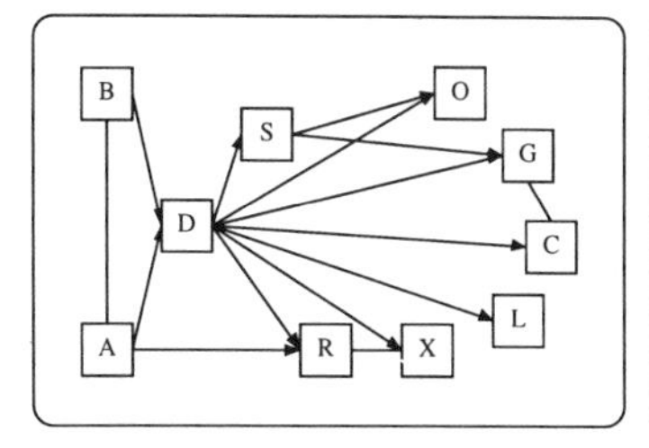

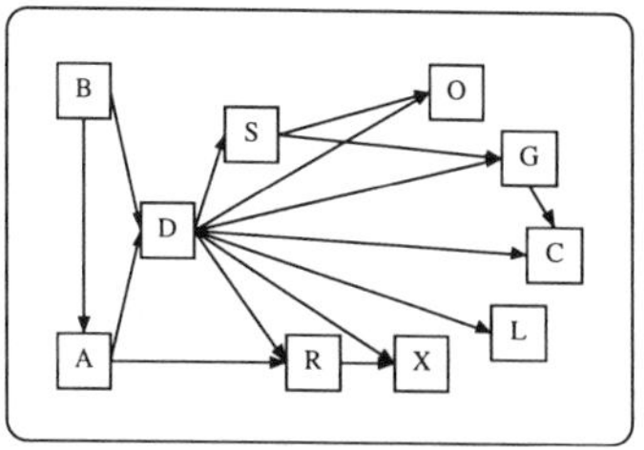

FIGURE 1: Graphical chain model with recursive structure $BA < D < S < LXROCG$ to the left, and HUGIN's DAG-model version to the right.

Two extreme examples of graphical chain models are undirected graphical models with only one block, and models where each variable is its own block, corresponding to a model with a directed acyclic graph (DAG-model in the following). In a DAG, A is said to be a *parent* of C if there is a directed edge from A to C. Two parents A and B of C are called *unmarried* if there is no edge between them.

The leftmost model in Figure 1 is an example of a graphical chain model with recursive structure $BA < D < S < LXROCG$. The meaning of the graph associated with a graphical chain model is that the joint distribution of the variables satisfy the *chain graph Markov property* [Frydenberg 89]. This means that any pair of variables that are not connected with an edge are conditionally independent [Dawid 79] given its set of *concurrent variables* which is the set of variables that are in the same block or previous blocks as any variable of the pair considered. For example, in the recursive structure $BAD < S < LXROCG$, a missing edge between A and D means that A is conditionally independent of D given B, written as $A \perp\!\!\!\perp D \mid B$, whereas in the recursive structure $BADS < LXROCG$, it means $A \perp\!\!\!\perp D \mid B, S$. Similarly, a missing edge BR means $B \perp\!\!\!\perp R \mid A, D, S, L, X, O, C, G$ in both of these recursive structures.

Some graphical chain models, called decomposable, have the property that they are equivalent — in terms of conditional independence properties — to DAG-models for suitably constructed DAGs. In particular a DAG model is equivalent to a specific graphical chain model if no pair of unmarried parents to a given variable in the DAG are both in the same block of the recursive structure as this variable. [LW 89b]

There are several advantages of only considering such models. In particular, most computations involved in any analysis simplify considerably. But also, DAG-models can easily be used in a decision support system, exploiting HUGIN [AOJJ 89]. In Figure 1 the DAG-model displayed to the right is equivalent to the model to the left.

It can be confusing to see such a DAG-model with arrows pointing against intuition. Some links in the graph would more naturally be undirected. But this is just an artifact of the graphical display and plays no role in the performance of the model itself.

Selection by BIFROST

BIFROST [HST 92, HT 93] is a model selection program that allows the user to exploit a database of complete observations on a number of cases for selecting graphical chain models. Any model selected can be automatically exported to a decision support system specification for direct use in HUGIN. The program exploits another contingency table program, CoCo

[Badsberg 91], for the computationally heavy parts of the model selection.

To launch BIFROST, the user must specify a recursive structure as described. If wanted, the user can specify some edges to be present and some edges absent. The user can then choose between various selection methods.

The variables in our data set were grouped in either 1, 2, 3 or 4 blocks corresponding to the trivial recursive structure *BADLXROCG*, *BADS* $<$ *LXROCG*, *BAD* $< S <$ *LXROCG*, and *BA* $< D < S <$ *LXROCG*. There were a number of reasons for choosing these. First, it was obvious that *LXROCG* were response variables to *BAD*. The variables B and A had a quite subtle relation to D, B being partly explanatory and partly response and A in reality having to do with a selection effect, see [SDLC 93] for a further discussion of this issue. The variable S was considered a symptom that potentially would influence the reporting of other symptoms. Apart from this, minimal prior knowledge was given to the system, i.e. all conditional independences were induced from data.

Two selection methods were considered, both based upon direct backward selection from the saturated model. The direct backward selection process begins by considering the model with all edges present, and tries to see how the quality of data description is affected by removing each edge in turn, i.e. by assuming specific conditional independences. If an edge can be removed according to some criterion, see below, the 'most removable edge' is removed. The process continues recursively from this model until no edge can be removed. The two criteria investigated was one of

AIC. Akaike's information criterion [Akaike 74] measures the goodness of a model by $AIC = \textit{Deviance} + 2 \times \textit{complexity}$. The complexity is the number of parameters (conditional probabilities) needed to represent the model and the *deviance* is twice the minimal entropy distance $\textit{Deviance} = 2\sum_{i \in \mathcal{I}} n(i) \ln \frac{n(i)}{n\hat{p}(i)}$, where $\mathcal{I}$ is the set of all combinations of configurations of the variables, $\hat{p}$ is the probability specified in the model which minimizes the entropy distance, $n(i)$ is the observed counts of cases having configuration i, and n is the total number of cases.

An edge is considered removable if AIC decreases when the edge is removed, i.e. when the deterioration of fit by assuming additional conditional independence pays off in terms of reduced complexity at the exchange rate of 2 units of fit per parameter. The edge which gives the largest decrease in AIC is considered to be the most removable.

Significance test. This is a more traditional statistical approach, where an edge is deemed removable at any stage, if the probability that the deviance is at least as high as the one observed — the p-value — is above a given level α. In our case we had $\alpha = .10$ and the p-values were determined in BIFROST by exploiting CoCo to perform Monte Carlo simulation as in [Patefield 81].

Using this criterion, the most removable edge is the edge with highest p-value.

When a model has been selected, it is translated to an equivalent DAG-model, and its conditional probability specifications are estimated by maximum likelihood, which in this case amounts to using observed relative frequencies.

The selected models are displayed by their equivalent DAG-models in Figure 2. The models selected by AIC with two and four blocks are identical, and this also holds in the testing case for two and three blocks.

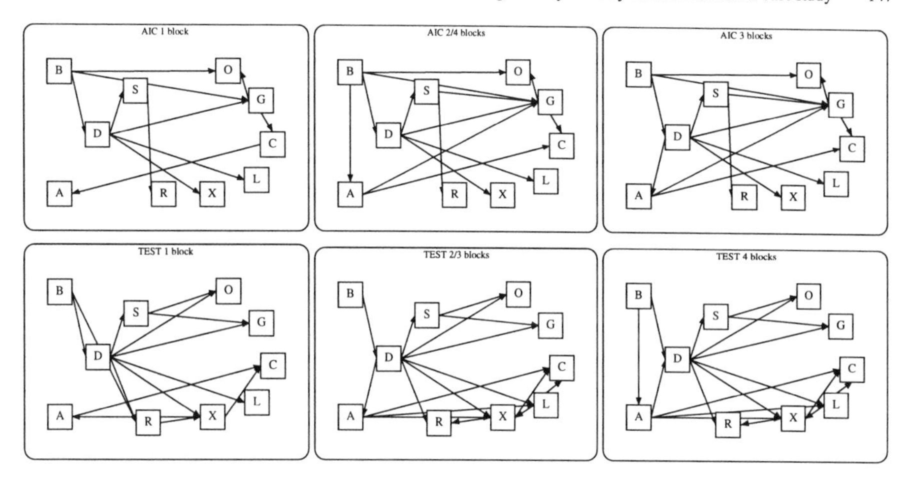

FIGURE 2: The six different models selected by BIFROST using AIC and testing.

Alternative models

The program CoCo [Badsberg 91] is an advanced program for analysing contingency tables of high dimension. With CoCo, a variety of model searches with different degrees of user interaction can be made. In the context here several of these were tried, including various versions of the model selection procedure introduced in [EH 85].

The latter methods select a list of models that are all acceptable descriptions of the data. Several user options include fixing of edges. Without going into details, these methods gave generally too large a list of acceptable models. This is natural from a scientific point of view in that it indicates that the data are really too sparse to identify a particular model with any certainty. Many models are consistent with the data. This still holds even if the simplifying assumption is made that interactions between three or more factors vanish, as e.g. suggested in [Whittaker 90]. None of the models selected in this way were evaluated as a basis for diagnostic systems.

Alternatively, highly user controlled strategies are possible. The graphical chain model in Figure 1 was selected by an attempt to make forward inclusion of edges, beginning with a simple model, having all manifestations conditionally independent, given the disease. While making the selections, data were inspected visually and as a rough guideline, a significance level of 5% was used for all tests. This, however, was not taken rigorously. After the forward selection had stopped, various attempts to remove edges were tried, unsuccessfully. In other words, the model is selected by a traditional statistical analysis of the data.

A completely different approach to establishing a model is through elicitation of subjective expert knowledge. The model in Figure 3 was established in this fashion, and includes unobserved pathophysiological variables, not in the database. Preliminary subjective specification of the probabilities was also made. The probabilities were given with an assessment of their imprecision, and then modified using the cases in the training set by maximizing the penalized likelihood. Here the data were used in their original incomplete representation, using the EM algorithm [Lauritzen 93] as implemented in [Thiesson 91]. See [SDLC 93] for further details of

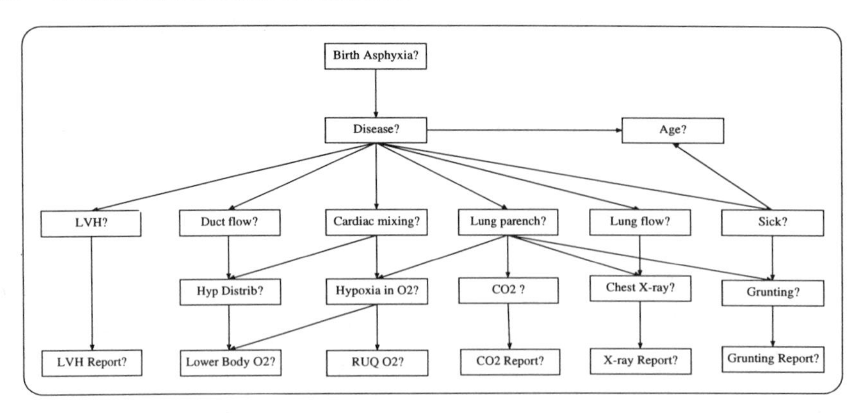

FIGURE 3: Network elicited from expert.

this procedure.

15.4 Model validation

After each model had been selected, it was exported to the shell HUGIN and the 87 cases in the test set were diagnosed using the model and the case data with the disease category masked as input information. That is, for each case HUGIN calculated the conditional probabilities of the six diseases, given the remaining observed variables on the given case. These probabilities are referred to as the *system probabilities* in the following. Various ways of measuring the performance were used.

Calibration plots: The general idea of calibration [Dawid 82, Dawid 86] compares the system probability of a given disease with the frequency of that disease being the correct diagnosis. The system is *well-calibrated* if the frequency of cases with system probability p and the given disease being correct, is about p.

In practice, cases with system probability in a given interval, say .10-.20 etc. must be grouped together and the disease frequency calculated. Calibration plots display the frequency among cases against the interval midpoint. Such plots were made for all six diseases and all selected models, but the large number of diseases and the small number of cases made it difficult to compare the models, due to random fluctuations. Table 15.1 gives examples of such calibration numbers. There is some tendency that the model obtained by modifying expert opinion is doing better, but it is hard to judge.

Scoring: An alternative quantitative approach is that of accumulating a penalty for giving small probabilities to events that happen [MW 77, MW 84]. A good scoring rule is based on the *Brier score,* which accumulates $\sum_{d \in diseases}(p(d) - \chi(d))^2$ over cases, where $p(d)$ is the system probability given to the disease d and $\chi(d)$ is the indicator function for the true disease.

Alternatively the *logarithmic score* accumulates $-\ln p$ over cases, where p is the system probability given to the disease that turned out to be true. In instances like the present, where the system probability is sometimes equal to zero, this will give infinite penalty. To make the comparison more interesting, we have only scored cases with strictly positive probability and

Disease	Model	0-.10	.10-.20	.20-.40	.40-.60	.60-.80	.80-.90	.90-1
		Probability intervals						
PFC	AIC 2/4	1/60=.02	6/15=.40	0/4=0	0/1=0	3/5=.60	1/1=1	0/1=0
PFC	Pen. Lik.	1/63=.02	0/5=0	1/5=.20	3/7=.43	2/3=.67	1/1=1	3/3=1
Fallot	AIC 2/4	4/57=.07	0/0=?	1/8=.13	0/0=?	1/3=.33	16/19=.84	0/0=?
Fallot	Pen. Lik.	3/51=.06	1/12=.08	1/3=.33	2/2=1	2/4=.50	4/5=.80	10/10=1

TABLE 15.1: Calibration tables for the diseases PFC and Fallot. The model with 2 (or 4) blocks selected by AIC is compared to the model obtained by modified expert opinion.

	BIFROST with testing			Own
	1 block	2 & 3 blocks	4 blocks	(1–4 blocks)
Brier	39.24	42.47	42.70	42.68
Log	37.60 (11)	42.37 (11)	42.70 (11)	45.00 (11)
Correct diagnoses (of 87)	63	58	58	59

TABLE 15.2: Performance measures of various models selected by significance testing. The numbers in parenthesis indicate the number of cases where the system probability for the correct disease was equal to zero.

simply counted the number of cases where the system probability for the true disease was equal to zero.

Diagnostic accuracy: A simple measure of performance is to choose the diagnosis with maximal system probability as the system diagnosis, and then count the number of correct diagnoses. This is slightly unfair to any probabilistic system since it does not take into account that a full probability distribution over diseases is given. When there are several candidates with reasonably high probabilities, a probabilistic system should not diagnose at all, but rather summarize its truly differential diagnosis, listing the possibilities and their probabilities.

15.5 Results

The scores and diagnostic accuracies are displayed in Table 15.2 and Table 15.3. From the tables it appears that the model exploiting expert opinion performs better than any of the automatically selected models, no matter which criterion is used to judge performance.

Among the models selected by using procedures involving significance testing, the 1 block model performs best and at a level which is close to the expert model. It was a surprise to the authors that this was the case, in particular that the crudely, automatically selected model

	BIFROST with AIC			Penalized
	1 block	2 & 4 blocks	3 blocks	Max. likelihood
Brier	37.33	38.08	39.72	33.91
Log	52.57 (7)	50.50 (8)	50.84 (8)	73.44
Correct diagnoses (of 87)	62	63	62	64

TABLE 15.3: Performance measures of various models. The numbers in parenthesis indicate the number of cases where the system probability for the correct disease was equal to zero. The BIFROST models are selected by Akaike's information criterion. The graph of the model in the rightmost column is constructed by expert knowledge alone. The probabilities were subsequently found by maximizing a likelihood function which is penalized by the expert's prior probabilities.

with one block and many repeated significance tests outperformed the model based on our own statistical analysis. If a lesson can be taken here, it must be that it is difficult to analyse data of this complexity, and the systematic approach used in model selection algorithms can be of good value. Alternatively, a significance level of about 5% seems to make the model too sparse when there is so little data available.

Among the models selected by AIC, the model with 1 block performs best in terms of everything else than diagnostic accuracy, where the model based on 2 or 4 blocks has one case more correctly diagnosed. The difference between the performance of the models selected by AIC is not so big as in the previous case, and in fact all perform on about the same level as the best model selected by significance testing.

It is of interest to note that in terms of simple diagnostic accuracy, the best models perform at the same level as the clinicians at Great Ormond Street, who in a similar evaluation had 114 out of 168 cases correctly diagnosed [FSMB 91].

15.6 Conclusion

As a consequence of the results, it seems appropriate to conclude that even a very crude approach to model selection seems capable of generating sensible baseline models. However, certain weaknesses are also apparent: (i) With data as sparse as here, the rough maximum likelihood estimates tend to give systems with too extreme behaviour, reflected, for example, through diagnostic probabilities equal to zero for true diseases. (ii) The blocking of variables in a recursive structure seems to have an unpredicted effect on the model selection. This aspect should be further investigated.

These weaknesses, combined with the fact that the model constructed from expert opinion outperformed all other models selected, suggest that it is worthwhile designing and using a model selection procedure which more explicitly uses expert knowledge.

The difference in performance between the best models selected by automatic methods and the model given by the expert is not so great. Hence it is conceivable that such a combined approach could prove quite effective in similar examples to the present.

15.7 References

[Akaike 74] Akaike, H. (1974). A new look at the statistical model identification. *IEEE Transactions on Automatic Control*, **19**, 716–723.

[AKA 91] Andersen, L. R., Krebs, J. H., and Andersen, J. D. (1991). STENO: An expert system for medical diagnosis based on graphical models and model search. *Journal of Applied Statistics*, **18**, 139–153.

[AOJJ 89] Andersen, S. K., Olesen, K. G., Jensen, F. V., and Jensen, F. (1989). HUGIN – A shell for building Bayesian belief universes for expert systems. In *Proc. 11th Int. Joint Conference on Artificial Intelligence*, 1080–1085.

[Badsberg 91] Badsberg, J. H. (1991). A guide to CoCo. Technical Report R-91-43, Dept. of Mathematics and Computer Science, Aalborg University.

[Dawid 79] Dawid, A. P. (1979). Conditional independence in statistical theory (with discussion). *Journal of the Royal Statistical Society, Series B*, **41**, 1–31.

[Dawid 82] Dawid, A. P. (1982). The well-calibrated Bayesian (with discussion). *Journal of the American Statistical Association*, **77**, 605–613.

[Dawid 86] Dawid, A. P. (1986). Probability forecasting. In Kotz, S. and Johnson, N. L., editors, *Encyclopedia of Statistical Sciences*, Volume 4, Wiley, New York.

[Dawid 92] Dawid, A. P. (1992). Applications of a general propagation algorithm for probabilistic expert systems. *Statistics and Computing*, **2**, 25–36.

[EH 85] Edwards, D. and Havránek, T. (1985). A fast procedure for model search in multidimensional contingency tables. *Biometrika*, **72**, 339–351.

[FSMB 89] Franklin, R. C. G., Spiegelhalter, D. J., Macartney, F., and Bull, K. (1989). Combining clinical judgements and statistical data in expert systems: over the telephone management decisions for critical congenital heart disease in the first month of life. *Int. J. of Clinical Monitoring and Computing*, **6**, 157–166.

[FSMB 91] Franklin, R. C. G., Spiegelhalter, D. J., Macartney, F., and Bull, K. (1991). Evaluation of a diagnostic algorithm for heart disease in neonates. *British Medical Journal*, **302**, 935–939.

[Frydenberg 89] Frydenberg, M. (1989). The chain graph Markov property. *Scandinavian Journal of Statistics*, **17**, 333–353.

[HST 92] Højsgaard, S., Skjøth, F., and Thiesson, B. (1992). *User's guide to BIFROST*. Dept. of Mathematics and Computer Science, Aalborg University.

[HT 93] Højsgaard, S. and Thiesson, B. (1993). BIFROST — Block recursive models Induced From Relevant knowledge, Observations and Statistical Techniques. To appear in *Computational Statistics and Data Analysis*.

[JLO 90] Jensen, F. V., Lauritzen, S. L., and Olesen, K. G. (1990). Bayesian updating in causal probabilistic networks by local computation. *Computational Statistics Quarterly*, **4**, 269–282.

[Kreiner 89] Kreiner, S. (1989). User's guide to DIGRAM - a program for discrete graphical modelling. Tech. Report 89-10, Statistical Research Unit, Univ. of Copenhagen.

[Lauritzen 89] Lauritzen, S. L. (1989). Mixed graphical association models (with discussion). *Scandinavian Journal of Statistics*, **16**, 273–306.

[Lauritzen 93] Lauritzen, S. L. (1993). The EM algorithm for graphical association models with missing data. To appear in *Computational Statistics and Data Analysis*.

[LS 88] Lauritzen, S. L. and Spiegelhalter, D. J. (1988). Local computations with probabilities on graphical structures and their application to expert systems (with discussion). *Journal of the Royal Statistical Society, Series B*, **50**, 157–224.

[LW 89b] Lauritzen, S. L. and Wermuth, N. (1989b). Graphical models for associations between variables, some of which are qualitative and some quantitative. *Annals of Statistics*, **17**, 31–57.

[MW 77] Murphy, A. H. and Winkler, R. L. (1977). Reliability of subjective probability forecasts of precipitation and temperature. *Applied Statistics*, **26**, 41–47.

[MW 84] Murphy, A. H. and Winkler, R. L. (1984). Probability forecasting in meteorology. *Journal of the American Statistical Association*, **79**, 489–500.

[Patefield 81] Patefield, W. M. (1981). AS 159. An efficient method of generating random $r \times c$ tables with given row and column totals. *Applied Statistics*, **30**, 91–97.

[SS 90] Shenoy, P. P. and Shafer, G. R. (1990). Axioms for probability and belief-function propagation. In Shachter, R. D. et al., editors, *Uncertainty in Artificial Intelligence IV*, pages 169–198, Amsterdam. North-Holland.

[SDLC 93] Spiegelhalter, D. J., Dawid, A. P., Lauritzen, S. L., and Cowell, R. G. (1993). Bayesian analysis in expert systems (with discussion). *Statistical Science*, **8**, 219–247 and 204–283.

[Thiesson 91] Thiesson, B. (1991). (G)EM algorithms for maximum likelihood in recursive graphical association models. Master's thesis, Dept. of Mathematics and Computer Science, Aalborg University.

[Wermuth 76] Wermuth, N. (1976). Model search among multiplicative models. *Biometrics*, **32**, 253–263.

[WL 90] Wermuth, N. and Lauritzen, S. L. (1990). On substantive research hypotheses, conditional independence graphs and graphical chain models (with discussion). *Journal of the Royal Statistical Society, Series B*, **52**, 21–72.

[Whittaker 90] Whittaker, J. (1990). *Graphical Models in Applied Multivariate Statistics*. John Wiley and Sons, Chichester.

16

A Survey of Sampling Methods for Inference on Directed Graphs

Andrew Runnalls

Computing Laboratory, The University, Canterbury CT2 7NF, UK.
arr@ukc.ac.uk

ABSTRACT We survey methods for performing inferences from measurement data to unknowns within a Bayesian network expressed as a directed graph, in which each node represents a scalar quantity capable of taking on values within a continuous range.

16.1 Introduction

We consider a system $\mathbf{X} = \{X_1, \ldots X_n\}$ of real-valued r.v.s (nodes) whose dependencies are expressed by means of a directed acyclic graph. With each node X_i in the system is associated a set $\mathbf{P}_i$, possibly empty, of *parent* nodes. Certain of the nodes, constituting a proper subset $\mathbf{Z}$ of $\mathbf{X}$, represent measurements or observations to hand, and have known values $\mathbf{Z} = \mathbf{z}$. We shall use $\check{\mathbf{X}}$ to denote the set of nodes that are *not* measurements, i.e. $\check{\mathbf{X}} = \mathbf{X} \setminus \mathbf{Z}$. $\check{\mathcal{X}}$ will denote the sample space spanned by $\check{\mathbf{X}}$. In general terms, the task is to determine the expected values of one or more (integrable) functions $T = T(x_1, \ldots x_n)$ of the node values, *conditional* upon the measured values.

Each X_i is a real-valued r.v., and we shall assume (*i*) that for each i we are given $f_i(x_i|\mathbf{P}_i)$, the pdf for node X_i conditional on the values of the parents of X_i,[3] and (*ii*) that the network as a whole satisfies the *conditional independence condition*, i.e. conditional on its parents, each node is independent of all other nodes apart from itself and its descendants (cf. [11, p. 120]). The conditional independence condition implies that the joint density of the node values is given by the product of the conditional densities f_i, i.e.

$$f(\mathbf{x}) = \prod_{i=1}^{n} f_i(x_i|\mathbf{P}_i)$$

We shall use the notation $\check{f}$ to denote the function which results from the joint density f by fixing the measurements $\mathbf{Z}$ to their known values $\mathbf{z}$. $\check{f}$ is an unnormalised form of the joint posterior density, $f(\check{\mathbf{x}}|\mathbf{Z} = \mathbf{z})$.

Some special cases (e.g. linear-Gaussian) can be solved by analytical means. Generally, however, numerical methods must be used. The following methods are of broad utility:

---

[1] *Selecting Models from Data: AI and Statistics IV.* Edited by P. Cheeseman and R.W. Oldford.

[2] This paper describes work sponsored by the Defence Research Agency, Farnborough, UK.

[3] In the problems which motivate this survey, the conditional densities f_i associated with the *non*-measurement nodes are generally of a tractable analytical form, and in particular it is usually straightforward to draw pseudorandom values from these conditional distributions. In contrast, the measurement nodes often have much more complicated conditional distributions, and frequently give rise to multimodal likelihood functions.

1. **Monte Carlo integration** using importance sampling. We have $E(T|\mathbf{Z}=\mathbf{z}) = \int_{\breve{\mathcal{X}}} T\breve{f}\,d\breve{\mathbf{x}} / \int_{\breve{\mathcal{X}}} \breve{f}\,d\breve{\mathbf{x}}$, so one approach is to evaluate these numerator and denominator integrals separately, by Monte Carlo integration, typically using importance sampling [6, 3, 14]. This approach has its merits, but space prevents us from exploring it further in this paper.

2. **Independent sampling:** Generate N *independent* pseudorandom samples from the posterior density $f(\breve{\mathbf{x}}|\mathbf{Z}=\mathbf{z})$, and use these to estimate the posterior expectation of T. Typically this could be achieved by generating random samples from some *proposal distribution* g, and then converting this into a sample from $f(\breve{\mathbf{x}}|\mathbf{Z}=\mathbf{z})$ by applying rejection sampling based on the ratio $\breve{f}/g$. Generally the criteria for choosing an appropriate proposal distribution g are the same as those for the choice of an importance sampling function in Method 1, and this technique will not be considered further in this paper.

3. **Monte Carlo Markov Chain:** Generate a Markov chain with states in the space $\breve{\mathcal{X}}$ whose limiting distribution is the desired posterior $f(\breve{\mathbf{x}}|\mathbf{Z}=\mathbf{z})$, and use one or more sample paths from this chain to estimate the posterior expectation of T. The method of Metropolis *et al.* [10] is one general method for generating such a Markov chain.

4. **Local updating:** This is a special case of Method 3, in which each of the transitions in the Markov chain is generated by a sequence of updates to single nodes in the directed graph. Each such update is performed in a manner which depends only on the current values of (a) the affected node $X_i \in \breve{\mathbf{X}}$, and ($b$) the neighbours of X_i in the moral graph derived from the directed graph. Gibbs sampling [2] is an example of this method.

16.2 Monte Carlo Markov chain (MCMC) methods

The idea here is to generate a Markov chain $\{\breve{\mathbf{X}}^{(1)}, \breve{\mathbf{X}}^{(2)}\ldots\}$ with states in the space $\breve{\mathcal{X}}$ whose limiting distribution is the desired posterior $f(\breve{\mathbf{x}}|\mathbf{Z}=\mathbf{z})$, and use one or more sample paths from this chain to estimate the posterior expectation of T. This requires us to find a *transition kernel* $P(\breve{\mathbf{x}}^{(n)}, \breve{\mathbf{x}}^{(n+1)})$ defining the probability distribution of $\breve{\mathbf{X}}^{(n+1)}$ conditional on $\breve{\mathbf{X}}^{(n)}$, with the following three properties:

- $f(\breve{\mathbf{x}}|\mathbf{Z}=\mathbf{z})$ must be an *invariant* distribution under this transition kernel, i.e. if $\breve{\mathbf{X}}^{(n)}$ has this distribution, then $\breve{\mathbf{X}}^{(n+1)}$ will too.

- The kernel must be *aperiodic*, i.e. it must not be possible to divide the space $\breve{\mathcal{X}}$ up into two or more subsets which the resulting Markov chain will visit in strict rotation.

- The chain must be *irreducible*: roughly, the chain must be capable of getting into any part of $\breve{\mathcal{X}}$ whatever its starting value. More precisely, for any starting value $\breve{\mathbf{X}}^{(1)} = \breve{\mathbf{x}}^{(1)}$ and any set A receiving positive probability under $f(\breve{\mathbf{x}}|\mathbf{Z}=\mathbf{z})$, there must be an integer n such that $\breve{\mathbf{X}}^{(n)}$ has non-zero probability of falling in A.

If these conditions are satisfied, the distribution of $\breve{\mathbf{X}}^{(n)}$ will converge to the desired distribution $f(\breve{\mathbf{x}}|\mathbf{Z}=\mathbf{z})$ as $n \to \infty$, and under very mild additional conditions the chain will be ergodic. (See particularly [16] for further mathematical details.)

Having chosen a suitable kernel P, the procedure for estimating $E(T|\mathbf{Z}=\mathbf{z})$ is: (1) Use P to generate one or more pseudorandom realisations $\{\breve{\mathbf{x}}^{(1)}, \breve{\mathbf{x}}^{(2)}, \ldots \breve{\mathbf{x}}^{(N)}\}_1$, $\{\breve{\mathbf{x}}^{(1)}, \breve{\mathbf{x}}^{(2)}, \ldots \breve{\mathbf{x}}^{(N)}\}_2$, etc. of the chain (preferably with different starting values). (2) Possibly identify and delete the transient initial portion from each of these realisations. (3) Calculate the mean value of $T(\check{\mathbf{x}})$ over the remaining portions of the sample paths. This forms the estimate of $E(T|\mathbf{Z} = \mathbf{z})$. (4) Further examine the sample paths to estimate the accuracy of this estimate. Stages 2 and 4 here pose unattractive statistical complications in comparison with Monte Carlo integration, or with other methods based on generating *independent* samples from $f(\check{\mathbf{x}}|\mathbf{Z} = \mathbf{z})$. Nevertheless the total effort in estimating $E(T|\mathbf{Z} = \mathbf{z})$ to within a given accuracy may be less using this Markov chain approach.

[16] and [17] present a useful survey of Monte Carlo Markov chain methods. See also [1], which particularly surveys applications to *un*directed graphs.

16.3 The Metropolis method

How can we construct a transition kernel with $f(\check{\mathbf{x}}|\mathbf{Z} = \mathbf{z})$ as its equilibrium distribution? One versatile approach is that due to Metropolis *et al.* [10] and generalised in [7]. In this approach, we choose some irreducible and aperiodic kernel $q(\check{\mathbf{x}}^{(n)}, \check{\mathbf{x}}^{(n+1)})$ defining a probability density for $\check{\mathbf{x}}^{(n+1)}$ conditional on $\check{\mathbf{x}}^{(n)}$. For the moment, q is otherwise arbitrary. Given the value $\check{\mathbf{x}}^{(n)}$ of the n-th term of a sample chain, the method generates the next term as follows:

1. Use the kernel q to generate a pseudorandom *candidate state* $?\check{\mathbf{x}}^{(n+1)}$.
2. Evaluate the criterion ratio
$$r = \frac{\check{f}(?\check{\mathbf{x}}^{(n+1)})q(?\check{\mathbf{x}}^{(n+1)}, \check{\mathbf{x}}^{(n)})}{\check{f}(\check{\mathbf{x}}^{(n)})q(\check{\mathbf{x}}^{(n)}, ?\check{\mathbf{x}}^{(n+1)})}$$
3. Accept the candidate state $?\check{\mathbf{x}}^{(n+1)}$ (i.e. put $\check{\mathbf{x}}^{(n+1)} = ?\check{\mathbf{x}}^{(n+1)}$) unconditionally if $r \geq 1$.
4. Otherwise accept $?\check{\mathbf{x}}^{(n+1)}$ with probability r.
5. If the candidate state $?\check{\mathbf{x}}^{(n+1)}$ is not accepted, then $\check{\mathbf{x}}^{(n+1)}$ is set equal to the preceding state $\check{\mathbf{x}}^{(n)}$.

Notice that the algorithm can work directly with the *un*normalised posterior $\check{f}$.

It can be shown (see for example [13]) that under very general conditions, this yields a Markov chain with the desired posterior as its equilibrium distribution. The efficiency of the method depends on choosing a q with the following properties: (*a*) Whatever the current state $\check{\mathbf{x}}^{(n)}$, there is a reasonably high probability of the candidate $?\check{\mathbf{x}}^{(n+1)}$ for the next state being accepted. This avoids long runs of identical states in the resulting chain. (*b*) Subject to the preceding condition, the distribution of candidate values should be as diffuse as possible, to reduce the correlation between successive terms in the Markov chain. [16] and [17] survey possible methods of constructing q. Among those applicable to directed graphs are:

Random Walk Chains In this method, the candidate state $?\check{\mathbf{x}}^{(n+1)}$ is generated by taking the current state $\check{\mathbf{x}}^{(n)}$ and *adding* to it a random vector (of appropriate size) drawn from some fixed distribution, e.g. a multivariate normal or t-distribution. In choosing the variance of this distribution, a trade must be made between factors (*a*) and (*b*) above. Since in this case $q(x, y) \equiv q(y, x)$, the criterion ratio r reduces to $\check{f}(?\check{\mathbf{x}}^{(n+1)})/\check{f}(\check{\mathbf{x}}^{(n)})$.

Independence Sampling In this method, the candidate state is drawn from some fixed distribution g, i.e. the distribution of $?\check{\mathbf{x}}^{(n+1)}$ does not in fact depend on the current state $\check{\mathbf{x}}^{(n)}$. In this case r reduces to

$$\frac{\check{f}(?\check{\mathbf{x}}^{(n+1)})/g(?\check{\mathbf{x}}^{(n+1)})}{\check{f}(\check{\mathbf{x}}^{(n)})/g(\check{\mathbf{x}}^{(n)})}$$

The criteria for choosing g are essentially the same as for choosing an importance sampling function.

16.4 The Gibbs sampler

The Gibbs sampler (the 'stochastic simulation' of [11]) generates a Markov chain by means of the following transition algorithm: each non-measurement node is visited in turn, and a new value is assigned to that node by drawing from the distribution of that node conditional on all the remaining nodes (including measurement nodes), using the values currently assigned to those other nodes. That is to say, $x_1^{(n+1)}$ is drawn from $f(x_1|x_2^{(n)},\ldots x_m^{(n)})$, $x_2^{(n+1)}$ is drawn from $f(x_2|x_1^{(n+1)},x_3^{(n)},\ldots x_m^{(n)})$, and so on. The new state $\check{\mathbf{x}}^{(n+1)}$ of the Markov chain is that resulting from a complete sweep of this process over the non-measurement nodes.

It can be shown (cf. [13]) that under very general conditions, this process will result in a chain whose equilibrium distribution is the posterior $f(\check{\mathbf{x}}|\mathbf{Z}=\mathbf{z})$.

Now in view of the conditional independence relationships implied in the directed graph, the distribution of a given node conditional on *all* the other nodes is identical to its distribution conditional on its neighbours in the corresponding moral graph.[4] In fact, for each non-measurement node X_i the new value of x_i must be drawn from a density proportional to

$$f_i^*(x_i) = f_i(x_i|\mathbf{P}_i) \prod_{j:X_i\in\mathbf{P}_j} f_j(x_j|\mathbf{P}_j)$$

where the product is over all the children of node X_i.

16.5 Sampling a node conditionally on its neighbours

A problem that frequently arises in applying Gibbs sampling to problems represented as directed graphs is this: although it may be quite straightforward to generate random samples from f_i, the distribution of X_i conditional on its *parents*, it may be very much less straightforward to draw random samples from the distribution of X_i conditional on its *neighbours* in the moral graph. There are two problems:

Problem 1 The distribution from which we wish to sample is given to us only in the form of an unnormalised density, f_i^*.

Problem 2 From each such distribution, only one sample is required, because next time around all of the conditioning variables will have changed.

[4] The moral graph is the undirected graph obtained by introducing arcs between any two parents of a common child, and dropping the directions of the arcs (cf. [9]).

On its own, Problem 1 suggests the use of methods such as rejection sampling, or—better—the ratio-of-uniforms method and its various generalisations [8, 18], which generate pseudorandom samples using an unnormalised density directly. However, these methods require certain parameters—dependent on the desired distribution—to be set up before sampling can take place, and this is where Problem 2 comes into play. If these parameters are too small the resulting samples have the wrong distribution, while if they are too big the sampling process is inefficient. Whilst there is a wide band of acceptable values between these extremes, the effort of choosing these parameters just to generate one sample may render these methods unattractive.

If f_i^* is log-concave, an attractive approach is to use adaptive rejection sampling [5], which constructs an efficient rejection sampling scheme on the fly, using a piecewise linear approximation to the logarithm of f_i^*. An alternative, not restricted to log-concave f_i^*, is the 'Griddy Gibbs sampler' approach, but since this does not sample f_i^* exactly, it is perhaps better considered in relation to the nodewise Metropolis method, discussed next.

16.6 Nodewise Metropolis

As we described it in §16.3, the Metropolis method was a *global* sampling technique, in that a complete new candidate state $?\breve{\mathbf{x}}^{(n+1)}$, embracing values for each of the non-measurement nodes, was generated before an accept-or-reject decision was made. However, it can be more efficient to apply the method on a node-by-node basis, analogously to the Gibbs sampler.

Specifically, each non-measurement node is visited in turn, and for each such node X_i a candidate new value $?x_i^{(n+1)}$ is generated from the current value $x_i^{(n)}$ by using a transition density $q_i(\breve{\mathbf{x}}^{(n)}, ?x_i^{(n+1)})$ which may depend also on the values currently assigned to the neighbours of X_i in the moral graph. We then evaluate the criterion ratio

$$r = \frac{f_i^*(?x_i^{(n+1)})q_i(?x_i^{(n+1)}, x_i^{(n)})}{f_i^*(x_i^{(n)})q_i(x_i^{(n)}, ?x_i^{(n+1)})}$$

and accept the candidate value with probability $\min(1, r)$, otherwise setting $x_i^{(n+1)} = x_i^{(n)}$. The new state $\breve{\mathbf{x}}^{(n+1)}$ of the Markov chain is that resulting from a complete sweep of this process over the non-measurement nodes.

As before, there are various ways of choosing the node transition kernels q_i, for example:

Random Walk Chains The candidate value $?x_i^{(n+1)}$ is generated by taking the current value $x_i^{(n)}$ and adding to it a random number drawn from some fixed distribution.

Forward Sampling Working through the nodes in a parents-before-children ordering, draw $?x_i^{(n+1)}$ from $f_i(x_i|\mathbf{P}_i)$, the distribution of X_i conditional on its parents, conditioning on the values currently assigned to its parents. In this case the criterion ratio reduces to the likelihood ratio

$$r = \prod_{j:X_i \in \mathbf{P}_j} \left(\frac{f_j(?x_j^{(n+1)}|\mathbf{P}_j)}{f_j(x_j^{(n)}|\mathbf{P}_j)} \right)$$

where the product is over the children of node X_i.

This method can be very easy to implement, but is apt to be inefficient if the likelihood function induced by the children of X_i is very much more concentrated than f_i.

Grid-based Sampling In this method [12, 15], f_i^* is first evaluated over a finite grid of values (e.g. the values $x_k = x_0 + k\Delta x$ for $-n \leq k \leq n$, where x_0 is the current value of the node and Δx and n are constants). One of the values x_k on the grid is then chosen with a probability proportional to $f_i^*(x_k)$, and then perturbed by an increment drawn from some fixed distribution. This amounts to drawing a random sample from a kernel-based approximation to the density defined by f_i^*.

16.7 Correlated nodes

Except in trivial cases, it is inevitable that the posterior distribution $f(\check{\mathbf{x}}|\mathbf{Z} = \mathbf{z})$ will involve couplings between the variables constituting $\check{\mathbf{X}}$. If nodes are very strongly correlated in this way, this will mean that local (nodewise) sampling procedures will be very slow to explore the posterior distribution, and statistics derived from early segments of the resulting Markov chain may be entirely misleading. In the extreme case, if two (or more) nodes have values that are functionally related (i.e. given the value of one, the value of the other is determined with probability 1), these nodes will remain stuck at the values they received after the first iteration of the sampling algorithm.

There are various ways of combatting this problem:

1. In some cases, where there is a group of tightly coupled nodes, it may be possible to redefine the network in such a way that this group is replaced by a single (scalar-valued) node. The problem then disappears.

2. If the preceding stratagem would not retaining adequate fidelity to the original problem, an alternative is to replace the group of nodes by a single vector-valued node, with the original nodes corresponding to components of the new node. (Cf. [11, p. 223]) A snag with either of these methods is that if the group of tightly coupled nodes is very large, then the graph resulting from their amalgamation may no longer make explicit the conditional independence relationships which made a directed graph representation attractive in the first place.

3. In general statistical problems, a change of variables may avoid the problem. However, it may not be easy to determine an appropriate transformation in advance, and in directed graph problems, the necessary transformation may, once again, destroy the conditional independence relationships that the graphical representation was intended to exploit.

4. The methods described below for combatting problems of multimodality will also alleviate problems with tightly coupled nodes.

16.8 Multiple modes

Another obstacle that local sampling techniques must overcome is illustrated in Fig. 1. In the Gibbs sampler, an iteration consists in sampling x_1 from its distribution conditional on the current value of x_2, and then sampling x_2 conditional on the new value of x_1. It is evident that, in the example above, the sample path will immediately be drawn into one of the two modal regions and then stay there indefinitely. Although the conditions for the convergence of the Gibbs sampler are formally satisfied, in that the probability of switching to the other modal

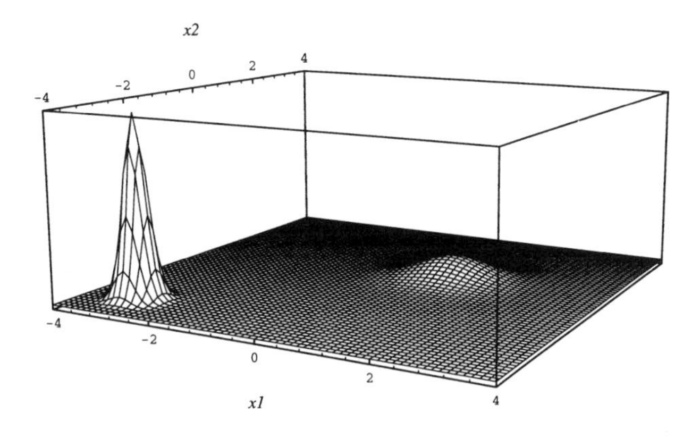

FIGURE 1: Example of a bimodal distribution that is problematic for nodewise sampling methods.

region is not zero, this probability is extremely low: about 5×10^{-7}, so you would need perhaps 100 million terms of the Markov chain before time averages along the sample path could even begin to be regarded as reliable estimates of the corresponding population statistics. Similar problems arise with the nodewise Metropolis method.

One idea would be to generate multiple sample paths with different starting values. This may well *reveal* the presence of multiple modal regions, but combining the statistics from the multiple paths may not present a correct picture overall, in that the frequency with which sample paths end up in a particular modal region will depend more on the distribution of starting points than on the relative probability volumes associated with the modes. For example, suppose that in the example we generate multiple paths with starting points uniformly distributed over the square region illustrated. About three-quarters of the paths will end up in the modal region centred on $(1, 1)$, and only a quarter of them around the taller mode at $(-3, -3)$. Yet these modal regions in fact have *equal* probability volume. To sum up, the 'catchment area' of a modal region may bear no relation to its volume.

16.9 Compound updating schemes

A good way of circumventing problems with multiple modes and correlated nodes is to adopt a compound updating scheme, in which each term of the Markov chain is generated by an iterative step which involves both local and global sampling methods. For example, on each iteration:

- First perform a sweep through the nodes using a local updating scheme, e.g. Gibbs sampling or nodewise Metropolis;
- Then perform a global update using the Metropolis method.

If the local and global updating steps are each characterised by a kernel which satisfies the conditions in §16.2, then the compound update certainly will be too,[5] so this method will still

[5] Indeed, the compound update may be irreducible even if its component steps aren't.

generate a Markov chain with the desired equilibrium distribution.

The philosophy here is to use the global updates to ensure that the sample path ranges over the full support of the desired posterior distribution, and in particular moves from one modal region to another with appropriate frequencies. This purpose will be fulfilled even if the candidate states generated at this stage are accepted (using the Metropolis criterion) only occasionally. The local updating scheme ensures that the states of the Markov chain will continue to vary, with the appropriate local distribution, even when the global updates are rejected (rather than remaining absolutely static, as would be the case if the global updating scheme were used on its own).

As before, there are various ways in which one can generate the candidate states for the global Metropolis update. One way is to draw the candidate state from some fixed distribution g representing a rough approximation to the posterior. This approximation can be constructed, for example, using the auxiliary network method of [14].

Another intriguing possibility is to generate it using a 'high temperature' version of the original network, using the 'auxiliary process' concept, discussed next.

16.10 The auxiliary process technique

In the Monte Carlo Markov chain approach as described in §16.2, the objective is to generate a Markov chain $\{\check{\mathbf{X}}^{(1)}, \check{\mathbf{X}}^{(2)} \ldots\}$ with states in the space $\check{\mathcal{X}}$, and whose limiting distribution is the desired posterior $f(\check{\mathbf{x}}|\mathbf{Z} = \mathbf{z}) \propto \check{f}(\check{\mathbf{x}})$. In the auxiliary process method, proposed in [4], we instead generate a Markov chain whose states are ordered pairs $(\check{\mathbf{X}}^{(i)}, \check{\mathbf{X}}'^{(i)})$, where $\check{\mathbf{X}}$ and $\check{\mathbf{X}}'$ each range over the space $\check{\mathcal{X}}$. As before, the intention is that the limiting distribution of $\check{\mathbf{X}}^{(i)}$ will be the desired posterior. At the same time, the *auxiliary process* $\check{\mathbf{X}}'^{(i)}$ will converge to some related distribution, with density $\check{f}'$ (not necessarily normalised). The purpose of the auxiliary process is to increase the 'fluidity' of the *target* process $\check{\mathbf{X}}^{(i)}$, thus avoiding problems of tight coupling and multimodality, this being achieved by periodically swapping the states of the target and auxiliary process.

As applied to directed graph problems, the target state $\check{\mathbf{X}}$ will represent an assignment of values to the non-measurement nodes of the original ('target') network, while the auxiliary state will typically represent an assignment of values to the non-measurement nodes in another network which—in analogy to statistical physics—we shall call the high-temperature network.[6] The high-temperature network will have identical topology to the target network, and corresponding measurement nodes will be assigned identical values, but the conditional probability distribution associated with a node in the high-temperature network will typically be more diffuse than for the corresponding node in the target network.

Each stage in the Markov chain is now generated *via* the following steps: (1) A new state $\check{\mathbf{x}}^{(n+1)}$ for the target process is generated by one or more sweeps of a local updating procedure. (2) A new state $\check{\mathbf{x}}'^{(n+1)}$ for the auxiliary process is generated by one or more sweeps of a local updating procedure. (3) A global Metropolis update is carried out in which the candidate state is that in which $\check{\mathbf{x}}^{(n+1)}$ and $\check{\mathbf{x}}'^{(n+1)}$ are interchanged. The criterion ratio reduces to:

$$r = \frac{\check{f}(\check{\mathbf{x}}')\check{f}'(\check{\mathbf{x}})}{\check{f}(\check{\mathbf{x}})\check{f}'(\check{\mathbf{x}}')}$$

[6] One might of course equally call it the *auxiliary network*, but this phrase was used in a different sense in [14].

A general way of constructing the high-temperature network is to set its conditional probabilities as follows:

$$f_i'(x_i|\mathbf{P}_i) \propto (f_i(x_i|\mathbf{P}_i))^{1/T_i}$$

where the 'temperature' T_i is ≥ 1, and is particularly high for the nodes chiefly responsible for tight coupling and multimodality problems in the target network, but the details are a promising field for future research.

16.11 References

[1] Julian Besag and Peter J. Green. Spatial statistics and Bayesian computation. *JRSS (B)*, 55, 1993.

[2] A. E. Gelfand and A. F. M. Smith. Sampling based approaches to calculating marginal densities. *JASA*, 85(410):398–409, 1990.

[3] John Geweke. Bayesian inference in econometric models using Monte Carlo integration. *Econometrica*, 57:1317–39, 1989.

[4] C. J. Geyer. Monte Carlo maximum likelihood for dependent data. In *Computer Science and Statistics: Proc. 23rd Symp. Interface*, 1992.

[5] W. R. Gilks. Derivative-free adaptive rejection sampling for Gibbs sampling. In J. M. Bernardo, J. O. Berger, A. P. Dawid, and A. F. M. Smith, editors, *Bayesian Statistics 4*, pages 641–9. Oxford University Press, 1992.

[6] J. M. Hammersley and D. C. Handscomb. *Monte Carlo Methods*. Methuen, 1964.

[7] W. K. Hastings. Monte Carlo sampling methods using Markov chains and their applications. *Biometrika*, 57(1):97–109, 1970.

[8] A. J. Kinderman and J. F. Monahan. Computer generation of random variates using the ratio of uniform deviates. *ACM Trans. Math. Software*, 3:257–60, 1977.

[9] S. L. Lauritzen and D. J. Spiegelhalter. Local computation with probabilities on graphical structures, and their application to expert systems. *JRSS (B)*, 50:157–244, 1988. With discussion.

[10] N. Metropolis, A. W. Rosenbluth, M. N. Rosenbluth, A. H. Teller, and E. Teller. Equations of state calculations by fast computing machines. *J. Chem. Phys.*, 21:1087–92, 1953.

[11] Judea Pearl. *Probabilistic Reasoning in Intelligent Systems*. Morgan Kaufmann, 1988.

[12] C. Ritter and M. A. Tanner. The griddy Gibbs sampler. Technical report, Division of Biostatistics, University of Rochester, 1991.

[13] G. O. Roberts and A. F. M. Smith. Simple conditions for the convergence of the Gibbs sampler and Metropolis-Hastings algorithms. Technical Report TR-92-12, Statistics Section, Department of Mathematics, Imperial College of Science, Technology and Medicine, 1992.

[14] Andrew R. Runnalls. Adaptive importance sampling for Bayesian networks applied to filtering problems. In David J. Hand, editor, *Artificial Intelligence Frontiers in Statistics: AI and Statistics III*. Chapman and Hall, 1993. Based on the proceedings of the Third International Workshop on AI and Statistics, Fort Lauderdale, 1991.

[15] M. A. Tanner. Tools for statistical inference: Observed data and data augmentation. *Lecture Notes in Statistics*, 67, 1991.

[16] Luke Tierney. Markov chains for exploring posterior distributions. Technical Report 560, School of Statistics, University of Minnesota, 1991.

[17] Luke Tierney. Exploring posterior distributions using Markov chains. In *Computer Science and Statistics: Proc. 23rd Symp. Interface*, 1992.

[18] J. C. Wakefield, A. E. Gelfand, and A. F. M. Smith. Efficient generation of random variates via the ratio-of-uniforms method. *Statistics and Computing*, 1:129–33, 1991.

17
Minimizing decision table sizes in influence diagrams: dimension shrinking

Nevin Lianwen Zhang, Runping Qi, and David Poole

Department of Computer Science
University of British Columbia
Vancouver, B.C. V6T 1Z2, CANADA

ABSTRACT One goal in evaluating an influence diagram is to compute an optimal decision table for each decision node. More often than not, one is able to shrink the sizes of some of the optimal decision tables without any loss of information. This paper investigates when the opportunities for such shrinkings arise and how can we detect them as early as possible so as to to avoid unnecessary computations. One type of shrinking, namely dimension shrinking, is studied. A relationship between dimension shrinking and what we call lonely arcs is established, which enables us to make use of the opportunities for dimension shrinking by means of pruning lonely arcs at a preprocessing stage.

17.1 Introduction

Influence diagrams were introduced by Howard and Matheson [3] as a graphical representation of decision problems. In comparison with decision trees (Raiffa [7]), influence diagrams are more intuitive and can be easily understood by people in all walks of life and degrees of proficiency. They are equally precise and more structured, thus allow more efficient treatment by computers.

An influence diagram is an acyclic directed graph of three types of nodes: random nodes, decision nodes, and value nodes, which respectively represent random quantities, decisions to make, and utilities (Howard and Matheson [3]).

Given an influence diagram, a decision table for a decision node is a table in which there is a cell for each array of values of its parents, and the value in that cell is the action to take when the parents do assume those values. One goal in evaluating an influence diagram is to compute an optimal decision table for each decision node.

The *size* of a decision table is the number of cells in it. It is the diagram itself, not the evaluation process that determines the size of a decision table. More often than not, after an optimal decision table is computed one realizes that it is possible to reduce the size of the table without losing any information.

There are at least two cases when the size of a decision table for a decision node d can be harmlessly reduced. First, there may exist a parent, say b, of d such that the b-dimension of the decision table is *irrelevant to* d in the sense that fixing the values of all the other parents of d, whatever value b takes the action for d remains the same. When it is the case, the b-dimension of the table can be deleted, resulting in a decision table of one less dimension. We call this deletion operation *dimension shrinking*.

A second case is when the real world situation that a cell represents never occurs in the sense

[0]Research is supported by NSERC Grant OGPOO44121. Comments from the reviewers' have been useful.

[1]*Selecting Models from Data: AI and Statistics IV.* Edited by P. Cheeseman and R.W. Oldford.

that its probability of occurring is zero. We call such a situation *unreachable*. When a situation is unreachable, the action in the corresponding cell can be deleted from the table, resulting in a decision table with one less cell. We call this deletion operation *cell shrinking*.

Cell shrinking has been studied under the term of asymmetry by Fung and Shachter [2], Covaliu and Oliver [1], Smith *et al* [12], Shenoy [11], and Qi and Poole [6]. This paper is concerned with dimension shrinking. We investigate when are dimension shrinkings possible? How can one detect them and make use of them to save computations?

A relationship between dimension shrinking and what we call lonely arcs is established (Section 17.4), which enables us to make use of the opportunities for dimension shrinkings by means of pruning lonely arcs at a preprocessing stage (Section 17.5). Since pruning arcs from influence diagrams leads to the violation of the so-called no-forgetting constraint, we need to work with a setup that is more general than influence diagrams. Introduced by Zhang *et al* [13], stepwise-decomposable decision networks (Section 17.3) prove to be a good setup for studying dimension shrinking.

Let us begin by reviewing the concepts of influence diagrams and decision networks.

17.2 Decision networks and influence diagrams

This section gives the formal definitions of Bayesian networks, decision networks and influence diagrams. The reader is referred to Zhang *et al* [13] for semantic details.

Before getting started, let us note that in this paper, standard graph theory terms such as acyclic directed graphs, parents (direct predecessors), children (direct successors), predecessors, descendants (successors), leaves (nodes with no children), and roots (nodes with no parents) will be used without giving the definitions. The reader is referred to Lauritzen *et al* [4] for exact definitions.

We shall use π_x to denote the set of parents of a node x in a directed graph. We shall use Ω_x to denote the frame of a variable x, i.e the set of possible values of x. For any set B, define $\Omega_B = \prod_{x \in B} \Omega$.

A *Bayesian network* $\mathcal{N}$ is a triplet $\mathcal{N} = (X, A, \mathcal{P})$, where

1. X is a set of variables (nodes) and A is a set of arcs over X such that (X, A) is an acyclic directed graph,

2. $\mathcal{P}$ is a set $\{P(x|\pi_x)|x \in X\}$ of conditional probabilities of the variables given their respective parents[3].

The *prior joint probability* $P_{\mathcal{N}}(X)$ of a Bayesian network $\mathcal{N} = (X, A, \mathcal{P})$ is defined by

$$P(X) = \prod_{x \in X} P(x|\pi_x). \tag{17.1}$$

In words, $P(X)$ the multiplication of all the conditional probabilities in $\mathcal{N}$. For any subset $B \subseteq X$, the *marginal probability* $P(B)$ is defined by

$$P(B) = \sum_{X-B} P(X), \tag{17.2}$$

[3]Note that when x is a root, π_x is empty. When it is the case, $P(x|\pi_x)$ stands for the prior probability of x.

where $\sum_{X-B}$ means summation over all the values of the variables in $X-B$.

A *decision network skeleton* is an acyclic directed graph $G = (Y, A)$, which consists of three types of nodes: *random nodes*, *decision nodes*, and *values nodes*, and where the value nodes have no children. Figure 1 (1) shows a decision network skeleton. Following the convention in the literature, decision nodes are drawn as rectangles, random nodes as cycles, and value nodes as diamonds.

A decision network skeleton is *regular* if there is a directed path that contains all the decision nodes. A regular decision network skeleton is *no-forgetting* if each decision node and its parents are parents to all subsequent decision nodes. The skeleton in Figure 1 (1) is not no-forgetting.

A *decision network* $\mathcal{N}$ is a quadruplet $\mathcal{N} = (Y, A, \mathcal{P}, \mathcal{F})$ where

1. (Y, A) is a decision network skeleton. We use C to denote the set of random nodes, D to denote the set of decision nodes, and V to denote the set of value nodes.

2. $\mathcal{P}$ is a set $\{P(c|\pi_c)|c{\in}C\}$ of conditional probabilities of the random nodes given their respective parents.

3. $\mathcal{F}$ is a set $\{f_v : \Omega_{\pi_v}{\rightarrow}R^1|v \in V\}$ of *value (utility) functions* for the value nodes, where R^1 stands for the set of real numbers.

We shall say that $\mathcal{N}$ is a decision network over the skeleton (Y, A) and that (Y, A) is the skeleton underlying $\mathcal{N}$.

A decision network is regular (or no-forgetting) if its underlying skeleton is. An *influence diagram* is a decision network that is regular and no-forgetting, and contains only one value node[4].

A *decision function (table)* for a decision node d_i is a mapping $\delta_i : \Omega_{\pi_{d_i}} \rightarrow \Omega_{d_i}$. The *decision function space* Δ_i for d_i is the set of all the decision functions for d_i. Let $D = \{d_1, \ldots, d_k\}$ be the set of all the decision nodes in a decision network $\mathcal{N}$. The Cartesian product $\Delta = \prod_{i=1}^{k} \Delta_i$ is called the *policy space* for $\mathcal{N}$, and a member $\delta = (\delta_1, \ldots, d_k) \in \Delta$ is called a *policy*. The relationship between a decision node d_i and its parents π_{d_i} as indicated by a decision function $\delta_i : \Omega_{\pi_{d_i}} \rightarrow \Omega_{d_i}$ is equivalent to the relationship as represented by the conditional probability $P_{\delta_i}(d_i|\pi_{d_i})$ given by

$$P_{\delta_i}(d_i{=}\alpha|\pi_{d_i}{=}\beta) = \begin{cases} 1 & \text{if } \delta_i(\beta) = \alpha \\ 0 & \text{otherwise,} \end{cases} \quad \forall\alpha{\in}\Omega_{d_i}, \forall\beta{\in}\Omega_{\pi_{d_i}}. \tag{17.3}$$

Since $\delta = (\delta_1, \ldots, \delta_k)$, we sometimes write $P_\delta(d_i|\pi_{d_i})$ for $P_{\delta_i}(d_i|\pi_{d_i})$. Because of equation (17.3), we will abuse the symbol δ by letting it also denote the set $\{P_\delta(d_i|\pi_{d_i})|d_i{\in}D\}$ of conditional probabilities of the decision nodes.

In a decision network $\mathcal{N} = (Y, A, \mathcal{P}, \mathcal{F})$, let $X = C{\cup}D$. Let A_X be the set of all the arcs of A that lie completely in X. Then the triplet $(X, A_X, \mathcal{P}{\cup}\delta)$ is a Bayesian network [5]. We shall refer to this Bayesian network as the *Bayesian network induced from $\mathcal{N}$ by the policy δ*, and write it as $\mathcal{N}_\delta$. The prior joint probability $P_\delta(X)$ is given by

$$P_\delta(X) = \prod_{x \in C} P(x|\pi_x) \prod_{x \in D} P_\delta(x|\pi_x). \tag{17.4}$$

---

[4] In the literature, influence diagrams do no have to satisfy all those three constraints. However, only influence diagrams that satisfy all the three constraints have been studied before Zhang *et al* [13].

[5] Remember that δ denotes a set of conditional probabilities for the decision nodes.

For any value node v, π_v contains no value nodes. Because if otherwise, there would be some other value nodes that has a child v. Hence $\pi_v \subseteq X$. The expectation $E_\delta[v]$ of the value function $f_v(\pi_v)$ of v under P_δ is given by

$$E_\delta[v] = \sum_{\pi_v} P_\delta(\pi_v) f_v(\pi_v).$$

The *expected value* $E_\delta[\mathcal{N}]$ of $\mathcal{N}$ under the policy δ is defined by

$$E_\delta[\mathcal{N}] = \sum_{v \in V} E_\delta[v]. \tag{17.5}$$

The *optimal expected value* $E[\mathcal{N}]$ of $\mathcal{N}$ is defined by

$$E[\mathcal{N}] = max_{\delta \in \Delta} E_\delta[\mathcal{N}]. \tag{17.6}$$

The optimal value of a decision network that does not have any value nodes is zero. An *optimal policy* δ^o is one that satisfies

$$E_{\delta^o}[\mathcal{N}] = E[\mathcal{N}]. \tag{17.7}$$

For a decision network that does not have value nodes, all policies are optimal.

In this paper, we shall only consider variables with finite frames. Hence there are only finite possible policies. Consequently, there always exists at least one optimal policy.

To evaluate a decision network is to (1) find an optimal policy, and (2) find the optimal expected value.

17.3 Stepwise-decomposable decision networks

This paper proposes a way for making use of the opportunities for dimension shrinking by means of arc removal. Since pruning arcs from an influence diagram leads to the violation of the no-forgetting constraint, we need a more general setup. One may suggests general decision networks; the problem with them is that they are hard to evaluate. Our choice is stepwise-decomposable decision networks (SDDN's) (Zhang *et al* [13]), which is briefly reviewed below.

Let $G = (X, A)$ be a directed graph. An arc from x to y is written as an ordered pair (x, y). The *moral graph* $m(G)$ of G is an undirected graph $m(G) = (X, E)$ whose edge set E is given by

$$E = \{\{x, y\} | (x, y) \text{ or } (y, x) \in A, \text{ or } \exists z \text{ such that } (x, z) \text{ and } (y, z) \in A\}.$$

In words, $\{x, y\}$ is an edge in the moral graph if either there is an arc between them or they share a common child. The term moral graph was chosen because two nodes with a common child are "married" into an edge [4].

In an undirected graph, two nodes x and y are *separated* by a set of nodes S if every path connecting x any y contains at least one node in S. In a directed graph G, x and y are *m-separated* by S if they are separated by S in the moral graph $m(G)$. Note that any node set m-separates itself from any other set of nodes.

Suppose $\mathcal{K} = (Y, A)$ is decision network skeleton and d is a decision node in $\mathcal{K}$. Let $Y_I(d, \mathcal{K})$, or simply Y_I, be the set of all the nodes that are m-separated from d by π_d, with the nodes in π_d

excluded. Let $Y_{II}(d,\mathcal{K})$, or simply Y_{II}, be the set of all the nodes that are not m-separated from d by π_d. We observe that Y_I, π_d, and Y_{II} forms a partition of Y.

The decision node d is a *stepwise-decomposability candidate node*, or simply a *candidate node*, if the downstream set $Y_{II}(d,\mathcal{K})$ contains neither any decision nodes other than d nor the parents of such decision nodes.

A decision network skeleton $\mathcal{K}$ is *stepwise-decomposable* if either it contains no decision nodes or there exists a candidate node d such that when d is regarded as a random node, $\mathcal{K}$ (now with one less decision node) is stepwise-decomposable. A decision network is *stepwise-decomposable* if its underlying skeleton is.

The reader is encouraged to verify that influence diagrams are SDDN's.

The remainder of this section is concerned with how to evaluate an SDDN. The decision network skeleton $\mathcal{K}$ is *smooth at d* if all the arcs between nodes in π_d and nodes in $Y_{II}(d,\mathcal{K})$ go from π_d to $Y_{II}(d,\mathcal{K})$. $\mathcal{K}$ is *smooth* if it is smooth at all the decision nodes. A decision network is *smooth* if its underlying skeleton is.

For technical simplicity, we shall restrict our attention only to smooth decision networks in this paper. However, all the results of this paper are true for non-smooth decision networks as well (see [14], Chapter 7).

Let d be a candidate node in a smooth decision network skeleton $\mathcal{K}$. The *tail* of $\mathcal{K}$ w.r.t d, denoted by $\mathcal{K}_{II}(d,\mathcal{K})$ or simply $\mathcal{K}_{II}$, is the decision network skeleton obtained by first restricting $\mathcal{K}$ to $\pi_d \cup Y_{II}$ and then removing those arcs that connect two nodes in π_d.

The *body* of $\mathcal{K}$ w.r.t d, denoted by $\mathcal{K}_I(d,\mathcal{K})$ or simply K_I, is the decision network skeleton obtained by first restricting $\mathcal{K}$ to $Y_I \cup \pi_d$ and then adding a node u and drawing an arc from each node in π_d to u. The node u is to be used to store the value of the tail, and is thus called the *tail-value node*.

Proposition 1 *(Zhang et al [13]) Suppose d is a candidate node in a smooth SDDN skeleton $\mathcal{K}$. Then the body $K_{II}(d,\mathcal{K})$ is also a smooth SDDN skeleton.*

Let $\mathcal{N}$ be a decision network over $\mathcal{K}$. The *tail* of $\mathcal{N}$ w.r.t d, denoted by $\mathcal{N}_{II}(d,\mathcal{N})$ or simply by $\mathcal{N}_{II}$, is a semi-decision network[6] over $\mathcal{K}_{II}$. The value functions for all the value nodes of $\mathcal{N}_{II}$ remain the same as in $\mathcal{N}$. The conditional probabilities of the random nodes of $\mathcal{N}_{II}$ that lay outside π_d also remain the same as in $\mathcal{N}$. The nodes in π_d, random or decision, are viewed in $\mathcal{N}_{II}$ as random nodes whose prior probabilities are missing.

Let X_t be the set of random and decision nodes in $\mathcal{N}_{II}$. Let $P_0(X_t)$ be the multiplication of all the conditional probabilities in $\mathcal{N}_{II}$. For any subset B of X_t, let $P_0(B)$ be obtained from $P_0(X_t)$ by summing out all the variables in $X_t - B$. We define the *evaluation functional* $e(d,\pi_d)$ of $\mathcal{N}_{II}$ by

$$e(d,\pi_d) = \sum_{v \in V_t} \sum_{\pi_v} P_0(\pi_v, \pi_d, d)\mu_v(\pi_v), \tag{17.8}$$

where V_t is the set of value nodes in $\mathcal{N}_{II}$.

The *body* of $\mathcal{N}$ w.r.t d, denoted by $\mathcal{N}_I(d,\mathcal{N})$ or simply by $\mathcal{N}_I$, is a decision network over $\mathcal{K}_I$. The conditional probabilities of all the random nodes remain the same as in $\mathcal{N}$. The values functions of the value nodes other than u also remain the same as in $\mathcal{N}$. The value function

[6] A semi-decision network is a decision network except the prior probabilities of some of the random root nodes are missing.

$\mu_u(\pi_d)$ of the downstream-value node u is defined by

$$\mu_u(\pi_d{=}\beta) = max_{\alpha\in\Omega_d} e(d{=}\alpha, \pi_d{=}\beta), \forall\beta \in \Omega_{\pi_d}. \tag{17.9}$$

Theorem 1 *(Zhang et al [13]) Suppose d is a candidate node in a smooth decision network $\mathcal{N}$. Let e be the evaluation functional of the tail $\mathcal{N}_{II}(d, \mathcal{N})$. Then*

1. *The optimal decision functions δ^o: $\Omega_{\pi_d} \rightarrow \Omega_d$ of d can be found through*

$$\delta^o(\beta) = arg\ max_{\alpha\in\Omega_d} e(d{=}\alpha, \pi_d{=}\beta), \forall\beta \in \Omega_{\pi_d}, \tag{17.10}$$

2. *And the optimal decision functions for decision nodes other than d are the same both in $\mathcal{N}$ and in the body $\mathcal{N}_I(d, \mathcal{N})$.*

17.4 Dimension shrinking and Lonely arcs

Before we can make use of opportunities for dimension shrinking, we need first detect them. To this end, we introduce the concept of independence for decision node. In a decision network, a decision node d is *independent* of one of its parents b if there exists an optimal decision table for d in which the b-dimension is irrelevant to d (and hence can be shrunken). In a decision network skeleton, a decision node d is *independent* of one of it parents b if d is independent of b in every decision network over the skeleton.

The remainder of this section introduces lonely arcs and show that they implies independencies for decision nodes, and hence opportunities for dimension shrinking.

Suppose $\mathcal{K}$ is a smooth decision network skeleton. Let d be a decision node in $\mathcal{K}$, and let b be a parent of d. The arc $b{\rightarrow}d$ is said to be *accompanied* if there exists at least one edge in the moral graph $m(\mathcal{K})$ of $\mathcal{K}$ that connects b and some nodes in the downstream set $Y_{II}(d, \mathcal{K})$. When it is the case, we say that the edge *accompanies* the arc $b{\rightarrow}d$. We say that the arc $b{\rightarrow}d$ is *lonely* if it is not accompanied by any edges of $m(\mathcal{K})$. In a decision network $\mathcal{N}$, an arc $b{\rightarrow}d$ is *lonely* if it is lonely in the underlying skeleton.

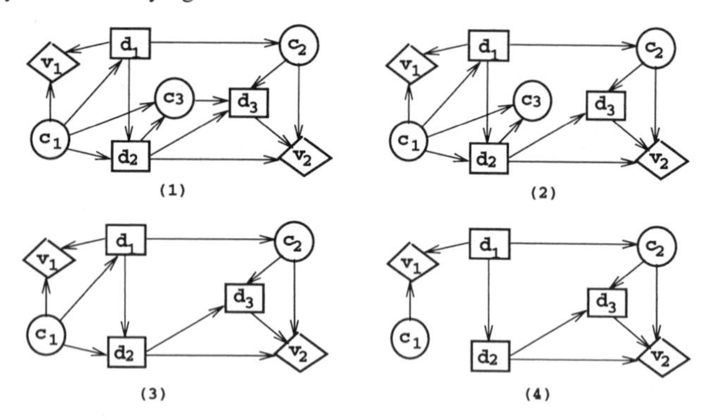

FIGURE 1: Removable arcs and removable nodes.

In the decision network skeleton shown in Figure 1 (1), the downstream set $Y_{II}(d_3, \mathcal{K})$ consists of a single node v_2. Since there is the arc $c_2{\rightarrow}v_2$ in $\mathcal{K}$, there is the edge (c_2, v_2) in $m(\mathcal{K})$, which accompanies the arc $c_2{\rightarrow}d_3$. However, the arc $c_3{\rightarrow}d_3$ is lonely.

Theorem 2 *Suppose $\mathcal{K}$ is a smooth SDDN skeleton. If an arc $b \rightarrow d$ into a decision node d is lonely, then d is independent of b.*

Proof: Let $\mathcal{N}$ be an arbitrary decision network over $\mathcal{K}$. We need to show that d is independent of b in $\mathcal{N}$.

It is easy to verify that for any candidate node $d' \neq d$, the arc $b \rightarrow d$ is lonely in $\mathcal{K}$ if and only if it is lonely in the body $\mathcal{K}_I(d', \mathcal{K})$. Because of this fact and because of Proposition 1 and Theorem 1, we can assume that d is a candidate node without losing any generality. Let $\mathcal{N}_{II}$ be the tail of $\mathcal{N}$ w.r.t d. Let P_0 be multiplication of the conditional probabilities in N_{II}. Since the arc $b \rightarrow d$ is lonely and $\mathcal{N}$ is smooth, none of the conditional probabilities in $\mathcal{N}_{II}$ involve b. So, neither does P_0.

Again because $b \rightarrow d$ is lonely, b can not be a parent to any value nodes in $Y_{II}(d, \mathcal{N})$. Hence, all the value functions $\mu_v(\pi_v)$ in $\mathcal{N}_{II}$ do not involve b.

Therefore, the evaluation functional $e(d, \pi_d)$ (see equation (17.8)) does not involve b. According to equation (17.10), the optimal decision functions of d is independent of b. The theorem is proved. □

In Figure 1 (1), the arc $c_3 \rightarrow d_3$ is lonely, hence d_3 is independent of c_3. From the proof of Theorem 2, we actually conclude that in all the optimal decision tables of d_3, the c_3-dimension is irrelevant to d_3 and hence can be deleted.

17.5 Utilizing opportunities for dimension shrinking

Now that we have learned how to detect opportunities for dimension shrinking, the natural next question is: How can we make use of them to avoid unnecessary computations? Our answer is to prune lonely arcs at a preprocessing stage. The justification is provided by the following corollary, which follows from the proof of Theorem 2.

Corollary 1 *Suppose $b \rightarrow d$ is a lonely arc in a smooth SDDN $\mathcal{N}$. Let $\mathcal{N}'$ be the decision network obtained from $\mathcal{N}$ by removing the arc $b \rightarrow d$. Then, $\mathcal{N}'$ and $\mathcal{N}$ have the same optimal decision tables for all the decision nodes other than d. Furthermore, the optimal decision tables for d in $\mathcal{N}'$ can be obtained from the optimal decision tables for d in $\mathcal{N}$ by shrinking the irrelevant b-dimension.* □

In order to repeatedly apply this corollary, we need the following theorem.

Theorem 3 *(Zhang et al [13]) The decision network skeleton resulted from pruning a lonely arc from a smooth SDDN skeleton is again a smooth SDDN skeleton.* □

17.5.1 Potential lonely arcs and barren nodes

In a decision network skeleton, a *barren node* is a random or a decision node that does not have any children. The node c_3 in Figure 1 (2) is a random barren node.

Proposition 2 *(Shachter [8]) A barren node may be simply removed from a decision network. If it is a decision node, then any decision function is optimal.*

Now we recursively define the concepts potential lonely arcs and potential barren nodes. A *potential lonely arc* is a lonely arc or an arc that becomes lonely after the removal of some barren and potential barren nodes, and the removal of some other lonely and potential lonely arcs. A *potential barren node* is a barren node or a node that becomes barren after the removal of some lonely and potential lonely arcs, and the removal of some other barren and potential barren nodes.

Going back to our example, after the removal of $c_3 \rightarrow d_3$, the the node c_3 becomes barren, and hence can be removed. After the removal of c_3, $c_1 \rightarrow d_2$ becomes lonely. After the removal of $c_1 \rightarrow d_2$, $c_1 \rightarrow d_1$ becomes lonely.

The corollary below follows from Corollary 1 and Theorem 3.

Corollary 2 *Suppose $\mathcal{K}$ is a smooth SDDN skeleton. If an arc $b \rightarrow d$ into a decision node d is a potential lonely arc, then d is independent of b.* □

17.5.2 An pruning algorithm

The following algorithm prunes all the potential lonely arcs and potential barren nodes from a smooth SDDN skeleton.

Procedure PRUNE-REMOVABLES($\mathcal{K}$)

- Input: $\mathcal{K}$ — a smooth SDDN skeleton.
- Output: A smooth SDDN skeleton that does not contain potential lonely arcs and potential barren nodes.

1. **If** there is no decision node in $\mathcal{K}$, output $\mathcal{K}$ and stop.
2. **Else**
   - Find and remove all the barren nodes;
   - Find a candidate node d of $\mathcal{K}$, and compute the downstream set $Y_{II}(d, \mathcal{K})$ of π_d.
   - Find and remove all the lonely arcs among the arcs from π_d to d.
   - Treat d as a random node and recursively call PRUNE-REMOVABLES($\mathcal{K}$).

Let $\mathcal{K}'$ be the output decision network skeleton of of PRUNE-REMOVABLES($\mathcal{K}$). How is $\mathcal{K}'$ related to $\mathcal{K}$ in terms of decision tables? Let $\mathcal{N}$ and $\mathcal{N}'$ be decision networks over $\mathcal{K}$ and $\mathcal{K}'$ respectively such that in both $\mathcal{N}$ and $\mathcal{N}'$ each variable has the same frame, each random variable has the same conditional probability, and each value node has the same value functions. By repeatedly applying Corollary 1, we can conclude that the optimal decision tables for $\mathcal{N}'$ can be obtained from those for $\mathcal{N}$ by deleting certain irrelevant dimensions.

Let us now consider applying PRUNE-REMOVABLES to the SDDN skeleton in Figure 1 (1). There are no barren nodes at the beginning; and d_3 is the only candidate node. The downstream set $\mathcal{K}_{II}(d_3, \mathcal{K})$ is the singleton $\{v_2\}$. One can see that $c_3 \rightarrow d_3$ is the only lonely arc. After the removal of $c_3 \rightarrow d_3$, the skeleton becomes as shown in Figure 1 (2), where c_3 is a barren node. After the removal of c_3, we get the skeleton in Figure 1 (3). Since d_3 is now treated as a random node, d_2 becomes a candidate node. The arc $c_1 \rightarrow d_2$ is lonely, and hence is removed. Thereafter, d_2 is also treated as a random node, rendering d_1 a candidate node. The arc $c_1 \rightarrow d_1$ is lonely and hence removed. The final skeleton is shown in Figure 1 (4).

Finally, let us note that [14] (Chapter 7) contains a more general version of PRUNE-REMOVABLES, which finds and prunes potentially lonely arcs in non-smooth SDDN's as well as in smooth SDDN's. In [14] (Chapter 8), we have also proved that the concept of potential lonely arcs captures all the opportunities of dimension shrinking that can be identified by using only graphical information, without resorting to numerical computations.

17.6 Conclusions

This paper has investigated how to detect opportunities for dimension shrinking in influence diagrams and how to make use of then to avoid unnecessary computations. A relationship between dimension shrinking and lonely arcs has been established, which enables us to make use of the opportunities for dimension shrinking by means of pruning lonely arcs at a preprocessing stage. Since pruning arcs from influence diagrams leads to the violation of the no-forgetting constraint, the investigation has been carried out in a more general setup of SDDN's.

The opportunities for dimension shrinking have been mentioned and to some extent made use of explicitly by Shachter [9], and implicitly by Shenoy [10]. This is paper is the first to propose a method of making use of opportunities for dimension shrinking at a preprocessing stage by means of pruning lonely arcs. The advantage of this method is that it simplifies influence diagrams or decision networks before the evaluation takes place. This method also leads to a divide and conquer strategy for evaluating influence diagrams [13].

17.7 References

[1] Z. Covaliu and R. M. Oliver (1992), Formulation and solution of decision problems using decision diagrams, Working paper, University of California at Berkeley.

[2] R. M. Fung and R. D. Shachter (1990), Contingent Influence Diagrams, Advanced Decision Systems, 1500 Plymouth St., Mountain View, CA 94043, USA.

[3] R. A. Howard, and J. E. Matheson (1984), Influence Diagrams, in *The principles and Applications of Decision Analysis*, Vol. II, R. A. Howard and J. E. Matheson (eds.). Strategic Decisions Group, Menlo Park, California, USA.

[4] S. L. Lauritzen, A. P. Dawid, B. N. Larsen, and H. G. Leimer (1990), Independence properties of directed Markov fields, *Networks*, Vol. 20, pp. 491-506.

[5] J. Pearl (1988), *Probabilistic Reasoning in Intelligence Systems: Networks of Plausible Inference*, Morgan Kaufmann, Los Altos, CA..

[6] R. Qi and D. Poole (1992), *Exploiting asymmetries in influence diagram evaluation*, Unpublished manuscript.

[7] H. Raiffa, (1968), *Decision Analysis*, Addison-Wesley, Reading, Mass.

[8] R. Shachter (1986), Evaluating Influence Diagrams, *Operations Research*, 34, pp. 871-882.

[9] R. Shachter (1988), Probabilistic Inference and Influence Diagrams, *Operations Research*, 36, pp. 589-605.

[10] P. P. Shenoy, (1992), Valuation-Based Systems for Bayesian Decision Analysis, *Operations research*, 40, No. 3, pp. 463-484.

[11] P. P. Shenoy, Valuation network representation and solution of asymmetric decision problems, work paper No. 246, Business School, University of Kansas.

[12] J. E. Smith, S. Holtzman, and J. E. Matheson (1993), Structuring conditional relationships in influence diagrams, *Operations Research*, 41, N0. 2, pp. 280-297.

[13] N. L. Zhang, R. Qi and D. Poole (1993), A computational theory of decision networks, *International Journal of Approximate Reasoning*, to appear.

[14] N. L. Zhang (1993), A computational theory of decision networks, Ph.D thesis, Department of Computer Science, University of British Columbia, Vancouver, B.C., V6T 1Z1, Canada.

18
Models from Data for Various Types of Reasoning

Raj Bhatnagar and Laveen N Kanal

University of Cincinnati
Cincinnati Ohio 45221
and
University of Maryland
College Park Md 20742

ABSTRACT Often the primary objective of constructing a model from the data is to model the phenomenon that produces the data in such a way that the model is useful for the desired type of reasoning objectives. In many graph models of probabilistic knowledge various aspects of these phenomena are represented by nodes while the edges represent the probabilistic dependencies among them. We demonstrate one type of reasoning objective that is better served by those models in which some qualitative relationships of the phenomena, are used as the edges and the hyperedges of the graph models. We have also outlined our methods for handling such models for reasoning and for learning the qualitative relationships of a domain from the empirical data.

18.1 Introduction

A *model* is intended to be a representation of a person's knowledge and beliefs about either a domain or a particular situation of a domain. The operations that manipulate the model and make inferences in its context are also expected to resemble the way the person deals and reasons with his knowledge and beliefs about the domain.

When constructing a model from a set of observed data the following aspects of the desired model are of interest :

- The intended use for the desired *domain model*, which may be either to make inferences in the context of the entire domain or to extract smaller *situation models* from the larger domain models.
- The structural primitives in terms of which the domain and the situation models are defined.
- The acquisition and learning of the relevant structural primitives from the available domain data.

The objective of the research presented here is to construct the *domain models* in such a way that the models retain the domain's probabilistic knowledge and are also structured in terms of the relevant qualitative relationships of the domain.

---

[1]*Selecting Models from Data: AI and Statistics IV.* Edited by P. Cheeseman and R.W. Oldford.

18.2 An Example

We describe below an example domain and then state two different types of reasoning objectives for which suitably structured models may be required. We then present our solution for constructing appropriate types of models and also for learning the relevant structural information from the empirical data.

The example domain is as follows. An organic compound with concentration *oc* biodegrades with a half-life of *hl*. The factors that affect the process, and thus the value of *hl*, include soil acidity *sa*, number of micro-organisms in the soil *nm*, and the soil humidity *hu*. The above represent five discrete and multivalued variables, described as follows:

```
oc   (high-conc, med-conc, low-conc)    hl   (10,20, . . .100) (days)
sa   (lev1, lev2, .... lev5)            nm   (10,100,1000)
hu   (l1,l2,l3,l4)
```

The available data for the domain consists of a database of recorded cases of biodegradation. In each case a value for each of the above five variables is recorded.

From the chemistry of the biodegradation processes we know that this organic compound breaks down via four different paths of chemical reactions. We call these paths R_1, R_2, R_3, and R_4. Each reaction runs at different rates in different soil conditions. For the complete domain it is true that a particular soil condition can support two, three, or four different types of reactions. However, in any individual instance typically only one of these reactions may have started and the others may have remained completely absent. Therefore, given a soil condition, there is some uncertainty about which reaction will start and there is also some uncertainty about the half-life of the process when a particular reaction is active. While recording an instance in our database of cases, it may also be possible to determine and record the type of the reaction active during the instance.

18.2.1 Type-1 Objective

In many reasoning and decision making situations we need a model in whose context the following type of inference may be made.

- **Given** : 1. The database of known cases from the past; 2. The observed events from the situation of interest, e.g., some soil attributes.
- **Determine** : Probability values for the unobserved variables such as the variable *hl*, evaluated in the context of the known cases.

The above type of objective has also been viewed as an instance of abductive reasoning [6, Pearl 88] in which the inferred probabilities for the unobserved variables form an *explanation* for the observed events. The premise of the abductive inference being that if the inferred probabilities were to hold for the unobserved variables then the observed events would have occurred as a matter of course.

18.2.2 Type-2 Objective

Another type of abductive reasoning in which we seek to extract smaller *situation models* from a larger domain model may be described as follows:

- **Given** : 1. The database of known cases from the past; 2. The observed events from the situation of interest, e.g., some soil attributes.

- **Determine** : The reaction which would result in the smallest expected value for the half-life variable hl, and the uncertainty associated with this reaction.

This type of objective is different from the previous one in two respects. The first is that the *situation model* it seeks needs to be structured in terms of the reactions and their associated probabilities. The second difference is that the hypothesized *situation model* must be optimal in some sense, e.g., the minimum expected value for the variable hl.

Given a particular soil condition, the type-1 objective seeks the probability values of hl events in the context of the entire domain, that is, we do not specify which particular reactions may become active. Therefore, our answer weighs in the relative probabilities of various reactions that may become active in the given soil condition. The type-2 objective seeks a particular reaction (= situation model) to be hypothesized and its associated uncertainties determined such that the hypothesis satisfies some optimality criterion. We now examine the structures of the domain models that are best suited for answering the above queries.

18.3 Structure of the Models

All the techniques for modeling the probabilistic knowledge aim to compress the information of the joint probability distribution into some structure of manageable size while not losing any essential parts of the information. A number of graph based models along with their associated inference procedures have been extensively studied in the literature [5, 6, Pearl 88]. Bayesian networks [6, Pearl 88] is one such graph representation. If the joint probability distribution is specified using k variables, $x_1, x_2, ...x_k$, then the joint probability distribution can be written as :

$$P[x_1, x_2, ...x_k] = P[x_1 \mid x_2, x_3..., x_k] * P[x_2 \mid x_3, x_4...x_k] \ldots P[x_{k-1} \mid x_k] * P[x_k] \quad (18.1)$$

This expression contains k product *terms*, each of which is a conditional probability of one variable conditioned on some other variables.

18.3.1 Type-1 Model

For constructing a graph model from the above distribution, each *term* is modified such that it retains only those conditioning variables on which the conditioned variable is directly dependent [6, Pearl 88]. The graph structure is then constructed by taking each node and making its conditioning variables its parent nodes and drawing an edge between every pair of parent-child nodes. We then associate with each node the probability values that are conditioned on all its parent nodes. One possible graph model for the biodegradation example may be structured as shown in Figure-1.

In this model we associate with each node n the conditional probability values $P[n \mid Parents(n)]$. The operations on this graph model let us determine the probabilities for the rest of the variables when evidence for some of them is available. Such inferencing can be performed by the operations described in either [5] or [Pearl 88].

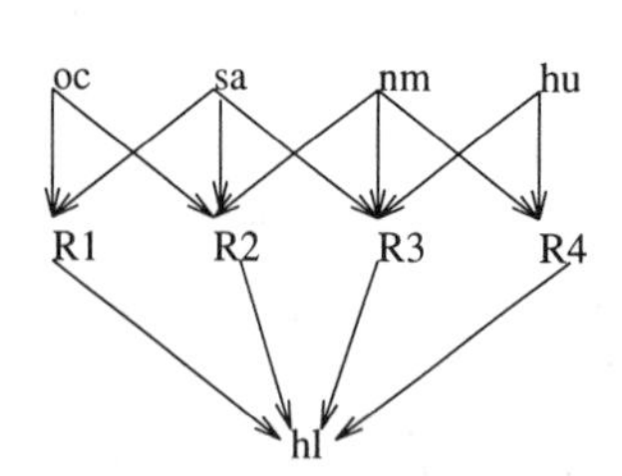

Figure-1 : A Model Containing All the Variables

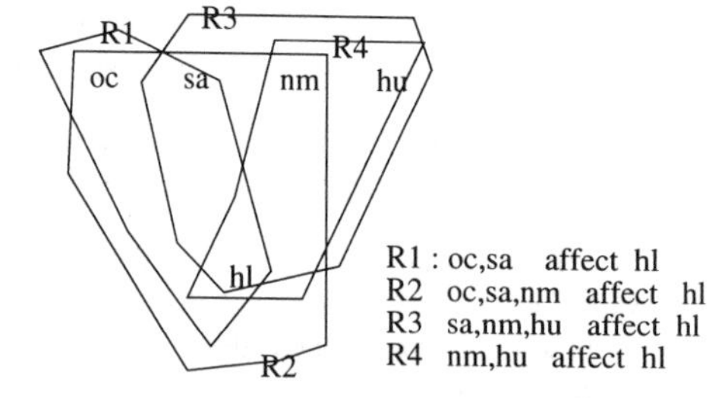

Figure-2 : Proposed Hypergraph Model

This model can easily facilitate the reasoning for the instances of type-1 objective. However, this model is not equally suitable when the following question of type-2 objective is encountered.

> The soil-acidity, and the number of micro-organisms for a soil are given. For which one of the four chemical reactions would the expected value for the variable *hl* be minimum?

In the context of the type-1 model it is not easily possible to hypothesize the unique optimal reaction according to this type of criterion. To answer a type-2 query using a type-1 model we need to partition the nodes into two different groups. The reaction nodes have to be assigned either a '0' or a '1' and the probabilities values computed for the rest of the nodes.

18.3.2 Type-2 Model

The model shown in Figure-2 has been constructed as follows. We assume that in any particular situation only one of the reactions may be active. We partition the database into four distinct groups, each corresponding to one of the possible reactions. Each polygon of Figure-2 is a hyperedge of the graph and corresponds to one of the possible reactions. Each Hyperedge includes the variable *hl*, called its *consequent node*, and the *antecedent nodes* which are known to causally affect *hl* when the corresponding reaction is active. The partition of the database cases corresponding to a hyperedge is used for determining the joint probability distribution to be associated with the hyperedge. Each hyperedge thus represents one of the many distinct qualitative *relationships* that are possible from the perspective of the causal influencing of the consequent node. The hypergraph of Figure-2 is a *domain model* in which the reaction variables are used as hyperedges and not nodes as is the case in Figure-1. This is done because reaction variables represent those structural relationships in terms of which we want to hypothesize the smaller situation models.

By saying that a particular hyperedge is the *situation model*, we are in effect hypothesizing that the situation (reaction here) corresponding to the hyperedge is active and all the other situations (reactions) are inactive. Using the joint probability distribution associated with the hyperedge and the observed events, we can determine the probability value for the event of interest in the context of the *situation model*. This can be done in exactly the same way as in [6, Pearl 88] but should be contrasted with the model of Figure-1 in which all probabilistic inferences are in the context of the *domain model*.

To answer the above type-2 query we need to find that *situation model* (hyperedge) from among the possible four in whose context the inferred probability for the variable *hl* is such that it has the minimum expected value.

18.4 Comparing the Models

Both the models discussed above have been derived from the same database of observations and can be related to the r.h.s. expression of equation #1 stated above. The node hl in Figure-2 is the consequent node of four different hyperedges. When we hypothesize that the situation being modeled is such that in it only R_2 is active then we use the joint probability distribution associated with the (reaction) hyperedge R_2 alone. Therefore, the consideration of other reactions being active in the given soil condition and their effect on the variable hl is completely excluded from the hypothesized situation model.

A more general form of the above model building exercise can be viewed as follows. In a graph of the type of Figure-1 resulting from the r.h.s. of expression #1, all the nodes are examined one at a time. Instead of associating a unique conditional probability function with it, as would be the case in a Bayesian network, it is made the consequent node of a number of distinct hyperedges each corresponding to a qualitative relationship of the domain. This step, however, may require us to do away with those variables in terms of which the hypothesization of situation models is desired to be done, and these variables get associated with the hyperedges. In the above example the reaction variables were such variables. After applying this transformation to every node of the graph of the type of Figure-1, we obtain a hypergraph which is the *domain model* of type-2.

When answering a query of the type-2 objective, we select a unique hyperedge for each consequent node of the graph, thus hypothesizing one of the possible situations. In the context of the complete hypergraph (= domain model) the problem of finding a situation model is the same as that of finding a sub-hypergraph of the complete graph. The problem of finding the optimal situation model is the same as that of finding the optimal subhypergraph of the hypergraph.

In the case discussed above the knowledge about the qualitative relationships (reactions) is available from the domain's theory. This knowledge is then embedded by us into the model along with the probabilistic knowledge available from the data. The structural primitives in terms of which a Bayesian network is constructed are the relationships of probabilistic dependence and independence. In the models of type-2 the qualitative relationships of the domain are the primitives in terms of which the models are structured. In case the knowledge of such qualitative relationships is not available from the domain it may be possible to infer it from the data itself to reasonably good extent. We have developed an entropy minimization based algorithm for learning the hyperedge structures from data.

18.5 Learning Qualitative Relationships

Algorithms for inducing relationships from databases have been presented in [2, 4, 9] and all these methods concern themselves with discovering the relationships of probabilistic dependence and independence so that graphs representing Bayesian networks for a domain could be induced.

We now examine a way for inducing structural relationships for the type-2 models from empirical data. We have tested this algorithm on a number of different databases with reasonable success. In the biodegradation example it is possible that we may only be able to measure the soil attributes and the half-lives of compounds and it may not be possible to know the active reaction. From a database of many recorded cases of biodegradation we attempt to identify the clusters of different behaviors from the perspective of each consequent node of the model graph.

Our assumption and hope is that such boundaries would be an approximation of the distinct qualitative relationships affecting the consequent node in question. Our algorithm for learning these boundaries is modeled after the following idea from Information Theory. Consider the situation in which some messages are transmitted across a noisy channel. Let us say that x_i's represent the messages sent from the source and the y_j's represent the signals observed at the receiving end. An observer can only see the received signals and the sent messages remain hidden from him. In an ideal situation when there is no noise introduced by the channel we should observe the signal y_k whenever the message x_k is sent. In a noisy channel, however, it is possible to observe different received signals for the same sent message. The following conditional entropy measures the uncertainty of the sender about the signal that will be received whenever a message is sent.

$$\sum_{i=1}^{K} \sum_{j=1}^{K} P[x_i] * P[y_j \mid x_i] * Log(P[y_j \mid x_i]). \tag{18.2}$$

The analogy between this example and the situation of hidden qualitative relationships and observed attributes is as follows. We equate the qualitative relationships with the *sent messages* and the domain attributes with the *received signals*. We consider the case of a single consequent node v_j of the hypergraph and assume that there are m distinct qualitative relationships in the domain that causally affect it. Each such relationship is represented by a distinct hyperedge h_i having v_j as its consequent node. There are also N binary valued attributes in the domain represented by v_1, v_2,. . . v_n. Each qualitative relationship h_i is a possible message that is sent by a sender, and each v_i represents a possible signal that can be observed at the receiver's end. In the context of the channel through which the effects of various qualitative relationships on v_j are observed, the conditional entropy computed for a database of recorded cases is:

$$E_D = \sum_{i=1}^{m} \sum_{k=1}^{K_j} P[h_i] * P[v_j = v_{jk} \mid h_i] * Log(P[v_j = v_{jk} \mid h_i]). \tag{18.3}$$

Our learning problem is formulated under the assumption that the database is a noiseless and true representation of reality. Therefore, we seek to discover that partitioning of the database cases which minimizes the above conditional entropy. Each of these partitions corresponds to one hyperedge. The conditional entropy value becomes zero when each database case is considered as corresponding to a different partition. We are seeking such distinct partitions to which significantly many database cases may belong. The rest of the cases, belonging to sparsely populated qualitative representations, may be merged into one hyperedge which forms the *rest-of-the-causes* between the antecedent nodes and the consequent node. The partitions thus discovered are considered the identifiable qualitative relationships. The *rest-of-the-causes* partition mentioned above corresponds to the "other unknown causes" and contains all those database cases which do not belong to other partitions.

A similar entropy based technique is used by Quinlan in [10] for inducing efficient decision trees from a database of cases. Our objective and approach are somewhat different. Given a database, Quinlan's method seeks to determine that attribute sequence which partitions the database most efficiently into categories. A similar technique is also employed in [3] to determine the attribute that should be measured to most efficiently partition the candidate space. We seek to partition the database from the perspective of each attribute separately, that is, once for each *term* of equation #1, to determine the identifiable distinct categories in which the variable gets influenced by other variables. Details of this algorithm have been presented in [1].

18.6 Conclusion

Often the primary objective of constructing a model from the data is to model the phenomenon that produces the data. In many graph models of probabilistic knowledge various aspects of these phenomena are represented by variables and edges represent the probabilistic dependencies among them. We have demonstrated one type of reasoning objective that is better served by those models in which some of the aspects of the phenomena correspond to the edges and the hyperedges of the graphs. We have also outlined our method for handling such models for reasoning and learning their characteristics from empirical data.

18.7 References

[1] Raj Bhatnagar and Laveen N Kanal. Structural and Probabilistic Knowledge for Abductive Reasoning, IEEE Transactions on Pattern Analysis and Machine Intelligence, vol. 15, No. 3, pp 233-245.

[2] G. F. Cooper and E. Herskovits. A Bayesian Method for Constructing Bayesian Belief Networks from Databases. *Proceedings of the Seventh Conference on Uncertainty in Artificial Intelligence*, 1991, pp. 86-94.

[3] Johan de Kleer and Brian C. Williams. Diagnosis With Behavioral Modes, *Proceedings of the IJCAI, 1989* Morgan Kaufmann, pp 1324-1330.

[4] E. Herskovits and G. Cooper. An Entropy-Driven System for Construction of Probabilistic Expert Systems from Databases. *Uncertainty in Artificial Intelligence 6*, P. P. Bonissone, M. Henrion, L. N. Kanal and J. F. Lemmer (eds), North Holland, 1991, pp 117 - 125.

[5] S. L. Lauritzen and D. J. Spiegelhalter, Local Computations with Probabilities on Graphical Structures and their Application to Expert Systems, *The Journal of Royal Statistical Society, Series B,* vol. 50, No. 2, pp 157-224, 1988.

[6] Richard E. Neapolitan *Probabilistic Reasoning in Expert Systems*, John Wiley and Sons, Inc. 1990.

[7] Judea Pearl. Fusion, Propagation, and Structuring in Bayesian Networks, *Artificial Intelligence* vol. 29, 1986, pp 241-288.

[8] Judea Pearl. *Probabilistic Reasoning in Intelligent Systems: Networks of Plausible Inference.*, Morgan Kauffman publishers, 1988.

[9] J. Pearl, T. S. Verma. A Theory of Inferred Causation. *Principles of Knowledge Representation and Reasoning: Proceedings of the Second International Conference*, Morgan Kaufmann. (April, 1991).

[10] J. R. Quinlan. Induction of Decision Trees, *Machine Learning* vol. 1, number 1, pp 81-106, 1986.

Part III

Causal Models

19
Causal inference in artificial intelligence

Michael E. Sobel

University of Arizona, Tucson, AZ. 85721

ABSTRACT In recent years, artificial intelligence researchers have paid a great deal of attention to the statistical literature on recursive structural equation models and graphical models, especially to the literature on directed acyclic independence graphs. Some researchers have argued that graphical models are useful in studying causal relationships among variables, and that causal relations can be read off the conditional independence graph. Some authors have even attempted to use the data to derive both a causal ordering and causal relationships. A number of others have refrained from imparting a causal relationship to graphical models. This paper discusses the subject of causation and causal inference in general, and then applies the discussion to the case of graphical models. Specifically, a distinction between causation and causal inference is made; the inferential activity consists of making statements about causation. Using a counterfactual account of the causal relation that derives from work in statistics, I show that the usual types of inferences made in the structural equations literature and the graphical models literature do not square with this account; the results are relevant because a number of authors have claimed otherwise.

19.1 Introduction

In recent years, as artificial intelligence researchers have paid increasing attention to the statistical literature on graphical models, (especially the literature on directed independence graphs), and to the related literature on recursive structual equation models (e.g., Wright [39], Duncan [8]), long-standing issues concerning the proper interpretation of these models have also been taken up.

In a recursive structural equation model, a finite set of variables is partitioned into subsets of exogenous and endogenous variables. The exogenous variables form a block, that is, a group of variables that are regarded (sometimes inappropriately) as unordered with respect to one another. The exogenous variables are allowed to be arbitrarily associated, and in the path diagram corresponding to the model, associations among exogenous variables are represented by curved bi-directional arrows. Finally, the exogenous variables are also taken to be causally prior to the endogenous variables. The endogenous variables are partitioned into one or more blocks, where blocks are ordered with respect to other blocks; for example, for any variable X in block A and any variable Y in block B, one of the variables is prior to the other. A particular endogenous variable is allowed to depend on the exogenous as well as prior endogenous variables, but not on variables within the same block. Such directed dependencies are represented in the path diagram by directed arrows (called "paths" in the literature) from prior variables to subsequent variables, or when a subsequent variable does not "depend" upon the prior variable, by the absence of a path. Finally, in a recursive structural equation model, the error terms are independent, and this feature is represented in the path diagram by the absence of curved bi-directional arrows linking the error terms. Thus, the independence graph for a recursive structural equation model is an example of a multivariate regression chain graph (Cox and Wermuth [6]), as opposed to a block

[0]The invited paper at the Fourth International Workshop on AI and Statistics, 1993.

[1]*Selecting Models from Data: AI and Statistics IV.* Edited by P. Cheeseman and R.W. Oldford.

recursive independence graph (Lauritzen and Wermuth [17]).

In the case where each endogenous variable constitutes a block, the set of endogenous variables is completely ordered, and the independence graph is a univariate recursive regression graph (Cox and Wermuth [6]), which is essentially the same as a directed acyclic graph. Note also that the recursive structural equation model corresponding to a multivariate regression chain graph is probabilistically equivalent to any univariate recursive regression graph derived from the multivariate graph by imposing any ordering on the endogenous variables within blocks, disallowing directed arrows between such variables in the new graph, and leaving the graph otherwise intact. However, in keeping with the view that the data should not generally be allowed to dictate a causal ordering of variables (at least not when substantive scientific questions are at stake), it is useful to distinguish between these probabilistically equivalent representations.

Researchers who use recursive structural equations associate each path with a real number (parameter) that describes the relationship of a prior variable to a subsequent variable. The parameters, which are ordinary regression coefficients, are often endowed with a "causal" interpretation. In particular, if X has a path to Y, i.e., there is a directed arrow from X to Y, the parameter linking X to Y is nonzero, and it is customary to say that X causes Y directly, with direct effect equal to the value of the parameter. When a parameter is zero, corresponding to the absence of a path, it is customary to say that the prior variable does not directly cause the subsequent variable, or equivalently, that the direct effect is 0 (e.g., see Sobel [30, 31]). The direct effect is not the only effect of interest, for while it describes the relationship of an independent variable to a response variable, net of all variables prior to the response, a description of the relationship of the independent variable to the response net only of variables prior to and concurrent with the cause is often desired. This regression coefficient, called a reduced form coefficient in econometrics, is often called a total effect (Sobel [32]) in psychology and sociology.

Recursive structural equation models with normally distributed errors are closely related to graphical models for directed independence graphs. Using elementary properties of the normal distribution, the direct effect of a prior variable X on a subsequent variable Y is 0 if and only if X and Y are independent, conditional on all other variables prior to Y. Thus, a direct effect of 0 is equivalent to a missing edge between the vertices corresponding to X and Y in a directed independence graph (see Whittaker [38] for a discussion of these graphs). In the graphical models literature, some authors use the same criterion (or highly related criteria) to state that X does not directly cause Y (Kiiveri and Speed [15], Kiiveri, Speed and Carlin [16], Pearl and Verma [21]). However, not all authors endow these conditional independence and dependence relationships with causal interpretations. For example, Cox and Wermuth [6] argue that while such relationships may be useful for the subsequent development of causal explanations, these relationships should not be equated with causation per se. Holland [13] and Sobel [32, 34, 35], who discuss structural equation models, adopt a counterfactual view (subsequently discussed) of the causal relationship, and conclude that causal interpretations of the parameters (so called direct effects) of structural equation models are often unwarranted. In the literature on graphical models, researchers have not paid a great deal of attention to causation that is not entirely direct. However, for the case when the independent variable is exogenous, a total effect of 0 corresponds to a missing edge in a multivariate regression graph.

The aim of this paper is to introduce to artificial intelligence researchers, who are already familiar with graphical models, a counterfactual account of the causal relation and a corresponding approach to causal inference that derives from statistical work (Cox [4],Neyman [18], Kempthorne [14], Rubin [25, 26, 27, 28]) on randomized and non-randomized studies. Both the account and the approach are relevant to researchers in artificial intelligence, some of whom have argued (e.g., Pearl and Verma [20]) that their results square with a manipulative account of causation. The approach to causal inference that derives from the counterfactual account (which subsumes a manipulative account) is then compared with the inferential strategies previously described, leading to the conclusion that such strategies do not square with a counterfactual account.

Throughout, basic ideas are emphasized using simple settings; for a more formal approach and proof of the results stated here, see Sobel [33]. No consideration is given to the question, studied by some artificial intelligence researchers, of how the data might be used to establish the direction of causation. For criticism of the notion that the data can be used in such a way, see Duncan [8] and Wermuth and Lauritzen [37], for example.

The paper proceeds as follows. Section two motivates and introduces the inferential strategy that derives from the counterfactual account, and compares, by way of example, this approach with the approach to causal inference discussed above. Section 3 formalizes the approach that derives from the counterfactual account and gives several results due to Sobel [33] that pin down the relationship between it and approaches taken in previous work on structural equation models and graphical models, as discussed above.

19.2 Causation and causal inference

The term "causal inference" is used in this paper to refer to the act of using evidence to make inferences about causation. The distinction between "causal inference" and causation, while generally overlooked, is critical, for it makes causation logically independent of the inferential strategies used to assess causal relationships. This is important, for just as different inferential strategies may be employed, different notions of causation may be held. It follows readily that the suitability of a particular inferential strategy hinges on the notion of causation under consideration, and different notions of the causal relation privilege alternative inferential strategies. Thus, I briefly discuss several notions of the causal relation. For more extended treatments see, for example, Cox [5], Holland [12], Sobel [34, 35].

Causation has often been equated with universal predictability. This is essentially the position taken by Hume. Under this view, the causal relation is deterministic at the general, as opposed to singular level, and probabilistic relations are not causal. Various attempts to weaken the universality requirement and introduce probabilistic relations have been made. For example, Basmann [1] draws upon a variant of the "same cause, same effect" notion, popular in mid-twentieth century phsyics, that equates causation with the idea that a closed system, under a given set of initial conditions, runs through the same sequence of states. In Basmann's formulation, the sequence is not determinate, for on different trials, with the same initial conditions, the sequence can vary somewhat.

In the context of deterministic causation, Hume's position has been criticized by those who argue that not all stable relationships exhibiting temporal precedence are causal. A relationship between a putative cause X and an effect Y may merely reflect the operation of a variable (called a common cause) that is prior to and affects both X and Y. Concern with this issue, in a non-deterministic context, seems to be the impetus for much of the work in the literature on probabilistic causation. Here (e.g., Suppes [36]), the essential idea is to classify an association between a putative cause and an effect as causal if conditioning on variables prior to the cause does not remove this association, and to define the relationship as non-causal (spurious) if the putative cause and the effect are conditionally independent, given the prior variable(s). In other words, causation is now equated with genuine predicability, as opposed to merely spurious predictability. Notice the similarity between the foregoing definition of spuriousness and the idea (in recursive structural equation models) that a prior variable X does not directly cause a response Y when the direct effect in the model is 0. Similar ideas are also featured in the econometric literature on Granger causation (Granger [10, 11], Geweke [9]) and in some of the literature on graphical models. Zellner [40], who notes that the essence of Granger causality is predictability, defines causation as predictability according to a law. Under this view, the causality tests featured in the economic literature only make sense when substantive considerations are introduced.

In the foregoing literature on probabilistic causation, causation (with the exception of Zellner) is more or less operationally defined by the criteria used to test for its presence or absence, i.e., causal inference and causation are apparently conflated. But many authors also supply informal motivation for these definitions, and it is natural to regard this motivation as a key to the notion of causation espoused. The definitions of spuriousness, etc, are consequently best regarded as part of the inferential strategy. In this vein, economists often assume that causes (in the Granger sense) are akin to manipulable inputs with the effect as output (Sims [29],Geweke [9]). Similarly, sociologists and psychologists typically interpret the "effects" in structural equation models as the unit (or average) amount of change in the effect variable due to manipulatively increasing the level of the cause one unit, while "holding constant" other relevant variables (Sobel [30]). More recently, in the artificial intelligence literature, Pearl and Verma [20] have argued that their inferential strategies square with the foregoing interpretations. For further material on this manipulative account of causation, see Sobel [34, 35].

Under the view of causation above, justification for the inferential step hinges on the argument that the effect variable does not change when a false cause is actually manipulated, even if these two variables are correlated. Further, if the true cause were known, conditioning on it's value would render the false cause and the effect independent. Sometimes this true cause is viewed as a common cause of both the false cause and the effect (Reichenbach [23]). By employing this inferential strategy, one presumably obtains the same inferences that would be obtained from a randomized experiment. For an argument along these lines, see Cartwright [2].

The idea that causes are events that are manipulated has a long history (Collingwood [3]). Under this view, events that precede the effect but are not results of manipulations are not causes per se, but part of the causal background; the background includes events prior to the manipulation as well as events that are subsequent, and therefore possibly affected by the manipulation. Consequently, the causal background and foreground are relative: in a particular context, the manipulation produces this partition. Thus, if a variable X caused (in some sense) a variable Y, with W as background, this refers to the setup in which X is manipulated and W is passively observed. In some other setup, if W were manipulated, and X passively observed, W would be the cause and X a part of the causal background.

The notion of the cause as a manipulation is compatible with either a deterministic or probabilistic treatment of causation, and it is compatible with either a singular or general account of the causal relation. The account espoused here is both singular and deterministic, but it is easily modified to be singular and probabilistic (Sobel [33]). A key element of a manipulative account is that because an agent assigns the value of the cause to the singular units, the unit can take on (before the assignment) any of the possible values of the cause. While the value x of the cause may be assigned to a singular unit, because the value x' could have been assigned, the manipulative account is readily fitted with the more general view (in some accounts of causation) that a causal relationship should sustain a counterfactual conditional, in that the latter view requires comparison between what occurs when the cause has value x and what would have occured had the cause been assigned value x'. Here, the manipulation, which in substantive research must be described, might best be viewed as giving content to a counterfactual that might otherwise remain vague (Sobel [35]).

The previous remarks on probabilistic causation indicate that the notion of the causal relation espoused in that literature (at least some portion thereof) is both manipulative and counterfactual. This raises the question of whether or not the inferential criteria used in that literature to assess causal relations support, as claimed by some, a manipulative account. The following example, which is used to informally introduce such an account, indicates the answer is no.

To keep matters simple, consider a simple medical experiment, in which the units (subjects) of a population $\mathcal{P}$ are either treated with a particular drug ($X = 1$) or not ($X = 0$), and their health status Y (good = 1, not good = 0) subsequently assessed.

Under a counterfactual account, each untreated unit in $\mathcal{P}$ could have been treated (but was not) and each treated unit could have been left untreated (but was not). Thus, it is plausible, in advance of treatment assignment (assignment by the investigator to either the treatment group or the control group) to associate two values with each unit, the value of the response when the unit is not treated (hereafter Y_0), and the value of the response when the unit is treated (hereafter Y_1). At this unit level, under a counterfactual account, comparison of the values of Y_0 and Y_1, denoted y_0 and y_1, respectively, evidences causation (lack of causation) when the unit effect $y_1 - y_0 \neq 0$ ($y_1 - y_0 = 0$).

Table 1 displays 16 patterns associated with the medical experiment. The patterns are created from the columns of the table labeled Y_0, Y_1, X and Z as follows: for any unit, there are two possible values of the response in the absence of treatment (Y_0), two possible values under treatment (Y_1), and two possible assignments (X). In addition, the investigator measures a binary variable Z on each unit, for a total of 16 patterns. Each unit must fit one and only one of these patterns, and the probability associated with each pattern is given in the last column of the table. For example, in pattern 1a, which occurs with probability P_{1a}, all the units have the value 0 on the binary covariate Z. Further, all units with this pattern have poor health whether treated or not, and thus the unit effect is 0, evidencing no causation. In addition, these units do not receive the treatment ($x = 0$). Thus, only the value of the response y_0 (underlined) is actually observed; y_1 is not observed, for this would require $x = 1$. The fact that causal inference requires comparison of two (in this case) responses on each unit, when in fact only one response can be observed, is called the fundamental problem of causal inference by Holland [12]. It is important to note that this problem raises no difficulties for a counterfactual account of causation, but only for inferences based on this account (as we shall see).

For the moment, attention focuses on the eight patterns obtained when the variable Z is collapsed; thus patterns 1a and 1b are combined into pattern 1, with probability $P_1 = P_{1a} + P_{1b}$, etc. The role of Z is discussed later.

Returning to the account, interest may center on the average effectiveness of treatment; this criterion, which is of little interest to a patient (who wants to know the efficacy of treatment in his/her case but cannot know this by virtue of the fundamental problem of causal inference), is typically of greater interest to scientists and policy makers. The average effectiveness is just the

TABLE 19.1: Possible patterns for the simple medical experiment

Pattern	Y_0	Y_1	X	Z	Unit Effect	Probability
1a.	$\underline{y_0 = 0}$	$y_1 = 0$	0	0	0	P_{1a}
1b.	$\underline{y_0 = 0}$	$y_1 = 0$	0	1	0	P_{1b}
2a.	$y_0 = 0$	$\underline{y_1 = 0}$	1	0	0	P_{2a}
2b.	$y_0 = 0$	$\underline{y_1 = 0}$	1	1	0	P_{2b}
3a.	$\underline{y_0 = 0}$	$y_1 = 1$	0	0	1	P_{3a}
3b.	$\underline{y_0 = 0}$	$y_1 = 1$	0	1	1	P_{3b}
4a.	$y_0 = 0$	$\underline{y_1 = 1}$	1	0	1	P_{4a}
4b.	$y_0 = 0$	$\underline{y_1 = 1}$	1	1	1	P_{4b}
5a.	$\underline{y_0 = 1}$	$y_1 = 0$	0	0	-1	P_{5a}
5b.	$\underline{y_0 = 1}$	$y_1 = 0$	0	1	-1	P_{5b}
6a.	$y_0 = 1$	$\underline{y_1 = 0}$	1	0	-1	P_{6a}
6b.	$y_0 = 1$	$\underline{y_1 = 0}$	1	1	-1	P_{6b}
7a.	$\underline{y_0 = 1}$	$y_1 = 1$	0	0	0	P_{7a}
7b.	$\underline{y_0 = 1}$	$y_1 = 1$	0	1	0	P_{7b}
8a.	$y_0 = 1$	$\underline{y_1 = 1}$	1	0	0	P_{8a}
8b.	$y_0 = 1$	$\underline{y_1 = 1}$	1	1	0	P_{8b}

average of the unit effects:

$$P_3 + P_4 - (P_5 + P_6) = Pr(Y_1 = 1) - Pr(Y_0 = 1) = E(Y_1) - E(Y_0), \tag{19.1}$$

and when this is 0, it would be natural to say that X does not cause Y in mean in $\mathcal{P}$. Note also that the average effect is 0 if and only Y_0 and Y_1 have the same distribution, in which case it would also be natural to say X does not cause Y in distribution in $\mathcal{P}$.

Now the inference step is discussed, where for the sake of simplicity, it is supposed that data are available for all units in the population. Because, for any given unit, the response is observed only under one of the treatments, the data on which inferences are based are pairs (x, y), where $y = y_0$ if $x = 0$, and $y = y_1$ if $x = 1$. Since the value of the response under the treatment that was not given is not known, this variable is collapsed over, precluding inference about the unit effects (unless additional assumptions are made). For example, the data can be used to discover that a unit has pattern 1 or 3, but not which of these two patterns.

From the bivariate data, the conditional probabilities are:

$$Pr(Y = 1 \mid X = 1) = Pr(Y_1 = 1 \mid X = 1) = (P_4 + P_8)/(P_2 + P_4 + P_6 + P_8), \tag{19.2}$$

$$Pr(Y = 1 \mid X = 0) = Pr(Y_0 = 1 \mid X = 0) = (P_5 + P_7)/(P_1 + P_3 + P_5 + P_7). \tag{19.3}$$

These probabilities are often used (e.g., in both randomized experiments and quasi-experimental studies) in place of $Pr(Y_1 = 1)$ and $Pr(Y_0 = 1)$, respectively, in (19.1). As such, the relationship between these sets of quantities is of interest:

$$Pr(Y_1 = 1) = Pr(Y_1 = 1 \mid X = 1)Pr(X = 1) + Pr(Y_1 = 1 \mid X = 0)Pr(X = 0). \tag{19.4}$$

The data can be used to compute the marginal probability distribution of X and the conditional probability $Pr(Y_1 = 1 \mid X = 1)$, but as Y_0 is always observed when $X = 0$, the data contain no information about $Pr(Y_1 = 1 \mid X = 0)$. Similarly, the data contain no information about the term $Pr(Y_0 = 1 \mid X = 1)$ appearing in the analogous expression for $Pr(Y_0 = 1)$. Therefore, the data cannot be used to compute (19.1) unless additional knowledge is brought to the study. However, additional subject matter knowledge is typically inadequate for ascertaining the value of these missing probabilities. Here is where the treatment assignment mechanism comes into play: if units are assigned to treatments at random, the distribution of the outcomes Y_0 and Y_1 will be the same in treatment and control groups, i.e., $Pr(Y_1 = 1 \mid X = 1) = Pr(Y_1 = 1 \mid X = 0)$, enabling computation of (19.1).

Several points are in order. First, the assumption of random assignment is:

$$X \perp\!\!\!\perp Y_j, j = 0, 1, \tag{19.5}$$

where the symbol "$\perp\!\!\!\perp$", due to Dawid [7] is used to denote independence; similarly "$\not\perp\!\!\!\perp$" is used to denote dependence. This assumption should not be confused with the statement $X \perp\!\!\!\perp Y$. In general X and Y will be associated, even when treatment assignment is random. Second, if X causes Y in distribution in $\mathcal{P}$, this does not entail that X and Y are associated (Sobel [33]), i.e., under a counterfactual account, causation does not imply association, despite some claims to the contrary (e.g., Pearl, Geiger and Verma [19]). Nor does (X, Y) association imply causation in distribution in $\mathcal{P}$.

Investigators typically measure variables in addition to X and Y, e.g., the binary variable Z in Table 1. Suppose Z is a covariate (a measurement taken prior to X) such as sex of subject.

In a randomized experiment, one might use Z to improve the precision with which (19.1) is estimated from sample data. Alternatively, one might wish to examine conditional average effects; for example, if women are coded 0,

$$Pr(Y_1 = 1 \mid Z = 0) - Pr(Y_0 = 1 \mid Z = 0) = (P_{3a} + P_{4a} - (P_{5a} + P_{6a}))/(\sum_{r=1}^{8} P_{ra}) = E(Y_1 \mid Z = 0) - E(Y_0 \mid Z = 0) \quad (19.6)$$

is the average effectiveness of treatment for women, and when this is 0 it is natural to say that X does not cause Y in mean (distribution) in the subpopulation of women.

Inferences about the conditional effects are based on triples (x, y, z), where $y = y_0$ if $x = 0$, $y = y_1$ if $x = 1$, and $z = 0$ or 1. As before, the unit effects cannot be ascertained; for example, the data can indicate that a woman has pattern 1a or 3a, but not which of these two patterns. From the data, the conditional probabilities

$$Pr(Y = 1 \mid X = 1, Z = 0) = Pr(Y_1 = 1 \mid X = 1, Z = 0) = (P_{4a} + P_{8a})/(P_{2a} + P_{4a} + P_{6a} + P_{8a}), \quad (19.7)$$

$$Pr(Y = 1 \mid X = 0, Z = 0) = Pr(Y_0 = 1 \mid X = 0, Z = 0) = (P_{5a} + P_{7a})/(P_{1a} + P_{3a} + P_{5a} + P_{7a}), \quad (19.8)$$

are readily obtained. These are often used (e.g., in both randomized and quasi-experiments) in place of $Pr(Y_1 = 1 \mid Z = 0)$ and $Pr(Y_0 = 1 \mid Z = 0)$ in (19.6). As such, the relationship between the quantities in (19.6) and the quantities in (19.7) and (19.8) are of interest:

$$Pr(Y_1 = 1 \mid Z = 0) = Pr(Y_1 = 1 \mid X = 1, Z = 0)Pr(X = 1 \mid Z = 0) + Pr(Y_1 = 1 \mid X = 0, Z = 0)Pr(X = 0 \mid Z = 0). \quad (19.9)$$

The conditional probability distribution of X, given Z, and the conditional probability $Pr(Y_1 = 1 \mid X = 1, Z = 0)$ can be computed from the data, but as Y_0 is always observed when $X = 0$, the data contain no information about $Pr(Y_1 = 1 \mid X = 0, Z = 0)$. Similarly, the data contain no information about the term $Pr(Y_0 = 1 \mid X = 1, Z = 0)$ in the analogous expression for $Pr(Y_0 = 1 \mid Z = 0)$.

Thus, the data cannot be used to compute (19.6) unless additional knowledge is brought to the study. As before, subject matter knowledge is typically inadequate for ascertaining the value of these missing probabilities. However, if units are randomly assigned to treatments, $X \perp\!\!\!\perp (Y_j, Z)$, $j = 0, 1$, and therefore $(X \perp\!\!\!\perp Y_j) \mid Z$, i.e.,

$$Pr(Y_j = 1 \mid X = j, Z = 0) = Pr(Y_j = 1 \mid Z = 0) \quad (19.10)$$

for $j = 0, 1$, allowing computation of (19.6). Equation (19.10) shows that random assignment is sufficient, but not necessary for computation of (19.6): the key to computing this quantity is the condition $(X \perp\!\!\!\perp Y_j) \mid Z$, $j = 0, 1$, as stated in (19.10). Here, random assignment holds conditionally, within levels of Z, but Z itself may be involved in the treatment assignment process, e.g., different fractions of men and women are assigned to the control and treatment groups.

The importance of the conditional random assignment assumption (which is essentially the same as the assumption of strongly ignorable treatment assignment in Rosenbaum and Rubin [24]) cannot be overstated. As previously demonstrated, in the absence of random assignment, the (x, y) pairs cannot be used to compute (19.1), unless other assumptions are made. In non-randomized studies, an investigator can attempt to name and measure the covariates that are sufficient for conditional random assignment, even though the study itself does not employ randomization. If treatment assignment is random, given the measured covariates, the investigator can then compute (19.6); if primary interest is in (19.1), this quantity is the weighted average (weighted by the distribution of the covariates) of the conditional average effects. However, if treatment assignment is not conditionally random, given the covariates measured, then (19.10) does not hold and the triples (x, y, z) cannot be used to compute either (19.6) or (19.1).

In non-experimental studies, while Z is sometimes thought of as a covariate that makes treatment assignment random, in many cases, Z is simply a variable that is deemed somehow "relevant"to the analysis. The manner in which Z is "relevant" is typically not made explicit, but discussions of the role of Z do not suggest it is viewed in the manner above. For example, when the causal relationship between X and Y is of interest, investigators will often statistically control for the prior variable Z, which is allegedly "relevant", and sometimes held to be causally "relevant", but the investigators never say how this is so; continuing, when the (X, Y) relationship vanishes, conditioning on Z, X is held to be a spurious cause of Y. Note that users of structural equation models and graphical models often employ this inferential strategy, albeit in a slightly more general setup (discussed in section 3).

Several additional points are in order. First, the assumption of conditional random assignment should not be confused with the statement $(X \underline{\|} Y) \mid Z$. Second, using (19.6) and (19.9), examples where the average conditional effects are 0, meaning X does not cause Y in distribution in either subpopulation (hence that X does not cause Y in distribution in $\mathcal{P}$), but $(X \not\!\underline{\|} Y) \mid Z$ are readily constructed by suitably choosing $Pr(Y_1 = 1 \mid X = 0, Z = j)$ for $j = 0, 1$. This means that the absence of causation (on average or in distribution) does not imply conditional independence (the spuriousness criterion). Similarly, causation (on average or in distribution) does not imply the absence of spuriousness. These results suggest that the inferential strategies used in the literature on probabilistic causation do not yield inferences that match up with a counterfactual account of the causal relation. For further details, see Sobel [33].

19.3 Generalization

This section briely formalizes and extends the introductory material in section 2 to the general case, after which applications to recursive structural equation models and graphical models are given. The setup here is similar to that in Pratt and Schlaifer [22]; these authors extended the setup in Rubin [25, 26, 27, 28]. For a slightly more general setup, see Sobel [33]. The key ingredients are: 1) the population $\mathcal{P}$, with units i (as in section 2); 2) the cause X, a vector taking values x in a set Ω_1 with more than one value (in section 2, $\Omega_1 = \{0, 1\}$) 3) the response Y, a vector taking values y in Ω_2 (in section 2, $\Omega_2 = \{0, 1\}$); 4) a function F defined on the Cartesian product $\mathcal{P} \times \Omega_1$, with values $F(x, i) = y_{xi}$. For fixed x, as i ranges over $\mathcal{P}$, the values of F generate the random vector Y_x (in section 2, binary variables Y_0 and Y_1 were considered). Sobel [33] defines four types of causation: 1) X causes Y for unit i if there are distinct values x and x' such that $y_{xi} \neq y_{x'i}$; otherwise X does not cause Y; 2) X causes Y in $\mathcal{P}$ if X causes Y

for some unit; otherwise X does not cause Y in $\mathcal{P}$; 3) X causes Y in distribution in $\mathcal{P}$ if there are distinct values x and x' such that $D(Y_x) \neq D(Y_{x'})$, where $D(Y_x)$ denotes the probability distribution associated with Y_x; otherwise X does not cause Y in distribution in $\mathcal{P}$; 4) X causes Y in mean in $\mathcal{P}$ if there are distinct values x and x' such that both $E(Y_x)$ and $E(Y_{x'})$ exist and are unequal; if, for all x, $E(Y_x)$ exists and is constant, X does not cause Y in mean in $\mathcal{P}$.

The foregoing definitions are readily extended to the case where a random vector Z is measured by replacing $\mathcal{P}$ with subpopulations $\mathcal{P}_z$. Thus, for example, X causes Y in distribution in $\mathcal{P}_z$ if there are distinct values x and x' such that $D(Y_x \mid Z = z) \neq D(Y_{x'} \mid Z = z)$; otherwise, X does not cause Y in distribution in $\mathcal{P}_z$. Similarly, X causes Y in distribution in $\mathcal{P}_Z$ if for some z, X causes Y in distribution in $\mathcal{P}_z$ (ignoring sets of measure 0). Causation in mean in subpopulations of $\mathcal{P}$ is handled similarly.

Parallelling section 2, attention centers first on inferences about whether X causes Y in distribution in $\mathcal{P}$. In the simplest case, such inferences are often based on the (x, y) pairs, where if $x = \tilde{x}$, the value $y = y_{\tilde{x}}$ is observed. From these pairs, the distributions

$$D(Y \mid X = x) = D(Y_x \mid X = x) \tag{19.11}$$

can be calculated, for all $x \in \Omega_1$ (compare (19.11) with (19.2) and (19.3)); these distributions are then used in place of $D(Y_x)$ to infer whether or not the (X, Y) relationship is causal. If treatment assignment is random, i.e., $X \| Y_x$ for all $x \in \Omega_1$ (compare with (19.5) in section 2), $D(Y \mid X = x) = D(Y_x)$ for all x and the observed data can be used to assess whether or not X causes Y in distribution in $\mathcal{P}$; if treatment assignment is not random, the observed data will yield inferences that are misleading.

For the example in section 2, causation in distribution and causation in mean were equivalent. In general, however, causation in mean implies causation in distribution, but the converse is not true. The absence of causation in distribution implies the absence of causation in mean, but again, the converse is not true. In the statistical literature, investigators have focused on causation in mean; here it is usual to examine the average effect (compare with (19.1)) of x vs. x', where x and x' are arbitrary elements of Ω_1:

$$E(Y_x) - E(Y_{x'}). \tag{19.12}$$

In practice, investigators often choose x' to be a substantively meaningful anchor point, allowing x to take values in $\Omega_1 - \{x'\}$. Also, in some instances, investigators will not be interested in the mean difference as a measure of effect, but this will not be taken up here.

Inferences about causation in mean are typically obtained by using the observed (x, y) pairs to calculate $E(Y \mid X = x) = E(Y_x \mid X = x)$, where E denotes expected value, and using these conditional expectations in place of $E(Y_x)$. When treatment assignment is random, $E(Y \mid X = x) = E(Y_x)$, justifying this inferential strategy, but when treatment assignment is not random, in general $E(Y \mid X = x) \neq E(Y_x)$.

Even in randomized experiments, it is typical for researchers to measure additional variables Z (here Z typically contains variables Z^* prior to the cause and variables Z^{**} subsequent to the cause, but prior to the effect). Inclusion of these variables can yield estimates of (19.12) that are more precise than the usual sample estimates based on (x, y) pairs only. Additionally, investigators often want to know if X causes Y in distribution in $\mathcal{P}_Z$ (or $\mathcal{P}_{Z^*}$), and how the average effects vary over theoretically relevant subpopulations: for example, the conditional average effect of x vs. x' at $Z = z$ is (compare with (19.6)):

$$E(Y_x \mid Z = z) - E(Y_{x'} \mid Z = z). \tag{19.13}$$

Inferences about the effects of interest are typically based on triples (x, y, z^*) (or (x, y, z)). Consider first the case where inferences are based on (x, y, z^*). >From the data, for all x and z^*, the conditional distributions $D(Y \mid X = x, Z^* = z^*) = D(Y_x \mid X = x, Z^* = z^*)$ and the conditional expectations $E(Y \mid X = x, Z^* = z^*) = E(Y_x \mid X = x, Z^* = z^*)$ can be computed and these are used in place of $D(Y_x \mid Z^* = z^*)$ and $E(Y_x \mid Z^* = z^*)$, respectively. This is akin to the strategy considered in section 2, where (19.7) and (19.8) are used to make inferences about (19.6). If treatment assignment is random, conditionally on Z^*, i.e.,

$$(X \perp\!\!\!\perp Y_x) \mid Z^*, \tag{19.14}$$

$D(Y \mid X = x, Z^* = z^*) = D(Y_x \mid Z^* = z^*)$, and the observed data can be used to make inferences about causation in causation in distribution in $\mathcal{P}_{\mathcal{Z}^*}$; but if treatment assignment is not random, conditionally on Z^*, the observed conditional distributions will yield misleading inferences about $D(Y_x \mid Z^* = z^*)$.

In the literature on probabilistic causation, and in both theoretical and empirical work in a number of disciplines (e.g., economics, sociology, epidemiology), the (X, Y) relationship is said to be spurious (i.e, X does not cause Y in some sense) when $(X \perp\!\!\!\perp Y) \mid Z^*$; sometimes Z^* is regarded as a common cause of X and Y. According to some authors, were it possible to conduct a randomized experiment, the spuriousness of the (X, Y) relationship would be evident, but insofar as it is not possible to conduct such an experiment, it is necessary to control for other variables to obtain the correct conclusions. >From the previous material, which parallels the treatment in section 2, it is evident that spuriousness (lack of spuriousness) does not entail the absence (presence) of causation in distrbituion in $\mathcal{P}_{\mathcal{Z}}$. As in section 2, inferences using the observed data typically do not match up with a counterfactual account. However, under essentially the condition of conditional random assignment, the absence of causation in distribution coincides with spuriousness (Sobel [33]).

In the literature on recursive structural equation models (with normal errors) and graphical models, the distributions $D(Y \mid X = x, Z = z)$ are used to infer whether or not the (X, Y) relationship is "directly"causal. In the former literature, the direct effect of X on Y is 0 if and only if for all x and z, $D(Y \mid X = x, Z = z) = D(Y \mid Z = z)$, i.e, $X \perp\!\!\!\perp Y \mid Z$. In the literature on graphical models, under these same conditions X is not a direct cause of Y. In general, if X causes Y directly (in one of the senses above), this does not imply X causes Y in distribution in $\mathcal{P}_{\mathcal{Z}}$ (or $\mathcal{P}$). Similarly, X may not cause Y in distribution in $\mathcal{P}_{\mathcal{Z}}$ (or $\mathcal{P}$), while X causes Y directly. Again, the usual inferences from the observed data do not square with a counterfactual account.

To develop conditions under which the notions of direct causation and causation in distribution coincide, suppose that treatment assignment is random, conditional on Z^*, i.e., $(X \perp\!\!\!\perp Y_x, Z_x^{**}) \mid Z^*$. In conjunction with mild regularity conditions, it can be shown (Sobel [33]) that X is not a direct cause of Y if and only if X does not cause Y in distribution in $\mathcal{P}_{\mathcal{Z}}$.

The foregoing shows that if Z^* is interpreted as a vector of covariates that makes treatment assignment conditionally random, inferences from the observed data square with a counterfactual account. Otherwise, this will not generally be true. Most writers have not interpreted Z^* in the fashion proposed here. In treatments of recursive structural equation models and graphical models, many view the random vector Z as containing other variables relevant to the analysis. However, it is not clear how these variables are relevant, except insofar as not taking these into consideration is held to lead to incorrect inferences, i.e., correlation is not causation. Further, the parameters corresponding to elements of Z in recursive structural equation models are also called direct effects, indicating that (X, Z) is viewed as a cause. Similar remarks apply to the literature on graphical models. Nevertheless, in the literature on recursive structural equation models, parameters are incorrectly interpreted as if they sustain a counterfactual account of the causal relation (Sobel [32]), and in a related vein, some authors in the literature on graphical models have argued that their results support a manipulative account of the causal relation (Pearl and Verma [20]).

19.4 REFERENCES

[1] Basmann, R. L. (1963) "The Causal Interpretation of Non-Triangular Systems of Economic Relations," *Econmtca*, **31**, 439-453.

[2] Cartwright, N. (1989) *Nature's Capacities and their Measurement.* Clarendon Press, Oxford.

[3] Collingwood, R. G. (1940) *An Essay on Metaphysics.* Clarendon Press, Oxford.

[4] Cox, D. R. (1958) *The Planning of Experiments.* Wiley, New York.

[5] Cox, D. R. (1992) "Causality: Some Statistical Aspects", *JRSS-A* **155**, 291-301.

[6] Cox, D. R., and Wermuth, N. (1993) "Linear Dependencies Represented by Chain Graphs", *StatSci* **8**, 204-218.

[7] Dawid, A. P. (1979) "Conditional Independence in Statistical Theory (with discussion)", *JRSS-B* **41**, 1-31.

[8] Duncan, O. D. (1975) *Introduction to Structural Equation Models.* Academic Press, New York.

[9] Geweke, J. (1984) "Inference and Causality in Economic Time Series Models," pp. 1101-1144 in *Handbook of Econometrics*, volume 2, eds. Z. Griliches and M. D. Intriligator, North Holland, Amsterdam.

[10] Granger, C. W. (1969) "Investigating Causal Relations by Econometric Models and Cross-Spectral Methods," *Econmtca*, **37**, 424-438.

[11] Granger, C. W. (1980) "Testing for Causality: A Personal Viewpoint," *Journal of Economic Dynamics and Control*, **2**, 329-352.

[12] Holland, P. W. (1986) "Statistics and Causal Inference (with discussion)," *JASA*, **81**, 945-970.

[13] Holland, P. W. (1988) "Causal Inference, Path Analysis, and Recursive Structural Equation Models (with discussion)," pp. 449-493 in *SocMethd, 1988*, ed. C. C. Clogg, American Sociological Association, Washington, D.C.

[14] Kempthorne, O. (1952) *The Design and Analysis of Experiments.* Wiley, New York.

[15] Kiiveri, H. and Speed, T. P. (1982) "Structural Analysis of Multivariate Data: A Review," pp. 209-289 in *SocMethd, 1982*, ed. S. Leinhardt, Jossey Bass, San Francisco.

[16] Kiiveri, H., Speed, T. P., and Carlin, J.B. (1984) "Recursive Causal Models," *JAstrMA*, **36**, 30-52.

[17] Lauritzen, S. L., and Wermuth, N. (1989) "Graphical Models for Associations Between Variables, Some of Which are Qualitative and Some Quantitative," *AnlsStat*, **17**, 31-57.

[18] Neyman, J.S. (1923) [1990] "On the Application of Probability Theory to Agricultural Experiments. Essay on Principles. Section 9. (with discussion)," *StatSci*, **4**, 465-80.

[19] Pearl, J., Geiger, D., and Verma, T. (1990) "The Logic of influence Diagrams (with discussion)," pp. 67-87 in *Influence Diagrams, Belief Nets and Decision Analysis*, eds. R. M. Oliver and J. Q. Smith, Wiley, New York.

[20] Pearl, J., and Verma, T. S. (1991) "A Theory of Inferred Causation", pp. 441-452 in *Principles of Knowledge Representation and Reasoning: Proceedings of the Second International Conference*, eds. J. A. Allen, R. Fikes and E. Sandewall, Morgan-Kaufmann, San Mateo, CA.

[21] Pearl, J., and Verma, T. S. (1992) "A Statistical Semantics for Causation," *StatComp*, **2**, 91-95.

[22] Pratt, J. W., and Schlaifer, R. (1988) "On the Interpretation and Observation of Laws," *JEconmtx*, **39**, 23-52.

[23] Reichenbach, H. (1956) *The Direction of Time*. University of California Press, Berkeley, CA.

[24] Rosenbaum, P. R., and Rubin, D. B. (1983) "The Central Role of the Propensity Score in Observational Studies for Causal Effects," *Biomtrka*, **70**, 41-55.

[25] Rubin, D. B. (1974) "Estimating Causal Effects of Treatments in Randomized and Non-Randomized Studies," *JEdPsych*, **66**, 688-701.

[26] Rubin, D. B. (1977) "Assignment to Treatment Groups on the Basis of a Covariate," *JEdStat*, **2**, 1-26.

[27] Rubin, D. B. (1978) "Bayesian Inference for Causal Effects: The role of randomization,"*AnlsStat*, **6**, 34-58.

[28] Rubin, D. B. (1980) Comment on "Randomization analysis of experimental data: The Fisher randomization test" by D. Basu, *JASA*, **75**, 591-593.

[29] Sims, C. A. (1977) "Exogeneity and Causal Ordering in Macroeconomic Models," pp. 23-43 in *New Methods in Business Cycle Research: Proceedings from a Conference*, ed. C. A. Sims, Federal Reserve Board, Minneapolis.

[30] Sobel, M. E. (1982) "Asymptotic Confidence Intervals for Indirect Effects in Structural Equation Models", pp. 290-312 in *SocMethd, 1982*, ed. S. Leinhardt, Jossey Bass, San Francisco.

[31] Sobel, M. E. (1987) "Direct and Indirect Effects in Linear Structural Equation Models", *SocMethR*, **16**, 155-176.

[32] Sobel, M. E. (1990) "Effect Analysis and Causation in Linear Structural Equation Models," *Psymtrka*, **55**, 495-515.

[33] Sobel, M. E. (1993) "Causation, Spurious Correlation and Recursive Structural Equation Models: A Reexamination," manuscript.

[34] Sobel, M. E. (1994a) "Causal Inference in Latent Variable Models", forthcoming in *Analysis of Latent Variables in Developmental Research*, eds. A. von Eye and C. C. Clogg, Sage, Los Angeles.

[35] Sobel, M. E. (1994b) "Causal Inference in the Social and Behavioral Sciences", forthcoming in *A Handbook for Statistical Modeling in the Social and Behavioral Sciences*, eds. G. Arminger, C. C. Clogg and M. E. Sobel, Plenum Press, New York.

[36] Suppes, P. (1970) *A Probabilistic Theory of Causality*, North Holland, Amsterdam.

[37] Wermuth, N., and Lauritzen, S. L. (1990) "On Substantive Research Hypotheses, Conditional Independence Graphs and Graphical Chain Models (with discussion)," *JRSS-B*, **52**, 21-72.

[38] Whittaker, J. (1990) *Graphical Models in Applied Mathematical Multivariate Statistics*. Wiley, New York.

[39] Wright, S. (1921) "Correlation and causation," *Journal of Agricultural Research*, **20**, 557-585.

[40] Zellner, A. (1984) "Causality and Econometrics," pp. 35-74 in *Basic Issues in Econometrics*, ed. A. Zellner, University of Chicago Press, Chicago.

20

Inferring causal structure among unmeasured variables

Richard Scheines

Carnegie Mellon University

ABSTRACT Linear structural equation models with latent variables are common in psychology, econometrics, sociology, and political science. Such models have two parts, a measurement model that specifies how the latent variables are measured, and a structural model which specifies the causal relations among the latent variables. In this paper I discuss search procedures for finding a 'pure' measurement model and for using this pure model to determine features of the structural model. The procedures are implemented as the Purify and MIMbuild modules of the TETRAD II program.

20.1 Introduction

Although quite a lot of attention is now being given to the problem of inferring the causal structure governing a set of measured variables,[3] much less has been paid to the problem of inferring the causal structure governing the set of latent, or unmeasured, variables that might affect those that are measured.[4] Models that employ latent variables are common in psychometrics, sociometrics, econometrics, and other domains. Investigators who use psychometric or sociometric tests or questionnaires often have hundreds of item responses which in themselves are of little interest. At issue is the causal relations among the latent variables, i.e., the features of persons or systems that the item responses measure. Even if simplifying assumptions such as linearity are made, it is often thought to be hopeless to extract from the data any reliable conclusions about the causal relations of the unmeasured causes. Researchers typically hypothesize and test a handful of causal models, but since the number of plausible alternatives is often astronomical, a hypothesized model or models that cannot be rejected on statistical grounds is little evidence for the substantive conclusions that follow from such models.

In this paper I describe procedures that under suitable assumptions give provably reliable information about the causal structure among a set of latent variables. The second procedure is able to test, using only correlations among measured variables, for independence and conditional independence relations among the latents, but only after the first procedure has located a 'pure' way to measure them. The procedures are implemented as the Purify and MIMbuild (Multiple Indicator Model Builder) modules of TETRAD II.[5] In what follows I describe the crucial features of these procedures, the assumptions under which they are reliable, state theorems concerning their reliability and informativeness, and discuss the limitations of the procedures as well as possible extensions and generalizations.

---

[0] I thank Peter Spirtes, Clark Glymour, and Steve Klepper for countless valuable suggestions.

[1] *Selecting Models from Data: AI and Statistics IV.* Edited by P. Cheeseman and R.W. Oldford. ©1994 Springer-Verlag.

[3] See, for example, [Pearl 88]; [Spirtes 93]; [Herskovits 90].

[4] An exception is [Pearl 88], section 8.3. Pearl's treatment, however, assumes that the causal structure among hidden variables is tree-like. The procedure I discuss in this paper makes no such assumption.

[5] TETRAD II will be distributed by Lawrence Erlbaum, NJ.

20.2 Structural equation models, causal structure and constraints.

This work applies to recursive linear structural equation models (RSEMs) with latent variables. Because space is limited I will not describe these models in detail. Several sources provide good introductions, e.g. [Bollen 89]. It is straightforward to represent the causal claims of such models with a directed acyclic graph (DAG). A **causal DAG** over a set of variables **V** contains a directed edge from A to B just in case A is a direct cause of B. Moving from a causal DAG to the corresponding system of structural equations involves specifying each variable in **V** as a linear combination of its immediate causes in the causal DAG plus an error variable [Spirtes 93]. If **V** includes all the common causes of variables in **V**, then **V** is **causally sufficient**. The error terms are assumed to be independent in a causally sufficient RSEM.

The joint distribution among the non-error variables V in an RSEM is determined by the triple $\langle G, D(\epsilon), \phi \rangle$, where G is the causal DAG over **V**, $D(\epsilon)$ is the joint distribution among the error terms, and ϕ the linear coefficients that correspond to each arrow in the DAG. By fixing G, we can parameterize all the distributions representable by an RSEM with a given causal structure by $\langle D(\epsilon), \phi \rangle$. A constraint on the joint distribution, e.g., independence, is **implied by the causal structure** if it holds in every parameterization of G.

I deal with two sorts of constraints on the population covariances: vanishing partial correlations and vanishing tetrad differences. The graph theoretic conditions under which an RSEM's causal structure implies both of these constraints is known. D-separation [Pearl 88] [Spirtes 93] characterizes the set of (partial) correlations that the causal structure of an RSEM implies must vanish, and the Tetrad Representation Theorem [Spirtes 93] characterizes the set of tetrad differences that the causal structure of an RSEM implies must vanish.

Tetrad differences are simply the determinant of a 2x2 sub matrix of the covariance or correlation matrix Σ. There are three possible tetrad differences among four variables y_1 - y_4:

$$\rho_{y1,y2} * \rho_{y3,y4} - \rho_{y1,y3} * \rho_{y2,y4}$$
$$\rho_{y1,y2} * \rho_{y3,y4} - \rho_{y1,y4} * \rho_{y2,y3}$$
$$\rho_{y1,y3} * \rho_{y2,y4} - \rho_{y1,y4} * \rho_{y2,y3}$$

Because a constraint holds in a population produced by an RSEM does not entail that the constraint is implied by the RSEM's causal structure G. The constraint might hold only for particular paramaterizations of G but not for others. Such a situation requires unusual parameterizations, however, and in what follows every vanishing partial correlation and vanishing tetrad difference that holds in a population is assumed to be implied by the causal structure of the RSEM that produced the population. This is the **Linear Faithfulness** assumption discussed in [Spirtes 93]

20.3 Pure measurement models

Structural equation models (e.g., figure 1) are typically divided into two parts, the measurement model and the structural model. Roughly, the structural model involves only the causal connections among the latent variables, and the measurement model the rest. Consider the graph in figure 1, in which the η variables are latent, the Y variables are measured, and the ϵ and ζ variables are error terms.[6]

[6] For purely illustrative purposes, one might imagine that this model applies to married, male Navy pilots. η_3 might express the pilots level of job satisfaction, η_4 how challenging he finds his career, η_1 how traditional a family the pilot comes from, and

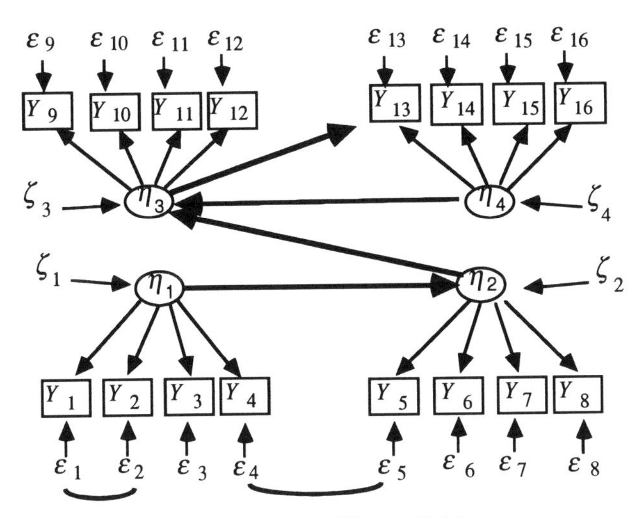

FIGURE 1: A Structural Equation Model.

In this case the structural model is the maximal subgraph involving only η and ζ variables, and the measurement model its complement, e.g., figure 2.

A measured variable is **pure** if it is a direct effect of exactly one latent and if its error term is independent of all other variables in the model, including other error terms. For example, Y_1, Y_2, Y_4, Y_5, and Y_{13} are the only impure indicators in the figures above. A variable V that measures latent η_i is **almost pure** if it is d-separated from every other measured variable by η_i.[7] A measurement model is **(almost) pure** just in case all of its measured variables are (almost) pure.[8]

Purity in the measurement model affords access to the relations among the latents, because if the measurement model is almost pure, then any causal connection between measured indicators is mediated by the latents they measure. Accordingly, the overall strategy is to first find pure measurement models (the Purify procedure in TETRAD II) and then use them to determine features of the structural model (the MIMbuild procedure in TETRAD II).

20.4 Starting points

The strategy begins with a covariance matrix over a set of measured variables **V**, a set of latents η, and an initial measurement model M_o that specifies each member of **V** as a measure of one latent in η, there being at least 3 measures for each latent. For example, **V** might be Y_1 through Y_{16}, and the initially specified measurement model for η_1 - η_4 as shown in figure 3.

η_2 how supportive the pilot's spouse is toward his Navy career. The Y variables might be questionaire responses.

[7] Peter Spirtes pointed out this definition, which is an more elegant but equivalent definition to the one in [Scheines 93].

[8] An almost identical definition is given in the structural equation modelling literature as a "unidimensional" measurement model. See [Anderson 82], [Anderson 88].

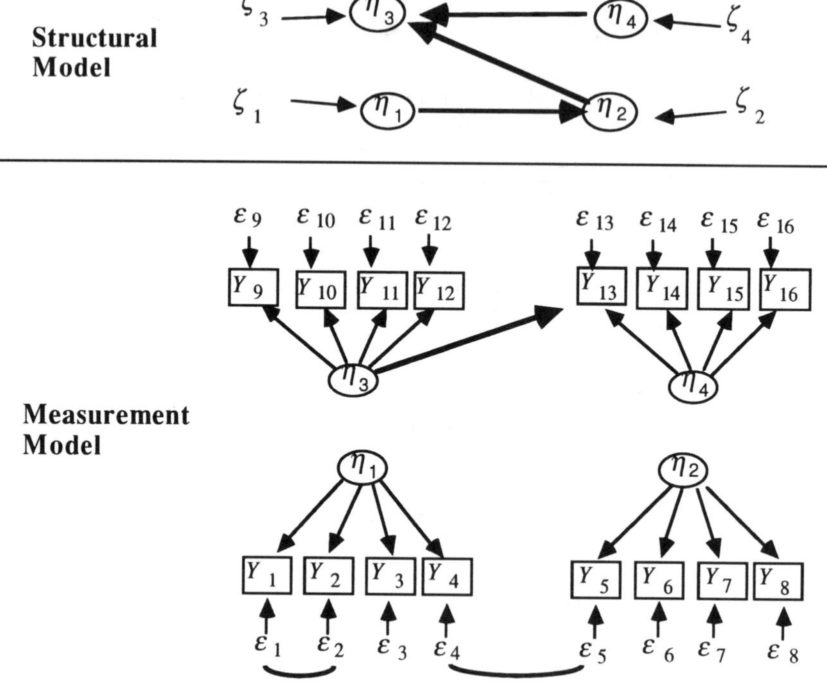

FIGURE 2: Measurement and structural models.

It is assumed that the initially specified measurement model is a sub-model of the actual RSEM. That is, each direct cause specified in the initial measurement model is in the actual RSEM, but several might have been left out. Put another way, it is not assumed that each initially specified measured indicator is pure, only that it does measure *at least* the latent it is specified to measure. This assumption is often quite easy to justify. For example, suppose that in the figures above η_3 is a latent interpreted as 'Job Satisfaction.' Further, suppose that Y_9 is a scale measuring agreement or disagreement with the statement 'I generally like going to work.' It is not controversial to assume that Y_9 measures η_3.

20.5 Purifying the measurement model

The vanishing tetrad difference can be used to locate which of the initially specified indicators are impure. Assuming that a tetrad difference vanishes in the population, it has an asymptotic normal distribution with mean 0 and variance given in [Bollen 90] in terms of the fourth moments of the observables (for the case of normally distributed variables the fourth moments specialize

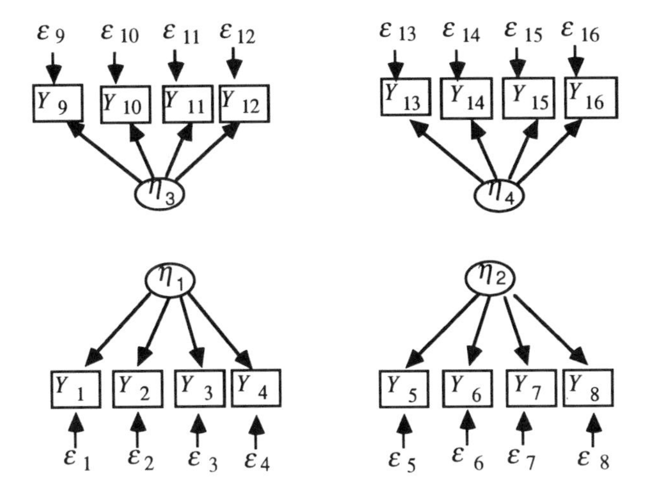

FIGURE 3: Initially specified measurement model.

to be functions of the covariances, so the asymptotic variance of a tetrad difference simplifies to a function of only the covariances [Wishart 28] and [Bollen 90]. Bollen's test is asymptotically distribution free and accounts for the dependence between covariances or correlations in a single tetrad difference by using Σ_{ss}, the covariance matrix of the covariances, the elements of which are defined in terms of fourth and second moments. The present version of Purify uses the Wishart formula, which assumes a multivariate normal distribution.

A **3x1 tetrad difference** involves three measures of η_i, (e.g. measures of η_1: Y_1, Y_2, and Y_3 in figure 3) and one measure of η_j, (e.g., Y_5 as a measure of η_2 in figure 3). 3x1 tetrads test for impurities *regardless of the connection between* η_i *and* η_j.

More precisely, let M_o be an initially specified measurement model over $\mathbf{V'} \cup \eta$ that satisfies the above assumptions, and let $\mathbf{V'}(\eta_i)$ = the elements of **V'** that measure η_i, and let $\mathbf{V} \subseteq \mathbf{V'}$. Then:

> **Theorem 1:** If M_o is a sub-model of an RSEM G over $\mathbf{V} \cup \eta$, and G's causal structure implies that every 3x1 tetrad difference vanishes among **V**, then for every $\eta_i \in \eta$, and every $V \in \mathbf{V}(\eta_i)$ such that $|\mathbf{V}(\eta_i)| \geq 3$, V is almost pure.[9]

Provided all the modeling assumptions are satisfied, this means that if we begin with an initially specified measurement model M_o, and find a subset **V** of M_o's indicators such that all 3x1 tetrad differences vanish among **V**, then all the indicators in **V** are almost pure, regardless of the connections among the latent variables.

The Purify procedure in TETRAD II uses 3x1 tetrad differences to locate and discard impure indicators in an initially specified measurement model. For example, suppose that the actual model G is the one shown in figure 1. Suppose further that the initially specified measurement

[9]This theorem is proved in [Scheines 93].

model is the one shown in figure 3. Indicators Y_1, Y_2, Y_4, Y_5, and Y_{13} are impure (figure 2). Purify might discard Y_1, Y_5, Y_{13}, arriving at a pure measurement model involving the remaining Y variables.[10]

20.6 The structural model

In an almost pure measurement model, an indicator of η_i and an indicator of η_j are d-separated given the empty set just in case η_i and η_j are d-separated given the empty set. Since vanishing correlations test for 0-order d-separation relations[11] in Faithful RSEMs [Spirtes 93], testing for 0-order d-separation among η_i and η_j reduces to testing for vanishing correlations among indicators of η_i and indicators of η_j. Assuming that each latent has at least two indicators, testing for an uncorrelated pair of latents means testing for several overlapping pairs of uncorrelated indicators. A simultaneous test for a set of vanishing correlations can be derived again by using Σ_{ss}.

Whereas 3x1 tetrad differences provide specification tests for impurity in the measurement model, another sort of vanishing tetrad difference provides a test for 1st-order d-separation among latent variables in an RSEM with an almost pure measurement model.

> **Theorem 2:** If G is an RSEM with a pure measurement model in which each latent has at least two measured indicators, then latent variables η_1 and η_3, whose measured indicators include J and L respectively, are d-separated given latent variable η_2, whose measured indicators include I and K, if and only if G implies $\rho_{JI}\rho_{LK} = \rho_{JL}\rho_{KI}$.[12]

After using vanishing correlations and tetrad differences to test for 0 and 1st order d-separation relations among the latents, MIMbuild uses the PC algorithm given in [Spirtes 93] to take these d-separation facts as inputs and return an equivalence class of structural models (a pattern). Since MIMbuild can at present only test for 0 and 1st order d-separation facts, the class of structural models returned is larger than it would be were MIMbuild able to test for d-separation facts of arbitrary order.

In the example used throughout, MIMbuild would proceed through the stages shown in figure 4 to constructing its output.

20.7 Reliability and limitations of the output

One further modification is made to the output to inform the user of potential uncertainties in the set of adjacencies due to the limitations on the order of d-separation information available. In the above example, 0 and 1st order d-separation facts are enough to uniquely identify the adjacencies, and MIMbuild can compute that they are enough. In cases where 2nd or higher order d-separation relations are needed to eliminate an adjacency, MIMbuild will not be able to do so. Nevertheless, it can identify the set of adjacencies that might have been eliminated had

---

[10] Y_2 and Y_4 need not be discarded, because once Y_1 and Y_5 are discarded, these two are no longer impure. For more detail, see TETRAD II: Tools for Discovery, forthcoming from Lawrence Erlbaum.

[11] D-separation is a relation between three sets, e.g., **X**, **Y**, are d-separated by **Z**. The order of the relation is just the cardinality of the separating set **Z**.

[12] This theorem is proved in [Scheines 93] and in [Spirtes 93].

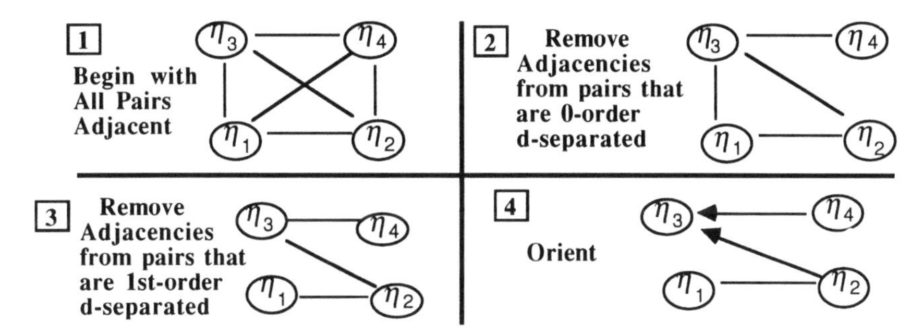

FIGURE 4: Trace of MIMbuild.

2nd or higher order d-separation relations been available. These adjacencies are marked with a '?.' The output is reliable in the following sense.[13]

> **Theorem 3:** If the input to MIMBuild is a list of all vanishing 0 and 1st-order d-separation relations among a set of causally sufficient latent variables implied by the causal structure of an RSEM G and Π is the output of MIMBuild then
>
> A-1) If X and Y are not adjacent in Π, then they are not adjacent in G.
>
> A-2) If X and Y are adjacent in Π and the edge is not labeled with a '?', then X and Y are adjacent in G.
>
> O-1) If $X \rightarrow Y$ is in Π, then every causal connection[14] in G between X and Y is into Y.
>
> O-2) If $X \rightarrow Y$ is in Π and the edge between X and Y is not labeled with a '?,' then $X \rightarrow Y$ is in G.

If the assumptions given above are satisfied, and the outcome of statistical tests are correct, i.e., hypotheses are rejected if and only if they are false, then MIMbuild will correctly identify the 0 and 1st order d-separation relations among the latents.

An early version of these procedures has been applied to a data set collected by the Navy Personnel Research and Development Center,[15] and its usefulness was found to compare favorably with factor analytic approaches. The procedure has also been tested extensively on Monte Carlo data generated from models with as many as 12 latent variables and 60 indicators. The procedure is quite reliable when its assumptions are satisfied and the sample is of reasonable size, e.g., over 200.

---

[13] Again, the proof is in [Spirtes 93].

[14] A causal connection between X and Y is either a directed path from X to Y, a directed path from Y to X, or a pair of directed paths from W to X and W to Y.

[15] See [Callahan 92]

Alternative strategies are available for building the structural model. One could, for example, purify the measurement model and then specify a 'theoretical model' in which each pair of latent variables is directly correlated. A maximum likelihood estimate of this structure will give a consistent estimate of the correlation matrix among the latent variables. This correlation matrix can then be used as input to the PC algorithm, which would then construct the equivalence class of faithful structural models.

20.8 REFERENCES

[Anderson 82] Anderson, J., and Gerbing, D., (1982) "Some Methods for Respecifying Measurement Models to Obtain Unidimensional Construct Measurement," *Journal of Marketing Research*, **XIX**, 453-60.

[Anderson 88] Anderson, J., and Gerbing, D., (1988). "Structural Equation Modeling in Practice: A Review and Recommended Two-Step Approach," *Psychological Bulletin*, **103**, 411-423.

[Bollen 89] Bollen, K. (1989) *Structural Equations with Latent Variables*. Wiley, New York.

[Bollen 90] Bollen, K. (1990). "Outlier Screening and a Distribution-free Test for Vanishing Tetrads," *Sociological Methods and Research*, **19**, 80-92.

[Callahan 92] Callahan, J., and Sorensen, S., (1992). "Using TETRAD II as an Automated Exploratory Tool," *Sociological Methods and Research*

[Glymour 87] Glymour, C., Scheines, R., Spirtes, P., and Kelly, K. (1987) *Discovering Causal Structure*. Academic Press, San Diego, CA.

[Herskovits 90] Herskovits, E., and Cooper, G. (1990). "Kutato: An Entropy-driven System for Construction of Probabilistic Expert Systems from Databases," *Proc. Sixth Conf. Uncertainty in AI. Association for Uncertainty in AI*, Mountain View, CA.

[Pearl 88] Pearl, J. (1988) *Probabilistic Reasoning in Intelligent Systems*. Morgan Kaufman, San Mateo, CA.

[Scheines 93] Scheines, R., (1993). "Linear Latent Variable Models," *Tech-Report PHIL-39*, Department of Philosophy, Carnegie Mellon University, Pittsburgh, PA, 15213.

[Scheines 92] Scheines, R., and Spirtes, P. (1992). "Finding Latent Variable Models in Large Data Bases," *International Journal of Intelligent Systems*, G. Piatetski-Shapiro, (ed.).

[Spirtes, Glymour, and Scheines 93] Spirtes, P., Glymour, C., and Scheines, R. (1993) *Causation, Prediction and Search*. Springer-Verlag, *Lecture Notes in Statistics*, New York.

[Wishart 28] Wishart, J. (1928). "Sampling Errors in the Theory of Two Factors," *British Journal of Psychology*, **19**, 180-187.

21
When can association graphs admit a causal interpretation?

Judea Pearl and Nanny Wermuth

Cognitive Systems Laboratory
University of California
Los Angeles, CA 90024
judea@cs.ucla.edu
and
Psychologisches Institut
Gutenberg Universität
55099 Mainz, Germany
wermuth@upymzc.sowi.uni-mainz.de

ABSTRACT
We discuss essentially linear structures which are adequately represented by association graphs called covariance graphs and concentration graphs. These do not explicitly indicate a process by which data could be generated in a stepwise fashion. Therefore, on their own, they do not suggest a causal interpretation. By contrast, each directed acyclic graph describes such a process and may offer a causal interpretation whenever this process is in agreement with substantive knowledge about causation among the variables under study. We derive conditions and procedures to decide for any given covariance graph or concentration graph whether all their pairwise independencies can be implied by some directed acyclic graph.

21.1 Introduction

Many of the most widely used statistical techniques center around the interpretation of the covariance matrix. The exponential family of the multivariate normal distribution provides a justification to reduce data to sample means, $\bar{y}_i$, $i = 1, \ldots, p$, and a covariance matrix, S, and for some purposes to sample means and a concentration matrix, S^{-1}, that is the inverse covariance matrix.

Under multivariate normality, a zero covariance, $\sigma_{ij} = 0$, means marginal independence and a zero concentration, $\sigma^{ij} = 0$, means conditional independence of variable pair Y_i, Y_j given all $p-2$ remaining variables. In general, these conditions just indicate linear independencies, since the standardized versions of covariance and concentration are the marginal correlation coefficient, ρ_{ij}, and the partial correlation coefficient, $\rho_{ij.k}$, given all the remaining variables $k = \{1, \ldots, p\} \backslash \{i, j\}$:

$$\rho_{ij} = \sigma_{ij}(\sigma_{ii}\sigma_{jj})^{-1/2}, \rho_{ij.k} = -\sigma^{ij}(\sigma^{ii}\sigma^{jj})^{-1/2}.$$

Thus, data reduction to covariance or concentration matrix implies that essentially linear

[0]The research was partially supported by Air Force grant #AFOSR 90 0136, NSF grant #IRI-9200918, and Micro grants #92-122/3. This work benefited from discussions with David Cox, Goeran Kauermann, and Thomas Verma.

[1]*Selecting Models from Data: AI and Statistics IV.* Edited by P. Cheeseman and R.W. Oldford.

relations among the variables under study are of interest. These may be relevant not only for continuous but also for ordinally scaled variables or for dichotomous variables. But, since this is a fairly strong assumption in some applications, checks for nonlinearity or special linearisations are needed, such as those discussed in [Cox-Wer 92a, Cox-Wer 92b] and in [Wermuth-Cox 92].

A covariance matrix with some zero off-diagonal entries can be represented by an undirected graph with the corresponding edges missing [Cox-Wer 93] called **covariance graph**, abbreviated by G_{cov} here. Similarly, a concentration matrix with some zero entries can be represented by an undirected graph with the corresponding edges missing [Kiiveri-Spe 82] called **concentration graph**, abbreviated by G_{con} here.

A **complete** covariance or concentration graph has no missing edges, that is, it has exactly one edge for each variable pair, $1/2\ p\ (p-1)$ in all. There can be a number of reasons for considering special incomplete undirected graphs. For instance, it could be that

- the linear independencies expressed thereby are themselves of substantive interest,
- reduction of dimensionality is needed to keep a reasonable balance between the number of parameters to be estimated and the number of available independent observations, or
- it is desired to decide whether the structure given by a covariance or concentration graph could have been generated by some stepwise process of univariate recursive regressions.

We concentrate here on the last aspect. This implies in particular that we consider covariance and concentrations graphs less as representing models of their own standing but, rather, a possible reflection of some directed acyclic graph, abbreviated as **dag**, which represents a recursive process for stepwise data generation. Such a process would admit a causal interpretation if the order of the variables involved in the process, especially their classification into response and explanatory variables, is in agreement with subject-matter knowledge about the causal process [Cox 92].

Each dag can be thought of as specifying dependence relationships between an ordered set of variables $\{Y_1, ..., Y_p\}$ where $j > i$ designates Y_j as potentially explanatory for Y_i; it becomes actively explanatory in case an arrow is drawn from Y_j to Y_i. A missing edge for variable pair Y_i, Y_j has then the interpretation: response Y_i is (linearly) independent of Y_j given the remaining potential explanatory variables of Y_i, $\{Y_{i+1} ..., Y_{j-1}, Y_{j+1}, ..., Y_p\}$. The set of missing edges defines the independence structure of the dag and permits[3] the stronger statement that Y_i is conditionally independent of all variables in $\{Y_{i+1}, ..., Y_p\}$ that are not linked to Y_i, given those that are linked to Y_i. Each dag **implies** a set of pairwise marginal (linear) independencies, i.e. a covariance graph, and a set of pairwise conditional (linear) independencies given all $p-2$ variables, i.e. a concentration graph. By "implied" we mean independencies that must hold in *every* distribution that fulfills all the missing-link conditions shown in the dag. This excludes independencies that are just introduced by accidental equalities among the numerical values of some parameters in the model. From a given dag, these implied independencies may be read off directly, using the graphical criterion of d-separation [Pearl 88, page 117] and [Geiger etal 90].

For example the following Markov chain

$$\begin{matrix} Y_1 & & Y_2 & & Y_3 & & Y_4 & & Y_5 \\ \circ & \longleftarrow & \circ & \longleftarrow & \circ & \longleftarrow & \circ & \longleftarrow & \circ \end{matrix}$$

[3] Provided S is non-singular.

is a dag which implies a complete covariance graph, i.e. no marginal independencies, and a chain-like concentration graph, obtained by removing the arrow heads. If r, s, t, u denote correlations between pairs $(Y_1, Y_2), (Y_2, Y_3), (Y_3, Y_4), (Y_4, Y_5)$, then the correlation matrix implied by the Markov chain has the form

$$\begin{matrix} 1 & r & rs & rst & rstu \\ \bullet & 1 & s & st & stu \\ \bullet & \bullet & 1 & t & tu \\ \bullet & \bullet & \bullet & 1 & u \\ \bullet & \bullet & \bullet & \bullet & 1. \end{matrix}$$

It can be used to explain why it may be difficult to decide on marginal linear independence merely from observed marginal correlations. Suppose $r = s = t = u$ and $r = 0.3$ then $r^2 = 0.09.$, $r^3 = 0.027$, $r^4 = 0.0081$ would be close to zero; even when $r = 0.6$, $r^4 = 0.13$ would still not be far from zero. Thus, if we know that the data are generated by the above Markov chain we can conclude that all marginal correlation are nonzero but a correlation is weaker the further the variables are apart in the chain. However, if the process by which the data actually are generated is not known, and we have to rely merely on data inspection or associated tests, we may judge some of the smaller correlations to be zero apart from sample fluctuations i.e. we come to erroneous conclusions about the covariance structure (represented by the covariance graph) or the generating process. Similar arguments apply to the concentration graph.

Formal tests for agreement of an observed covariance matrix or concentration matrix with a corresponding hypothesized association graph are available. An arbitrary pattern of zeros in a covariance matrix is a special case of hypotheses linear in covariances [Anderson 73] and an arbitrary pattern of zeros in a concentration matrix is a covariance selection model [Dempster 72]. Maximum likelihood estimates and associated likelihood ratio tests are available under the assumption of multivariate normality and may for instance be computed with the help of LISREL [Jöreskog-Sör 78] for the former and with the help of MIM [Edwards 92] for the latter. The same estimates are called quasi-likelihood estimates if just the assumption of linearity is retained.

The problem of finding a causal explanation for a general distribution was treated in [Pearl-Wer 91] and, in the case of normal distributions, might require the processing of an exponential number of submatrices (i.e., all the majors of G_{cov}). When an explicit list of *all* conditional and marginal independencies is available, the problem can be solved in time polynomial in the length of the list [Verma-Pea 92a] [Dori-Tar 92]. Still, the length of this list can be very large indeed. The current paper attempts to find a causal interpretation on the basis of a more limited information, assuming that the only vanishing dependencies available to the analyst are those corresponding to the zero entries of the covariance and concentration matrices.

To this end, we pose the following set of problems:

• Given a covariance graph, G_{cov}, decide whether it could have been generated by a directed acyclic graph

(i) with the same nodes and edges,

(ii) with the same nodes but fewer edges,

(iii) with some additional nodes (representing latent or hidden variables).

• Given a concentration graph, G_{con}, decide whether it could have been generated by a directed acyclic graph

(iv) with the same nodes and edges,
(v) with the same nodes but fewer edges,
(vi) with additional (latent) nodes.
• Given a covariance and a concentration graphs, decide whether
(vii) both could have been generated by the same directed acyclic graph.

Theorems 1 through 7 of the next section present declarative and procedural solutions to these seven problems, respectively. Proofs are omitted, and will be included in an expanded version of this paper.

21.2 Main results

Definition 1 *A directed acyclic graph D is said to imply an independence I if I holds in every probability P that fulfills all the missing-link conditions of D, that is, choosing any variable ordering consistent with the direction of arrows in D, P renders each variable Y_i independent on its inactive explanatory variables (nonadjacent predecessors), given its active explanatory variables (adjacent predecessors). If D fails to imply an independence for pair (i, j) we say that D implies a dependence for that pair which, in turn, will induce an edge ij in G_{cov} or G_{con}.*

Lemma 1 *D implies an independence I iff I satisfies the d-separation criterion in D* [Pearl 88]. *In particular, D implies a marginal independency $\sigma_{ij} = 0$ if every path between nodes i and j in D traverses at least one pair of converging arrows (i.e., $\longrightarrow \star \longleftarrow$). Likewise, D implies a conditional independence given all remaining variables, $\sigma^{ij} = 0$, if every path between nodes i and j in D traverses at least one pair of non-converging arrows (i.e., either $\longrightarrow \star \longrightarrow$ or $\longleftarrow \star \longrightarrow$).*

Definition 2 *(generate and coincide) A dag D is said to (a) generate a covariance graph G_{cov} if the set of marginal independencies implied by D matches the set of missing edges of G_{cov} and (b) to coincide with G_{cov} if D not only generates G_{cov} but also has the same edges as G_{cov}.*

Similar definitions of *generate* and *coincide* apply to G_{con}. For example, the Markov chain discussed in Section 21.1 represents a dag which coincides with its G_{con} but not with its G_{cov}, since some edges in G_{cov} do not appear in the chain.

Definition 3 *($\vee$-configuration) A triplet of nodes (a, b, c) in an undirected graph G is said to be a $\vee$-configuration (pronounced as vee-configuration), if ab and bc are edges in G while ac is not.*

Definition 4 *(sink orientation) Given an undirected graph G, the sink orientation of G results from assigning arrows $a \longrightarrow b \longleftarrow c$ to each $\vee$-configuration (a, b, c) in G.*

Note that a sink orientation may contain bi-directed edges, since two opposing arrows may be assigned to the same edge. For example, the sink orientation of the 4-chain a—b—c—d is given by $a \longrightarrow b \longleftrightarrow c \longleftarrow d$, since both (a, b, c) and (b, c, d) are $\vee$-configurations. We mark that the sink orientation of a graph with n nodes can be formed in $O(n^3)$ steps.

Theorem 1 *The following statements are equivalent. A covariance graph G_{cov} coincides with a dag D if and only if (i) no edge in the sink orientation of G_{cov} is bi-directed; (ii) G_{cov} contains no chordless 4-chain. Moreover, if the sink orientation of G_{cov} contains no bi-directed edges,* *then every acyclic orientation of the remaining undirected edges constitutes a dag that coincides with* G_{cov}.

Example 1 Consider the graph

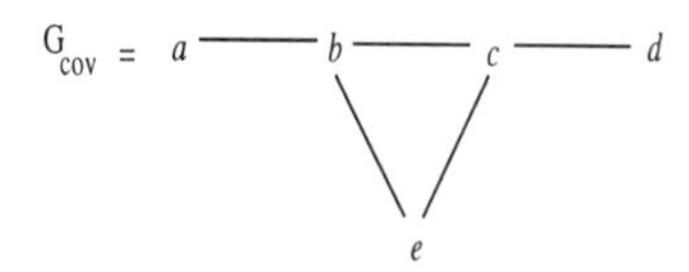

The 4-chain a—b—c—d is chordless, hence, there is no dag that coincides with G_{cov}. Indeed, the $\vee$-orientation of G_{cov} is

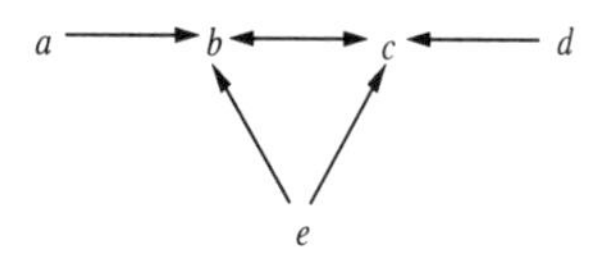

which renders the edge $b \longleftrightarrow c$ bi-directed.

Turning to problem (ii), we first define the notions of **sink completion** and **exterior cliques**.

Definition 5 ($sink\ completion$) *A sink completion of an undirected graph* G_{cov} *is any dag obtained from* G_{cov} *as follows:* a) *the sink orientation of* G_{cov} *is formed,* b) *all bi-directed edges are removed,* c) *the remaining undirected edges are oriented to form a dag.*

Lemma 2 *Every undirected graph has at least one sink completion.*

This follows from the facts that (1) every sink orientation is acyclic and (2) any acyclic partial orientation of a graph can be completed to form a dag.

Definition 6 ($exterior\ cliques$) *A clique in an undirected graph is said to be exterior if it contains at least one node that is adjacent only to nodes in that clique. We call such a node "extremal".*

Note that finding an extremal node in a graph with n nodes and E edges takes at most nE^2 steps; for each node i we test whether every two neighbors of i are adjacent to each other. The same procedure can be used to enumerate all exterior cliques of a graph, since an extremal node must reside in one and only one clique.

Theorem 2 *The following statements are equivalent:* (i) G_{cov} *is generated by a directed acyclic graph;* (ii) *a sink completion of* G_{cov} *implies all edges of* G_{cov}; (iii) *every edge of* G_{cov} *resides in an exterior clique.*

Example 2 Consider again the graph G_{cov} of Example 1. The sink completion of G_{cov} is given by the dag $D = a \longrightarrow b \longleftarrow e \longrightarrow c \longleftarrow d$ which implies all edges of G_{cov}. In particular, the edge b—c, missing from D, is implied by the path $b \longleftarrow e \longrightarrow c$, which does not contain converging arrows (see Lemma 1.) Hence D generates G_{cov}. Indeed, following condition (iii), the exterior cliques of G_{cov} are (a, b), (b, c, d) and (c, d), which contain every edge of G_{cov}.

A simple example of a graph G_{cov} that cannot be generated by a dag is the 4-chain a—b—c—d. The sink completion of G_{cov} is the disconnected dag $a \longrightarrow b \quad c \longleftarrow d$ which does not imply the edge b—c in G_{cov}. Likewise, G_{cov} has only two exterior cliques, (a, b) and (c, d), thus violating condition (iii) relative to edge b—c.

Turning to Problem (iii), we have:

Theorem 3 *Every covariance graph G_{cov} can be generated by a directed acyclic graph with additional nodes by replacing each edge ab of G_{cov} with $a \longleftarrow (*) \longrightarrow b$, where $(*)$ represents an unobserved variable.*

Remark: To reduce the number of latent nodes, it is not necessary to replace each edge of G_{cov}, but only those which are not implied by the sink completions of G_{cov}. For example, the 4-chain a—b—c—d is generated by the dag $a \longrightarrow b \longleftarrow(*)\longrightarrow c \longleftarrow d$, where the replaced edge bc is the only one that is not implied by the sink completion of this chain (see Example 2).

We now turn to problems (iv) - (vi) where we seek a dag that agrees with the structure of a given concentration matrix.

Theorem 4 *A concentration graph G_{con} coincides with some dag D if and only if G_{con} is a chordal graph, i.e. it contains no chordless n-cycle, $n > 3$. (Effective tests for chordality are well known in the literature* [Tarjan-Yan 84].)

Theorem 5 *A concentration graph G_{con} can be generated by some dag if and only if there exists an ordering O such that the following procedure removes all edges of G_{con}:*

1. *Find an exterior clique, remove all its extremal nodes and their associated edges, mark all remaining edges in that clique, and add them to a list L of marked edges, in some order O.*

2. *Repeat the process on the reduced graph until either all nodes have been eliminated or, in case the subgraph contains no exterior clique, choose the first marked edge on L, and remove it from the graph.*

3. *Repeat step 1 and 2.*

Example 3 To see how the marking procedure helps identify a generating dag, consider the graph G_{con} below.

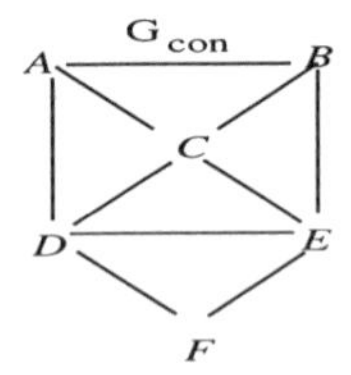

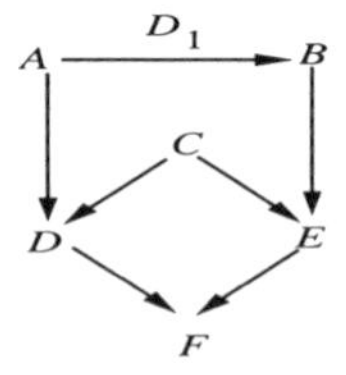

The cliques are: ABC, ACD, BCE, CDE and DEF. Only DEF is exterior, with extremal vertex F. Removing F together with edges DF and FE, leaves a subgraph with no exterior clique. However, edge DE is marked, so it can be removed, which creates new exterior cliques, allowing the process to continue till all edges are removed. Indeed, directing arrows toward each node removed in step 2, yields the dag D_1 above, which generates G_{con}.

Example 4 *The following example (due to Verma) shows that the order of edge removals may be crucial, that is, failure to eliminate all edges in one ordering does not imply that no elimination ordering exists. Consider the following concentration graph:*

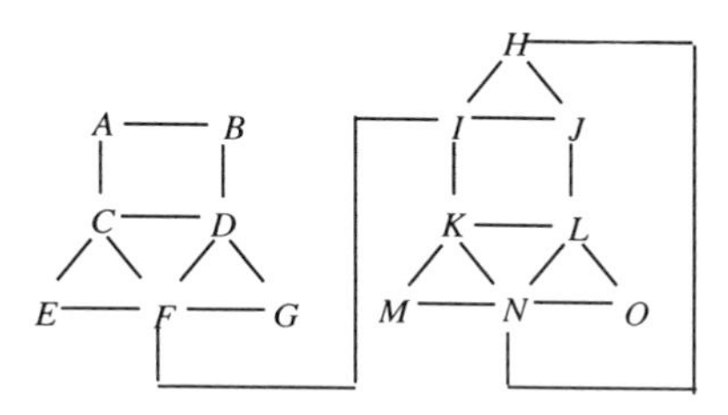

The only extremal vertices are E, G, M, and O. If we remove these (in any order) it will result in a graph with no exterior cliques and with the following marked links: CF, DF, KN, and LN. If we first remove the marked edges FC and FD, the process will halt (because the cycle $A - B - C - D - A$ cannot be eliminated.) However, if we first remove the marked edges LN and KN, then we will find a good elimination ordering: ..., N, H, J, L, K, I, F, ...

The fact that it is impossible by local means to decide which of the marked edges should be removed first renders the decision problem a difficult one. While in the example above it is clear that one should postpone the removal of FC and FD, because it leads to an impasse (the cycle $A - B - C - D - A$), such local clues are not available in the general case. Indeed, Verma and Pearl have shown that the problem is NP-Complete [Verma-Pea 92b]. Nevertheless, effective necessary conditions are available, which makes the decision problem feasible when the number of nodes is not too large.

Lemma 3 *The following are necessary conditions for a concentration graph G_{con} to be generated by a dag. (1) every chordless n-cycle of G_{con}, $n > 3$, must have at least one edge that resides in some k-clique, $k > 2$. (2) G_{con} must have at least one exterior clique.*

Example 5 The graph G_{con} below shows that these two conditions are not sufficient; it satisfies both, yet it cannot be generated by any dag.

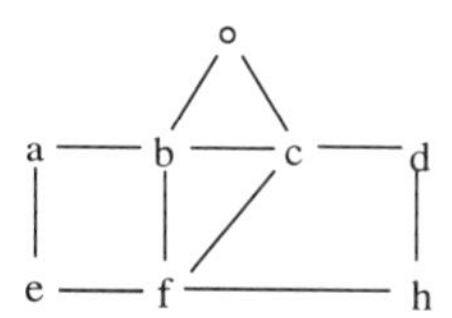

The edge removal procedures of Theorem 5 can easily be converted into a backtrack algorithm that guarantees finding a generating dag, if such a dag exists. Whenever the procedure reaches an impasse, the algorithm backtracks to the state of last decision point, and selects another edge for removal. The efficiency of this algorithm can be improved by insisting that at any decision point we consider for removal only edges such that the reduced graph will satisfy the necessary conditions of Lemma 3. In general, we should also try to remove marked edges that belong to chordless n-cycles, $n > 3$, as early as possible, because we know that at least one such edge must eventually be removed if the process is to come to a successful end.

Theorem 6 *A concentration graph G_{con} can be generated by a dag with additional latent nodes if and only if it can be generated by a dag without latent nodes.*

In other words, latent variables do not improve the power of dags to generate richer patterns of vanishing concentrations. This is in sharp contrast to patterns of vanishing covariances (Theorem 3); every such pattern can be generated by a dag with latent nodes.

Finally, turning to problem (vii) we have:

Theorem 7 *A covariance graph G_{cov} and a concentration graph G_{con} can both be generated by the same dag only if the following three conditions hold:*

1. *G_{cov} and G_{con} must each be generated by a dag.*
2. *The set of nodes of every exterior clique of G_{cov} must induce a subgraph of G_{con} that can be generated by a dag.*
3. *Every edge of G_{con} that does not reside in a larger clique must also be an edge of G_{cov}.*

We are not sure at this point whether the conditions above are sufficient; so far we were not able to find a pair of undirected graphs that fulfill these three conditions and yet cannot be generated by a dag.

21.3 Conclusions

This paper provides conditions and procedures for deciding if patterns of independencies found in covariance and concentration matrices can be generated by a stepwise recursive process represented by some directed acyclic graph. If such an agreement is found, we know that one or several causal processes could be responsible for the observed independencies, and our procedures could then be used to elucidate the graphical structure common to these processes, so as to evaluate their compatibility against substantive knowledge of the domain.

If we find that the observed pattern of independencies does not agree with any stepwise recursive process, then there are a number of different possibilities. For instance,
– some weak dependencies could have been mistaken for independencies and led to the wrong omission of edges from the covariance or concentration graphs.
– the sampling procedure used in data collection was not properly randomized. Selection bias tends to induce symmetrical dependencies that cannot be emulated by a recursive process [Pearl 88, page 130].
– the process responsible for the data is non-recursive, involving aggregated variables, simultaneous reciprocal interactions, or mixtures of several recursive processes.
– some of the observed linear dependencies reflect accidental cancellations or hide actual non-linear relations.

In order to distinguish accidental from structural independencies it would be helpful to conduct several longitudinal studies under slightly varying conditions. These fluctuating conditions would tend to affect the numerical values of the covariances but not the basic structural properties of the process underlying the data. Under such conditions, if the data were generated by a recursive process represented by some directed acyclic graph, those independencies that are "implied" by the dag (see Definition 1) would persist despite the fluctuating conditions, while accidental independencies would be perturbed and hence discarded. We regard this as a possibility to pursue Cochran's suggestion [Cochran 65, page 252] that "when constructing a causal hypothesis one should envisage as many different consequences of its truth as possible and plan observational studies to discover whether each of these consequences is found to hold."

21.4 References

[Anderson 73] Anderson, T.W.(1973) "Asymptotically Efficient Estimation of Covariance Matrices With Linear Structure," *Ann. Statist.* **1**, 135-141.

[Cochran 65] Cochran, W.G. (1965) "The Planning of Observational Studies of Human Populations," *J. Roy. Statist. Soc. A* **128**, 234-265.

[Cox 92] Cox, D.R. (1992) "Causality: Some Statistical Aspects," *J. Roy. Statist. Soc.* Ser. A. **155**, 291-301.

[Cox-Wer 92a] Cox, D.R. and Wermuth, N. (1992) "Tests of Linearity, Multivariate Normality and the Adequacy of Linear Scores," Berichte zur Stochastik und verw. Gebiete, Universität Mainz **92-6**. To appear in: *Applied Statistics, Journal of the Royal Statistical Society*, Series C.

[Cox-Wer 92b] Cox, D.R. and Wermuth, N. (1992) "Response Models for Mixed Binary and Quantitative Variables," *Biometrika* **79**, 441-461.

[Cox-Wer 93] Cox, D.R. and Wermuth, N. (1993) "Linear Dependencies Represented by Chain Graphs," *Statistical Science* **8**, 204-218.

[Dempster 72] Dempster, A.P. (1972) "Covariance Selection," *Biometrics* **28**, 157-175.

[Dori-Tar 92] Dori, D. and Tarsi, M. (1992) "A Simple Algorithm to Construct a Consistent Extension of a Partially Oriented Graph," Computer Science Department, Tel-Aviv University. Also *Technical Report R-185*, UCLA, Cognitive Systems Laboratory.

[Edwards 92] Edwards, D. (1992) "Graphical Modeling With MIM," Manual.

[Geiger etal 90] Geiger, D., Verma, T.S., and Pearl, J. (1990) "Identifying Independence in Bayesian Networks," *Networks* **20**, 507-534. John Wiley and Sons, Sussex, England.

[Jöreskog-Sör 78] Jöreskog, D.G. and Sörbom, D. (1978) *LISREL IV, Analysis of Linear Structural Relationships by Maximum-Likelihood, Instrumental Variables and Least Squares Methods*, 3rd Edition, Mooresville: Scientific Software.

[Kiiveri-Spe 82] Kiiveri, H. and Speed, T.P. (1982) "Structural Analysis of Multivariate Data: A Review," in *Sociological Methodology*, S. Leinhardt ed., 209-289. Jossey Bass, San Francisco.

[Pearl 88] Pearl, J. (1988) *Probabilistic Reasoning in Intelligence Systems*. Morgan Kaufmann, San Mateo, CA.

[Pearl-Wer 91] Pearl, J. and Verma, T. (1991) "A Theory of Inferred Causation," in *Principles of Knowledge Representation and Reasoning: Proceedings of the Second International Conference*, Allen, J.A., Fikes, R., and Sandewall, E. eds., 441-452. Morgan Kaufmann, San Mateo, CA. Short version in *Statistics and Computing* **2**, 91-95. Chapman and Hall, London.

[Tarjan-Yan 84] Tarjan, R.E. and Yannakakis, M. (1984) "Simple Linear-Time Algorithms to Test Chordality of Graphs, Test Acyclicity of Hypergraphs and Selectively Reduce Acyclic Hypergraphs," *SIAM Journal of Computing* **13**, 566-579.

[Verma-Pea 92a] Verma, T. S., and Pearl, J. (1992) "An Algorithm for Deciding if a Set of Observed Independencies has a Causal Explanation," in *Proceedings, 8th Conference on Uncertainty in Artificial Intelligence, Stanford*, 323-330. Morgan Kaufmann, San Mateo, CA.

[Verma-Pea 92b] Verma, T.S. and Pearl, J. "Deciding Morality of Graphs is NP-complete," in *Proceedings of the Ninth Conference on Uncertainty in Artificial Intelligence, Washington, D.C.*, D. Heckerman and A. Mamdani eds., 391-397.

[Wermuth-Cox 92] Wermuth, N. and Cox, D.R. (1992) "On the Relation Between Interactions Obtained With Alternative Codings of Discrete Variables," *Methodika* **VI**, 76-85.

22
Inference, Intervention, and Prediction

Peter Spirtes and Clark Glymour

Department of Philosophy
Carnegie Mellon University

It is with data affected by numerous causes that Statistics is mainly concerned. Experiment seeks to disentangle a complex of causes by removing all but one of them, or rather by concentrating on the study of one and reducing the others, as far as circumstances permit, to comparatively small residium. Statistics, denied this resource, must accept for analysis data subject to the influence of a host of causes, and must try to discover from the data themselves which causes are the important ones and how much of the observed effect is due to the operation of each.
–G. U. Yule and M. G. Kendall,1950

George Box has [almost] said "The only way to find out what will happen when a complex system is disturbed is to disturb the system, not merely to observe it passively." These words of caution about "natural experiments" are uncomfortably strong. Yet in today's world we see no alternative to accepting them as, if anything, too weak.
–G. Mosteller and J. Tukey, 1977

What can be predicted when the causal structure and the joint distribution among a set of random variables is known that cannot be predicted when only the joint distribution over the set of random variables is known? The answer is that with the former we can predict the effects of intervening in a system by manipulating the values of certain variables, while with only the latter we cannot. For example, if we know only the joint distribution of smoking and lung cancer, we cannot determine whether stopping people from smoking will reduce the rates of lung cancer. On the other hand, if we also know that smoking causes lung cancer (and that there is no common cause of smoking and lung cancer), we can predict that stopping smoking will reduce lung cancer, and by how much. As the quotations at the beginning of the article show, there is a debate within the statistics community about whether predicting the effects of interventions from passive observations is possible at all. In this paper we will describe some recent work for predicting the effects of interventions and policies given passive observations and some background knowledge. While addressing some of the claims just considered, the results we describe unify two traditions in statistics–one, beginning at least as early as Sewell Wright ([Wright 34]; [Simon 77]; [Blalock 61]; [Kiiveri 82]; [Wermuth 83]; [Lauritzen 84]; [Kiiveri 84]; [Wright 34]; [Glymour 87]; [Pearl 88]), connects directed acyclic graphs with probability distributions through constraints on conditional independence relations, while the other ([Neyman 35]; [Rubin 77]; [Rubin 78]; [PearVerm 91]) connects causal claims

---

[0]This article is partially based on "Prediction and Experirmental Design with Graphical Causal Models," by P. Spirtes, C. Glymour, R. Scheines, C. Meek, S. Fienberg, and E. Slate, Carnegie Mellon Technical Report CMU-Phil-32, September, 1992. Our research was supported grant N00014-91-J-1361 from ONR and NPRDC.

with "counterfactual distributions" and offers rules for predicting the distribution of a variable that will result if other variables are deliberately manipulated ([Rubin 77]; [Pratt 88]). We consider the following two cases.

Case 1: We know the causal structure, the joint sample distribution over the variables in the causal structure, which variables will be directly manipulated, and what the direct manipulation will do to those variables. We want to predict the distribution of variables that will not be directly manipulated.

Case 2. A more realistic case is when we do not know the causal structure and latent variables may or may not be present, but we do know a joint sample distribution over a set of variables, which variables will be directly manipulated, and what the direct manipulation will do to those variables. We want to predict the distribution of variables that will not be directly manipulated.

In many areas of applied statistics, causal relations among variables are often represented by directed acyclic graphs (DAGs) over a set of variables $\mathbf{V}$ in which the vertices represent variables and a directed edge $A \rightarrow B$ indicates that A is a direct cause of B relative to $\mathbf{V}$. We say a set $\mathbf{V}$ of variables is *causally sufficient* for a population if for all variables W such that W directly causes two or more members of $\mathbf{V}$ but W itself is not in $\mathbf{V}$, W has the same value for each unit in the population. Given a population of systems each with a causal structure represented by the same DAG, G, whose vertex set $\mathbf{V}$ is causally sufficient, we assume the distribution P of values of variables in $\mathbf{V}$ satisfies two conditions: (i) (Markov Condition) conditional on its parents (that is, its direct causes) in G, each variable V is independent of all variables that are not its descendants (that is, effects of V); (ii) (Faithfulness or DAG Isomorph Condition) all conditional independence relations true in the distribution P follow from satisfying the Markov Condition applied to G. The significance and consequences of these assumptions are described at length in [Pearl 88] and [Spirtes 93]. [Pearl 88] also states a graphical condition named d-separation such that if a distribution P satisifes the Markov and Faithfulness Conditions for G, then $\mathbf{X}$ is independent of $\mathbf{Y}$ given $\mathbf{Z}$ in P if and only if $\mathbf{X}$ is d-separated from $\mathbf{Y}$ given $\mathbf{Z}$ in G (where boldface variables represents sets of random variables). The interest (if not the truth) of the conditions is established by the fact that empirical statistical models with causal interpretations very often satisfy them. For systems in which each variable in a graph is written as a function of its parents plus independent error, the Markov Condition is *necessarily* satisfied; for linear systems the condition is also necessarily satisfied if zero correlations and zero partial correlations are substituted for independence and conditional independence relations in the definition of the Markov Condition. And in the linear case it is intuitively clear, and has been proved [Spirtes 93] that given a system of simultaneous linear functionals the Faithfulness Condition holds for all but a set of values of the linear parameters with zero Lebesgue measure.

Let us now consider the case where the causal structure G_U is known (figure 1.) We will call the population actually sampled (or produced by sampling and some experimental procedure) the **unmanipulated** population, and the hypothetical population for which smoking is banned the **manipulated** population. We suppose that it is known that if the policy of banning smoking were put into effect it would be completely effective, stopping everyone from smoking, but would not affect the value of any other variable in the population except through its effect on *Smoking*; in that case we say that *Smoking* has been manipulated. If *Smoking* is manipulated, the distribution of *Drinking* in the population does not change. The causal graph for the hypothetical manipulated population will be different than for the unmanipulated population, and the distribution of *Smoking* is different in the two populations. The unmanipulated and manipulated causal graphs (G_U and G_M respectively) are shown in figure 1.

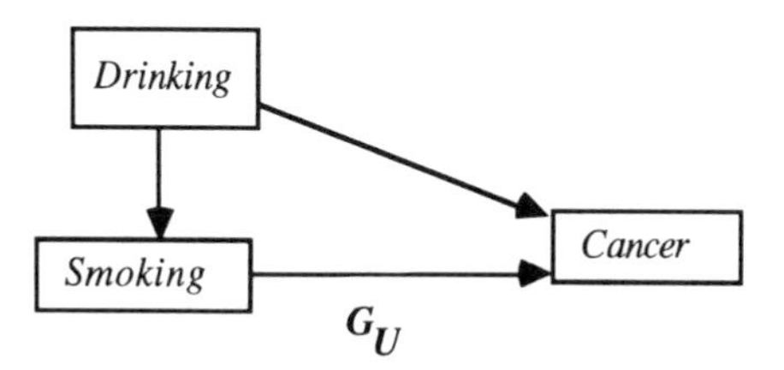

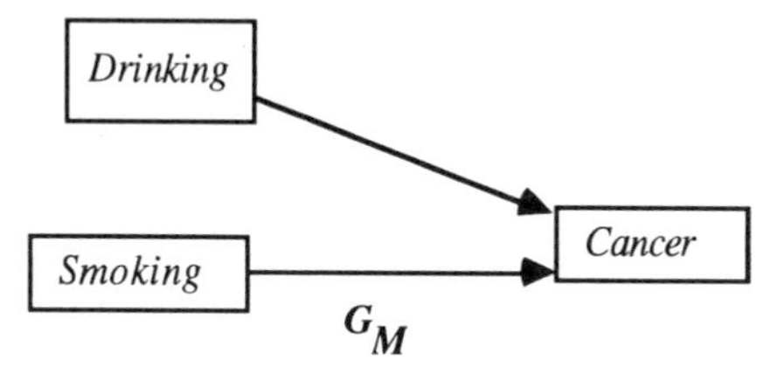

FIGURE 1:

The difference between the unmanipulated graph and the manipulated graph is that some vertices that are parents of the manipulated variables in G_U may not (depending upon the precise form of the manipulation) be parents of manipulated variables in G_M and vice-versa.

How can we describe the change in the distribution of *Smoking* that will result from banning smoking? One way is to note that the value of a variable that represents the policy of the federal government is different in the two populations. So we could introduce another variable into the causal graph, the *Ban Smoking* variable, which is a cause of *Smoking*. In the actual unmanipulated population the *Ban Smoking* variable is *off*, and in the hypothetical population the *Ban Smoking* variable is *on*. In the actual population we measure *P*(*Smoking* | *Ban Smoking* = *off*); in the hypothetical population that would be produced if smoking were banned *P*(*Smoking* = 0 | *Ban Smoking* = *on*) = 1. The full causal graph, G_C, represents the causal structure in the combined unmanipulated (*Ban Smoking* = 0) and (hypothetical) manipulated (*Ban Smoking* = 1) populations. It combines the edges from both G_U and G_M, includes the new variable representing smoking policy, and is shown in figure 2. For any subset **X** of **V** = $\{Smoking, Drinking, Cancer\}$ in the causal graph, let $P_U(\mathbf{X})$ be *P*(**X** | *Ban Smoking* = *off*) and let $P_M(\mathbf{V})$ be *P*(**V** | *Ban Smoking* = *on*) .

We can now ask if $P_U(Cancer \mid Smoking) = P_M(Cancer \mid Smoking)$ (for those values of *Smoking* for which $P_M(Cancer \mid Smoking)$ is defined, namely *Smoking* = 0)? Clearly the answer is affirmative exactly when *Cancer* and *Ban Smoking* are independent given *Smoking*; but if the distribution is faithful this just reduces to the question of whether *Cancer* and *Ban Smoking* are d-separated given *Smoking*, which they are not in this causal graph. Further, $P_U(Cancer) \neq P_M(Cancer)$ because *Cancer* is not d-separated from *Ban Smoking* given the empty set. But in contrast

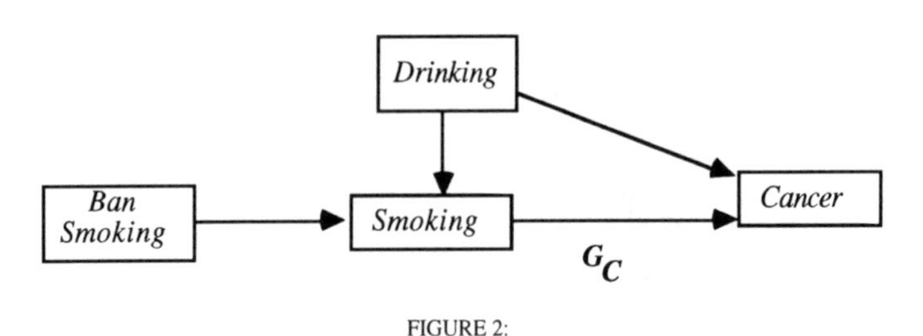

FIGURE 2:

$$P_U(Cancer \mid Smoking, Drinking) = P_M(Cancer \mid Smoking, Drinking)$$

(for those values of *Smoking* for which $P_M(Cancer \mid Smoking, Drinking)$ is defined, namely *Smoking* = 0), because *Ban Smoking* and *Cancer* are d-separated by the pair of variables $\{Smoking, Drinking\}$. The importance of this invariance is that we can predict the distribution of cancer if smoking is banned by by the following formula (remembering that $P_U(Smoking = 0) = 1$):

$$P_M(Cancer) =$$

$$\vec{\sum_{Smoking, Drinking}} P_M(Cancer \mid Drinking, Smoking) \times P_M(Drinking) \times P_M(Smoking) =$$

$$\vec{\sum_{Drinking}} P_U(Cancer \mid Drinking, Smoking = 0) \times P_U(Drinking)$$

which in general is not equal to $P_U(Cancer \mid Smoking = 0)$.(The special summation symbol indicates that we are summing over all values of the random variables listed under the summation sign for which the conditional distributions in the scope of the summation symbol are defined.) Note that this formula which predicts the distribution of *Cancer* in the manipulated population uses only quantities that can be measured from the unmanipulated population and the supposition that $P_U(Smoking = 0) = 1$.

One of the inputs to our conclusion about $P_M(Cancer)$ is that the ban on smoking is completely successful and that it does not affect *Drinking*; this knowledge does not come from the measurements that we have made on *Smoking*, *Drinking* and *Cancer*, but is assumed to come from some other source. Of course, if the assumption is incorrect, there is no guarantee that our calculation of $P_M(Cancer)$ will yield the correct result. In [Spirtes 93] a general theorem stating how to calculate $P_M(\mathbf{Y} \mid \mathbf{Z})$ when a set of variables $\mathbf{X}$ is manipulated is given.

We will now consider the case where the causal structure is not known and latent variables may be present, but we do know a sample joint distribution over a set of variables, which variables will be directly manipulated, and what the direct manipulation will do to those variables. We have developed the Prediction Algorithm, that in some cases can make predictions about the distribution of $P(\mathbf{Y} \mid \mathbf{Z})$ in these circumstances. The details are in [Spirtes 93]. Here we will simply sketch an example of the major steps in the algorithm, without going in to detail about how each step is carried out.

Suppose we measure *Genotype* (G), *Smoking* (S), *Income* (I), *Parents' smoking habits* (PSH) and *Lung cancer* (L) and the unmanipulated distribution is faithful to the graph shown in figure 3(i). The first step of the Prediction Algorithm is to infer as much as possible about the causal structure from the sample distribution. We have developed an algorithm that can be used on either discrete or linear models, called the Faster Causal Inference (FCI) Algorithm for inferring some features of a causal structure from sample data. The algorithm is described in [Spirtes 93]. (For other causal inference algorithms, see [Wright 34], and [Cooper 91].) The input is either raw data or a correlation matrix, and optional background knowledge about the causal relations among the variables. The output is a graphical structure called a partially oriented inducing path graph (POIPG) that contains valuable information about the causal structure among the variables. The procedure has been implemented in the TETRAD II program, available from the authors. The POIPG for the graph in figure 3(i) is shown in figure 3(ii). The small "o"s at the ends of some edges indicate that the presence or absence of an arrowhead is undecided. The *Smoking* $\rightarrow$ *Lung cancer* edge indicates that *Smoking* is a cause of *Lung cancer*. However, the POIPG does not tell us whether *Income* causes *Smoking* or *Income* and *Smoking* have an unmeasured common cause, and so on. While the POIPG narrows down the class of graphs that could have produced the observed distribution, it does not uniquely identify such a graph. For example, from the POIPG shown in figure 3(ii) we are unable to determine whether the graph that produced the observed distribution was the one shown in figure 3(i) or the graph shown in figure 3(iii) (or others besides.) (In figure 3(iii) T_1 and T_2 are unmeasured).

If we directly manipulate *Smoking* so that *Income*, *Parents' smoking habits*, *Genotype*, and any latent variable other than *Ban Smoking* are not parents of *Smoking* in the manipulated graph, then no matter which of the graphs compatible with the POIPG produced the marginal distribution, the POIPG and the Manipulation Theorem tell us that if *Smoking* is directly manipulated that in the manipulated population the resulting causal graph will look like the graph shown in figure 4.

In this case, we can determine the distribution of *Lung cancer* given a direct manipulation of *Smoking*. Three steps are involved. Here, we simply give the results of carrying out each step. First, from the partially oriented inducing path graph we find a way to factor the joint distribution in the manipulated graph. Let P_U be the distribution on the measured variables and let P_M be the distribution that results from a direct manipulation of *Smoking*. It can be determined from the partially oriented inducing path graph that

$$P_M(I, PSH, S, G, L) = P_M(I) \times P_M(PSH) \times P_M(S) \times P_M(G) \times P_M(L \mid G, S)$$

where I = *Income*, PSH = *Parents' smoking habits*, S = *Smoking*, G = *Genotype*, and L = *Lung cancer*. This is the factorization of P_M corresponding to manipulating *Smoking* in the graph of figure 3(i) by breaking the edges into *Smoking*.

Second, we can determine from the partially oriented inducing path graph which factors in the expression just given for the joint distribution are needed to calculate $P_M(L)$. In this case $P_M(I)$ and $P_M(PSH)$ prove irrelevant and we have:

$$P_M(L) = \sum_{G,S}^{\rightarrow} P_M(S) \times P_M(G) \times P_M(L \mid G, S)$$

Third, we can determine from the partially oriented inducing path graph that $P_M(G)$ and $P_M(L \mid G,S)$ are equal respectively to the corresponding unmanipulated probabilities, $P_U(G)$

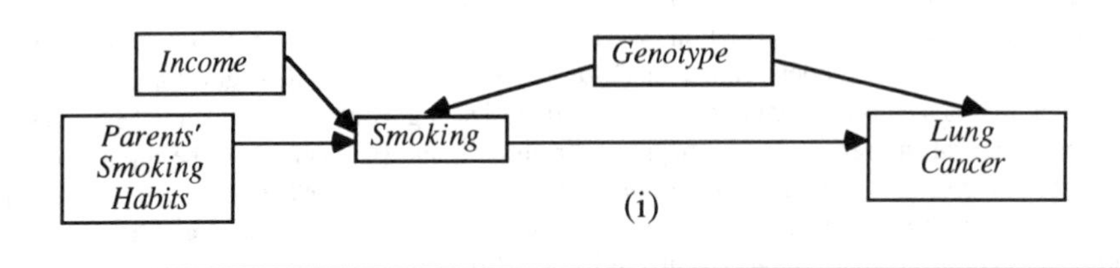

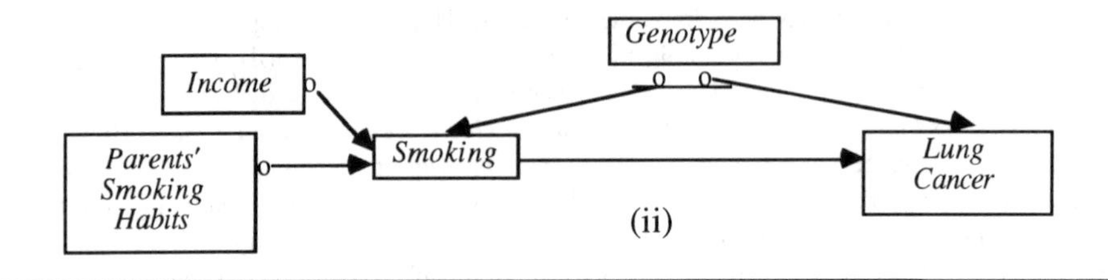

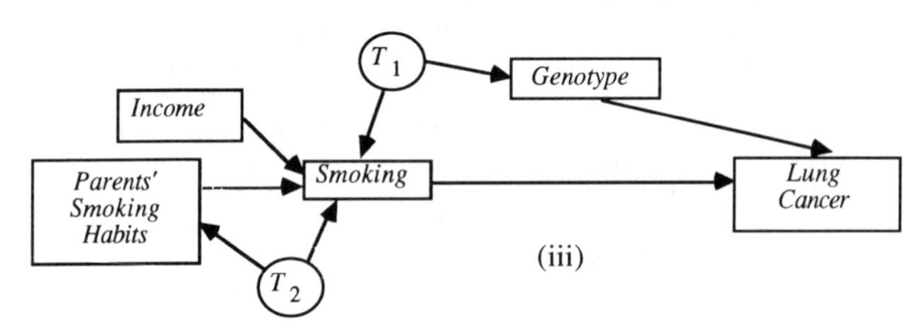

FIGURE 3:

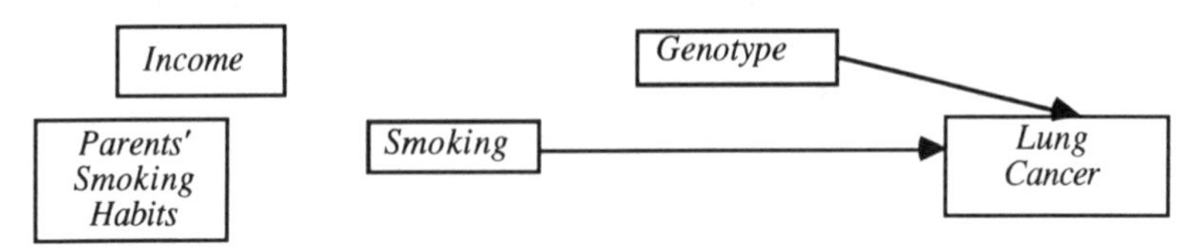

FIGURE 4:

and $P_U(L \mid G,S)$. Furthermore, $P_M(S)$ is assumed to be known, since it is the quantity being manipulated. Hence, all three factors in the expression for $P_M(L)$ are known, and $P_M(L)$ can be calculated. Note that $P_M(L)$ can be predicted even though $P(L)$ is most definitely not invariant under a direct manipulation of S.

22.1 REFERENCES

[Blalock 61] Blalock, H. (1961) *Causal Inferences in Nonexperimental Research*. University of North Carolina Press, Chapel Hill, NC.

[Cooper 91] Cooper, G. and Herskovitz, E. (1991) "A Bayesian Method for the Induction of Probabilistic Networks from Data," *Journal of Machine Learning* **9**, 309-347.

[Glymour 87] Glymour, C., Scheines, R., Spirtes, P., and Kelly, K. (1987) *Discovering Causal Structure*. Academic Press, San Diego, CA.

[Holland 86] Holland, P. (1986) "Statistics and Causal Inference," *JASA* **81**, 945-960.

[Kiiveri 82] Kiiveri, H. and Speed, T. (1982) " Structural Analysis of Multivariate Data: A Review," in *Sociological Methodology*, S. Leinhardt, ed. Jossey-Bass, San Francisco.

[Kiiveri 84] Kiiveri, H., Speed, T., and Carlin, J. (1984) "Recursive Causal Models," *Journal of the Australian Mathematical Society* **36**, 30-52.

[Lauritzen 84] Lauritzen, S., and Wermuth, N. (1984) "Graphical Models for Associations Between Variables, Some of Which are Quantitative and Some of which are Qualitative," *Ann. Stat.* **17** 31-57.

[Mosteller 77] Mosteller, F., and Tukey, J. (1977) *Data Analysis and Regression, A Second Course in Regression*. Addison-Wesley, Massachusetts.

[Neyman 35] Neyman, J. (1935) "Statistical Problems with Agricultural Experimentation," *J. Roy. Stat. Soc. Suppl. 2* 107-180.

[Pearl 88] Pearl, J. (1988) *Probabilistic Reasoning in Intelligent Systems*. Morgan Kaufman, San Mateo, CA.

[Pratt 88] Pratt, J. and Schlaifer, R. (1988) "On the Interpretation and Observation of Laws," *Journal of Econometrics* **39**, 23-52.

[Rubin 77] Rubin, D. (1977) "Assignment to Treatment Group on the Basis of a Covariate," *Journal of Educational Statistics* **2**, 1-26.

[Rubin 78] Rubin, D. (1978) "Bayesian Inference for Causal Effects: The Role of Randomizations" *Ann. Stat.* **6**, 34-58.

[Simon 77] Simon, H. (1977) *Models of Discovery*. D. Reidel, Dordrecht, Holland.

[Spirtes 93] Spirtes, P., Glymour, C., and Scheines, R. (1993) *Causation, Prediction and Search*. Springer-Verlag, *Lecture Notes in Statistics*, New York.

[Verma 90] Verma, T. and Pearl, J. (1990) "Equivalence and Synthesis of Causal Models," in *Proc. Sixth Conference on Uncertainty in AI*. Association for Uncertainty in AI, Inc., Mountain View, CA.

[Wermuth 83] Wermuth, N. and Lauritzen, S. (1983) "Graphical and Recursive Models for Contingency Tables," *Biometrika* **72**, 537-552.

[Wright 34] Wright, S. (1934) "The method of Path Coefficients," *Ann. Math. Stat.* **5**, 161-215.

[Yule 37] Yule, G. and Kendall, M. (1937) *An Introduction to the Theory of Statistics*. Charles Griffin, London.

23
Attitude Formation Models: Insights from TETRAD

Sanjay Mishra and Prakash P. Shenoy

School of Business, University of Kansas,
Summerfield Hall, Lawrence, Kansas 66045-2003
sanjay@ukanvm.cc.ukans.edu and pshenoy@ukanvm.cc.ukans.edu

ABSTRACT The goal of this paper is to examine the use of TETRAD to generate attitude formation models from data. We review the controversy regarding models of attitude formation in the psychology literature, and we re-analyze the data using different subroutines from TETRAD to shed further light on this controversy.

23.1 Introduction

The goal of this paper is to examine the use of TETRAD to generate causal models about attitude formation from existing data. TETRAD is a computer program that discovers directed acyclic graph (DAG) probabilistic models by examining conditional independence relations in data [Glymour *et al.* 1987]. Since the directed edges in a DAG are often interpreted in terms of causality, this program is helpful in searching for "causal" models underlying the available data. As structural equation modelers are aware of, the definition and the imputation of causality are highly controversial. These discussions have been regularly activated in most fields of sciences, liberal arts, and humanities. Thus, when we talk about a "causal" model, our statements and inferences about causality are only as strong as those of other researchers facing this dilemma. This is especially true when the data is collected using non-experimental methods.

There are numerous possible explanations for observed patterns in non-experimental data. It is practically impossible to develop and evaluate possible explanations. Generally, a few explanations are expounded upon and tested for their viability. In this process, one hopes that the explanation that appears most viable is the best and "true" explanation of the data. As researchers have shown by fitting alternative models to the same data, this may not always be the case. For example, the one-factor and the two-factor models of attitude formation have been extensively debated in the psychology literature [Moreland and Zajonc 1979,Zajonc 1980,Zajonc 1981,Birnbaum and Mellers 1979,Birnbaum 1981,Mellers 1981,Wilson 1979]. There are other examples of such controversies in the social sciences. If one could do a *systematic* search for alternative causal models using available data in the field, then our faith in the conclusions would be greater. TETRAD offers one such option.

We review the controversy regarding models of attitude formation in the psychology literature and re-analyze the data using TETRAD. We find some support for the two-factor theory that states that people do not always have to cognitively process a stimulus before developing a liking

[0]This work is based upon work supported in part by the National Science Foundation under Grant No. SES-9213558, and in part by the General Research Fund of the University of Kansas under Grant Nos. 3605-XX-0038 and 3077-XX-0038. We are grateful to Peter Spirtes and Richard Scheines for providing the latest version of TETRAD II and for their comments and discussions.

[1]*Selecting Models from Data: AI and Statistics IV.* Edited by P. Cheeseman and R.W. Oldford. ©1994 Springer-Verlag.

for it. However, TETRAD suggests models that are different from either the one-factor or the two-factor model proposed in the psychology literature.

23.2 TETRAD

The TETRAD program is a semi-automated method for finding a DAG model for continuous variables [Glymour *et al.* 1987,Spirtes *et al.* 1992]. TETRAD assumes a normal linear model holds for the data set. The linearity assumption enables the program to check for conditional independencies using partial correlations. This is also referred to as the search for vanishing partial correlations or tetrads. TETRAD has been evaluated and compared to statistical techniques such as LISREL and EQS [Spirtes, Scheines, and Glymour 1990]. Both LISREL and EQS have modification indices or a similar number to aid in model development. However, the method of using modification indices is computationally less efficient and has been found to be less reliable than the method used by TETRAD. In structural equation models, the problem of improper solutions can be particularly severe for complex models with small sample sizes [Boomsma 1985]. A likely cause of improper solutions is a misspecified model. Using the TETRAD approach, improper specification of the model is less likely to occur. Thus, one can infer that the misfits are be due to sampling variation only. In case of improper solutions, researchers have been also investigating other approaches [Bagozzi and Yi 1991].

The TETRAD program uses artificial intelligence techniques—graph algorithms and heuristics—to suggest alternative models in a systematic manner and test for these models. The goal of TETRAD is to discover models that are consistent with the correlational data. Other common techniques like factor analysis and regression analysis are also driven by similar goals, such as factor rotations in factor analysis, and stepwise regression procedures in regression analysis. The question however remains: On what basis should we decide which is the best model? Theoretical justifications and the argument for parsimony have been used for resolving this issue. As suggested earlier, the strength of TETRAD lies in its use of artificial intelligence techniques to systematically explore for the possible configurations that could be a representation of the observed data. The "best" elaborations of the initial models are defined as those that imply patterns or constraints that are judged to hold in the population, that do not imply patterns judged not to hold in the population, and that are as "simple as possible."

To suggest possible modifications of the input model, TETRAD uses the following principles [Glymour *et al.* 1987, pp. 93–95]: (1) *Spearman's principle:* Other things being equal, prefer those models that for values of their free parameters, entails the constraints judged to hold in the population. (2) *Thurstone's principle:* Other things being equal, a model should not imply constraints that are not supported by the sample data. (3) *The Simplicity principle:* Simpler models are those that have less causal connections. Other things being equal, select models that suggest the least number of causal connections.

It is not always that these principles lead to identical suggestions. For example, Spearman's principle is more likely to hold for simple models. Simple models imply most constraints, but seldom are constraints supported by data—Thurstone's principle. Thus, the fewer the model constraints implied, the better Thurstone's principle holds. As we incorporate more and more parameters into a model, the principle of simplicity is less likely to be satisfied. In spite of the apparent conflict between the principles, TETRAD has performed admirably in generating models, identifying "correct" models, and suggesting modifications to existing models [Glymour *et al.* 1987].

Past research with TETRAD. Spirtes *et al.* compare the reliability of the computer aided model specifications using TETRAD [Spirtes, Scheines, and Glymour 1990], and the other commonly available software such as EQS [Bentler and Chou 1990] and LISREL [Joreskog and Sorbom 1990]. On the basis of simulation studies, the authors conclude that the TETRAD software is better than the others in model specification and search. The main difference is that EQS and LISREL suggest a single model, whereas TETRAD suggests a small list of such "models." Whether the models suggested in the small list are interpretable is not considered in TETRAD. Others [Bentler and Chou 1990,Joreskog and Sorbom 1990] raise interesting questions about the brute force search method and the computer aided model search philosophy underlying TETRAD.

Scheines has extended the use of TETRAD in modeling the measurement of latent variables [Scheines 1993]. The emphasis is more on developing the "best" measurement model rather than the structure of latent variables. The MIMBUILD subroutine, in TETRAD, is used to develop a "pure" measurement model—a model where each observed variable is an indicator of a single latent variable.

Notwithstanding the objections of the classical theorists, we think that research in the use of TETRAD for inducing models from data has been inadequate. This is highlighted considering the fact that TETRAD gives us a systematic approach to evaluating a host of alternative representations of the correlational (or covariance) data. Applications of TETRAD in theory development in the applied as well as the basic sciences should be welcomed.

23.3 Models of attitude formation

The purpose of the current study is to demonstrate the use of TETRAD in theory development and conflict resolution in a research area. We will focus on the models of attitude formation in psychology for this demonstration. Do preferences need inferences? Is cognition a necessary precondition for liking to develop?

Two different models of attitude formation have been suggested in the literature [Birnbaum and Mellers 1979,Moreland and Zajonc 1977,Moreland and Zajonc 1979]. The two models are shown in Figures 1 and 2. In these figures, latent variables are shown as elliptic nodes, and observed variables are shown as rectangular nodes. Also, latent variables are given upper-case labels, and observed variables are given lower-case labels. The model in Figure 1 is called the *one-factor* model, and the model in Figure 2 is called the *two-factor* model. The one-factor model supports the traditional view that to prefer or like something, one has to first recognize it (Figure 1). The two-factor model supports Zajonc's claim that recognition of a stimulus does not have to precede liking it (Figure 2).

Cognition and affect controversy. Till the late sixties and the early seventies, practically all research in stimulus processing assumed cognition to be an intervening stage for liking, i.e., subjects responded to a stimulus only after a cognitive processing cycle [Zajonc 1980, p. 153]. It was only later that this paradigm was challenged by Zajonc and his associates [Zajonc 1980,Zajonc and Markus 1982,Moreland and Zajonc 1979]. This resulted in a controversy among cognitive psychologists. This is evidenced by the extensive rebuttals, and re-analysis of these studies [Birnbaum and Mellers 1979,Lazarus 1982,Lazarus 1984]. Experimental evidence suggests that subjects do not appear to recognize certain stimuli but still have positive feelings

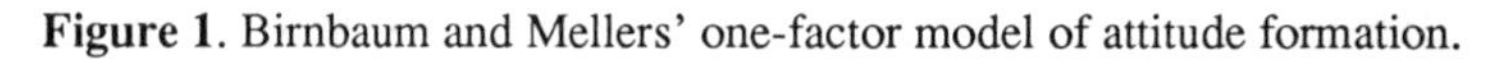

Figure 1. Birnbaum and Mellers' one-factor model of attitude formation.

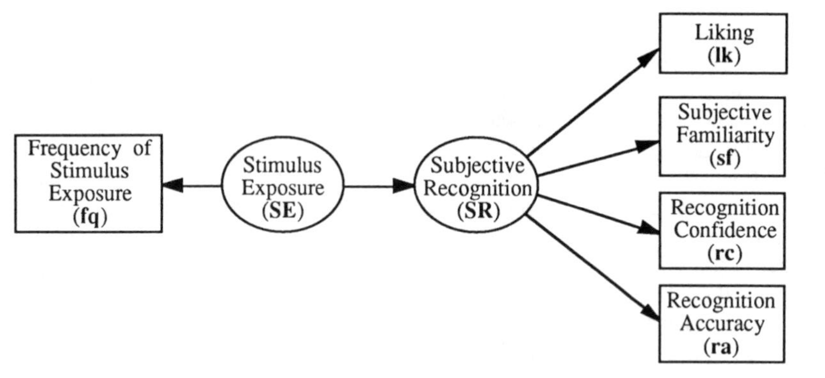

Figure 2. Moreland and Zajonc's two-factor model of attitude formation.

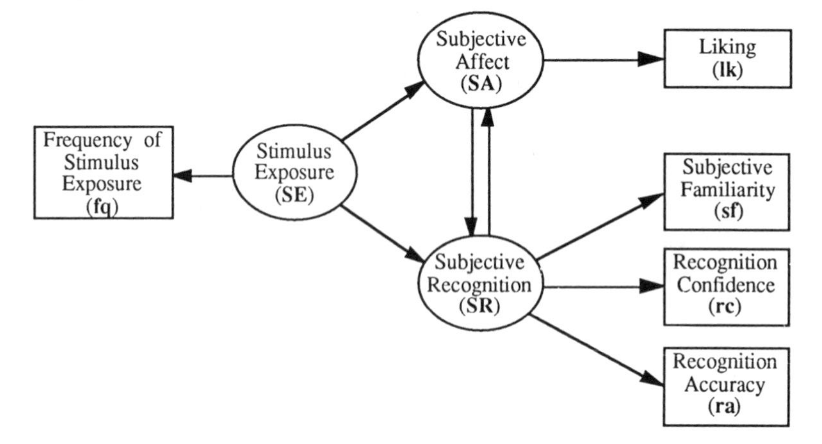

Table 1. The correlation matrix.

	fq	lk	sf	rc	ra
fq	1.000				
lk	0.663	1.000			
sf	0.579	0.533	1.000		
rc	0.340	0.252	0.291	1.000	
ra	0.413	0.302	0.555	0.417	1.000

about them if they have been exposed to them. This led Zajonc to model "feeling without thinking" or "preferences without inferences." The affectists argue that affective reactions to stimuli are often the first reactions and can occur without extensive perceptual and cognitive encoding. These reactions are made with more confidence than cognitive judgments and can be made sooner. This is also referred to as the two-factor theory (Figure 2). The cognitivists believe that "thought is a necessary condition of emotion" and are proponents of the one-factor theory (Figure 1).

Moreland and Zajonc conducted two experiments to investigate the recognition and affect theory [Moreland and Zajonc 1977]. They conclude that subjective recognition is not a necessary condition for the occurrence of the observed exposure effects. This is in contrast to the one-factor theory that considers affect to be post-cognitive. However, a re-analysis of the same data (Table 1) led others to conclude that the data could have been generated from a one-factor model by incorporating correlated errors for the recognition measures [Birnbaum and Mellers 1979].

23.4 Further insights into the controversy

The debate over the data and the re-evaluation of the conclusions of the Moreland and Zajonc study make this an appropriate test of TETRAD's capacity to induce models from data. There are numerous ways in which we can use TETRAD. Most of the differences are determined by how much of the hypothesized model one accepts. As stated earlier, TETRAD has an inherent bias for simpler models hence an edge specified in the model is less likely to be eliminated. The program suggests *elaborations* to a hypothesized model. The deletion of edges can be based on results from EQS.

Assuming properly specified measurement model. We shall first consider the results from TETRAD, assuming that the measurement model as developed by Moreland and Zajonc [Moreland and Zajonc 1977] is appropriate, i.e. stimulus exposure is measured by frequency, subjective affect is measured by liking, and subjective recognition is measured by subjective familiarity, recognition accuracy, and recognition confidence.

The SEARCH subroutine is used to suggest enhancements to the model. The input includes the correlation matrix, a hypothesized graph, significance levels for the tests. (See [Mishra and Shenoy 1993] for details of the various models run and subroutines used.) Extra knowledge about the phenomenon under study is incorporated in the "/Knowledge" section. Since exposure of the individual to the stimulus must occur before any affect or recognition of the stimulus, we set temporal precedence of **fq** over the other variables. We also prevented any observed variables from affecting the latent variables or other observed variables. However, latent variables could cause observed variables and other latent variables. TETRAD suggested three edges be added: **SA→sf**, **SE→ra**, and **SA→ra**. The tetrad score of 100 implies that every constraint passes the program's statistical tests and none fail. The model, denoted by M1, is presented in Figure 3. The dashed arrows represent hypothesized relations and the solid arrows represent elaborations suggested by TETRAD. We conclude from M1 that stimulus exposure affects recognition accuracy. Also, subjective affect affects recognition accuracy and subjective familiarity. However, M1 says nothing about the relationships between the latent variables.

Figure 4 shows the model suggested by TETRAD, denoted by M2, when only edges between latent variables are permitted. The tetrad score for the added edges is 92.50. The difference from the earlier model, M1, is that relationships between the latent variables and measured variables not included in the graph statement are forbidden. We conclude from M2 that **SE** causes **SR**, and **SR** causes **SA**. M2 is unlike the two-factor model shown in Figure 2 because there is no direct path from **SE** to **SA**. This means that **SE** does not directly influence **SA**. M2 appears to support the one-factor theory. However, M2 is not the one-factor model because **SR** does not directly influence the liking variable even when this is permitted.

Figure 3. Model M1 suggested by the SEARCH subroutine in TETRAD with elaborations from latent variables to observed variables permitted. Dashed arrows were assumed, and solid arrows were suggested by TETRAD.

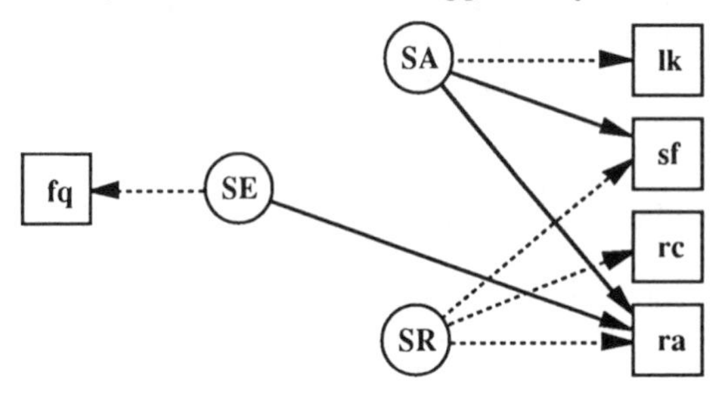

Figure 4. Model M2 suggested by the SEARCH subroutine in TETRAD with elaborations permitted among latent variables only. Dashed arrows were assumed, and solid arrows were suggested by TETRAD.

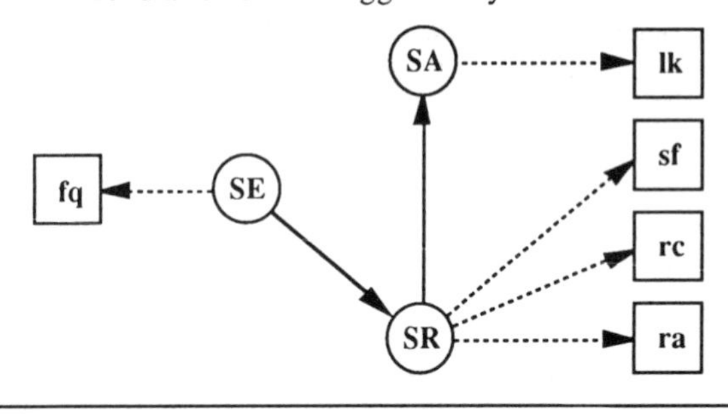

We conclude that the models suggested by TETRAD are neither the one- nor the two- factor models suggested by past researchers. A major assumption for these sets of models is that the measurement model is appropriate. A closer look at the past results [Birnbaum and Mellers 1979,Moreland and Zajonc 1979] shows that the factor loadings for the recognition accuracy and confidence measures may not be adequate. Normally, loadings should be greater than 0.707 (which means that at least 50% of the variance in measured variables is attributable to the latent variables). The loadings for the recognition accuracy and confidence variables are below this accepted norm. Therefore, we used the MIMBUILD subroutine to check the purity of the measures.

Purity of the measurement model. A measurement model is pure if each indicator variable is directly caused by only one latent variable, and the error terms are uncorrelated [Spirtes *et al.*

1992]. The MIMBUILD routine stopped when there were only two indicators of both **SA** and **SR**. This is likely to happen when a pure model is absent because no further purification of the model is possible. Based on the EQS results and our understanding of the data, we believe that the specified measurement model is not a pure one. In fact, model M1 suggests **SA** as a direct cause of **ra** and **sf** even when these were already being caused by **SR**.

Using the BUILD subroutine. If we are not sure about the validity of the measurement model proposed by the earlier researchers (and there are some reasons to doubt the validity), TETRAD has a subroutine, BUILD, which suggests possible relations among the observed variables. The two major classes are: (1) Models that assume causal sufficiency, i.e. the measured variables do not have any latent causes, and (2) Models that do not make this assumption, i.e. the measurement aspect of the model is also considered.

In the input to BUILD, we fixed exposure to the stimulus to precede everything else. Also, because the debate involves whether liking for a stimulus is always preceded by cognitive action, we made no designation for the temporal order for liking with respect to the other variables.

When causal sufficiency is assumed, the suggested model, M3, includes the following directed edges: **fq→sf**, **fq→rc**, **lk→sf**, **ra→sf**, and **rc→ra** (as shown in Figure 5). The edge **fq—lk** is undirected, but based on the temporal precedence of **fq** over **lk**, we could direct the edge from **fq** to **lk**. This supports the two–factor theory that exposure frequency directly impacts liking. Contrary to the one–factor theory that liking is post-cognitive, we do not have a directed edge from any of the cognitive measures to **lk**. In fact, the only suggested path between the two sets is *from* **lk** *to* **sf**. Liking appears to be causing subjective familiarity.

Critics may argue that we did not consider any latent causes for the measured variables in this model, hence it is a misspecified model. Therefore, we also used the BUILD routine without assuming causal sufficiency. The resultant model, M4, is shown in Figure 6.

The routine directs the edges **fq→sf**, **fq→rc**, **ra→sf**, and **rc→ra**. The routine suggests that there could be a common cause for **fq** and **lk**, or the order of causality is undetermined. The edge o—o indicates that **fq** and **lk** are statistically dependent conditional of every set of measured variables. However, based on temporal precedence we can direct the edge from **fq→lk**. What is more interesting is that the routine does not suggest a common cause for the cognitive measures: **sf, rc,** and **ra**. This questions the validity of the measurement model suggested in one–factor and two–factor theory, i.e., **SR→sf**, **SR→rc**, and **SR→ra**. The edge **lko→sf** indicates that **lk** is either a cause of **sf** or there is a common unmeasured (or latent) cause of **lk** and **sf** or both. This means that the only *suggested* common cause is between **sf** and **lk**.

Based on the models suggested by the BUILD subroutine, we conclude that the measurement model for the subjective recognition variable suggested by the one–factor and two–factor models is not valid. Also, we find no evidence for liking being post-cognitive.

23.5 Conclusions

We used TETRAD to shed further light on the controversy between the one-factor and the two-factor models of attitude formation. Based on the BUILD results, we conclude that recognition accuracy, recognition confidence, and subjective familiarity do not appear to be measurements of a common cause, subjective recognition. Models M1 and M2 suggested by TETRAD confirm this conclusion. Also, the only directed edge is not from one of these measures to **lk**, but the other way around (liking to subjective familiarity). This suggests that liking is not post cognitive.

Figure 5. Model M3 suggested by the BUILD subroutine of TETRAD assuming causal sufficiency.

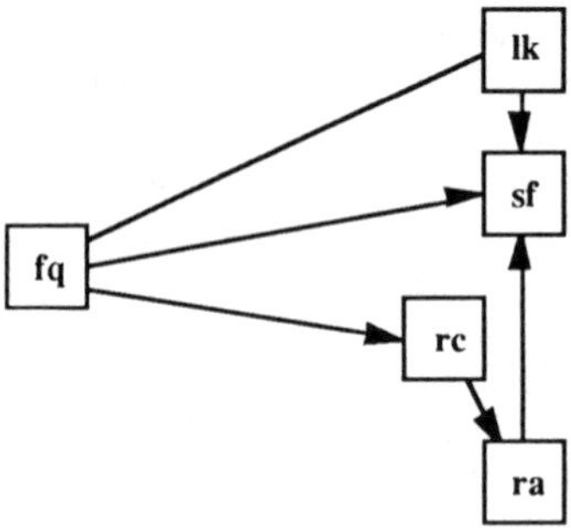

Figure 6. Model M4 suggested by the BUILD subroutine of TETRAD not assuming causal sufficiency.

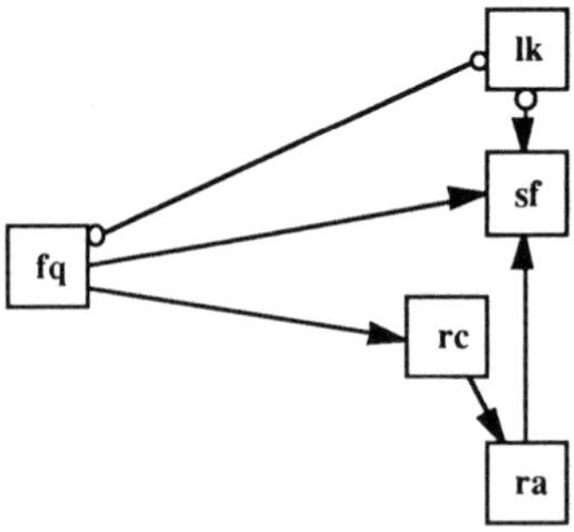

This supports the two-factor theory that people do not always have to cognitively process the stimulus before developing a liking for it: mere exposure is enough. Thus, although we find some support for the two-factor theory, neither the one-factor nor the two-factor model is supported by TETRAD. Instead, TETRAD suggests several different models. It would be interesting to estimate the fit of these models using LISREL or EQS.

The one-factor and two-factor models are relatively simple models with only five observed variables. Ideally, we would have liked to have multiple measures for the stimulus exposure and subjective affect variables. This would have helped purify the model better. The small sample size (n = 400) could be a problem as well. We did not vary the numerous control parameters of the TETRAD subroutines and for the most part worked with the defaults. Also, the results from EQS suggest very good fits for the models suggested in past research. This reduces the scope for improvement. Coupled with the fact that the data set has very few observed variables, the power of the TETRAD is not fully utilized. Future applications of TETRAD should consider these issues.

23.6 REFERENCES

[Bagozzi and Yi 1991] Bagozzi, R. P. and Y. Yi (1991), "Multitrait-Multimethod Matrices in Consumer Research," *J. Cons. Res.*, **17(4)**, 426–439.

[Bentler and Chou 1990] Bentler, P. and C. Chou (1990), "Model Search in TETRAD II and EQS," *Soc. Methods and Res.*, **19(1)**, 67–79.

[Birnbaum 1981] Birnbaum, M. (1981), "Thinking and Feeling: A Skeptical Review," *Amer. Psych.*, **36**, 99–101.

[Birnbaum and Mellers 1979] Birnbaum, M. and B. Mellers (1979), "One Mediator Model of Exposure Effects is Still Viable," *J. Pers. and Soc. Psych.*, **37(6)**, 1090–1096.

[Boomsma 1985] Boomsma, A. (1985), "Nonconvergence, Improper Solutions, and Starting Values in LISREL Maximum Likelihood Estimation," *Psychometrika*, **50(2)**, 229–242.

[Glymour *et al.* 1987] Glymour, C., R. Scheines, P. Spirtes and K. Kelly (1987), *Discovering Causal Structure*, Academic Press, New York, NY.

[Joreskog and Sorbom 1990] Joreskog, K and D. Sorbom (1990), "Model Search with TETRAD II and LISREL," *Soc. Methods and Res.*, **19(1)**, 93–106.

[Lazarus 1982] Lazarus, R. S. (1982), "Thoughts on the Relation Between Emotion and Cognition," *Amer. Psych.*, **37(9)**, 1019–1024.

[Lazarus 1984] Lazarus, R. S. (1984), "On the Primacy of Cognition," *Amer. Psych.*, **39(2)**, 124–129.

[Mellers 1981] Mellers, B. (1981), "Feeling More Than Thinking," *Amer. Psych.*, **36**, 802–803.

[Mishra and Shenoy 1993] Mishra, S. and P. Shenoy (1993), "Inducing Attitude Formation Models Using TETRAD," School of Business Working Paper No. 247, University of Kansas.

[Moreland and Zajonc 1977] Moreland, R. L. and R. Zajonc (1977), "Is Stimulus Recognition a Necessary Condition for Exposure Effects?" *J. Pers. and Soc. Psych.*, **35(4)**, 191–199.

[Moreland and Zajonc 1979] Moreland, R. and R. Zajonc (1979), "Exposure Effects May Not Depend on Stimulus Recognition," *J. Pers. and Soc. Psych.*, **37(6)**, 1085–1089.

[Scheines 1993] Scheines, R. (1993), "Inferring Causal Structure Among Unmeasured Variables," *Prelim. Papers of the 4th Intern. Workshop on A.I. and Stat.*, 151–162, Fort Lauderdale, FL.

[Spirtes, Scheines, and Glymour 1990] Spirtes, P., R. Scheines and C. Glymour (1990), "Simulation Studies of the Reliability of Computer-Aided Model Specification using the Tetrad II, EQS, and LISREL Programs," *Soc. Methods and Res.*, **19**, 3–66.

[Spirtes *et al.* 1992] Spirtes, P., R. Scheines, C. Glymour, and C. Meek (1992), *TETRAD II: Tools for Discovery*, Version 2.1 User's Manual, Department of Philosophy, Carnegie Mellon Univ., Pittsburg, PA.

[Wilson 1979] Wilson, W. (1979), "Feeling More Than We Can Know: Exposure Effects Without Learning," *J. Pers. and Soc. Psych.*, **37(6)**, 811–821.

[Zajonc 1980] Zajonc, R. (1980), "Feeling and Thinking: Preferences Need No Inferences," *Amer. Psych.*, **35**, 151–175.

[Zajonc 1981] Zajonc, R. (1981), "A One-Factor Mind About Mind and Emotion," *Amer. Psych.*, **36**, 102–103.

[Zajonc and Markus 1982] Zajonc, R. and H. Markus (1982), "Affective and Cognitive Factors in Preferences," *J. Cons. Res.*, **9(2)**, 123–131.

24

Discovering Probabilistic Causal Relationships: A Comparison Between Two Methods

Floriana Esposito, Donato Malerba, and Giovanni Semeraro

Dipartimento di Informatica
Università degli Studi di Bari
via Orabona, 4
70126 Bari
Italy
{esposito | malerbad | semeraro}@ vm.csata.it

ABSTRACT This paper presents a comparison between two different approaches to statistical causal inference, namely Glymour *et al.*'s approach based on *constraints on correlations* and Pearl and Verma's approach based on *conditional independencies*. The methods differ both in the kind of constraints considered while selecting a causal model and in the way they search for the model which better fits the sample data. Some experiments show that they are complementary in several aspects.

24.1 Introduction

Causal knowledge is a fundamental part of any intelligent system. In second generation expert systems, a causal model of the domain is used in order to perform a deep reasoning when heuristic knowledge is inadequate, so leading to a graceful performance degradation instead of an abrupt failure when there is no heuristic rule covering a particular situation [Steels 85]. In the middle eighties, the exigency of disposing of deep knowledge on the application domain has been raised in machine learning as well. In particular, a domain theory, often including a causal model, is used in explanation-based learning while generalizing concepts [Pazzani 90]. When the domain theory is incomplete, the automated discovery of causal relations in a database can help to recover part of it.

Causal relationships can be deterministic or probabilistic. The former can be established when the descriptions of all the variables involved in a phenomenon are sufficiently detailed, while the latter exist when some relevant variables cannot be measured or when some measurements are not sufficiently accurate. As already pointed out in [Suppes 70], most of the causal relationships that humans use in their reasoning are probabilistic.

In this paper, we compare two approaches to discovering probabilistic causal relations, namely Glymour *et al.*'s approach based on constraints on correlations [GlShSpKe 87] and Pearl and Verma's approach based on conditional independencies [PearVerm 91]. Preliminary experiments indicate that these methods are complementary since they show very different results for diverse causal models. Moreover, these methods can only discover a causal dependence between variables in the model and not the causal law. When the variables in the model are numerical (interval or ratio level measurements), coefficients of the linear models can be estimated by

[1] *Selecting Models from Data: AI and Statistics IV.* Edited by P. Cheeseman and R.W. Oldford. ©1994 Springer-Verlag.

means of regression analysis. However, the method based on conditional independencies can also be applied to nominal or ordinal variables, in which case symbolic inductive learning algorithms can help to find the causal rules of the phenomenon. Thus, inductive learning algorithms together with statistical causal inference systems can be exploited in order to synthesize a causal theory.

24.2 The method of constraints on correlations

A *causal model* can be represented by a directed graph, whose nodes are random variables and whose edges denote direct causal relationships between pairs of random variables. Figure 1 shows a causal model concerning electrical faults in a car. Henceforth, we will assume that the reader is familiar with some basic notions on directed graphs. We limit ourselves to recall only some concepts.

An *open path* is a path with no cycle. A *trek* between two distinct nodes v and w is a pair of open paths, p and q, from a node u to v and w respectively, such that they intersect only in u. A trek is denoted by $< p, q >$ and u is called *source* of the trek. Bold arrows in Figure 1 show a trek between the nodes "ignition does not start" and "battery voltage is abnormal." The source of the trek is the variable "battery is dead."

Definition 1 *A* stochastic causal theory *(SCT) is a triple ($\langle$V,E$\rangle$, (Ω,P), X) where:*

1. *$\langle$V,E$\rangle$ is a directed graph (causal model) having no edges from a node to itself.*

2. *(Ω,P) is a probability space for the random variables in V. Every variable in V has a non-zero variance. If there is no trek between two variables $u, v \in V$, then u and v are statistically independent.*

3. *$ErrVar$ is a subset of V containing error variables. Such variables have $indegree = 0$ and $outdegree = 1$. Moreover, for each variable $v \in V$ such that $indegree(v) > 0$ there exists only one error variable e_v with an edge in v. Edges connecting error variables with any other variable $v \in V - ErrVar$ are not taken into account when the indegree of v is computed. There is no trek between any two error variables.*

4. *X is a set of independent equations in V. For each $v \in V$ such that $indegree(v) > 0$ the equation $v = f_v(w_{v_1}, w_{v_2}, ..., w_{v_n}, e_v)$ is a member of X, where w_{v_1}, w_{v_2}, ..., w_{v_n} are variables adjacent to v.*

According to this definition, stochastic causal theories may differ in the form of the relationship between a set of causes (variables adjacent to v) and an effect (v itself), that is in the form of f_v. In some theories the functions f_v are linear.

Definition 2 *A* stochastic linear causal theory *(SLCT) is a 4-tuple ($\langle$V,E$\rangle$, (Ω,P), X, L) where:*

1. *($\langle$V,E$\rangle$, (Ω,P), X) is a stochastic causal theory;*

2. *L is a labelling function for the edges in E taking values in the set of non-null real numbers. For any edge g from u to v, we will denote $L(g) = a_{vu}$. The label of a path p, $L(p)$, is defined as the product of the labels of each edge in p. The label of a trek t=<p,q> is defined as $L(t) = L(p) \cdot L(q)$.*

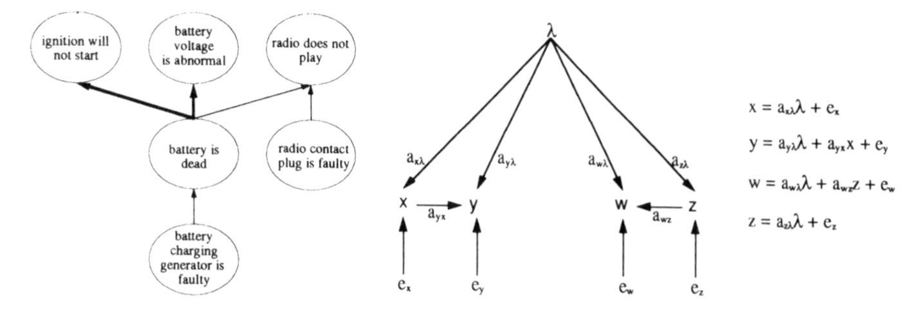

Figure 1: A causal model of electrical faults in a car.

Figure 2: A linear causal model and the corresponding set of equations.

3. *X is a set of independent homogeneous linear equations in V. For each $v \in V$ such that $indegree(v) > 0$ the equation: $v = \sum_{w \in Adj(v)} a_{vw} \cdot w + e_v$ is a member of X where $Adj(v) = \{w | \ w$ is adjacent to $v\}$.*

A variable v is said to be *independent* iff $indegree(v) = 0$, otherwise it is *dependent*. Figure 2 shows an example of causal model and its related set of equations. Here, λ is a *latent* variable representing a non-measured cause of the phenomenon described by the causal model. The other variables are *measured* or *observed*, since a sample is available for them.

Explaining observed data is the main reason for which a theory is built. In particular, for stochastic linear causal theories, the goal is that of discovering causal relationships by exploiting information on the correlations between measured variables. However, the following question raises: how can correlations (or covariances), which are symmetric information, provide hints on the direction of causality? Of course, by using a simple correlation between two variables no one can discriminate a genuine cause from a spurious association, i.e. the concomitant variations due to a common (hidden) cause, or the direction of causality in the case of genuine causal dependencies. On the contrary, the causal relationship between two variables can be determined by examining the covariances of a larger set of variables. For instance, if we are given two variables, x="battery is dead" and y="radio does not play", we can conclude that x causes y only if we consider a further control variable z, say "radio contact plug is faulty", such that z is correlated to y but not to x (see Figure 1). Symmetrically, we can also conclude that z causes y by taking x as control variable.

Given a SLCT T =(⟨V,E⟩, (Ω,P),X,L), Glymour *et al.* [GlShSpKe 87, pages 285-286] have proven the following formula for the covariance γ_{xy} between two variables x and y:

$\gamma_{xy} = \sum_{t \in T_{xy}} L(t)\sigma^2_{s(t)}$

where T_{xy} is the set of treks between x and y and $s(t)$ is the source of a trek t.

Covariances (correlations) computed according to this formula are said to be *implied* by the theory, in order to distinguish them from the sample covariances (correlations). Unfortunately, the computation of γ_{xy} for any two variables x and y requires knowledge about the edge labels and the variances of the independent variables. Rather than estimate coefficients and variances in order to compare the *values* of implied correlations with the corresponding values of sample correlations, Glymour *et al.* prefer to compare *constraints* on covariances implied by the model with constraints on covariances which hold in the data. Thus, the problem is that of *searching for a causal model such that constraints on implied correlations best fit constraints on sample correlations.*

There are several kinds of constraints on correlations which could be considered, but some considerations on computational complexity drive Glymour *et al.* to contemplate only two of them: *tetrad* and *partial equations*. Such constraints are enough to choose among several alternative causal models and can be easily tested as well.

Definition 3 *Let $M = \langle V, E \rangle$ be a causal model, and x, y, w, z four distinct measured variables in V. A* tetrad *equation between x, y, w,z is one of the following equations:*

$$\rho_{xy} \cdot \rho_{wz} = \rho_{xw} \cdot \rho_{yz} \qquad \rho_{xy} \cdot \rho_{wz} = \rho_{xz} \cdot \rho_{yw} \qquad \rho_{xw} \cdot \rho_{yz} = \rho_{xz} \cdot \rho_{yw}$$

where ρ_{ij} is the correlation between i and j.

Definition 4 *Let $M = \langle V, E \rangle$ be a causal model, and x, y, z three distinct measured variables in V. A partial equation between x, y, z is the following equation:*

$$\rho_{xz} = \rho_{xy} \cdot \rho_{yz} \qquad \text{(or equivalently } \rho_{xz.y} = 0\text{).}$$

Such constraints on correlations are particularly interesting since it can be shown that in linear causal models their satisfaction depends *only on the causal model*, that is the directed graph, and not on the probability space [GlShSpKe 87, pages 265-288, 293-310]. Thus, it is not necessary to estimate either the coefficients or the variances of the independent variables, since we can establish which equations are satisfiable (or implied) by a SLCT T =(⟨V,E⟩, (Ω,P), X, L) by simply looking at the causal model ⟨V,E⟩. In particular, the problem of checking whether a constraint (tetrad or partial equation) is satisfied by a SLCT is reduced to a search problem for some trek sets. Then, the problem is moved to verifying constraints which hold in the data. In order to establish whether a tetrad equation holds, an asymptotic test is performed on the *tetrad difference*:

H_0: $\hat{\rho}_{uv} \cdot \hat{\rho}_{wx} - \hat{\rho}_{ux} \cdot \hat{\rho}_{vw} = 0$

where $\hat{\rho}_{ij}$ is the sample correlation between the variables i and j. As to the partial correlation, the following hypothesis has to be tested:

H_0: $\hat{\rho}_{xy.z} = 0$.

In [Anderson 58], the asymptotic tests for both the tetrad difference and the partial correlation are reported. When different alternative causal models are hypothesized, they can be compared on the ground of the following three criteria:

1. $H - I$ (or *incompleteness* criterion): the number of constraints which hold in the data but are not implied by the model;
2. $I - H$ (or *inconsistency* criterion): the number of constraints implied by the model but not holding in the data;
3. *simplicity* criterion: the previous criteria being equal, prefer the simplest model, that is the causal model with the lowest number of edges.

It should be observed that a specialization of a causal model by adding an edge can increment the model incompleteness ($H - I$) and reduce the model inconsistency ($I - H$).

TETRAD's automatic search procedure [GlShSpKe 87] starts with an initial causal model and suggests the addition of sets of treks to the *initial* causal model in order to provide an *extended* model which better fits to the data according to the above criteria. The initial model must be an acyclic graph such that:

1. there are no edges between measured variables;

2. all measured variables are effect of a latent cause;
3. all latent variables are connected to each other.

For each foursome of measured variables, a particular subgraph of the causal model is selected. Such subgraph consists of:

1. the foursome of measured variables;
2. all latent variables, named parents, which are adjacent to any variable in the foursome;
3. all edges from the parents to any variable in the foursome;
4. all treks, including latent variables which are not parents, between the parents.

According to the type of subgraph, TETRAD suggests the addition of treks defeating the implication of constraints which do not hold in the data at a significance level α. However, some of the suggested treks may also increase the incompleteness of the model, in which case they will be excluded from further processing. The remaining suggested treks form a *locally minimal* set LM_i, since their addition to the model is evaluated by considering only a subgraph of the causal model. As search proceeds from foursome to foursome, it forms *globally minimal* sets GM_i of suggested treks such that GM_i will defeat as many implied constraints involving all foursomes considered so far as is possible without increasing incompleteness. At the end of the automatic search, TETRAD outputs the sets GM_i and the first significance level at which these sets are non empty.

24.3 The method of conditional independencies

This method builds the causal model by analysing the conditional independencies which hold in the sample data. It is based on a clear theory of causality which classifies several types of causal relationships. This is done in the light of Reichenbach's principle [Reichenb 56] according to which every dependence between two variables x and y has a causal explanation, so either one causes the other or there is a third variable z influencing both of them (z is the source of a trek between x and y). In this approach temporal information is not considered essential, but it can be easily exploited when available.

Let V be a set of random variables with a joint probability distribution P, and x, y, z_1, z_2, ..., $z_n \in V$. Then x and y are said to be *(conditionally) independent in the context* $\{z_1, z_2, ..., z_n\}$if for each n-tuple $(\zeta_1, \zeta_2, ..., \zeta_n) \in z_1 \times z_2 \times ... \times z_n$ it happens that: $f_{x,y|z_1=\zeta_1,...,z_n=\zeta_n} = f_{x|z_1=\zeta_1,...,z_n=\zeta_n} \cdot f_{y|z_1=\zeta_1,...,z_n=\zeta_n}$. Henceforth, $I(P)$ will denote the set of conditional independencies between the variables in V when P is the joint probability distribution. Moreover, we will write $I(x, y|z_1, z_2, ..., z_n)$ if x and y are conditionally independent in the context $\{z_1, z_2, ..., z_n\}$, otherwise we will write $\neg I(x, y|z_1, z_2, ..., z_n)$.

Given a sample distribution $P^\star$ on a set O of observed variables, the goal is that of finding a causal model $M = \langle V, E\rangle$ such that $O \subseteq V$ and there exists T =(⟨V,E⟩, (Ω,P), X) such that $P_{[O]} = P^\star$, where $P_{[O]}$ denotes the marginal distribution for the variables in O. In other terms, we aim at finding a causal model which may possibly contain some latent variables (when $O \subset V$) and is consistent with sample data. If that model exists, we say that the *latent structure* $S = \langle M, O\rangle$ *implies* the joint probability distribution $P^\star$ on O.

Actually, there could be an infinite number of causal models M satisfying the conditions above, so we need some constraints in order to simplify the problem. Pearl and Verma reduce the problem to *finding a causal model that is consistent with the set of conditional independencies which hold in the data.* More precisely, let P_S be the set of probability distributions on O

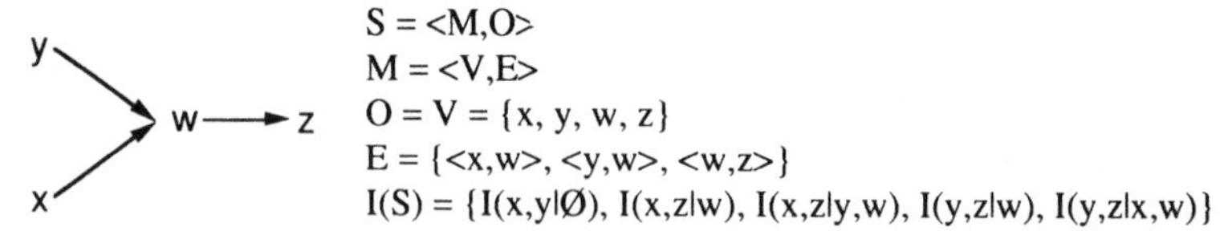

Figure 3: Latent structure and corresponding set of conditional independencies.

implied by a latent structure $S = \langle$M,O$\rangle$. Then, it is possible to provide the following definition of *implied conditional independence*:

Definition 5 *A latent structure $S = \langle M,O \rangle$ implies a conditional independence I between some variables in O, if I is a conditional independence for each probability distribution implied by the latent structure, i.e. $\forall P \in P_S$: $I \in I(P)$.*

Henceforth, $I(S)$ will denote the set of conditional independencies implied by S, that is $I(S) = \bigcap_{P \in P_S} I(P)$. It is worthwhile to observe that the set $I(S)$ can be obtained by simply looking at the causal model. Indeed, it can be proven that there is a conditional independence between two variables x and y in a given context S iff $(\forall t \in T_{xy}\ \exists z \in S : z \in t) \wedge [\forall z \in S\ (\forall t \in T_{zx}$ such that t does not contain variables in $S - \{z\}$ then t is a path from z to $x) \vee (\forall t \in T_{zy}$ such that t does not contain variables in $S - \{z\}$ then t is a path from z to $y)]$. Figure 3 shows a latent structure and the corresponding set of conditional independencies which could be derived according to the given rule.

Now, the search problem can be formally stated as follows: find a latent structure $S = \langle$M,O$\rangle$ such that $I(S) = I(P^\star)$. It is possible to make at least two considerations. Firstly, there is no guarantee that independencies observed on the sample distribution $P^\star$ are not *accidental* but they actually reflect the true independencies in the underlying causal model. For this reason, Pearl and Verma assume that $P^\star$ is a *stable distribution*, that is, $I(P^\star)$ is the set of *structural* independencies due to the underlying causal model. Secondly, there could be an infinite number of dependency-equivalent latent structures $S = \langle$M,O$\rangle$ such that $I(S) = I(P^\star)$, each of which having a different set of latent variables. Pearl and Verma restrict their search to particular latent structures called *projections*.

Definition 6 *A latent structure $S = \langle\langle V, E\rangle, O\rangle$ is a* projection *on O of another latent structure $S' = \langle\langle V', E'\rangle, O\rangle$ iff $I(S) = I(S')$ and each latent variable $\lambda \in V$ is a common cause of exactly two measured variables $v, w \in O$ which are not connected by any edge.*

Fortunately, it can be proven that for each latent structure there is at least one dependency-equivalent projection, thus it is reasonable to search a causal model only in the space of projections. For projections it is convenient to use hybrid graphs as representation formalism. A *hybrid graph* is a couple $G = \langle V, E\rangle$ where V is the set of nodes and $E = \langle E_N, E_U, E_B\rangle$ is a triple of three disjunct sets of edges, namely a set of *non-directed* edges $(u - v)$, E_N, a set of *uni-directed* edges $(u \rightarrow v)$, E_U, and a set of *bi-directed* edges $(u \leftrightarrow v)$, E_B. In hybrid graphs of projections, a latent cause λ for two observed variables u and v is represented by eliminating the variable λ and by adding a bi-directed edge between u and v. Thus the graph of a projection will have only observed variables. The intersection of the hybrid graphs representing all the possible projections of latent structures $S = \langle$M,O$\rangle$ for which $I(S) = I(P^\star)$ is called *core* of $P^\star$.

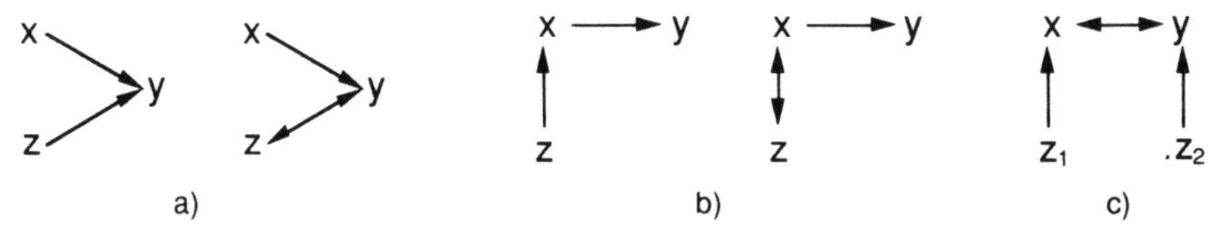

Figure 4: a) the two cases of potential causal influence; b) the two cases of genuine causal influence; c) a case of spurious association.

The algorithm of Inductive Causality proposed by Pearl and Verma [PearVerm 91] builds the core of a sample distribution $P^{\star}$. The input to the program is the set of conditional independencies which hold in $P^{\star}$. The output core has four kinds of edges:

1. *marked uni-directed edges* representing genuine causal relationships;
2. *unmarked uni-directed edges* representing potential causal relationships;
3. *bi-directed edges* representing spurious associations;
4. *non-directed edges* representing unclassifiable relationships.

Below, the definitions of genuine and potential causal relationships as well as the definition of spurious associations are provided.

Definition 7 *A variable x has a* potential causal influence *on another variable y if x and y are dependent in every context (for each context S : $\neg I(x,y|S)$) and there exists a variable z and a context S such that: $I(x,z|S) \wedge \neg I(z,y|S)$.*

Definition 8 *A variable x has a genuine causal influence on y if there exists another variable z such that:*

1. *x and y are dependent in every context and there is a context S such that z is a potential cause of x, z and y are dependent in S ($\neg I(z,y|S)$), and z and y are independent in $S \cup \{x\}$ ($I(z,y|S \cup \{x\})$), or*

2. *x and y are in the transitive closure of rule 1.*

If V contains only three variables, the only cases satisfying definition 7 are those shown in Figure 4a, while the the only two cases satisfying condition 1) of definition 8 are shown in Figure 4b.

Definition 9 *Two variables x and y have a* spurious association *if they are dependent in some context S ($\neg I(x,y|S)$) and there exist two variables z_1 and z_2 such that: $\neg I(z_1,x|S) \wedge I(z_1,y|S)$ $\wedge \neg I(z_2,y|S) \wedge I(z_2,x|S)$.*

In this definition, the first two conditions prevent x from causing y, while the last two conditions prevent y from causing x, so that the dependence between x and y can only be explained by a spurious association (see Figure 4c).

24.4 Empirical comparison between the two methods

In this Section, the two approaches to causal inference are empirically compared. The first three experiments have been performed by using models in which the assumptions of linearity, normality, acyclicity and stability are made.

Experiment 1 Let us consider the model in Figure 5a. All independent variables have a standard normal distribution, $N(0,1)$, the significance level for all the test is $\alpha = 0.05$ and the sample size is 1000. >From some tests on the sample correlations we derive the following conditional independencies:

$$I(x,w|\emptyset), \qquad I(x,z|\emptyset), \qquad I(y,z|\{w\})$$

In Figure 5b, the output of the Inductive Causality Algorithm is shown. The connection $w - z$ cannot be explained due to the lack of control variables, while the connections $x \rightarrow y$, $w \rightarrow y$ labelled as potential causes can be considered genuine causes if we assume that there are no latent variables.

By trying to analyse this sample with TETRAD, we soon meet a problem: what is the initial model? In Figure 5c the simplest initial model is presented. Such a model is evidently wrong since the latent variable τ does not exist in reality. However, we will be interested in finding causal relationships between observed variables, so ignoring connections with τ. The automatic search procedure suggests an addition of either a trek between x and y or a trek between w and z at the significance level $\alpha = 0$, but when we add the edges $x \rightarrow y$ and $w \rightarrow z$ to the initial model we get a model in which no further improvement is possible. Therefore, the method of constraints on correlations is not able to discover the causal relation $w \rightarrow y$.

Experiment 2 When there are latent variables, final results strongly depend on the topology of the initial causal model. For instance, let us consider the true causal model in Figure 6a, where τ is a latent variable. Once again, all independent variables in the sample to have a standard normal distribution. >From the correlations on a sample of 1000 observations we derive the following independencies:

$$I(x,w|y), \qquad I(x,z|\emptyset), \qquad I(y,z|\emptyset)$$

By applying the Inductive Causality Algorithm, the model in Figure 6b is built. This result is undoubtedly better than that obtained by TETRAD when the initial causal model is that shown in Figure 6c. Indeed, both a trek between x and y and a trek between z and y are suggested. Such treks could be realized by adding the following connections: $x \leftrightarrow y$, $z \rightarrow y$. However, the final model is not better than the initial one, neither can we discover the causal relationship $y \rightarrow w$.

Experiment 3 In this experiment we generated a sample of 5000 observations for the model in Figure 7a and we ran TETRAD with the initial model of Figure 7b. Correlations are given in Table III. In this case the addition of a trek between x_2 and x_5 is suggested and by asking a detailed analysis of the initial model augmented with the edge $x_2 \rightarrow x_5$ we are provided with further suggestions, namely:

$$x_4 \rightarrow x_5, \qquad x_4 \leftrightarrow x_5, \qquad \tau_1 \rightarrow x_4, \qquad x_6 \rightarrow x_5, \qquad \tau_1 \rightarrow x_6, \qquad x_6 \leftrightarrow x_5$$

By testing all these additions to the augmented model, we find that the best models are:

initial model + $x_2 \rightarrow x_5 + \tau_1 \rightarrow x_4$

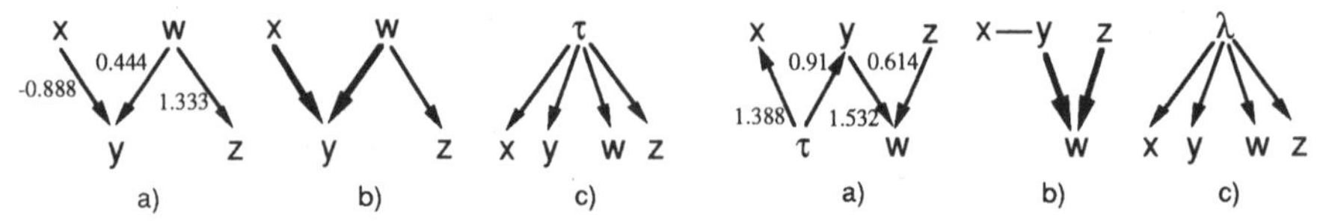

Figure 5: Causal models of experiment 1.

Figure 6: Causal models of experiment 2.

initial model + $x_2 \rightarrow x_5 + x_6 \rightarrow x_5$
initial model + $x_2 \rightarrow x_5 + x_6 \leftrightarrow x_5$
among which there is the true causal model.

The method of conditional independencies provides poor results since no independencies hold in the data. In fact, it is possible to discover independencies only between observed variables, but having excluded latent variables from consideration all couples of variables appear to be dependent (see Figure 7c). Therefore, the method of conditional dependencies is not suitable for studying those phenomena in which there are some latent variables controlling many observed variables which have no spurious association. This is due to the fact that this method can only discover projections of the real model, where a latent variable can affect exactly two observed variables.

Experiment 4 In this experiment we generated a sample of 1000 observations concerning the model in Fig. 8. Each independent observed variable has an ordinal domain with three possible values: low, medium, high. In this case there is no linear dependence between the variables, even if the causal relationships are still probabilistic. Conditional independencies are determined by a χ^2 test for each pair of variable. For $\alpha = 0.05$ the following independencies are significant:

$$I(x_1, x_2|\emptyset), \quad I(x_1, x_4|\emptyset), \quad I(x_1, x_5|\{x_3\}), \quad I(x_2, x_4|\emptyset), \quad I(x_2, x_5|\{x_3\}), \quad I(x_3, x_4|\emptyset)$$

The Inductive Causality Algorithm builds the original model with the only difference that all the causal relationships but $x_3 \rightarrow x_5$ are potential and not genuine. At this point, having discovered the causal dependencies between the variables, we used a learning system in order to induce the *causal rules* from the data. In particular, two learning problems were defined: the first for learning the causal law which relates the study level and the intelligence of a student with his/her preparation and the second for learning how the preparation and the leniency of the examiner can affect the final score. In both cases, the output of the learning system was a probabilistic decision tree [Quinlan 90].

24.5 Conclusions

In this paper, an empirical comparison between two different approaches to statistical causal inference has been presented. They differ both in the kind of constraints considered while selecting a causal model and the way they search for the model which better fits the sample data.

In Glymour *et al.*'s approach, constraints on correlation implied by the model are compared to those that hold in the data. Two kinds of constraints are considered: tetrad and partial equations. TETRAD, a program that finds causal models by testing vanishing tetrad differences and vanishing partial correlations, uses an automatic heuristic search strategy in order to recover the causal structure. Some limitations of this approach are the applicability to datasets with only

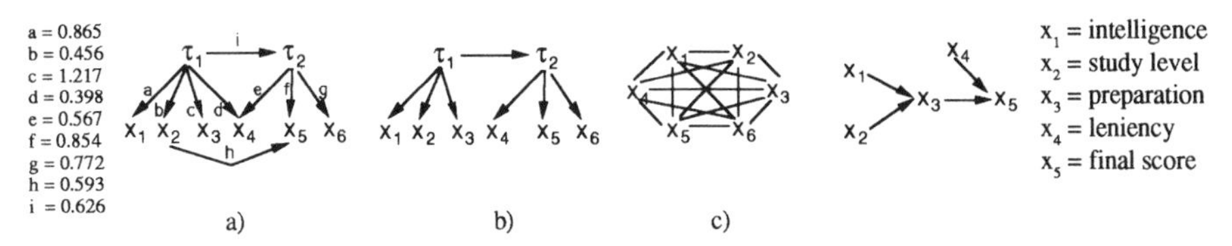

Figure 7: Causal models of experiment 3. Figure 8: Causal model of experiment 4.

interval or ratio level measurements, and the necessity to dispose of a complete initial model involving latent variables.

These limitations do not concern the approach based on conditional independencies which exploits constraints determined by conditional independencies in order to detect genuine causal influences among observed variables. There are a bias and a basic assumption in this method: the former is towards projections while the latter concerns the stability of the underlying distribution.

The complementarity of the two methods is also confirmed by some preliminary experimental results presented in the paper.

24.6 References

[Anderson 58] Anderson, T.W. (1958) *An Introduction to Multivariate Statistical Analysis* John Wiley and Sons, New York, NY.

[GlShSpKe 87] Glymour, C., Scheines, R., Spirtes, P. and K. Kelly (1987) *Discovering Causal Structure* Academic Press, Orlando, FL.

[Pazzani 90] Pazzani, M.J. (1990) *Creating a Memory of Causal Relationships: An Integration of Explanation-Based and Empirical Methods* Erlbaum, Hillsdale, NJ.

[PearVerm 91] Pearl, J. and T.S. Verma (1991) "A theory of Inferred Causation," in *Principles of Knowledge Representation and Reasoning: Proceedings of the Second International Conference*, J.A. Allen, R. Fikes, and E. Sandewall, eds., 441-452, Morgan Kaufmann, San Mateo, CA.

[Quinlan 90] Quinlan, J. R. (1990) "Probabilistic Decision Trees," in *Machine Learning: An Artificial Intelligence Approach, vol III*, Y. Kodratoff and R.S. Michalski, eds., 140-152, Morgan Kaufmann, San Mateo, CA.

[Reichenb 56] Reichenbach, H. (1956) *The Direction of Time*, University of California Press, Berkley, CA.

[Steels 85] Steels, L. (1985) "Second Generation Expert Systems," *Future Generation Computer Systems* **1**, 213-221.

[Suppes 70] Suppes, P. (1970) *A Probabilistic Theory of Causation* North Holland, Amsterdam, Holland.

25

Path Analysis Models of an Autonomous Agent in a Complex Environment

Paul R. Cohen, David M. Hart, Robert St. Amant, Lisa A. Ballesteros and Adam Carlson

Experimental Knowledge Systems Laboratory
Department of Computer Science
University of Massachusetts
Amherst, MA 01003

ABSTRACT We seek explanatory models of how and why AI systems work in particular environments. We are not satisfied to demonstrate performance, we want to understand it. In terms of data and models, this means we are not satisfied with descriptive summaries, nor even with predictive models. We want causal models. In this brief abstract we will present descriptive, predictive and causal models of the behavior of agents that fight simulated forest fires. We will describe the shortcomings of descriptive and predictive models, and summarize path analysis, a common technique for inducing causal models.

25.1 Phoenix

Phoenix is a simulated environment populated by autonomous agents. It is a simulation of forest fires in Yellowstone National Park and the agents that fight the fires [Cohen et al. 89]. Agents include watchtowers, fuel trucks, helicopters, bulldozers and, coordinating (but not controlling) the efforts of all, a fireboss. Fires burn in unpredictable ways due to wind speed and direction, terrain and elevation, fuel type and moisture content, and natural boundaries such as rivers, roads and lakes. Agents behave unpredictably, too, because they instantiate plans as they proceed, and they react to immediate, local situations like encroaching fires.

Phoenix *appears* to work very well, in that bulldozers are expeditiously dispatched to fires and appear to coordinate their efforts, and fires are usually contained. But as with any complex system, one cannot tell much from a demonstration. (In fact, the demo looked fine even though, as we later discovered, a critical algorithm was projecting events minutes instead of hours in the future!) For this reason we turned to an extensive statistical analysis of specific aspects of behavior and the factors that affect them.

25.2 Modeling Phoenix

We created a straightforward fire fighting scenario that controlled for many of the variables known to affect the planner's performance. In each of 343 trials, one fire of a known initial size

---

[0]This research was supported by ARPA under contracts F30602-91-C-0076 and F30602-93-C-0100; and by AFOSR under the Intelligent Real-time Problem Solving Initiative, contract AFOSR-91-0067. The US Government is authorized to reproduce and distribute reprints for governmental purposes notwithstanding any copyright notation hereon.

was set at the same location (an area with no natural boundaries) at the same time (relative to the start of the simulation). Four bulldozers were used to fight it. The wind's speed and direction were set initially and not varied during the trial. Thus, in each trial, the Fireboss receives the same fire report, chooses a fire-fighting plan, and dispatches the bulldozers to implement it. A trial ends when the bulldozers have successfully surrounded the fire or after 120 simulation-time hours without success.

For exploratory purposes we collected over 40 measurements on each trial, but we will concentrate on eight here.

Success. If the fire is contained in less than 120 simulated hours, the trial is a success, otherwise it is a failure.

Wind speed. We set the wind speed to be either 3, 6 or 9 kilometers per hour and didn't change it during a trial.

Realtime knob. This parameter controls the ratio of the fireboss's "thinking time" to the elapsed "environment time." The normal setting allows the fireboss one simulated minute of thinking time for one minute of environment time, during which fires burn. Halving this parameter means the fireboss takes twice as long, relative to the environment, to think.

Utilization. The proportion of a trial that the fireboss spends thinking, as opposed to idle, or waiting for requests, etc.

First plan. Phoenix adopts one of three general plans to fight a fire.

Number of plans. Phoenix sometimes replans when its first plan doesn't work. If number of plans = 1, then no replanning occurred.

Fireline built. This measures the number of meters of fireline cut by the bulldozers in a successful trial.

Shutdown time. This measures the number of hours, from the first sighting to the return to base of the last bulldozer, required to contain the fire.

25.2.1 Descriptive models

We wanted to know how these factors interacted; specifically, we wanted to know how success and shutdown time depended on the other factors. (We will focus on shutdown time, here, but see [Hart and Cohen 92]). A descriptive model is simply a correlation matrix of the factors. We don't recommend this kind of model. We suggest it only to point out its shortcomings, the most glaring of which is that a correlation represents the endpoints of a causal story, not the story itself. For example, we were very surprised to find in our data that wind speed was essentially uncorrelated with shutdown time ($r = -.053$). We knew that higher wind speed increased the spread rate of the fire, which would require more fireline, which would take more time to cut. If this causal story is right, then $r = -.053$ must represent the sum of two influences, the one we just described and another that cancels it out and also depends on wind speed. We traced the problem to a sampling bias in our experiment: For high wind speeds, if a fire isn't contained relatively quickly, then it might not be contained at all. For example, if a fire has been burning for 60 hours or more, and wind speed is 3, then the probability of the fire being contained eventually is .375. But if wind speed is 6, the probability of eventually containing an old fire is only .2, and if wind speed is 9, the probability drops to .13. We measured shutdown time for successful trials only. But successful containment of old fires is relatively unlikely at higher wind speeds, so as wind speed increases, we see fewer older fires contained, thus fewer high values of shutdown time. This results in a negative relationship between wind speed and shutdown time. Note that this relationship represents an effect of missing data, not a true negative causal relationship.

The two influences are shown in a *path diagram* in Figure 1. The path from wind speed to fireline built to shutdown time is the first causal story. We hypothesized another path, corresponding to the second story, from wind speed to shutdown time. By the rules for estimating correlations from path diagrams (described later) the estimated correlation between wind speed and shutdown time is the sum of the two stories: first, the effect of wind speed on fireline built multiplied by the effect of fireline built on shutdown time ($.363 \times .892$); second, the effect that the sampling bias must have ($-.377$) to offset the first story to the extent that the overall correlation is $-.053$. This model does not tell us that the second story is correct—only how strong the hypothesized sampling bias must be. Still, we prefer path models such as Figure 1 to raw correlations because they can tease apart the causal stories that are often confounded in correlations.

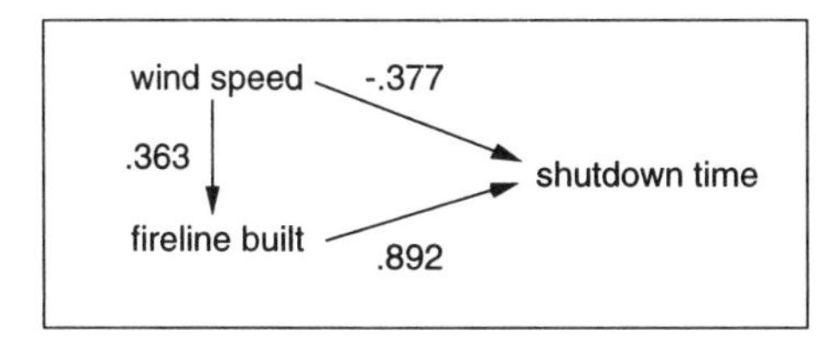

FIGURE 1: A path model of the influences of wind speed and fireline built on shutdown time.

25.2.2 *Predictive regression models*

We wanted a predictive model of how factors influence shutdown time in Phoenix, so we turned to multiple linear regression. Again, we do not recommend this technique, but present it to highlight its limitations. Here is a regression equation for shutdown time; the coefficients are standardized:

$$\begin{aligned} est.Shutdown &= .762 FirelineBuilt + .144 NumPlans + .036 FirstPlan - \\ &\quad .192 RealTimeKnob - .283 WindSpeed \end{aligned}$$

Unfortunately, while predictive, this model is hardly explanatory. It is a completely "flat" causal story. All it says is that five factors influence the dependent variable to different degrees. If our goal is to understand the interactions of factors in a complex environment (i.e., wind speed and real time knob) with decisions and actions of a complex agent (e.g., which plan to use first, how often to replan, and fireline built), then this model is useless. It is also unparsimonious, as described in [Hart and Cohen 92].

25.3 Causal path models

Path analysis is an exploratory modeling technique that has helped us incrementally develop a causal understanding of Phoenix. It is based on multiple regression, but it can be used to fit "multilevel" models in which some predictors point to the dependent variables and are in turn predicted by other variables. One way to run a path analysis is to specify a path model and see how well data fit it; another approach is to let the data suggest path models. The latter approach is causal induction, similar to the work of [Pearl and Verma 91], and [Spirtes et al. 93]. An algorithm is described later. Here, let us focus on the former approach, where the analyst specifies a path model. Figure 2 is a model we derived from the experiment described above. Note that it is neither flat nor unparsimonious: some causal influences are mediated by one or two endogenous variables, and the model includes only a fraction of the influences that are implicit in a multiple regression model. Consequently, the coefficient of determination R^2 of this model is lower than for the regression model. The trick in path analysis is to get a good degree of fit to the data while preserving the causal structure that you think exists, and elucidating causal structure that you didn't suspect. Path analysis has three steps:

1. Propose a path model. The model represents causal influences with directed arrows (e.g., $FirelineBuilt \rightarrow ShutdownTime$) and correlations with undirected links (see Figure 2).

2. Derive path coefficients, the magnitude of which are interpreted as a measure of causal influence.

3. Estimate the correlation between two factors by multiplying path coefficients along paths between the factors and summing the products over all legal paths between the factors.

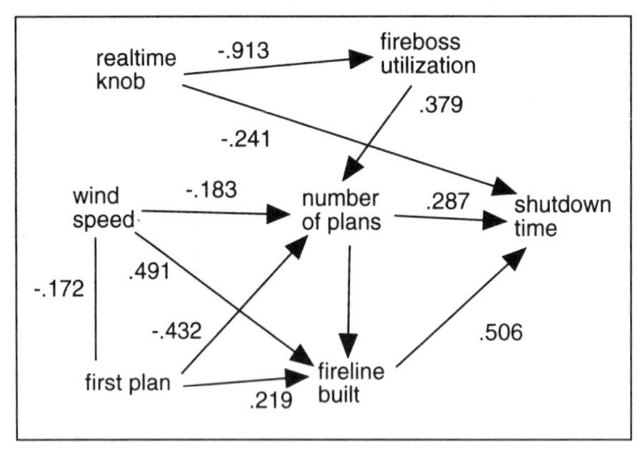

FIGURE 2: A path model derived from Pheonix data.

The three steps are repeated, modifying the model each time, until we have a model that is good by some criteria (more on this shortly). The second step is accomplished by multiple regression: A variable that is pointed to by one or more arrows is regressed on the variables at the tails, and the standardized regression coefficients are the path coefficients. For example, the regression of fireline built on first plan, wind speed and number of plans produces the following standardized regression model: $est.FirelineBuilt = .219(FirstPlan) + .491(WindSpeed) + .849)(NumPlans)$.

Two common statistical criteria of goodness are that the model accounts for a lot of the variance in the dependent variable and the model does a good job of estimating empirical correlations. The first criterion is the familiar R^2; the second requires some explanation. From the normal equations that underlie regression, we can derive a chain rule that says the estimated correlation between two variables is the sum of the products of the path coefficients on all the legal paths that connect the variables. Legal paths are those that don't visit a node more than once, don't follow arrows backward after they have gone forward, and don't include more than one undirected arc (see [Asher 83];[Li 75]; [Loehlin 87]; [Cohen 94] for details). For example, the estimated correlation between wind speed and shutdown time is $(-.183)(.287)+(-.183)(.843)(.506)+(.491)(.506)+(-.172)(.219)(.506) + (-.172)(-.432)(.287) + (-.172)(.432)(.843)(.506) = .089$. We have already seen that the empirical correlation between wind speed and shutdown time is -.053. This is the largest disparity between estimated and empirical coefficients in the entire model, so, by and large, the model does a pretty good job of explaining the empirical correlations. At the same time, the model accounts for 73% of the variance in shutdown time, 69% of the variance in fireline built, and 31% of the variance in the number of plans. This latter figure seems low and makes us wonder whether the variables in the model are "connected right." Perhaps another model does as good a job estimating correlations and also accounts for more of the variance in the dependent variables. We haven't found it.

25.4 An algorithm for automating path analysis

Estimating the fit of a path model in terms of R^2 and estimated correlations is straightforward, but proposing path models is not. Or rather, it is easy to propose models, but because the number of ways to connect N nodes with single- or double-headed arrows is 2^{N^2-N}, it is wise to propose them intelligently. We have developed an algorithm called PA that automatically searches the space of path models for those that fit the data and satisfy other heuristic criteria. The algorithm explores the search space by gradually refining models until a satisfactory model is reached or no more possibilities remain. Model refinement is by means of modifications. A modification to a model adds a direct causal influence (i.e., a new directed edge) between two unrelated variables, or a correlation between two directly related variables. The algorithm builds complex causal models incrementally, by modifying simpler models. Through successive modifications the algorithm can generate and evaluate all possible models, although this is to be avoided in most cases.

In general terms, the algorithm maintains a "best model" and a pool of further possibilities, in the form of potential modifications to models that have already been examined. The algorithm selects the best of these potential modifications and executes it, producing a new model. All the legal modifications of this new model are generated, scored, and returned to the pool of modifications. The new model itself is scored and compared to the best model. If the new model scores higher, it becomes the best model. The process repeats until the modification pool is empty. The best model encountered is finally returned.

Because of the size of the search space (e.g., 1,048,576 for five variables) the algorithm must be constrained. We limit consideration of modifications to those that are syntactically correct and meet particular heuristic criteria. Further, we guide search by considering the most promising modifications first. Syntactic criteria ensure that a model meets the requirements entailed by structural equations. There are several: We specify exactly one of the model variables as a dependent variable, which has no outgoing edges. Some model variables may be specified as independent variables, which have no incoming edges (bidirectional or correlation edges between independent variables and other variables are allowed, however.) The model must consist of a single connected component, not two or more unconnected substructures. A model may not contain a cycle. These constraints reduce the search space a good deal, but still only by a polynomial factor.

If a modification to a model will produce a new model that meets these syntactic requirements, it is then tested against heuristic criteria. There are two distinct kinds of heuristics: those that apply to the introduction of an new edge in a model, and those that evaluate the model as a whole.

Consider a modification that involves introducing an edge in a model from variable S to variable D. First, if variables S and D have some common ancestor, i.e., a common cause, a correlation edge directly between S and D is not allowed. Second, if the partial correlation between S and D with some third variable V held constant is below a threshold, then a direct edge between S and D is not allowed. The rationale for this heuristic is that the effect of S on D may be mediated by an intermediate variable, which is indicated by a low partial. Variations on this second heuristic constrain those variables considered as intermediate variables.

The second set of heuristic criteria deals with the model produced by a prospective modification. First, we may specify an upper limit on the number of edges in a model. Second, we may similarly specify an upper limit on the number of correlation edges in a model. Third, we may set a maximum on the branching factor of outgoing edges from any variable. All these heuristics limit the complexity of the models our algorithm produces; they are one means of encouraging parsimony.

What remains from this pruning process is a set we call the legal modifications to a model. These modifications are then evaluated for comparison with other possible modifications in the pool. The algorithm evaluates each modification by summing two component scores:

- $R^2_{new} - R^2_{old}$ is the difference between the determination coefficient for the existing model and the model produced by the modification. This component of the score prefers modifications that produce the largest increase in the statistical explanatory power of the model.

- $\sum_{i,j \in \text{vars}}(|r_{i,j} - \text{e\_old}_{ij}| - |r_{i,j} - \text{e\_new}_{ij}|)$ is more complex. For any two variables in a model we can calculate the estimated correlation based on the topology of the model, and the actual correlation produced by the data. If the difference between these two values is small, then this is a good model with respect to those two variables. The sum of this difference over all variable combinations in the model is one measurement of how well the model fits the data. The total value is the difference between the value of the existing model and the value of the model produced by the modification.

Once the highest scoring modification is selected and applied, the algorithm must evaluate the model that is produced. This model is scored on two counts: the determination coefficient for the dependent variable, and the summed difference between actual and estimated correlations. The highest scoring model is save d.

Besides limiting the search space, the algorithm must deal with two other efficiency concerns, the calculation of betas and of estimated correlations. Because any partial regression coefficient may be called for in evaluating a model, it may be necessary in the worst case to calculate all possible regressions during the course of the search. This requires an exponential number of calculations. Estimated correlations pose a larger problem. They are calculated by following paths in the model, and again in the worst case the process is exponential. The algorithm relies on the pruning process, in particular on the limitations on branching factor and the number of edges, to reduce these calculations to polynomial time.

25.4.1 *Illustration of PA*

The PA algorithm has been tested with artificial data [Cohen et al. 93] and also with the Phoenix dataset. We are currently conducting a study in which it is compared with Pearl's algorithm [Pearl and Verma 91] and the TETRAD II algorithm [Spirtes et al. 93]. We are also improving the algorithm. Thus, detailed evaluation of PA is forthcoming, and this section must be read more as an illustration than an evaluation.

We ran PA on a subset of the Phoenix dataset, leaving out wind speed and fireboss utilization. We told PA that realtime knob and first plan were independent variables, and shutdown time was a dependent variable. These facts reduced the size of PA's search space to 4096 possible models. Syntactic and heuristic pruning reduced the space further to 311 models. The desired model (the one that we built ourselves as a good model of Phoenix) is shown in Figure 3. When we ran PA on what we will call the *full* dataset, its best-ranked model was almost identical to Figure 3, missing only the link between first plan and fireline built. This link was pruned by the partial correlation heuristic, mentioned earlier.

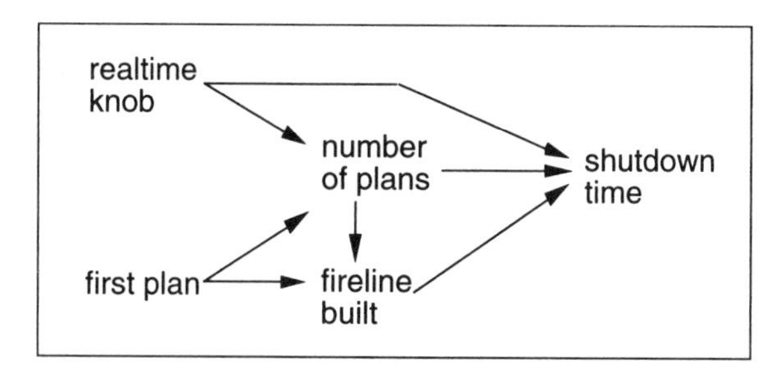

FIGURE 3: The desired model.

We were concerned that we had somehow tuned PA to perform well with the Phoenix dataset, so we partitioned this dataset by wind speed into three smaller sets of 85, 67 and 63 trials at wind speed of 3, 6 and 9 kph, respectively. We thought that PA ought to produce "essentially the same" model in these three conditions, recognizing, of course, that the covariances between wind speed and the other variables would be somewhat different in each condition. At the same time, we would be disappointed if the models were very different. After all, the "core" of the Phoenix planner should operate in pretty much the same way whether the wind speed is low, medium or high. For instance, the realtime knob should still influence the number of plans, which should in turn influence the amount of fireline built, and so on. If data from the three wind speed conditions produced very different models, it might mean that PA is too sensitive to unimportant differences in the covariance structure among the variables.

As you can see in Figure 4, many aspects of the models are common, and the differences are interesting "near misses." First, all the models have shutdown time directly influenced by realtime knob, number of plans, and fireline. All of the models have realtime knob influencing fireline and number of plans: in the 3 kph model, realtime knob influences fireline indirectly through number of plans; in the 6 kph model, both influences are direct; in the 9 kph model, fireline is directly influenced and in turn influences number of plans. The influence of first plan is hardest to pin down: it influences shutdown time, number of plans and fireline, respectively, in the three conditions. It appears that as wind speed increases, the influence of the first plan on shutdown time becomes more indirect. In retrospect, this is exactly what we should have expected: at low wind speed, Phoenix replans very infrequently, hence the influence of the first plan on shutdown time is greatest. At higher wind speeds, replanning is generally necessary, unless Phoenix's first plan happened to be one called "model," which was very conservative and built such a large perimeter of fireline that replanning was rarely required. Thus, at higher wind speed, the choice of the first plan should directly influence either the number of plans or the amount of fireline built.

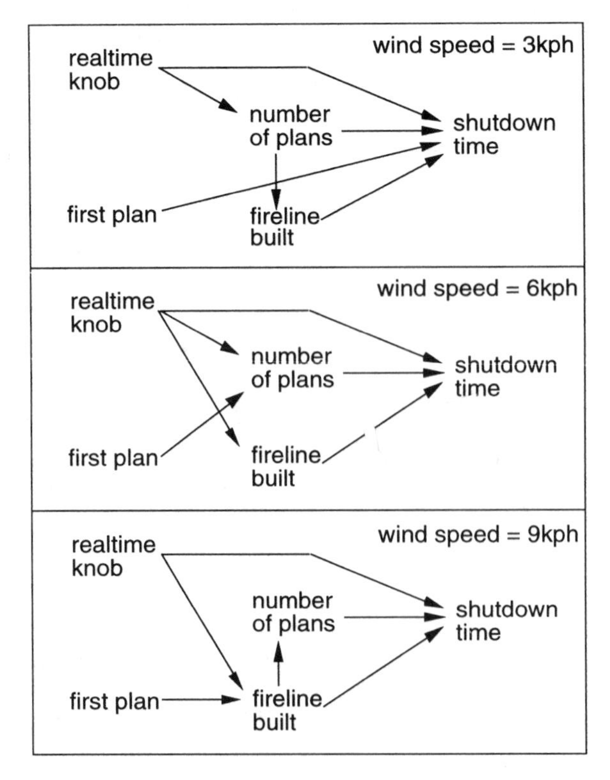

FIGURE 4: The top-ranked models produced by PA for partitions of the Phoenix dataset by wind speed.

25.5 Conclusion

Path analysis appears to be a promising technique for modeling complex AI systems, although our studies to date have been small and our results preliminary. Although one can build and estimate path models by hand, it would be better to give the job to an intelligent algorithm. Ours appears to work well on the Phoenix dataset, and also on artificial datasets, although a larger, more discriminating study is clearly needed.

25.6 References

[Asher 83] Asher, H.B. (1983) *Causal Modeling*. Sage Publications.

[Cohen et al. 89] Cohen, P.R., Greenberg, M.L., Hart, D.M., and Howe, A.E. (1989) "Trial by Fire: Understanding the Design Requirements for Agents in Complex Environments", *AI Magazine* **10(3)**, 32-48.

[Cohen et al. 93] Cohen, P.R., Carlson, A., and Ballesteros, L.A. (1993) "Automating Path Analysis for Building Causal Models from Data", *Proc. 10th Intl. Conf. on Machine Learning*, Amherst, MA, in press.

[Cohen 94] Cohen, P.R. (1994) *Empirical Methods for Artificial Intelligence*. The MIT Press, forthcoming.

[Hart and Cohen 92] Hart, D.M. and Cohen, P.R. (1992) "Predicting and Explaining Success and Task Duration in the Phoenix Planner", *Proc. 1st Intl. Conf. on AI Planning Systems, College Park, MD*, 106-115.

[Li 75] Li, C.C. (1975) *Path Analysis - A Primer*. Boxwood Press.

[Loehlin 87] Loehlin, J.C. (1987) *Latent Variable Models*. Lawrence Erlbaum Associates.

[Pearl and Verma 91] Pearl, J. and Verma, T.S. (1991) "A Theory of Inferred Causation", *Principles of Knowledge Representation and Reasoning: Proc. of 2nd Intl. Conf*, Morgan Kaufmann, 441-452.

[Spirtes et al. 93] Spirtes, P., Glymour, C. and Scheines, R. (1993) *Causation, Prediction and Search*. Springer-Verlag, New York.

Availability of Code

CLASP, our Common Lisp statistics package, is available. The path analysis code is part of CLASP. CLASP uses CLIM and will run on most platforms. Please contact David Hart (dhart@cs.umass.edu) if you want any of this code.

Part IV

Particular Models

26
A Parallel Constructor of Markov Networks

Randy Mechling and Marco Valtorta

SCT Utilities Systems Inc.
Columbia, SC 29205
and
Department of Computer Science
University of South Carolina
Columbia, SC 29208
mgv@usceast.cs.scarolina.edu.

ABSTRACT PaCCIN is a program to build Markov networks from data, implemented in C on an nCube/ten MIMD parallel computer with 1024 nodes. We describe the algorithm on which the program is based, the program itself, and some experimental results.

26.1 Introduction

PaCCIN (Parallel Constructor of Causal Independence Networks) is a parallel program to build Markov networks. The networks constructed by PaCCIN are useful for reasoning performed by belief network-based expert systems, such as MUNIN [AnWoFaAn 87,LaurSpie 88,AnOlJeJe 89]. Elsewhere, one of us discussed the feasibility of assisting the knowledge engineer in synthesis and refinement of numerical parameters for independence networks whose structure is designed by the knowledge engineer [ValtLove 92]. While we still feel that refinement is important, here we describe a program that builds Markov networks from data.

26.2 Markov Networks

The PaCCIN system follows the CONSTRUCTOR algorithm [FungCraw 90,CrawFung 91]. The original feature of PaCCIN is that it is implemented on a parallel computer, the nCube/ten hypercube available at the University of South Carolina. PaCCIN builds Markov networks from data.

A *Markov network* is a graph $G = (V, E)$, where $V = \{v_1, v_2, \ldots, v_n\}$ is a set of n events and v_i is adjacent to v_j if and only if v_j is in the Markov boundary of v_i. Given a set of events $V = \{v_1, v_2, \ldots, v_n\}$, the *Markov boundary* of v_i is the set of events $S = \{v_{i1}, v_{i2}, \ldots, v_{ik}\}$ such that

[0]We thank Dr. Robert Fung for sending us the LED data set, Dr. Caroline Macera for help in obtaining the National Health Survey database, Dr. Stephen Durham for discussion on statistical aspects, and Ken Sallenger, Beverly Huntsberger, and Dr. Terry Huntsberger for help with the nCube/ten hardware and software. The first author carried out the work described in this paper while at the Department of Computer Science, University of South Carolina. The second author was the discussant for Dr. Fung's presentation at the 1991 Third International Workshop on AI and Statistics and especially appreciates Dr. Fung's cordial and generous attitude.

1. $I(v_i, S, V - S - v_i)$

2. There is no subset of S that has property (1).

Here, $I(v_i, S, V - S - v_i)$ is the *independence relation*, indicating that v_i is independent of all the other events apart from itself and those in S when given set S. A set of events $S = \{v_{i1}, v_{i2}, \ldots, v_{ik}\}$ that has property (1), but not necessarily property (2) is called a *Markov blanket*. (See [Pearl 88, Chapter 3] for a discussion of the relationship of undirected Markov and directed Bayesian networks.)

26.3 Parallel implementation

Real world data that is useful for intelligent applications often involves large sets of information. PaCCIN was designed to eliminate as much as possible coding that would place limits on the size of the network it constructs. The data structures used are mostly one- and two-dimensional linked lists. Memory is allocated as needed and is limited only by the system memory. Another feature of PaCCIN that allows it to be used on large data sets is the parallel implementation.

Many of the routines involved in determining the conditional independence of a node given a set of other nodes are computationally intensive. Fortunately, however, much of the computation is mutually exclusive and can be easily parallelized. PaCCIN is implemented to run on the nCube/ten massively parallel supercomputer. This MIMD computer is composed of 1024 independent processors. that are linked by a hypercube configuration of communication channels. The number of node processes that are started is dependent upon the dimension of the hypercube. For the PaCCIN project, a node process is started for each of the events in the observations up to a maximum of 1024 processes. In this way, each node is able to concentrate on determining one Markov boundary in parallel with the other nodes.

The PaCCIN system operates in two large steps. In the first step, a data set is read and each event in the set is modeled as a node in the network, and the pair-wise dependencies of these nodes are determined using the Chi-square test for independence. (A different independence test could be used.) The second step it to determine the Markov boundary of each node, which, according to the definition given above, is the set of nodes adjacent to it (*neighbors*) in the Markov network. The various tasks at a more detailed level are divided between the host computer, in our case a Sun4/280, and node processors in the nCube/ten parallel computer.

The most computationally intensive portion of PaCCIN is finding neighbors. The Chi-square test is used as a test for conditional independence. Each event is tested against a set of other events, while considering all possible values of the remainder of the events. Each event can take one of a finite set of values defined in a configuration file. The *find neighbor* algorithm in PaCCIN begins with the candidate neighbor set consisting of single nodes and continues with pairs of nodes, and so on, if necessary. Nodes are entered in the candidate neighbor set in order of dependence, according to the results of the pairwise dependency tests, as done in CONSTRUCTOR. Given this candidate neighbor set and all its possible values, the target node is tested for independence against all remaining nodes. By assuming composability, again as in CONSTRUCTOR, this test is done by checking the target node against each node in the remainder set individually, given the candidate set. PaCCIN accomplishes this by giving each parallel process the target event, a different remainder event, and the candidate set. In this way the entire remainder set can be tested in parallel. If all parallel processes show that the target and the given remainder events are independent when the values of the candidates are known, the candidate set is in fact a neighbor set. An outline of the program is given in Figures 1, 2, and 3.

26.4 Model Construction with PaCCIN

The PaCCIN system has been tested with three data sets. One set was kindly supplied by Dr. Robert Fung and contains 199 observations of a faulty LED display [FungCraw 90,CrawFung 91]. Each observation contains a record of the state of each of the seven LED segments of a display and of which key was pressed. The data set describes a faulty display in which the state of the first LED segment (i.e., whether the segment is lit or not) is determined by the state of the second and third segments, rather than directly by the depressed key. In other words, the state of the first segment is conditionally independent of the state of the keys, given the state of the second and third segments. PaCCIN uncovered this conditional independence.

The other two data sets used to test PaCCIN contain data obtained from a national health survey conducted by the National Center for Health Statistics, carried out in 1971-1975. The two data sets differ only in their size: one hundred and 250 observations, respectively. Each observation corresponds to a patient and contains 147 fields, each one for a variable or event such as *Pulse*, *Ever Had Yellow Jaundice*, and *Past Week People Disliked Me*. We are not ready to discuss the results of our experiment with this data set from a domain-specific point of view, partly because we have not finished discussing them with the domain expert, Dr. Caroline Macera of the Department of Epidemiology and Biostatistics of the University of South Carolina, who kindly obtained the data for us.

Our experiments indicate a speed-up of a factor of four on the LED data set and a factor of 30 on the medical data sets. In contrast to the work by Gammerman and Luo [GammeLuo 91] on medical data analysis using belief networks, PaCCIN is not restricted to constructing Polytrees.

26.5 Conclusions

We are working on extensions to PaCCIN in different dimensions. In all cases, parallel design and implementation involve new work on the conversion of Markov networks to Bayesian networks (cf. [VermPear 93]), graphical interface (cf. [LeviKuba 92]), parallel belief computation (cf. [DamFouLi 92]), software engineering issues, analysis of additional data sets, and other algorithms for belief network construction (for the last three issues, cf. [SinghVal 93]).

26.6 References

[AnOlJeJe 89] Andersen, Stig K., Kristian G. Olesen, Finn V. Jensen, and Frank Jensen. "HUGIN—a Shell for Building Bayesian Belief Universes for Expert Systems." *Proceedings of the Eleventh International Joint Conference on Artificial Intelligence*, Detroit, 1080–1085, 1989 .

[AnWoFaAn 87] Andreassen, S., M. Woldbye, B. Falck, and S.K. Andersen. "MUNIN—A Causal Probabilistic Network for Interpretation of Electromyographic Findings." *Proceedings of the Tenth International Joint Conference on Artificial Intelligence (IJCAI-87)*, Milan, 366–372, 1987.

```
HOST BUILD DEPENDENCY TABLE

Determine size of hypercube
Start and load the hypercube
Send global data
Send data set
Receive results
Complete lower triangle of pairwise dependency table


NODE BUILD DEPENDENCY TABLE

Get node environment
Compute procindx
Receive global data
Receive data set
If procindx < number of events - 1
  Build a contingency table for each event > procindx
  Calculate Chi-square value from table
Return results
```

FIGURE 1: Outline of PaCCIN: Host and Node Programs (Part 1)

```
HOST FIND NEIGHBORS

Order events in ascending order by number of events
Loop A--all ordered events
  Choose target event from ordered list
  Loop B--until a neighbor set is found
    If this is the first pass
      Candidate set = known neighbors of target
    Else
      Choose events with highest degree of dependence
       on target to add to known neighbors of target
      Candidate set = known neighbors + extra event(s)
    Remainder set = dependents of target - candidate set
    If remainder set is empty
      Candidate set is neighbor set
      Break loop B
    Send target event
    Send candidate set
    Receive results
    If all results are successful
      Candidate set is neighbor set
      Break Loop B
    End Loop B
    Update target event's neighbor set
End Loop A
```

FIGURE 2: Outline of PaCCIN: Host and Node Programs (Part 2)

```
NODE FIND NEIGHBORS

Receive target event
Receive candidate set
Receive remainder set
If procindx > size of remainder set
  Remainder event pulled from remainder set,
   using procindx as index
  Loop A--each event in the candidate set
    Build 3D contingency table using target event,
     remainder event, and current candidate event
    Calculate Chi-square value for 3D contingency table
    Accumulate Chi-square values
    Choose next candidate event
  End Loop A
    Get Chi-square value from table
    If table value > accumulated Chi-square value
      success = 1
    Else
      success = -1
Else
  success = -1
Send success results
```

FIGURE 3: Outline of PaCCIN: Host and Node Programs(Part 3)

[CrawFung 91] Crawford, S.L. and R.M. Fung. "An Analysis of Two Probabilistic Model Induction Techniques." *Preliminary Papers of the Third International Workshop on Artificial Intelligence and Statistics*, Fort Lauderdale, Florida, January 1991 (pages not numbered).

[DamFouLi 92] D'Ambrosio, Bruce, Tony Fountain, and Zhaoyu Li. "Parallelizing Probabilistic Inference: Some Early Explorations." *Proceedings of the Eighth Conference on Uncertainty in Artificial Intelligence*, Stanford, CA, 59–66, 1992.

[FungCraw 90] Fung, R.M. and S.L. Crawford. "Constructor: A System for the Induction of Probabilistic Models." *Proceedings of the Eighth National Conference on Artificial Intelligence*, Boston, 762–769, 1990.

[GammeLuo 91] Gammerman, A. and Z. Luo. "Constructing Causal Trees from a Medical Database." Technical Report TR91002 (draft 2), Department of Computer Science, Heriot-Watt University, United Kingdom.

[LaurSpie 88] Lauritzen, S.L. and D.J. Spiegelhalter. "Local Computations with Probabilities on Graphical Structures and their Applications to Expert Systems." *Journal of the Royal Statistical Society, Series B (Methodological)*, 50 (1988), 2, 157-224.

[LeviKuba 92] Levine, David and Ted Kubaska. "Using the X Window System with the iPSC/860 Parallel Supercomputer." Typescript, available from Intel Corporation, Supercomputer System Division, 15201 N.W. Greenbrier Parkway, Beaverton, OR 97006.

[Pearl 88] Pearl, J. *Probabilistic Reasoning in Intelligent Systems: Networks of Plausible Inference*. San Mateo, California: Morgan-Kaufmann, 1988.

[SinghVal 93] Singh, Moninder, and Marco Valtorta. "An Algorithm for the Construction of Bayesian Network Structures from Data." *Proceedings of the Ninth Conference on Uncertainty in Artificial Intelligence*, Washington, DC, July 1993, 259–265.

[ValtLove 92] Valtorta, Marco and Donald W. Loveland. "On the Complexity of Belief Network Synthesis and Refinement." *International Journal of Approximate Reasoning*, 7, 3–4 (October-November 1992), 121–148.

[VermPear 93] Verma, T.S. and J. Pearl. "Deciding Morality of Graphs is NP-Complete." *Proceedings of the Ninth Conference on Uncertainty in Artificial Intelligence*, Washington, DC, July 1993, 391–397.

27

Capturing observations in a nonstationary hidden Markov model

Djamel Bouchaffra and Jacques Rouault

Cristal-Gresec - Université Stendhal
B.P. 25 - 38040 Grenoble Cedex 9 - France

ABSTRACT This paper is concerned with the problem of morphological ambiguities using a Markov process. The problem here is to estimate interferent solutions that might be derived from a morphological analysis. We start by using a Markov chain with one long sequence of transitions. In this model the states are the morphological features and a sequence correponds to a transition from one feature to another. After having observed an inadequacy of this model, one will explore a nonstationary hidden Markov process. Among the main advantages of this latter model we have the possibility to assign a type to a text, given some training samples. Therefore, a recognition of "style" or a creation of a new one might be developped.

27.1 Introduction

27.1.1 Automatic analysis of natural language

This work lies within a textual analysis system in natural language discourse (French in our case). In most systems used today, the analysis process is divided into *levels*, starting from morphology (first level) through syntax, and semantics to pragmatics. These levels are sequentially activated, without backtracking, originating in the morphological phase and ending in the pragmatic one. Therefore, the i-th level knows only the results of preceding levels. This means that, at the morphological level, each word in the text (*a form*) is analyzed autonomously out of context. Hence, for each form, one is obliged to consider all possible analysis.

Example : let's consider the sequence ot the two forms *cut* and *down* :

- *cut* can be given 3 analyses : verb, noun, adjective ;
- *down* can be a verb, an adverb or a noun.

The number of possible combinations based upon the independance of the analysis of one form in relation with the others implies that the phrase *cut down* is liable to *nine* interpretations, independently on the context.

These multiple solutions are transmitted to syntactic parsing which doesn't eliminate them either. In fact, as a syntactic parser generates its own interferent analyses, often from interferent morphology analysis, the problems with which we are confronted are far from being solved. In order to provide a solution to these problems, we have recourse to statistical methods. Thus the result of the morphological analysis is filtered when using a Markov model.

[1] *Selecting Models from Data: AI and Statistics IV.* Edited by P. Cheeseman and R.W. Oldford.

27.1.2 *Morphological analysis*

A morphological analyser must be able to cut up a word form into smaller components and to interpret this action. The easiest segmentation of a word form consists in separating word terminations (inflexional endings) from the rest of the word form called *basis*. We have then got a *inflexional morphology*. A more accurate cutting up consists in splitting up the basis into affixes (*prefixes, suffixes*) and *root*. This is then called *derivational morphology*.

The interpretation consists in associating the segmentation of a word form with a set of informations, particulary including :

- the general morphological class : verb, noun-adjective, preposition, ...
- the values of relevant morphological variables : number, gender, tense, ...

Therefore, an interpretation is a class plus values of variables ; such a combination is called a *feature*. Note that a word form is associated with several features in case where there are multiple solutions.

27.1.3 *Why statistical procedures?*

Because of the independance of the analysis levels, it is difficult to provide contextual linguistic rules. This is one of the reasons why we fall back on statistical methods. These latter method possess another advantage : they reflect simultaneously language properties, e.g. the impossibility to obtain a determinant followed directly by a verb, and properties of the analysed corpus, e.g. a large number of nominal phrases.

Some researchers used Bayesian approaches to solve the problem of morphological ambiguities. However, these methods have a clear conceptual framework and powerful representations, but must still be knowledge- engineered, rather than trained. Very often in the application of these methods, researchers have a good observation of the individuals of the population, *because the observation is a relative notion*. Therefore, we have difficulty in observing possible transitions of the individuals. The way of "capturing" the individuals depends on the environment encountered.

27.2 A morphological features Markov chain

27.2.1 *The semantic of the model*

Let m be the number of states, T the length of state sequence and $\{f_i/1 \leq i \leq m\}$ the states or morphological features ; we have only one individual ($n = 1$) for each transition time $t = 1, 2, \ldots, T$. A first order m-states Markov chain is defined by an $m \times m$ state transition matrix P, an $m \times 1$ initial probability vector Π, where :

$$P = (P_{f_i, f_j})$$

$$i, j = 1, 2, \ldots, m$$

$$P_{f_i, f_j} = Prob[e_{t+1} = f_j / e_t = f_i]$$

$$\Pi_{f_i} = Prob[e_1 = f_i]$$

$$i = 1, 2, \ldots, m$$

By definition, we have :

$$\sum_{j=1}^{j=m} P_{f_i, f_j} = 1 \qquad\qquad pour\ i = 1, 2, \ldots, m$$

$$\sum_{k=1}^{k=m} \Pi_{f_k} = 1$$

The probability associated to a realization E of this Markov chain is :

$$Prob[E/P, \Pi] = \Pi_{e_1} \times \prod_{t=2}^{t=T} P_{e_{t-1}, e_t}$$

27.2.2 *Estimation of transition probabilities*

As pointed out by Bartlett in Anderson and Goodman [AG57] the asymptotic theory must be considered with respect to the variable *number of times of observing the word form in a single sequence of transitions*, instead of the variable *number of individuals in a state when T is fixed.* However, this asymptotic theory was considered because the number of times of observing the word form increases ($T \rightarrow +\infty$). Furthermore, we cannot investigate the stationary properties of the Markov process, since we only have one word form (one individual) at each transition time. Therefore, we assumed stationarity. Thus, if N_{f_i, f_j} is the number of times that the observed word form was in the feature f_i at time $t-1$ and in the feature f_j at time t, for $t \in \{1, 2, \ldots, T\}$, then the estimates of the transition probabilities are :

$$\hat{P}_{f_i, f_j} = \frac{N_{f_i, f_j}}{N_{f_i +}}$$

where $N_{f_i +}$ is the number of times that the word form was in state f_i. The estimated transition probabilities are evaluated on one training sample. We removed the morphological ambiguities by choosing the sequence E of higher probability.

27.3 A Markov model with hidden states and observations

The inadequacy of the previous model to remove certain morphological ambiguities has led us to believe that some unkown hidden states govern the distribution of the morphological features. Instead of passing from one morphological feature to another, we focused only on the surface of one random sample, i.e. an observation was a morphological feature. As pointed out in [ROU88], this latter entity cannot be extracted without a context effect in a sample. In order to consider this context effect, we have chosen criteria like *the nature of the feature, its successor feature, its position in a sentence, the position of the sentence in the text.* An observation o_i is then a *known hidden vector* whose components are values of the criteria presented here. Of course, one can explore other criteria.

Définition 1 *A hidden Markov model (HMM) is a Markov chain whose states cannot be observed directly but only through a sequence of observation vectors.*

A HMM is represented by the state transition probability P, the initial state probability vector Π and a $T \times K$ matrix V (K is the number of states) ; the elements of V are the conditional densities $v_i(o_t) =$ *density of observation* o_t *given* $e_t = i$. Our aim is the determination of the optimal model estimate $\mathcal{V}^* = (\Pi^*, P^*, V^*)$ given a certain number of samples : this is the *training problem.*

Theoreme 1 *The probability of a sample* $S = \{o_1, o_2, \ldots, o_T\}$ *given a model* $\mathcal{V}$ *can be written as :*

$$Prob(S/\mathcal{V}) = \sum_E \Pi_{e_1} v_{e_1}(o_1) \times \prod_{t=2}^{t=T} P_{e_{t-1},e_t} v_{e_t}(o_t)$$

Proof : For a fixed state sequence $E = (e_1, e_2, \ldots, e_T)$, the probability of the observation sequence $S = \{o_1, o_2, \ldots, o_T\}$ is :

$$Prob(S/E, \mathcal{V}) = v_{e_1}(o_1) \times v_{e_2}(o_2) \times \ldots \times v_{e_T}(o_T)$$

The probability of a state sequence is :

$$Prob(E/\mathcal{V}) = \Pi_{e_1} \times P_{e_1,e_2} \times P_{e_2,e_3} \times \ldots \times P_{e_{T-1},e_T}$$

Using the formula :

$$Prob(S, E/\mathcal{V}) = Prob(S/E, \mathcal{V}) \times Prob(E/\mathcal{V})$$

and summing this joint probability over all possible states sequences E, one demontrates the theorem.

The interpretation of the previous equation is : initially at time $t = 1$, the system is in state e_1 with probability Π_1 and we observe o_1 with probability $v_{e_1}(o_1)$. The system then makes a transition to state e_2 with probability P_{e_1,e_2} and we observe o_2 with probability $v_{e_2}(o_2)$. This process continues until the last transition from state e_{T-1} to state e_T with probability P_{e_{T-1},e_T} and then we observe o_T with probability $v_{e_T}(o_T)$.

In order to determine one of the estimate of the model $\mathcal{V} = (\Pi, P, V)$, one can use the maximum likehood criterionn (or a max entropy) for a certain family S_i where $i \in \{1, 2, \ldots, L\}$ of training samples. Some methods of choosing representative samples of fixed length are presented in [BOU92]. The problem is expressed mathematically as :

$$\max_{v_i} f(S_1, S_2, \ldots, S_L/\mathcal{V}) = \max_{v_i}\{\prod_{j=1}^{j=L}[\sum_E \Pi_{e_1} \times v_{e_1}(o_1^j) \times \prod_{t=2}^{t=T} P_{e_{t-1},e_t} v_{e_t}(o_t^j)]\}$$

There is no known method to solve this problem analytically, that is the reason why we use iterative procedures. We start by determining first the optimal path for each sample. An optimal path E^* is the one which is associated to the higher probability of the sample. Using the well-known Viterbi algorithm, one can determine this optimal path. The different steps for finding the single best state sequence in the Viterbi algorithm are :

step 1 : initialization

$$\delta_1(i) = \Pi_i v_i(o_1) \qquad (1 \le i \le K)$$

$$\psi_1(i) = 0$$

step 2 : recursion for $2 > t > T$ and $1 \leq j \leq K$:

$$\delta_t(j) = \max_{1 \leq i \leq K}[\delta_{t-1}(i)P_{i,j}]v_j(o_t)$$

$$\psi_t(j) = \arg\max_{1 \leq i \leq K}[\delta_{t-1}(i)P_{i,j}]$$

step 3 : termination

$$P^* = \max_{1 \leq i \leq K}[\delta_T(i)]$$

$$e_T^* = \arg\max_{1 \leq i \leq K}[\delta_T(i)]$$

state sequence backtracking for $t = T-1, T-2, \ldots, 1$:

$$e_T^* = \psi_{t+1} e_{t+1}^*$$

P^* is the state-optimized likelihood function and $E^* = \{e_1^*, e_2^*, \ldots, e_T^*\}$ is the optimal state sequence. Instead of tracking all possible paths, one successively tracks only the optimal paths E_i^* of all samples. Thus, this can be written as :

$$g(o_1, o_2, \ldots, o_T; E^*, \mathcal{V}) = \max_E\{\Pi_{e_1} \times v_{e_1}(o_1) \times \prod_{t=2}^{t=T} P_{e_{t-1},e_t} v_{e_t}(o_t)\}$$

This computation has to be done for all the samples. Among all the v_i ($i \in \{1, 2, \ldots, L\}$) associated to optimal paths, we decide to choose as best model estimate the one which maximizes the probability associated to a sample. It can be written as :

$$\mathcal{V}^* = arg\{\max_{v_i} g(o_1^i, o_2^i, \ldots, o_T^i; E^*, \mathcal{V}_i)$$
$$i \in \{1, 2, \ldots, L\}$$

27.4 The different steps of the method

We present an interactive method which enables us to obtain an estimator of the model V. This method is suitable for direct computation.

First step : one has to cluster the sample with respect to the chosen criteria. two possibilities are offered : a *classification* or a *segmentation*. In this latter procedure, the user may structure the states ; operating in this way, the states appear like unknown hidden states. However, in a classification the system structures its own states according to a suitable norm. Thus, the states appear like unkown hidden ones. The clusters formed by one of the two procedures represent the first states of the model, they form *the first training path*.

Second step : one estimate the transition probabilities using the following equations and the probability of each training vector for each state $v_i(o_t)$. This is the first model $\mathcal{V}_1$. Let $i, i \in \{1, 2, \ldots, K\}$ and $t \in \{1, 2, \ldots, T\}$.

- Let $Nb(o_1, i)$ be the number of times the observation o_1 belongs to the state i and Nbp the number of training paths, then :

$$\hat{\Pi}_i = \frac{Nb(o_1, i)}{Nbp}$$

- Let $Nb(o_{t-1}, i; o_t, j)$ be the number of times the observation o_{t-1} belongs to the state i and the observation o_t belongs to the state j , then :

$$\hat{P}_{i,j}(t) = \frac{Nb(o_{t-1}, i; o_t, j)}{Nb(o_{t-1}, i)}$$

- The previous estimation formula can be written as :

$$\hat{P}_{i,j}(t) = \frac{N_{i,j}(t)}{N_i(t-1)} = \frac{N_{i,j}(t)}{N_{i+}(t)}$$

where $N_{i,j}(t)$ is the number of transitions from state i at time $t-1$ to state j at time t and $N_i(t-1)$ the number of times the state i is visited at time $t-1$.

- Let $Nbex(o_1, i)$ be the expected number of times of being in state i and observing o_t and $Nbex(i)$ the expected number of times of being in state i, then :

$$\hat{v}_i(o_t) = \frac{Nbex(o_1, i)}{Nbex(i)}$$

Third step : one computes $f(o_1, o_2, \ldots, o_T; \mathcal{V}_1)$ and determines the next training path, or clustering, necessary to increase $f(o_1, o_2, \ldots, o_T; \mathcal{V}_1)$. We apply the second step to this training path. The procedure is repeated until we reach the maximum value of the previous function. At this optimal value, we have E_1^* and v_1 of the first sample. This step uses Viterbi algorithm.

This algorithm is applied to a family of samples of the same text, so we obtain a family of E_i^* and $\mathcal{V}_i$. As mentioned previously, one decides reasonably to choose the model $\mathcal{V}^*$ whose probablitiy associated to a sample is maximum. This last model makes the sample the most representative, i.e. *we have a good observation in some sense.* This optimal model estimate is considered as *a type of the text processed.*

27.5 Test for first-order stationarity

As outlined by Anderson ang Goodman [AG57] the following test can be used to determine whether the Markov chain is first-order stationary, or not. Thus, we have to test the null hypothesis (H) :

$$P_{i,j}(t) = P_{i,j} \qquad (t = \{1, 2, \ldots, T\}$$

The likehood ratio with respect to the null and alternate hypothesis is :

$$\lambda = \prod_{t=1}^{t=T} \prod_{i=1}^{i=K} \prod_{j=1}^{j=K} \frac{P_{i,j}^{N_{i,j}(t)}}{P_{i,j}^{N_{i,j}(t)}(t)}$$

We now determine the confidence region of the test. In fact, the expression $-2 \log \lambda$ is distributed as a Chi-square distribution with $(T-1) \times K \times (K-1)$ degrees of freedom when the null hypothesis is true. As the distribution of the statistic $S = -2 \log \lambda$ is χ_2, one can compute a β point ($\beta = 95, 99.95$ %, etc.) as the threshold S_β. The test is formulated as :
If $S < S_\beta$, the null hypothesis is accepted, i.e. the Markov chain is first-order stationary. Otherwise, the null hypothesis is rejected at $100\% - \beta$ level of signifiance, i.e. the chain is not a first order stationary and one decides in favour of the nonstationary model.

27.6 How to solve the morphological ambiguities

This is the most important phase of our application. Let's consider an example of nine possible paths encountered in a test. Among these paths, the system has to choose the most likely according to the probability measure (see the third figure). Our decision of choosing the most likely path comes from the optimal model $\mathcal{V}^*$ obtained in the training phase. We show in this example how to remove the morphological ambiguities.

If the optimal state sequence obtained in the training phase is the one which corresponds to the figure :

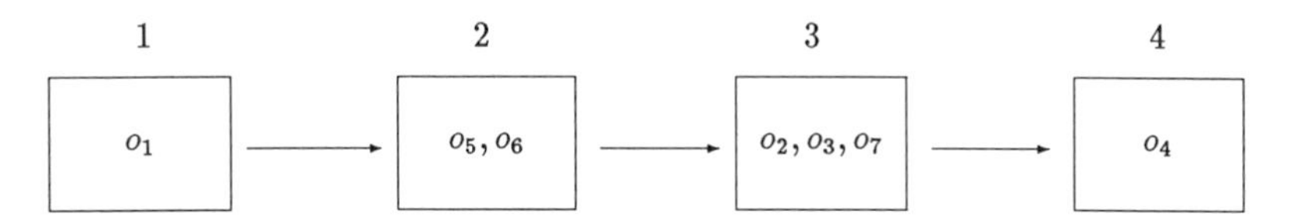

K=4 : the optimal state sequence is $1, 3, 3, 4, 2, 2, 3$

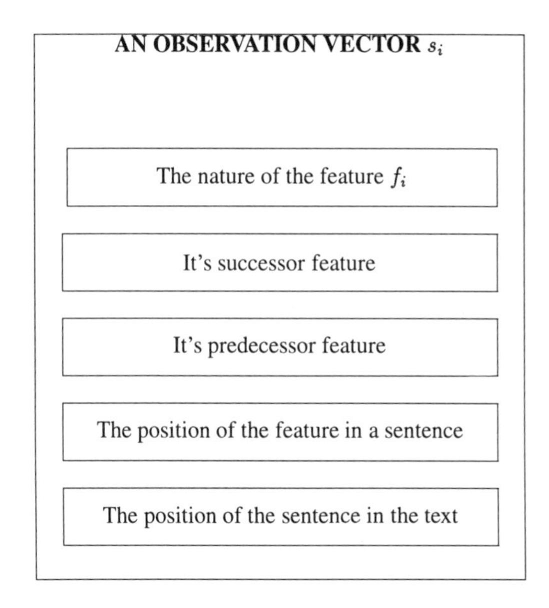

the one for example can choose between the two following paths of the figure :

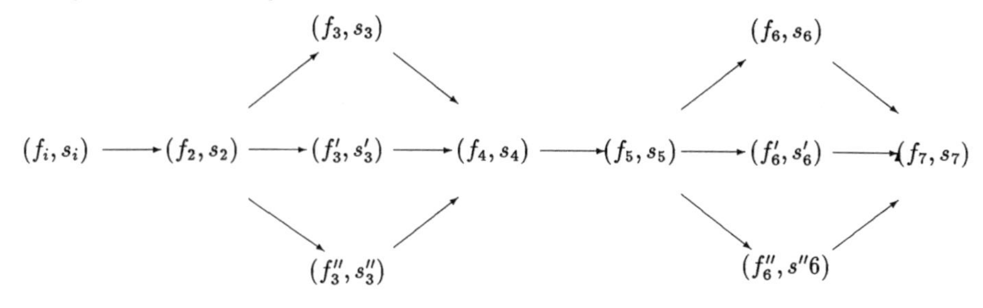

Path 1	s_1	s_2	s_3	s_4	s_5	s_6	s_7
Path 2	s_1	s_2	s'_3	s_4	s_5	s'_6	s_7

One compute the probabilities of these two realizations of the observations o_i $(i = 1, 2, \ldots, 7)$ using the formula :

$$Prob(o_1, o_2, \ldots, o_7/\mathcal{V}^*) = \Pi_{e_1} v_{e_1}(o_1) \times \prod_{t=2}^{t=7} P_{e_{t-1}, e_t} v_{e_t}(o_t)$$

The first figure shows that each s_i belongs to a state e_1 and, using the optimal model $\mathcal{V}^* = (\Pi, P, V)$ one can compute the probability of a path. Our decision to remove the morphological ambiguities is to choose the path with the highest probability.

27.7 Conclusion

We have presented a new approach for solving the morphological ambiguities using a hidden Markov model. This method may also be applied to other analysis levels as syntax. The main advantage of the method is the possibility to assign many different classes of criteria (fuzzy or completely known) to the training vectors and investigating many samples. Furthermore, we can define a "distance" between any sample and a family of type of texts called models. One can choose the model which gives the higher probability of this sample and conclude that the sample belongs to the specific type of texts. We can also develop a proximity measure between two models $\mathcal{V}_1^*$ and $\mathcal{V}_2^*$ through representative samples. However, some precautions must be taken in the choice of the distance used between the training vectors in the cluster process. In fact, the value of the probability associated to a sample may depend on this norm and, therefore, the choice of the best model estimated can be affected.

So far, we supposed that the criteria described the observations and the states are completely known (hard observation). Very often, when we want to make deep investigations, fuzziness or uncertainty due to some criteria or states are encountered, what should be done in this case? How can we cluster the observations according to some uncertainty measure? What is the optimal path and the best estimate model according to the uncertainty meausure? We are working in order to propose solutions to those questions in the case of a probabilistic [BOU91,BOU] and fuzzy logic.

27.8 References

[AG57] T.W. ANDERSON and L. A. GOODMAN. Statistical inference about markov chains.

Annals of Mathematical Statistics, 28, 1957.

[BOU] D. BOUCHAFFRA. *A relation between isometrics and the relative consistency concept in probabilistic logic*. Elsevier, North Holland.

[BOU91] D. BOUCHAFFRA. A relation between isometrics and the relative consistency concept in probabilistic logic. In *13th IMACS World Congress on Computational and Applied Mathematics*, juillet 1991.

[BOU92] D. BOUCHAFFRA. *Echantillonnage multivarié de textes pour les processus de Markov et introduction au raisonnement incertain dans le Traitement Automatique de la Langue naturelle*. PhD thesis, Université des Sciences Sociales, novembre 1992.

[ROU88] J. ROUAULT. *Linguistique automatique : applications documentaires*. P. Lang, 1988.

28

Extrapolating Definite Integral Information

Scott D. Goodwin, Eric Neufeld, and André Trudel

Dept. of Computer Science,
University of Regina,
Regina, SK, Canada, S4S 0A2,

Dept. of Computational Science,
University of Saskatchewan,
Saskatoon, SK, Canada, S7N 0W0,
and
Jodrey School of Computer Science,
Acadia University,
Wolfville, NS, Canada, B0P 1X0.

ABSTRACT We propose a probability-logic based representation of temporal knowledge that relates point and interval information through temporal integration. We illustrate the use of deduction and (probabilistic) induction to formally specify rational inference under uncertainty. In particular, we consider a previously unexplored aspect of the frame problem: *action and time invariance of statistical knowledge*.

28.1 Introduction

Difficult problems in temporal representation and reasoning arise even for simple stories such as

> Jake ran ten kilometers between 8am and 6pm. After the first hour, he rested for 15 minutes. After running six kilometers, he rested for 30 minutes.

This story entails complex relationships between propositions, intervals, and points on the time line. To represent this example and others, we use a first order temporal logic whose underlying temporal ontology is the real numbers. The logic is point based and information at the interval level is represented in terms of what is true at the point level through the use of the Riemann integral. Point based information is represented with a function $\beta(t)$ that equals unity if the property represented by the function β is true at time point t and equals zero otherwise. This unorthodox representation of point-based information facilitates the representation of interval based information. We can integrate functions such as β over an interval to get the duration of truth of β over the interval. Suppose Jake is running continuously between times 10 and 12:

$$\forall t.\ 10 \leq t \leq 12 \rightarrow \mathit{running}(t) = 1.$$

[0]Research supported by the Institute for Robotics and Intelligent Systems (IRIS) and NSERC grant OGP0122131, by IRIS and NSERC grant OGP0099045, and by NSERC grant OGP0046773.

[1]*Selecting Models from Data: AI and Statistics IV.* Edited by P. Cheeseman and R.W. Oldford. ©1994 Springer-Verlag.

If we integrate the running function over the interval [10,12], we get the length of time (in this case 2 time units) that Jake is running:

$$\int_{10}^{12} running(t)\, dt = 2.$$

This can be viewed as integrating "truth" to get "duration of truth". The value of the integral need not equal the length of the interval. For example, "Jake ran for 6 hours during some 10 hour period" is represented as:

$$\int_{t_0}^{t_0+10} running(t)\, dt = 6. \tag{28.1}$$

Note that running may not have occurred for 6 consecutive hours. It may have been interrupted many times.

In a uniform manner, the logic can represent temporal information of a quantitative nature (e.g., velocity or temperature). For example, "Jake ran ten kilometers between 8am and 6pm" is represented as:

$$\int_{8}^{18} velocity(running,t)\, dt = 10.$$

where *velocity(running,t)* equals the velocity due to running at time *t* and 18 equals 6pm on a twenty four hour clock. We integrate velocity to get displacement. We can easily represent the remaining sentences from the story about Jake's run. For the sentence, "After the first hour, he rested for 15 minutes," we have:

$$\int_{9}^{9.25} running(t)\, dt = 0.$$

where the .25 in 9.25 represents a quarter of an hour. "After running six kilometers, he rested for 30 minutes" is represented as:

$$\exists sixkm\, .\ \int_{8}^{\text{sixkm}} velocity(running,t)\, dt = 6 \ \ \& \ \int_{\text{sixkm}}^{\text{sixkm+0.5}} running(t)\, dt = 0.$$

A complete syntax and semantics for our logic appears in [GNT 92b].

The *integral* formalism has the advantage that it offers a uniform representation of both qualitative and quantitative information. We do not directly associate temporal information with an interval. Instead, we represent what is true at the point level and use this to describe intervals via the integral. The logic is based on the assumption that what is true at every point in an interval completely determines what is true over the whole interval.

Another advantage of using the *integral* is that examples such as formula (28.1) which involve temporal uncertainty (i.e., we do not know exactly when the running occurred) are easily represented. Temporal information of the type represented by formula (28.1) is called *definite integral information.*

Definition 1 (DEFINITE INTEGRAL INFORMATION)
Interval based information is called *definite integral information* if it can be represented in terms of a definite integral:

$$\int_{t_1}^{t_2} \beta(t)dt = \alpha$$

where $0 \leq \alpha \leq (t_2 - t_1)$ and β is a 0–1 function.

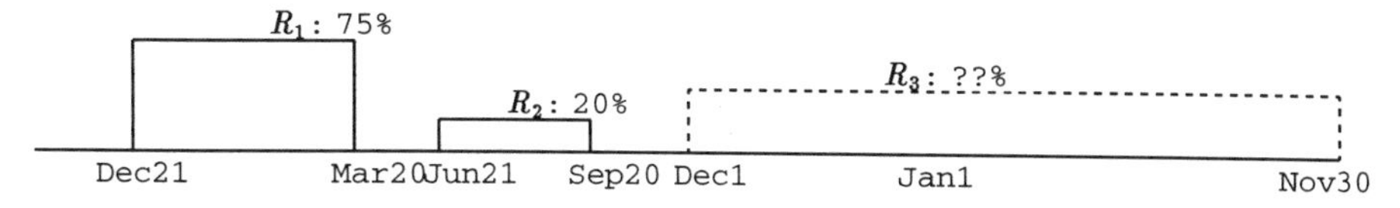

FIGURE 1: Rainfall Example

We are not aware of any other logic in AI that deals with the representation of definite integral information. In our logic, the representation is simple. But, reasoning with definite integral information is difficult because of the inherent uncertainty involved. We discuss this issue in the remainder of the paper.

28.2 Reasoning with Definite Integral Information

Another example of definite integral information is given in Figure 1 which depicts a situation where we know that it rained 75% of the time in the winter (region R_1):

$$\int_{R_1} \text{raining(t)}\, dt = 0.75 \times |R_1|$$

and 20% of the time during the summer (region R_2):

$$\int_{R_2} \text{raining(t)}\, dt = 0.20 \times |R_2|.$$

Reasonable questions to ask are "What is the rainfall during January?" and "What is the rainfall at noon on January 1st?" Note that both January and January 1st fall within winter (region R_1). We have information about interval R_1 but we cannot deduce anything useful about its sub-intervals. We call the problem of making reasonable inferences about sub-intervals and interior points of an interval for which we have definite integral information the problem of *interpolating definite integral information.*

Other reasonable questions to ask are "What will be the rainfall during the following year (December 1 to November 30; region R_3 in Figure 1)?" and "What will be the rainfall at noon on Janurary 1st of next year?". Both these questions involve predicting the future value of definite integral information based on current values. We call this the problem of *extrapolating definite integral information.*

We review the problem of interpolation and our proposed solution in the next section. The complete solution appears in [GNT 92a]. We then discuss approaches to the extrapolation problem in Section 28.4. In particular, we show how to extrapolate definite integral information in the special case where the temporal information is time over the entire interval.

28.3 Interpolating Definite Integral Information

Previous work shows how to draw reasonable inferences about interior points and subintervals from vague temporal knowledge [GNT 92a]. A (probabilistic) inductive inference rule identifies temporal intervals with reference classes and bases conclusions on the most relevant knowledge. This requires nonmonotonic "jumping to conclusions" about subintervals and points based on interval information. In the absence of "other information," we assume every subinterval has an equal chance of having some property. Taking this to the limit, we assign the same probability to individual points. But this is contingent on the interval being the most specific (most relevant) interval containing the point. Applying definite integral information to subintervals and points is related to the idea of *direct inference* in statistical reasoning [Pollock 84]. Direct inference is the application of statistical information about a reference class to a member of the class. When multiple reference classes are applicable, we must choose between them—this is the *reference class problem* [Reich 49,Kyburg 83]—we adopt the reference class selection policy proposed in [Goodwin 91].

Definite integral information acts as a set of probabilistic constraints that determine a set of possible interpretations. In general, we can only deduce weak constraints on subintervals and points from definite integral information. But the plausible inferences we wish to make hinge on an assumption of epistemological randomness (or, roughly, a principle of indifference); that is, each possible interpretation is assumed to be interchangeable (i.e., equally expected). From this we can infer an expected value for β at a particular point. We can compute this from the definite integral information by first determining an interval of points, that all have the same expected value for β, and using definite integral information about the interval to determine the expected value at the particular point. This interval of points having the same expected value for β is a *maximally specific reference interval*. For example, interval R_1 (Figure 1) is a maximally specific reference interval for raining during January (a sub-interval of R_1). Using the maximally specific reference interval, our approach would infer that it rains 75% of the time during January and that there is a 75% chance of rain at noon on New Year's Day. More complex situations involving conflicting reference intervals are addressed in [GNT 92a].

28.4 Extrapolating Definite Integral Information

Persistence (the frame problem) has been viewed in two ways: 1) *action-invariance of a property*: whether a property that is true before an action or event will remain true afterwards, cf. temporal projection [HanksMcD 87]; or, 2) *time-invariance of a property*: if a property is true at some point in time, how long is it likely to remain true [DeanKan 88]. Under these views, a property such as raining at a given point in time is highly action-invariant (few actions cause raining to stop) and slightly time-invariant (it rains for a while and then stops). Here we consider a previously unexplored aspect of the frame problem: **action and time invariance of statistical knowledge**.

Given statistical information (i.e., definite integral information) about various time intervals, we wish to make reasonable inferences about past or future intervals. For example, Figure 1 depicts a situation where we know that it rained 75% of the time in the winter, and 20% of the time during the summer. We have no statistical information about the coming year (December 1 to November 30) so the interpolation technique in the previous section is not applicable. The temporal projection technique of Hanks and McDermott is also inappropriate. We cannot determine from the statistical information whether it was raining at the last moment on September 20. Even if we knew it was raining at that time, it does not make sense to allow raining to persist indefinitely. We have no information about actions or events that may affect raining. Finally, Dean and Kanazawa's probabilistic temporal projection cannot be used as it requires the construction of a survivor function for *raining* based on many observations of raining changing from true to false. In our example, we have no observations of raining at particular points. We only have interval statistics.

Instead of considering persistence at the level of individual time points, we can view it at the interval level and describe the persistence of statistical information. If we take the observed statistics to be samples of raining over time, we can base our inferences for other intervals on these samples. For instance, we can infer an expected value for the statistic for R_3 in Figure 1 using R_1 and R_2 as follows:

$$E\left(\frac{\int_{R_3}\text{raining(t)}\,dt}{|R_3|}\right) = \frac{\int_{R_1}\text{raining(t)}\,dt + \int_{R_2}\text{raining(t)}\,dt}{|R_1| + |R_2|} = \frac{0.75\times 90 + 0.2\times 92}{182}$$
$$\approx 47\%.$$

Traditionally, persistence is modelled as the default or assumption that everything stays the same unless acted upon. Perhaps this assumption is valid in planning, where reasoning is done at a coarse grain size or in microworlds where a complete description of the world may be manipulated by a program, but generates unexpected results when uncertain knowledge is part of the domain. For example, *sunshine* in some climates might be expected to persist until acted upon by *rainfall*. However, at that point we do not really expect *rainfall* to persist indefinitely. If we were to make a prediction about the future, given no information to the contrary, we are much more likely to predict *sunshine*, because it is a normal state of affairs, than to predict *rainfall* because of a persistence axiom.

Within traditional discrete temporal formalisms such as the situation calculus, there are ways of circumventing the persistence axiom when it seems unnatural. If the time line has some representation as well, we simply introduce an unobservable micro-action that puts an end to the *rainfall* at about the expected time and causes *sunshine* to appear again and persist indefinitely. It may even be possible to parameterize these forces so that after some time the micro-action has acquired the critical mass needed to change the temporal information. But this seems like a roundabout way of introducing the sort of probabilistic information that the situation calculus was intended to avoid. In this work, we include an explicit representation of probabilistic information as well as techniques for making inferences from that knowledge.

There are at least two probabilistic ways to represent persistence: implicitly and explicitly. Although both approaches share intuitions with logicist approaches, both differ in striking ways. Implicit persistence means that we do not represent directly the conditions that give rise to a particular observable phenomenon. Rather, we represent these conditions as probabilities that give the conditional probability of some observable given it has occurred recently.

For example, consider rainfall. Certain climatic conditions have a certain "average" behaviour that is inclined to produce *sunshine*. For simplicity, assume *sunshine* simply occurs or not during an interval t and that $\neg sunshine_t \equiv rainfall_t$. However, the conditions may change such that they are inclined to produce *rainfall*. There is some random variation in either case, but this suggests it is reasonable to assume that a sequence of discrete temporal intervals forms a reversible Markov process. Let a_t denote the temporal information that property a holds in time interval t, and we assume that the probability of an event at t_{n+1} given knowledge of interval t_n is independent of earlier intervals. This yields the recursion

$$p(a_n|a_k) = p(a_n|a_{n-1}) \cdot p(a_{n-1}|a_k) + p(a_n|\neg a_{n-1}) \cdot p(\neg a_{n-1}|a_k),$$

that for $k = \infty$ yields the prior

$$p(a_n) = p(a_n|a_\infty) = \frac{p(a_n|\neg a_{n-1})}{2 - p(a_n|a_{n-1}) - p(\neg a_n|\neg a_{n-1})}\;.$$

To return to the weather example, we wish to represent the knowledge that conditions giving rise to *sunshine* persist, but that *rainfall* indicates that conditions have changed and that underlying conditions are likely to result in future rainfall. Letting a_t denote sunshine at time t, we might let

$p(\neg a_n|\neg a_{n-1}) = 0.7$, rainfall persists somewhat, but
$p(a_n|a_{n-1}) = 0.95$ sunshine persists even more.

Solving, $p(a_t) = 0.86$. This is a simple way of saying that the conditions that give rise to sunshine prevail overall, yet we are subject to rainy spells. We can model the transition of these conditions by letting the parameters vary over different values of t. So, for example, for Mar22 $\leq t <$ Jun22, $p(a_t) = 0.50$, but for Jun22 $\leq t <$ Sep22, $p(a_t) = 0.90$, says that spring is rainier than summer.

All parameters have strong mutual dependencies. Note that conditions giving rise to a persist equally and bidirectionally. That is, in general $p(a_n|a_{n-1}) = p(a_{n-1}|a_n)$. We later let parameters vary over the time line. If negative conditions persist as strongly as positive conditions (e.g. $p(a_n|a_{n-1}) = p(\neg a_n|\neg a_{n-1})$), then $p(a_t) = 1/2$, as expected.

In this example, *rainfall* is an event without known extent that occurs during some interval. It is straightforward to generalize our observable to proportions, so that the probability of *rainfall* during an interval varies as a function of some k previous intervals. If we assume all proportions are unity, we get a version of the situation calculus that includes an interesting probabilistic persistence axiom. We illustrate this idea with the Yale Shooting Problem [HanksMcD 87].

We recall that the Yale shooting scenario assumes that a gun is *loaded*, that someone *waits*, then *shoots*. The typical hearer of this scenario expects that someone is dead after being shot. But one logicist representation explains the target still being alive because the unloadedness of the gun persists backwards in time until just before the shooting. In a crude sense, this is related to the notion of equal bidirectional persistence.

This anomaly does not completely disappear if we use a simple Markov model to represent persistence, as shown in [Pearl 88]. Although many implausible scenarios result in the gun becoming unloaded while one waits, Pearl illustrates other syntactic features of this sort of inference, which we adapt here. Suppose we model this problem with a probabilistic persistence axiom that says the gun stays loaded (unloaded) given it is loaded (unloaded) already. We have the equations

$$p(d_n|s_{n-1} \wedge l_{n-1}) = 1.0, \quad p(l_n|l_{n-1}) = 0.9, \quad p(\neg l_n|\neg l_{n-1}) = 0.9,$$
$$p(a_n|a_{n-1}) = 0.99, \quad a_n \equiv \neg d_n, \quad l_0 \wedge s_1 \wedge a_1.$$

The first axiom says that if you are shot, you are dead; the gunner's aim is true. Notice, however, that there are at least two ways we can calculate the probability $p(a_2|l_0 \wedge s_1 \wedge a_1)$. Given a_1, we might ignore the effect of $l_0 \wedge s_1$, and conclude a_2 is most likely. Although this disturbs our intuitions about life and death, the inference makes sense if, as Pearl suggests, we interpret a_1 as meaning "wears bulletproof armour". Alternately, we might ignore a_1 and use the first statistic and the assumption that the value of a variable at time t_n, given the value of a variable at t_{n-1} is independent of its value at any previous time, we can write:

$$p(d_2|s_1 l_0) = p(d_2|s_1 l_1)p(l_1|l_0) + p(d_2|s_1 \neg l_1)p(\neg l_1|l_0)$$

which yields $p(d_2|s_1 l_0) = 0.9$.

By using Bayes' rule, if we observe the subject is alive at time 2 ($\neg d_2$), then

$$p(l_1|\neg d_2 l_0 s_1) = \frac{p(\neg d_2|l_1 l_1 s_1) \cdot p(l_1|l_0 s_1)}{p(\neg d_2|l_0 s_1)}$$

gives the probability the gun was loaded before shooting is zero.

To automatically effect the intended inference, we need to change both the representation as well as the inference rule. We change the representation by explicitly stating that loading and shooting overrule the effect that currently being alive has on death with an axiom of the form

$$p(d_n|a_{n-1} \wedge s_{n-1} \wedge l_{n-1}) = 1.0.$$

We augment our Markov model with a "nonmonotonic" inference rule that states when making an inference about a proposition, we use the probability with the narrowest condition on the right hand side of the conditioning bar.

But note that the entire inference can now be accomplished using only the nonmonotonic inference rule and the "Jeffrey's rule" and there is no need to worry about independence as pointed out in [Neufeld 93].

This suggests we should investigate the use of approaches based on statistical inference. Extrapolation from statistical data about intervals is more difficult than interpolation because in a sense it requires interpolation from the entire time line, for which we have no information. Thus, another approach to persistence is the explicit methodology we have described earlier as probabilistic regions of persistence.

"Arbitrary-duration" persistence is too crude an approximation. Instead, temporal information persists for some finite period of time into the future and/or past. Generally, we cannot say how long information persists. For example, if John is currently alive, we cannot determine with certainty the time of his death (assuming we do not murder John). But, from actuarial tables, we can estimate his life expectancy. This is true of most temporal reasoning: although we do not know exactly how long something should persist, we can make reasonable statistical estimates.

We approximate the truth values of a piece of information over time with *regions of persistence*. For example, let running be true at time 100. Assume that a person usually runs for 30 minutes and may sometimes run for up to 50 minutes. We expect running to be true for some 30 minute interval that contains time 100. For simplicity, we assume 100 is located in the center of the interval. We then expect running to persist over the interval (100-15,100+15) and we expect running not to persist outside (100-25,100+25). Over the intervals $(100 - 25, 100 - 15)$ and $(100 + 15, 100 + 25)$ we are unwilling to predict whether running persists. The general form of the *rop* relation is $rop(\beta,\ -t_1,\ -t_2,\ t_3,\ t_4)$ where β is a point-based item of information, although we easily generalize to intervals. If β is true at time X and nothing is known to affect β, then β is expected to persist throughout the interval $(X - t_2, X + t_3)$, β may or may not persist over $(X - t_1, X - t_2)$ and $(X + t_3, X + t_4)$, and β is expected to be false before $X - t_1$ and after

$X + t_4$. So, we predict β is true over $(X - t_2, X + t_3)$, we predict β is false before $(X - t_1)$ and after $(X + t_4)$ and otherwise we make no prediction. The regions are not necessarily symmetric around X. It may be that $t_2 \neq t_3$ and/or $t_1 \neq t_4$.

We give the *rop* relation a simple statistical semantics. Assume the duration of β is normally distributed with typical duration *(mean)* μ and variation *(variance)* σ^2 about that mean. Thresholding yields the desired logical inferences. Suppose we are satisfied to predict β remains true if the probability of β remaining true is greater than 50% and we wish to predict β is *false* if the probability is less than 5% (approximately) and otherwise we make no prediction. In this case, the relation $rop(\beta,\ -t_1,\ -t_2,\ t_3,\ t_4)$ holds if and only if $t_2 + t_3 = \mu$, and $t_1 + t_4 = \mu + 2\sigma$. This statistical semantics subtly changes the meaning of persistence; rather than stating that we can be reasonably sure β persists over $(X - t_1, X + t_4)$ it states that we can be reasonably sure it does not persist *beyond* the interval. This is consistent with the usual interpretation of a continuous probability distribution function. For example, if *running* truly has a normal distribution, the duration of a run is less than the mean 50% of the time. Thus at time X we expect the run to end within t_3 minutes with probability 0.5.

The semantics of the *rop* relation may vary with the problem domain. For example, if we must be 95% sure that running is true to predict that it is true, we let $t_2 + t_3 = \mu - 2\sigma$ and $t_1 + t_4$ is unchanged.

This simple relation offers solutions to several interesting inference problems. Suppose we observe John running at time 100 and we believe that, on average, he runs for 30 minutes with a standard deviation of 10 minutes. In this case, $rop(running, -25, -15, 15, 25)$ holds and we expect running to be true over the interval (85,115) and false before 75 and after 125.

We can extrapolate from definite integral information as well as point information. Suppose instead that running is known to be true over (80,120), that is, John has run for 40 minutes (10 minutes longer than the mean of 30) and we wish to estimate the time at which running will end. We can now use the idea of regions of persistence to reason about the expected endpoint. Suppose, as in the previous instance, we wish to predict running will persist if we are 50% sure it will continue, and we wish to predict it would not if we are 97.5% sure it would not and we make no prediction otherwise. As before, we assume that the duration of *running* is normally distributed, *but we simply consider the expected remaining time for those runs longer than 40 minutes.* By conventional methods, we find that about 50% of runs longer than 40 minutes last about another 4 minutes and 97.5% are completed within about 16 minutes. This inference can be viewed as nonmonotonic inference over an infinite set of reference classes, where a reference class in general consists of the set of runs longer than k minutes. We apply thresholds to the relevant part of the density function.

If we know that the run started at time 80, we conclude running is false before that time, we know it is true up to time 120, we predict it is true up to time 124 (about), we are uncertain about whether it is true up to time 136 and predict that it is false after that point. If we have no information about when the run started, we assume that time 100 is the midpoint of the run (the expected midpoint) and distribute the 4 minutes and the 16 minutes in equal amounts at each end of the run.

Assume the same parameters for *running*, but we know that John stopped running. We estimate that John began his run after time 70 with probability 0.5. With probability 0.025 (approximately), we expect he did not start the run before time 50. Here there are only four regions of persistence since we know running to be false after time 100.

Note in the case where we know that running started at 90 and ended at 100, the *rop* relation is not applicable, since we have complete information about running, there is nothing to predict.

The examples so far rely on the fact that the definite integral information represents continuous action, but this need not be the case. The *rop* formalism can be adapted to deal with other kinds of temporal information. Given that some temporal information occurred at only a portion of the points in an interval, it may be specified that the proportion of points will gradually decay, or that the proportion of points will persist until some point in the future and then stop. This depends on whether we are modelling an action that begins with some intensity and then winds down, or an action that is by its nature intermittent during its existence and then stops (for example, a leisurely walk).

When we have statistics for interrelated properties, the problem of extrapolation of statistical information can become complex. For instance, our statistical information could have a structure similar to the Yale Shooting Problem (YSP) [HanksMcD 87]. Suppose we have the definite integral knowledge:

$$\int_0^1 l(t)\,dt = 0.75 \quad \& \quad \int_1^2 a(t)\,dt = 0.9 \quad \& \quad \int_1^2 s(t)\,dt = 1 \quad \&$$

$$\forall t.[s(t) = 1 \;\&\; l(t) = 1] \rightarrow a(t+1) = 0$$

(a, l, and s could be interpreted as the outputs of three electronic gates, e.g., gate $a(t) = 1$ if gate a is on at time t). Just as in the YSP, there is a conflict between extrapolating l and a, i.e., if the integral of l over (1,2) is 0.75 by extrapolation, then a over (2,1) must be less than 0.25 (and not 0.9 by extrapolation). Resolving the conflict by extrapolating the earlier definite integral information first suffers the same problem as the chronological preference methods (e.g., [Shoham 86]). An even simpler difficulty is apparent in the statistical version of the Russian Roulette Problem [Goodwin 91]:

$$\int_0^1 a(t)\,dt = 0.9 \quad \& \quad \int_0^1 s(t)\,dt = 1 \quad \& \quad \forall t.[s(t) = 1 \;\&\; l(t) = 1] \rightarrow a(t+1) = 0$$

Here extrapolating the a information to the interval (1,2) entails $\int_0^1 l(t)\,dt < 0.1$ since $a(t+1) \;\&\; s(t) \rightarrow \neg l(t)$. This inference is faulty because a in the interval (1,2) should be contingent on l in the interval (0,1). We need the ability to express temporal conditionals; that is, the language must be expressive enough to encode statements expressing instances of the Markov property, that the future is independent of the past, given a sufficiently detailed present.

28.5 Conclusion

Our probability-logic based representation of (continuous) temporal knowledge connects point and interval information through temporal integration. This lets us automatically encode complex relationships between point and interval information. The reasoning component uses deduction and (probabilistic) induction to permit complex patterns of reasoning such as interpolation and extrapolation of definite integral information. The latter is a previously unexplored aspect of the frame problem, namely, the persistence of statistical knowledge.

28.6 REFERENCES

[DeanKan 88] Dean, T. and Kanazawa, K. (1988) Probabilistic Temporal Reasoning. *Proceedings of the Seventh National Conference on Artificial Intelligence*, 524–528.

[Goodwin 91] Goodwin, S.D. (1991) *Statistically Motivated Defaults.* Ph.D. dissertation, Department of Computer Science University of Alberta. Edmonton, Alberta. [also University of Regina, Computer Science TR-91-03].

[GNT 92a] Goodwin, S.D., Neufeld, E. and Trudel, A. (1992a) Definite Integral Information. *Proceedings of the Ninth Biennial Conference of the Canadian Society for Computational Studies of Intelligence*, 128–133.

[GNT 92b] Goodwin, S.D., Neufeld, E. and Trudel, A. (1992b) Temporal Reasoning with Real-valued Functions. *Proceedings of the Second Pacific Rim International Conference on Artificial Intelligence*, 1266–1271.

[HanksMcD 87] Hanks, S. and McDermott. D. (1987) Nonmonotonic Logic and Temporal Projection. *Artificial Intelligence* **33**(3), 379–412.

[Kyburg 83] Kyburg, H.E., Jr. (1983) The Reference Class. *Philosophy of Science* **50**(3), 374–397.

[Neufeld 93] Neufeld, E. (1993) Randomness and independence in nonmonotonic reasoning. In *Artificial Intelligence Frontiers in Statistics*, D.J. Hand, editor, Chapman and Hall, 362-369.

[Pearl 88] Pearl, J. (1988) *Probabilistic Reasoning in Intelligent Systems.* Morgan Kaufmann.

[Pollock 84] Pollock, J. (1984) Foundations of Direct Inference. *Theory and Decision* **17**, 221–256.

[Reich 49] Reichenbach, H. (1949) *Theory of Probability.* University of California Press.

[Shoham 86] Shoham, Y. (1986) Chronological Ignorance: Time, Nonmonotonicity, Necessity and Causal Theories. Proceedings of the Fifth National Conference on Artificial Intelligence, 389–393.

29
The Software Reliability Consultant

George J. Knafl and Andrej Semrl

Institute for Software Engineering
Department of Computer Science & Information Systems
DePaul University

ABSTRACT
In this paper, we provide a concise survey of software reliability modeling procedures as an indication of the complex variety of such procedures. We then provide a preliminary design for the Software Reliability Consultant (SRC), describing the current state of its development as well as future plans. SRC provides guidance in the choice of an appropriate performance measure and an effective modeling procedure for an arbitrary software reliability dat set.

29.1 Introduction

Software undergoing system testing based on random test cases generated according to the system's operational profile produces failure data that may be used in the planning and controlling of test resources. Such failure data consist of occurrence times for all observed failures together with the total amount of elapsed time over which the software has been exercised. Failure data may also be reported in grouped form and/or collected during operations for use in problem report management.

Although failure data sets have a relatively simple form, expert knowledge is required to navigate through the extensive variety of modeling procedures as well as the possible predictive performance schemes that may be used as the basis for comparing such procedures. Many modeling procedures have been developed for analyzing software reliability data. No one approach is appropriate for all such data sets [AdbChaLi 86] and so there is a need to choose a suitable procedure with which to analyze a given data set. Elementary modeling procedures have two basic components, a probabilistic model together with a parameter estimation procedure. Several criteria, for example, maximum likelihood, least squares, best past performance, and linear combinations, may be used to choose between, or weight, such procedures. The predictive performance of these composite procedures may be used to identify an effective procedure. Different composite modeling procedures may be appropriate under different performance measures and so an appropriate predictive performance measure needs to be selected on which to choose among alternative modeling procedures. The Software Reliability Consultant (SRC) is currently under development in order to provide guidance in the choice of an appropriate performance measure and an effective modeling procedure for an arbitrary software reliability data set.

In this paper, we provide a concise survey of software reliability modeling procedures as an indication of the complex variety of such procedures. Then we provide a preliminary design for the Software Reliability Consultant system, describing the current state of its development as

[1] *Selecting Models from Data: AI and Statistics IV.* Edited by P. Cheeseman and R.W. Oldford. ©1994 Springer-Verlag.

well as future plans. Much work of a prototyping nature still remains before the design of such a system is finalized.

29.2 Software reliability modeling procedures

Many software reliability models fall in the category of nonhomogeneous Poisson process (NHPP) models [Goel 85]. For NHPP models, the failure count for any time interval is distributed as a Poisson random variable and is independent of counts for other disjoint intervals. The random cumulative failure counts N(t) for the first t time units have mean values $\mu(t)$ that are nonlinear as a function of time t. This means that the failure intensity function $\lambda(t)$, defined as the derivative of the mean value function with respect to t, is nonconstant in t, and hence these models are called nonhomogeneous. The failure data consist of the observed cumulative failure counts n(t) (possibly in some adjusted form) which constitute a realization of the stochastic process N(t) based on actual execution of the system. Three of the more popular NHPP models are the exponential (EXP or 1-STAGE), logarithmic (LOG), and power (PWR) models with

EXP:	$\mu(t) = \alpha(1 - e^{-\beta t})$	$\lambda(t) = \alpha\beta e^{-\beta t}$	$\lambda(\mu) = \alpha\beta(1 - \mu/\alpha)$,
LOG:	$\mu(t) = \alpha \ln(1 + \beta t)$	$\lambda(t) = \alpha\beta/(1 + \beta t)$	$\lambda(\mu) = \alpha\beta e^{-\mu/\alpha}$,
PWR:	$\mu(t) = \alpha t^{\beta}$	$\lambda(t) = \alpha\beta t^{\beta-1}$	$\lambda(\mu) = \alpha^{1/\beta}\beta\mu^{(\beta-1)/\beta}$.

Many of the popular models have just two parameters, $\alpha > 0$ and $\beta > 0$, but models with more than two parameters, usually just three parameters, are also in use. Other relatively simple NHPP models include the delayed S-shaped (DSS or 2-STAGE) model [YamaOsa 85], the inverse linear (INVL) model [MusIanOk 87], the log power (LPWR) model [ZhaoXie 92], and the K-Stage Erlangian with fixed K (K-STAGE) models [KhosWoo 91]. Individually, these are all two-parameter models, but the family of K-STAGE models is a three parameter model. Some are finite failures models like EXP in the sense that $\mu(t)$ converges to the finite bound α as t converges to ∞. Others are infinite failures models like LOG and PWR.

Estimates of the model parameters, α and β, may be obtained through maximum likelihood estimation (MLE) or through least squares estimation. There are several variants of least squares estimation that may be employed. The most commonly employed variant of least squares estimation [MusIanOk 87] is based on estimating the failure intensity function $\lambda(t)$ at a series of time points, transforming these estimates, fitting a straight line regression model to the transformed estimates, and then transforming the parameter estimates of the regression model back into estimates of the parameters, α and β, of $\mu(t)$. The failure intensity estimates are obtained by partitioning the interval over which data are available into consecutive intervals. Usually the partition is random (RANDOM(k)) since it is based on skipping every kth failure with k=5 a recommended amount [MusIanOk 87]. Deterministic choices of the partition, for example, k equal sized intervals (EQUAL(k)), may also be employed. However, intervals with zero failures will have to be merged with adjacent intervals since transforms, especially the most commonly used natural logarithmic transform, would not be computable for intervals with zero failures. Other transforms like reciprocal and squart root transforms are appropriate for certain models, for example, the reciprocal transform applied to the failure intensity function of the logarithmic model produces a straight line function. The failure intensity estimates may be considered as estimates of $\lambda(t)$ at any point in the associated intervals with the left endpoint (LEFT) being a common choice [Okum 85]. However, for situations like the power model, where $\ln[\lambda(t)]$ is a straight line function of $\ln(t)$, this latter function will be undefined for the first interval since its left endpoint is at t=0. So other choices like the midpoint (MID) or the right endpoint (RIGHT) may be justified and the allowable choices are restricted in certain situations.

The regression models that are applied to transformed failure intensity estimates will depend on the transform that was used and on whether failure intensity is treated as a function $\lambda(t)$ of time (the LSTIME approach) or as a function $\lambda(\mu)$ of mean cumulative failures (the LSMEAN approach). For example, one popular approach utilizes the natural logarithmic transform together with the LSMEAN approach no matter what the model is [MusIanOk 87]. For the case of the logarithmic model, this approach generates ordinary least squares (OLS) estimates for the parameters of $\ln[\lambda(\mu)] = \phi + \psi\mu$ where $\phi = \ln(\alpha\beta)$ and $\psi = -1/\alpha$. These equations are inverted to obtain estimates of α and β from the OLS estimates of ϕ and ψ. For models other than the logarithmic model, $\ln[\lambda(\mu)]$ is nonlinear in μ and so nonlinear least squares (NLS) estimates of the associated parameters, ϕ and ψ, may be required. In certain situations, OLS in terms of a transform of μ will suffice. Unweighted least square procedures are often employed, but ordinary weighted least squares (OWLS) and nonlinear weighted least squares (NWLS) procedures are more appropriate due to the fact that the failure intensity estimates have nonconstant variance under the assumption that N(t) is a Poisson process [EhrlStam 90].

Note that if the least squares estimate of ψ for the above example based on the logarithmic model is positive, it produces a negative estimate of $\alpha = -1/\psi$ which is meaningless for software reliability data. Such situations occur infrequently but do occur for data sets found in the literature. There are other situations when a specific estimation procedure may be unallowable with a specific model, for example, failure intensity for the DSS model and, more generally, for the K-STAGE (with fixed K>1) models cannot be expressed in the form of an explicit function of μ. For these models, estimation by the MLE or LSTIME procedures is allowable, but not estimation by the LSMEAN procedure. More precisely, we rule out LSMEAN in this case because its computation is not straightforward, although actually possible through routines that invert $\mu(t)$.

Estimation procedures besides those mentioned above may be allowable for specific models. The cumulative failure counts n(t) may be fit with an unweighted nonlinear least squares model (NLSCUM) based on the form of the mean value function $\mu(t)$ or through weighted alternatives (NWLSCUM) [Schn 75,XieZhao 92]. More generally, least squares procedures may be based on fitting models to transforms of n(t). Models may also be based on hazard functions for the times between failures variables with or without Bayesian priors on the parameter(s) of the hazard function [AdbChaLi 86,MusIanOk 87]. For example, the Jelinski-Moranda geometric de-eutrophication model is a simple hazard function model which is not Bayesian while the Littlewood-Verrall model is the most popular Bayesian model. It is also possible to fit ARIMA models to the failure time variables with correlated rather than independent errors.

29.3 Composite modeling procedures

In practice, future predictions based on currently available failure data should be obtained through a composite modeling procedure [AdbChaLi 86,KeilMill 91]. Reliance on a single elementary procedure (or perhaps even a small number of them) will be highly inefficient for certain data sets in comparison to composite procedures. Whether the current data set is one of those problem data sets or not is not clear at the outset, and so it can only be classified as such through consideration of composite procedures. Composite modeling procedures depend on a modeling set M of elementary procedures emp=(m,ep), each determined by a specific model m and an estimation procedure ep. For example, emp=(LOG,MLE) utilizes maximum likelihood estimation applied to the logarithmic NHPP model. Some composite procedures generate predictions using a single elementary procedure chosen from a modeling set through some selection criterion. For one such approach, the ML approach, the choice will be the elementary procedure that generates the largest individual maximum likelihood value. For the LS approach, the choice will be the elementary procedure that generates the smallest individual least squares value. Different varieties of LS procedures correspond to how the least squares criterion is computed. For example, it might be based on the least squares fit to $\mu(t)$, to $\lambda(t)$, or to transforms of these two functions. Different modeling sets M would be appropriate for these different composite modeling procedures. For example, MLE is the natural elementary estimation procedure to utilize with ML while varieties of LSTIME and LSMEAN would be appropriate for LS based on fitting failure intensity. The NLSCUM estimation procedure or its weighted version NWLSCUM generates alternative forms of LS based on fitting cumulative failures. The best past performance (BPP) approach selects the elementary procedure with the smallest value of some predictive performance measure. It may be applied to any modeling set M and so may be employed in an attempt to improve the performance of other types of composite procedures. Different performance measures generate different BPP procedures.

The composite procedures considered so far select a single elementary procedure from the modeling set. Such procedures are called supermodels [KeilMill 91]. LC procedures, on the other hand, predict with linear combinations of the predictions of the elementary procedures in the modeling set [LuBroLit 92,LyuNiko 91]. Fixed weights may be used in forming the linear combinations or the weights may be chosen dynamically, for example, as inversely proportional to values of a predictive performance measure or proportional to some other transform (for example, the negative exponential transform) of predictive performance.

29.4 Predictive performance measures

Predictive performance may be measured in a variety of ways. Suppose that cumulative failure counts are available for all times t within some modeling interval (0,t']. The subset of the data in the first p% of total time t', that is, in the interval (0,pt'/100], may be used together with some modeling procedure mp to generate a prediction $\hat{x}(p,mp)$ of the number x(p) of failures occurring in some future time interval. The form of this future time interval is often taken as the remainder of the modeling interval, that is, (pt'/100,t']. This approach utilizes predictions of remaining failures and so will be designated as the RF approach. Alternately, the prediction interval may be the future interval whose length is a fixed percentage p' of the length of the current interval into the future, that is, (pt'/100,pt'(1+p'/100)/100] (the p'% AHEAD procedure), the future interval of constant length L (the CONSTANT(L) approach), or the future interval containing the next k failures after time pt'/100 (the NEXT k procedure). The utility of $\hat{x}(p,mp)$ as a prediction of x(p) will be scored through some function $w(\hat{x},x)$, for example, through absolute (ABS) errors with $w(\hat{x},x) = |\hat{x} - x|$, through relative (REL) errors [KeilMill 91], or negative log likelihood (NLL) errors [Knaf 91]. Under these scoring functions, smaller predictive performance corresponds to better performance. The choices for the percentages p% will vary from some starting percentage q% to some ending percentage r% by some step size d%. The choices of q, r, and d determine the set P of percentages p over which the performance scores $w(p, mp) = w(\hat{x}(p, mp), x)$ will either be averaged (AVG) or maximized (MAX) to determine the predictive performance PP(mp) of the modeling procedure mp. The prequential likelihood criterion [AdbChaLi 86] can be converted to a performance measure of the type discussed above through the negative logarithmic transform.

Best predictive performance composite modeling procedures choose a modeling procedure mp from some modeling set M with smallest predictive performance PP(mp). Composite modeling procedures cmp, including BPP approaches, also have predictive performance values PP(cmp) as do adaptive schemes for choosing between composite modeling procedures.

29.5 Structure of SRC and development plans

SRC is under development in three software layers. The lowest or computational layer supports a variety of modeling procedures, both elementary and composite. Future development will include support for a more complete set of elementary procedures and an expanded set of composite procedures. In particular, linear combination procedures still need to be implemented. A variety of predictive performance measures are currently available and are computable for all supported modeling procedures. The second or user interface layer together with the computational layer constitutes the Software Reliability Modeler (SRM). It supports both an interactive and a batch execution mode. SRM running in interactive mode provides capability similar to existing systems like SMERFS [Farr 91] and CASRE [LyuNiko 91,LyuNikFa 92], that is, direct manipulation of supported modeling procedures and predictive performance measures. The next section describes how SRM is used to perform a software reliability analysis. SRC will be implemented on top of SRM using expert system technology.

SRC may be run in closed mode in which case it will generate a sequence of composite procedures which SRM will analyze with the goal of computing an appropriate prediction (for example, of the time required to achieve a failure intensity objective). SRC will also provide an explanation upon demand of the predictive performance measure it used and the rules employed in its selection of a modeling procedure. These rules and the currently selected performance measure may be altered and a revised analysis requested. SRC may also be executed in an open, step-by-step fashion in conjunction with interactive user access to SRM. In this way, the user will be able to request a recommendation for improving the modeling procedure most recently considered within SRM. This latter feature will assist users in adjusting the rules encoded in SRC as well as in becoming more experienced in software reliability modeling.

Some preliminary heuristics for identifying effective composite procedures as an alternative to exhaustive search have been determined at this time. For example, the ML procedure appears to be a safe approach, that is, it is optimal in many cases and not too inefficient otherwise, and so it is a natural choice to compute first. Also, LS procedures which use some form of least squares parameter estimation as elementary modeling procedures may be distinguished either by comparing least squares criterion values or through their past performance. The former approach may be computed quickly. If the result is better or no worse than some level (say 80%) efficient compared to ML, then computing the more expensive best past performance alternative seems warranted, but perhaps not otherwise.

Some composite modeling procedures only work well for selected data sets. It is feasible to consider techniques that distinguish data sets based on project features, like development methods, process, and tools employed, results of inspections at stages prior to system test, or a variety of complexity measures. When such features are unavailable, as they are for data sets reported in the literature, techniques also need to be developed to classify data sets using only observed failure times. One possible, simple criterion could be based on having close average normalized failure times. In any case, the behavior of composite procedures for the data set currently under analysis may be predicted to be similar to that of past "close" data sets. Composite procedures that were effective (for example, with relative efficiencies no lower than 80%) in analyzing some close data set are reasonable to compute first for the current data set.

29.6 Software reliability analyses using SRM

A software reliability analysis consists of the following steps:

1. specify a failure data set,
2. specify a composite modeling procedure,
3. specify a predictive performance measure,
4. request a computation,
5. display and interpret the results.

The last four steps are repeated for as many composite modeling procedures as are deemed appropriate. For example, a comparison might be made between maximum likelihood estimation and the various forms of least squares estimation. Sensitivity analyses might be conducted, for example, to determine the effect of using an alternative predictive performance measure. One of these composite modeling procedures will be considered the most appropriate, perhaps because it has the best predictive performance under some preferred measure. The results from this composite modeling procedure will be used in making engineering decisions, for example, whether to stop testing. The Software Reliability Consultant (SRC) will provide guidance on what are appropriate analyses to consider.

The Software Reliability Modeler (SRM), on the other hand, provides a platform for conducting one software reliability analysis at a time. Its primary command level supports the analysis steps described above and has the following structure.

```
RELIABILITY
     FAILURE
     SPECIFICATION
     PREDICTIVEPERFORMANCE (ALIAS: PP)
     COMPUTE
     RESULTS
```

Each SRM command may be abbreviated or in some cases aliases are also provided. The execution of most commands invokes a lower command level which recognizes its own specific menu of commands. SRM is built with the Modeling Package MP [Knaf 87] and so its command structure inherits all the features of that tool. SRM fully supports a software reliability modeling language that may be invoked one command at a time through a simple user interface, in short command sequences to speed up interactive use, or in complete batch programs to bypass the user interface. SRC will invoke SRM through scripts executed by SRM's batch command facility. SRM will eventually be equipped with a graphical user interface to make it more convenient to use interactively.

A composite modeling procedure consists of a set of elementary modeling procedures together with a rule for selection of the "best" such procedure or a weighting scheme for combining results from all elementary procedures into a single linear combination result. The specification of a composite modeling procedure involves the following steps:

1. specify a set of elementary modeling procedures,
2. decide whether to
   (a) select the best elementary modeling procedure or
   (b) use a linear combination of elementary modeling procedures.

The SPECIFICATION command level within SRM is consequently structured as follows.

```
SPECIFICATION
     MODELS
     SELECTBY
     LINEARCOMBINATION (ALIAS: LC)
           not yet fully implemented
     EXCLUSIVE: SELECTBY with LINEARCOMBINATION
```

Whenever the user specifies one or the other of the SELECTBY or LC options, the other option is automatically cleared. The user need not worry about such consistency issues. Certain modeling options like the LC option are included in the interface even though its computational layer code is not fully implemented yet. The system will notify the user whenever a request is made for such an option.

An elementary modeling procedure consists of a model, represented by a mean value function, together with a parameter estimation technique. The user specifies which elementary modeling procedures to include within the current composite modeling procedure through the MODELS command level. One or more estimation techniques must be associated with each of the elementary models desired in the composite modeling procedure. The user may specify that certain aspects of estimation procedures be common to all the mean value functions included in the composite modeling procedure. The process procedes as follows:

```
specify all elementary modeling procedures to consider,
  1. indicate models and special estimation requests for them,
     specify any set of the following:
           1.1  delayed S-shaped model,
           1.2  exponential model,
           1.3  inverse linear model,
           1.4  K-stage Erlangian model,
           1.5  logarithmic model,
           1.6  log power model,
           1.7  power model,
     or
           1.8  all of the above,
```

```
2. indicate common aspects of estimation procedures,
   2.1  least squares estimation
        2.2.1 analyze cumulative failures,
        2.2.2 analyze estimates of failure intensity,
   2.2  maximum likelihood estimation (Poisson processes).
```

Elementary modeling procedures are specified in SRM as follows.

```
MODELS
    DELAYEDSSHAPED (ALIAS: DSS)
    EXPONENTIAL
    INVERSELINEAR (ALIAS: INVLINEAR)
    KSTAGE
    LOGARITHMIC
    LOGPOWER
    POWER
    ALL
    EXCLUSIVE:
        ALL with DSS, EXP, INVL, KSTAGE, LOG, LOGP, and POW
    COMMONESTIMATION
        LSE
            CUMFAILURES
                currently under test
            ESTFAILINTENSITY (ALIAS: EFI)
        MLE
```

Other command levels are similarly structured. The following is a sample command procedure for specifying a composite modeling procedure and predictive performance measure, compute and list the results, and save the reliability request and its results in a disk file.

```
RELIABILITY
    FAILURES
        ...         -- specify the failure data set to analyze
        EXIT
    SPECIFICATION
        MODELS      -- all with maximum likelihood estimation
            DELAYEDSSHAPED MLE ON EXIT
            EXPONENTIAL MLE ON EXIT
            INVERSELINEAR MLE ON EXIT
            LOGARITHMIC MLE ON EXIT
            LOGPOWER MLE ON EXIT
            POWER MLE ON EXIT
            EXIT
        SELECTBY    -- distinguish models using maximum likelihood
            ML ON
            EXIT
        EXIT
```

```
PREDICTIVEPERFORMANCE
     ...          -- specify a predictive performance measure
     EXIT
COMPUTE
RESULTS
     LIST ...
     EXIT
SAVE 'ANALYSIS1' --  save the current request in the workspace
FILE SAVE 'ALLWORK.SRM' EXIT -- save the workspace to disk
EXIT
```

The ANALYSIS1 reliability request may now be reused, for example, to change to least squares estimation and model selection by least squares. After executing the revised request, the predictive performance of the two samples may be compared to decide which composite modeling procedure is preferable.

29.7 References

[AdbChaLi 86] Adbel-Ghally, A. A., Chan, P. Y., and Littlewood, B., (1986) *Evaluation of Competing Software Reliability Predictions*, IEEE Trans. Software Eng., **SE-12**, 950-967, Sept. 1986.

[EhrlStam 90] Ehrlich, W. K., Stampfel, J. P., and Wu, J. R. (1990) *Applications of Software Reliability Modeling to Product Quality and Test Process*, Proc. 12th Int. Conf. Software Eng., 108-116, Mar. 1990.

[Farr 91] Farr, W. H. (1991) *Statistical Modeling and Estimation of Reliability Functions for Software (SMERFS) Users Guide*, Naval Surface Weapons Center, NSWC TR 84-373, Revision 2, Mar. 1991.

[Goel 85] Goel, A. L. (1985) *Software Reliability Models: Assumptions, Limitations, and Applicability*, IEEE Trans. Software Eng., **SE-11**, 1411-1423, Dec. 1985.

[KeilMill 91] Keiller, P. A., and Miller, D. R. (1991) *On the Use and the Performance of Software Reliability Growth Models*, Rel. Eng. and System Safety, **32**, 95-117, 1991.

[KhosWoo 91] Khoshgoftaar, T. M., and Woodcock, T. G. (1991) *The Use of Software Complexity Metrics in Software Reliability Modeling*, Proc. 1991 Int. Symp. Software Rel. Eng., 183-191, May 1991.

[Knaf 87] Knafl, G. J. (1987) *A Prototype Model Management System*, Proc. 19th Symp. on the Interface, 559-564, 1987.

[Knaf 91] Knafl, G. J. (1991) *Software Reliability Model Selection*, Proc. 15th Int. Comp. Software and Appl. Conf., G. J. Knafl, ed., 597-601, Sept. 1991.

[LuBroLit 92] Lu, M., Brocklehurst, S., and Littlewood, B. (1992) *Combination of Predictions Obtained from Different Software Reliability Growth Models*, 10th Ann. Software Rel. Symp. Proc., 24-33, June 1992.

[LyuNiko 91] Lyu, M. R., and Nikora, A. (1991) *Software Reliability Measurements through Combination Models: Approaches, Results, and a CASE Tool*, Proc. 15th Int. Comp. Software and Appl. Conf., G. J. Knafl, ed., 577-590, Sept. 1991.

[LyuNikFa 92] Lyu, M. R., Nikora, A., and Farr, W. H. (1992) *Making Better Software Reliability Estimations by a Practical Tool*, Proc. 2nd Bellcore/Purdue Symp. Issues in Software Rel. Estimation, 11-27, Oct. 1992.

[MusIanOk 87] Musa, J. D., Iannino, A., and Okumoto, K. (1987) *Software Reliability: Measurement, Prediction, Application*, McGraw-Hill, 1987.

[Okum 85] Okumoto, K. (1985) *A Statistical Method for Software Quality Control*, IEEE Trans. Software Eng., **SE-11**, 1424-1430, Dec. 1985.

[Schn 75] Schneidewind, N. F. (1975) *Analysis of Error Processes in Computer Software*, Sigplan Notices, **10**, 337-346, 1975.

[YamaOsa 85] Yamada, S., and Osaki, S. (1985) *Software Reliability Growth Modeling: Models and Applications*, IEEE Trans. Software Eng., **SE-11**, 1431-1437, Dec. 1985.

[XieZhao 92] Xie, M., and Zhao, M. (1992) *The Schneidewind Software Reliability Model Revisited*, Proc 3rd Int. Symp. Software Rel. Eng., 184-192, Oct. 1992.

[ZhaoXie 92] Zhao, M., and Xie, M. (1992) *Applications of the Log-Power NHPP Software Reliability Model*, Proc. 3rd Int. Symp. Software Rel. Eng., 14-22, Oct. 1992.

30
Statistical Reasoning to Enhance User Modelling in Consulting Systems

Paula Hietala

University of Tampere
Department of Mathematical Sciences, Statistics Unit
P.O. Box 607
FIN-33101 Tampere, Finland

ABSTRACT This paper describes methods that combine statistical reasoning with qualitative reasoning in the process of creating and updating of stereotypes of users. If a consulting system is equipped with stereotype based user modelling techniques it is able to predict a tentative user model before a more complete individual user model is available. It is hoped that this kind of a system is more capable to adjust its actions to the user. Moreover, if information about the users of the system has been cumulated during the sessions this information can be utilised in updating the stereotypes. In this paper these ideas are illustrated in the context of statistical consulting systems.

30.1 Introduction

The need for more adaptive helping and advising mechanisms in computer systems is commonly agreed. Several researchers have recently emphasized this aspect also with respect to statistical software (see e.g. [de Greef 91]; [MolePass 91]; [Popping 92]; [SautMade 90]). A promising approach to more user-adaptive systems is to apply techniques developed in the area of user modelling, for example so called stereotypes (see e.g. [KobsWahl 89]). This approach is based on the following idea: empirically, facts about people are not statistically independent events. This suggests that facts can be clustered into groups that frequently co-occur [Rich 89], i.e. into so called stereotypes. An essential advantage of this approach is that using these stereotypes the system is now able to predict an individual user model based on scarce or incomplete empirical data about the user. And when having an individual model, even a partial one, the system can better adjust its actions and output to the corresponding user. Of course, these stereotypes must include some kind of uncertainty measures. However, these measures are mostly qualitative (see e.g. [Chin 89]) or ad hoc numeric values (see e.g. [Rich 89]). We propose that by combining elements of statistical reasoning with the qualitative reasoning (especially into the process of creating and updating stereotypes) it is possible to make stereotypes more reliable.

In this paper we describe how this stereotype based approach can be made use of in the context of consulting systems. Throughout the paper we consider a specific statistical consulting system (an expert system for time series analysis) as an example to illustrate the structure of stereotypes and the possibilities offered by statistics in user modelling.

---

[1] *Selecting Models from Data: AI and Statistics IV.* Edited by P. Cheeseman and R.W. Oldford. ©1994 Springer-Verlag.

30.2 Stereotypes in a statistical consulting system

We focus here on systems in which the input from the user is based on selection (i.e. direct manipulation or menu driven systems). In this situation we can not analyze concepts used in textual commands as most of the earlier user modelling systems have done (see e.g. [Rich 89]; [Chin 89]). Instead, we must now monitor user actions (selections) during a session with the system and especially how he uses the advising facilities of the system. We also assume that we are able to divide the task of the user (e.g. analysis of one time series) into subtasks (e.g. input, graphical representation of data, calculation of simple statistics, transformations, etc.). Finally, we assume that we are able to count (reasonable) upper limits for each type of help (help concerning operations of the system, dictionary, etc.) in each subtask.

In our example system stereotypes are composed of two components: the knowledge the user possesses concerning statistics (statistical knowledge) and the knowledge the user has regarding system's behaviour (system knowledge). These components are related to each other in the following way. If the user possesses good statistical knowledge we can not infer anything about his knowledge about the system, but, on the contrary, if the user has obtained good system knowledge he probably knows most of the statistical concepts (we assume that the system operates only on one domain). So, these two components of the stereotype are not independent. Moreover, we attach a qualitative measure of difficulty (easy, moderate, difficult) to the statistical concepts. For example, we can say that the concepts "mean" and "variance" are both on the same level of difficulty, but the concept "partial autocorrelation" is located on a higher (i.e. more difficult) level. This means that all statistical concepts employed by the system are classified according their difficulty and we suppose that if the user knows a particular concept he is likely to know all concepts that are on the same or easier level of the classification (see a quite similar definition of double-stereotypes in [Chin 89]). Figure 1 clarifies our classification of statistical concepts. The above mentioned measure of the difficulty is used only in connection to the statistical knowledge, but, instead, we assume that the difficulty of concepts related to system's operations does not vary.

The selection of an appropriate stereotype for each user is based on observations about user's actions with the system. In other words, certain actions of the user are treated as activation events (so called triggers) of a stereotype. In our example system, the activation events are actions dealing with the advising facilities, such as:

- becoming familiar with the system,
- help concerning operations available in the system,
- help concerning statistical concepts (dictionary), and
- help concerning navigation in the system.

An outline of a stereotype "Occasional user with moderate knowledge of statistics" can be seen in Figure 2.

At the beginning of this section we made an assumption that we are able to divide the task of the user into subtasks. In this division the main problem is to find an optimal size for each subtask. Small subtasks cause "overloading" of the user modelling system. On the other hand, large subtasks are too rigid for the system's flexible adaptation to the user. Moreover, we must carry out the division without any overlapping in the user actions. This overlapping problem

	Difficulty		
	Easy	**Moderate**	**Difficult**
Statistical concepts	mean variance seasonality trend ...	autocorrelation level shift outlier periodicity ...	partial autocorrelation transformation differencing filtering ...

FIGURE 1: An example of qualitative measure of difficulty

arises especially if the user is quite freely allowed to utilize the advising facilities of the system. Furthermore, upper limits for each type of help in each subtask are easier to discover if subtasks satisfy the above requirements.

In our example system (an expert system for time series analysis) we can specify subtasks as follows (the task being the analysis of one time series):

1. specification of the new data,
   e.g. defining name, starting point and periodicity of data,

2. specification of the data transformation,
   e.g. differencing,

3. modification of the data specifications,
   e.g. changing name or starting point of the data,

4. data input,

5. modification of data,
   e.g. changing data value(s),

6. graphical representation of data,

7. calculation of statistics from the data,
   e.g. mean, variance, etc.

8. requesting advises concerning the data.

Finally, in each of the above mentioned eight subtasks upper limits can be defined for each type of help (i.e. help concerning operations available in the system, help concerning three classes of statistical concepts - easy, moderate and difficult).

30.3 Statistical and qualitative reasoning in user modelling

In the previous chapter we have described the general idea of stereotypes. However, in Figure 2 we did not yet apply any uncertainty measure to user's knowledge. We can now add (ad hoc) numeric values to describe how certainly the user knows a particular concept. For example, we may add a probability, say 0.9, to the sentence "user knows most of easy statistical concepts" (see

Statistical knowledge
- user knows most of easy statistical concepts
- user knows most of moderate statistical concepts
- user does not know most of difficult statistical concepts

System knowledge
- user knows a few system concepts

Triggers
- user does not read the introduction to the system but selects the help mode of the system
- user does not look for explanations of easy concepts but looks for explanations of more difficult concepts from the dictionary

FIGURE 2: Stereotype "Occasional user with moderate knowledge of statistics"

Figure 2) to define that the user probably knows 90% of all easy statistical concepts (employed by the system). Using this approach we can now summarize a simplified structure of the stereotypes (Figure 3).

In Figure 3 the value p_k (k = easy, moderate or difficult) denotes the probability that the user knows any given statistical concept and p_{sys} denotes probability that the user knows any given system concept.

Before we elaborate more closely on the user modelling we describe how the system estimates the p-values of the individual user model and how the system selects the stereotype to which the partial user model is best fitted. For the estimation of p-values we can assume that the user utilizes all the help he really needs (e.g. asks definitions for all statistical concepts he does not know). We know upper limits for each type of help in each subtask (say $L_{i,j}$, where i = the type of help and j = the subtask) and the system can count for each type of help how many times a particular user really has utilized them in any given subtask (say $U_{i,j}$). Now, we can estimate after each subtask the individual p-values using the formula $p_i = (L_{i,j} - U_{i,j})/\Sigma L_{i,j}$, where i = the type of help and summations are over all j subtasks that the user has completed. One possible solution to the selection of stereotype is to think p-values as a point in four dimensional space and to calculate the distances between p-values of the individual user model and p-values of each stereotype. The best stereotype is the nearest one. If a part of the p-values of an individual user model is missing, then the dimension of the space is the number of known p-values.

In our opinion the user modelling process should explicitly utilize and maintain structures that are associated with individual users as well as the stereotype itself. We envisage the user modelling to be composed of three phases:

1. creating and updating stereotypes,
2. selecting an appropriate stereotype for the individual user, and
3. creating an individual user model based on the above selected stereotype and updating this user model when more information of the user is obtained.

		Statistical Knowledge		
		Low	Moderate	High
System Knowledge	Low	$p_e = 0.9$ $p_m = 0.1$ $p_d = 0.1$ $p_{sys} = 0.1$	$p_e = 0.9$ $p_m = 0.9$ $p_d = 0.1$ $p_{sys} = 0.1$	$p_e = 0.9$ $p_m = 0.9$ $p_d = 0.9$ $p_{sys} = 0.1$
	Mod	$p_e = 0.9$ $p_m = 0.1$ $p_d = 0.1$ $p_{sys} = 0.5$	$p_e = 0.9$ $p_m = 0.9$ $p_d = 0.1$ $p_{sys} = 0.5$	$p_e = 0.9$ $p_m = 0.9$ $p_d = 0.9$ $p_{sys} = 0.5$
	High	-	$p_e = 0.9$ $p_m = 0.9$ $p_d = 0.1$ $p_{sys} = 0.9$	$p_e = 0.9$ $p_m = 0.9$ $p_d = 0.9$ $p_{sys} = 0.9$

FIGURE 3: Description of stereotypes

Phases (2) and (3) concern individual users and the reasoning is mainly based on qualitative information. On the other hand, the first phase (1) is mainly associated with the structure of stereotype and the reasoning here is typically statistical.

In our approach the individual user modelling process (phases 2 and 3) proceeds as follows. The first guess of appropriate stereotype for an individual user is based on triggers. After each subtask the system checks if the previously selected stereotype is valid by calculating new p-values for the individual user model and selecting the nearest stereotype (phase 2). An individual user model is at least partly based on stereotypes until all p-values of an individual user model are calculated. If the individual user model has all its own p-values, then the selected stereotype is not needed any more. However, the system continues to calculate new p-values for the individual user model after each subtask in order to adapt more closely (dynamically) to the individual user (phase 3). So, the system is able to adjust its advising facilities to suit the user who learns all the time more about the domain of the system as well as about the system itself.

Stereotypes are created (phase 1) by rule of thumb (see the descriptions of stereotypes outlined in Figure 3). So they must be updated to form more accurate descriptions according to the information received from the real users of the system. Our system stores p-values and the fitted stereotype of the user after an individual model is completed. After several sessions (and several users) with the system, this stored data can be utilized in modifying also the stereotypes. "Too large" frequency in a certain category (e.g. most of the users are located into the category "moderate statistical knowledge - moderate system knowledge") indicates that this category needs refinement or the classification of statistical concepts needs repairing. On the other hand, distributions of individual p-values reveal whether the default p-values should be changed (the measure of location differs significantly from the default p-value) or whether the classification of statistical concepts is not appropriate (the measure of variation of a certain p-value is significantly large).

It would be tempting to modify the stereotypes automatically after a set of sessions with the users. However, in our opinion, we can not assume that our current knowledge about the structure of stereotypes and about the possible findings of data is so complete that we are able to automate the updating procedure. The special features of data (e.g. the difficulty in making exact assumptions about distributions and small sample sizes) effect to the selection of proper statistical methods. Therefore, at least for the present, the updating of stereotypes must be done independently using nonparametric methods and human knowledge about real users.

30.4 Conclusions

The advantages of the approach outlined above are the following:

(a) the user model can be adapted to an individual user during the session (the model "lives and learns" with the user) and

(b) the information cumulated in each and every session with the users can be exploited in the development of the structure and content of stereotypes.

Usefulness of dynamic individual user models (advantage a) is commonly agreed (see e.g. [Rich 89]). We hope to have shown in this paper that evolution of the stereotypes themselves (advantage b) is also a very essential part of successful user modelling.

30.5 References

[Chin 89] Chin, D.N. (1989). "KNOME: Modeling What the User Knows in UC," in *User Models in Dialog Systems*, Kobsa, A. and Wahlster, W. eds. (1989), 74-107.

[de Greef 91] de Greef, P. (1991). "Analysis of Co-operation for Consultation Systems." *Journal of Applied Statistics*, Vol 18, 175-184.

[KobsWahl 89] Kobsa, A. and Wahlster, W. (eds.) (1989). *User Models in Dialog Systems*. Springer-Verlag, New York.

[MolePass 91] Molenaar, I.W. and Passchier, P. (1991) "The AGREE CONSULTANT: More Agreeable Software." *Journal of Applied Statistics*, Vol 18, 107-120.

[Popping 92] Popping, R. (1992). "More Agreeable Software?" *Computational Statistics & Data Analysis, (Statistical Software Newsletter)* Vol 13, 221-227.

[Rich 89] Rich, E. (1989). Stereotypes and User Modelling. in *User Models in Dialog Systems*, Kobsa, A. and Wahlster, W. eds. (1989), 35-51.

[SautMade 90] Sauter, V. L. and Madeo, L. A. (1990). "Using Statistics to Make Expert Systems 'User-Acquained'." *Annals of Mathematics and Artificial Intelligence*, Vol 2, 309-326.

31

Selecting a frailty model for longitudinal breast cancer data

D. Moreira dos Santos and R. B. Davies

Centre for Applied Statistics
Lancaster University, U.K.

ABSTRACT Different units often exhibit different failure rates. Frailty models allow for heterogeneity across units over and above the variation attributable to covariates included in the analysis. This paper reviews frailty model methods, using parametric and semiparametric specifications, for longitudinal breast cancer data.

31.1 Introduction

Most frailty models considered in the literature have a proportional hazard specification with a scalar and time-invariant frailty variable. Lancaster [11] was one of the first to generalize the proportional hazard model by introducing a positive multiplicative disturbance to represent unobserved heterogeneity (or frailty, in medical applications). Elbers and Ridder [6] showed that it is possible to distinguish between time dependence and frailty when covariates are included in the model. This paper addresses the main analysis issues that arise in these contexts, and reviews three models, illustrated with breast cancer data. The results demonstrate the misleading consequences of not taking into account the heterogeneity effect.

31.2 Heterogeneity in survival analysis

When considering cancer incidence, the tendency for malignant cell transformation differs between individuals. This may be due to different exposures to carcinogens, different genetic dispositions, and so on. Even for cancers when almost all patients succumb eventually, the speed with which the disease develops may vary considerably. These are heterogeneity effects.

Because individuals have different frailties, those who are most frail will tend to die (or experience an adverse event, e.g. recurrence of the disease) earlier than the others. In this way there is a progressive selection of individuals that are less frail than the rest. The observed survival curve (or hazard rate) is therefore not representative of any single individual, but is influenced by this progressive sample selection effect. In such a situation, heterogeneity is not only of interest in its own right, but actually distorts what is observed. Of course, some of the heterogeneity may be explainable in terms of observed covariates, but we may always anticipate an unexplained residual variation between individuals.

Mortality from cancer typically declines as a function of time from diagnosis, except for an initial increase. The decline has been interpreted as a biological phenomenon, reflecting a "regenerative process operating in cancer". Another natural explanation is that the decline is

[1] *Selecting Models from Data: AI and Statistics IV.* Edited by P. Cheeseman and R.W. Oldford. ©1994 Springer-Verlag.

Table 1: Modelling the length of time to first recurrence

VARIABLE	ESTIMATE	S.E.	MULTIPLIER
AGE	-0.0061	0.0038	
STAGE (2)	0.82	0.11	2.3
STAGE (3)	1.57	0.18	4.8

due to the selection effect associated with heterogeneity. Some patients are at lower risk of dying from their disease than others, and after some time the low risk group will dominate the survivors; the observed time dependence is a spurious consequence of unobserved heterogeneity.

31.3 Data

The data analysed in this paper were provided by a large U.K. hospital. They cover 917 breast cancer patients from initial diagnosis between 1980 and 1985 with details of survival and any recurrence up to July 1991. The explanatory variables used are AGE, STAGE (factor with 3 levels measuring severity of the disease at diagnosis - patients in the most severe category, level 4, were excluded from the analysis), and COUNT (factor with 3 levels distinguishing between durations from primary treatment to first recurrence, from first to second recurrence, and from second recurrence to death). It should be noted that defining recurrence is problematic. The results given in this paper are intended to illustrate statistical modelling issues; substantive conclusions would require more extensive analysis.

31.4 Basic inferences

The proportional hazards model is given by

$$h(t) = h_0 \, exp(\beta' x)$$

where h and h_0 are the overall and baseline hazards, respectively; β is the parameter vector; x is the covariate vector and t is the duration.

This model is straightforward to interpret because the explanatory variables have a multiplicative effect upon the baseline hazard. Moreover, by using the Cox [3] partial likelihood approach, it is possible to estimate the β parameters without making any parametric assumptions about the baseline hazard. The results below use an equivalent model fitting method (the piecewise exponential; see Breslow [1]) which also provides estimates of the baseline hazard.

The results were obtained using GLIM4 (Francis et al. [8]) and the maximum likelihood estimates (ESTIMATE), standard errors (S.E.), and the corresponding multiplicative effects (MULTIPLIER) are given in Table 1. We took each tenth distinct recurrence time for the piecewise representation.

From this table, we conclude that the hazard decreases with AGE and that STAGE is highly significant. Also, the hazard for women with STAGE (2) and STAGE (3) severity is 2.3 and 4.8 times greater, respectively, than the hazard for women with STAGE (1) severity.

The overall pattern for total survival time (see Table 2) is very similar to that for first recurrence but the relative effects of AGE and disease severity are more pronounced.

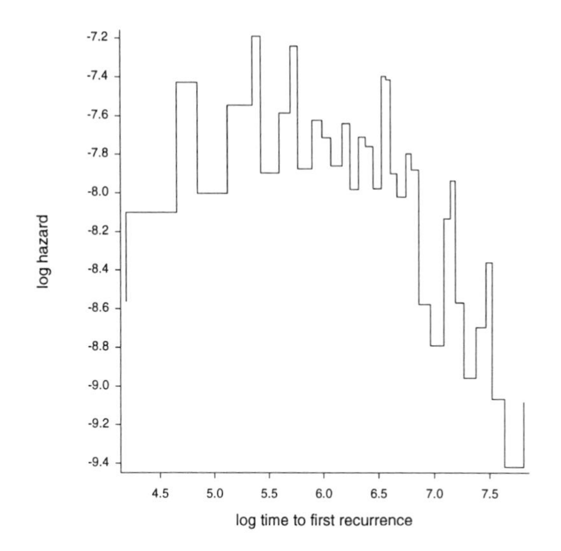

FIGURE 1: Baseline hazard for first recurrence

Table 2: Modelling the total survival time

VARIABLE	ESTIMATE	S.E.	MULTIPLIER
AGE	-0.0083	0.0047	
STAGE (2)	1.05	0.13	2.9
STAGE (3)	1.80	0.21	6.1

Figures 1 and 2 show the baseline hazard functions for first recurrence and total survival time. In both cases the hazard appears to decline after an initial increase, as expected.

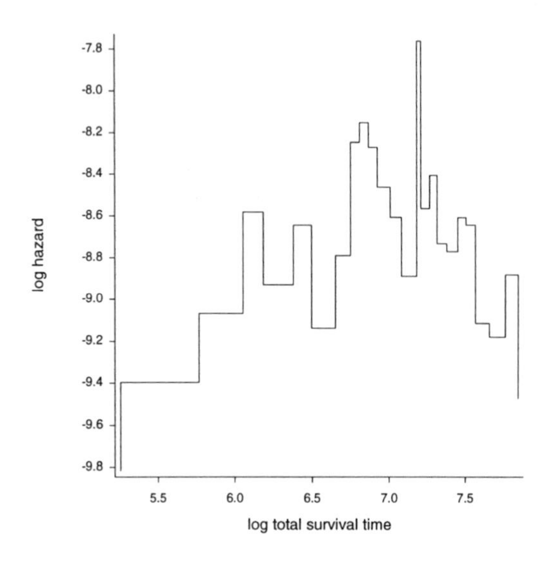

FIGURE 2: Baseline hazard for survival time

31.5 More complex models

The proportional hazards model with frailty (cf. Lancaster [11]; Elbers and Ridder [6]) is given by

$$h(t) = h_0 \; exp(\beta' x) \; \nu$$

where ν is the frailty variable.

The likelihood for each duration is given by

$$L(\beta, \nu) = [f(t)]^c \; [S(t)]^{1-c}$$

where $f(t)$ and $S(t)$ are the probability density function and survival function, respectively, corresponding to $h(t)$ and c is the censoring indicator ($c = 0$ if there is right censoring; $c = 1$, otherwise). Maximum likelihood inference is based upon the integrated likelihood given by

$$L(\beta) = \int L(\beta, \nu) g(\nu) \, d\nu$$

where $g(\nu)$ is the probability density function of the frailty ν over the population. This approach is readily extended to multiple durations for each individual.

It is now necessary to model both the baseline hazard and the frailty distribution. Consider the possible combinations of parametric and nonparametric representations:

HAZARD $[h_0]$	FRAILTY $[g(\nu)]$	MODEL TYPE	
parametric	parametric	parametric	(1)
nonparametric	parametric	semiparametric	(2)
parametric	nonparametric	semiparametric	(3)
nonparametric	nonparametric	nonparametric	(4)

Most researchers have used the first model, (1), because it is comparatively simple to fit. A common approach is to assume a gamma distribution for the frailty and a Weibull distribution for the hazard; these are conjugate distributions which lead to an analytically tractable model specification. Here we provide some results using the semiparametric formulations (2) and (3) in addition to (1).

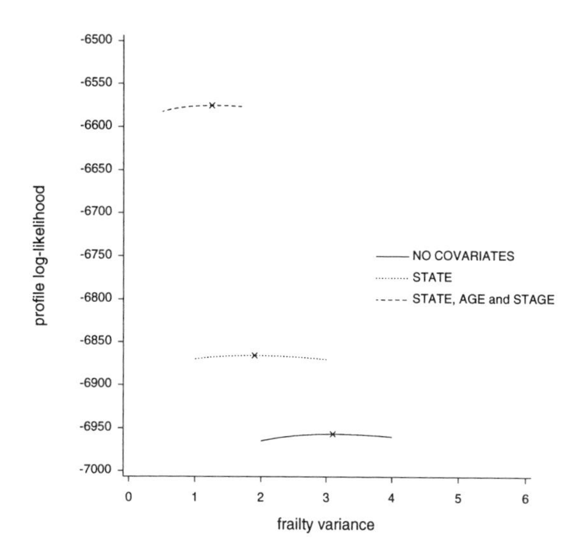

FIGURE 3: Profile log-likelihood for the frailty variance

Widespread application of these "mixture" models has been inhibited by at least two factors. First, there has been some concern that, in empirical applications, the frailty distribution is often estimated to have zero variance and the frailty model reverts to a conventional proportional hazard specification. Using GLIM macros for the model (2), nonparametric hazard and parametric frailty, we plotted the graphs of the profile log-likelihood for the frailty variance, as suggested by Clayton [2], for a variety of specifications. Here, frailty is assumed to be gamma distributed with mean 1 (without loss of generality because there is a constant term in the model) and the profile log-likelihood is obtained by varying the values for the variance. The results shown in Figure 3 were obtained for the full dataset with up to three durations for each individual as explained in Section 3. The results indicate that there is a substantial frailty variance when no covariates are included in the model. Moreover, including covariates reduces but does not eliminate the frailty variance. Our experience suggests that the zero frailty variance problem

Table 3: Conventional Weibull model
log-likelihood = -7596.86 (no. of iterations =10)

VARIABLE	ESTIMATE	S.E.	MULTIPLIER
COUNT (2)	1.56	0.07	4.8
COUNT (3)	1.30	0.10	3.7
AGE	-0.0086	0.0023	
STAGE (2)	0.63	0.06	1.9
STAGE (3)	1.02	0.11	2.8
weib. param.	0.77	0.03	
intercept	-6.26	0.24	

Table 4: Adding heterogeneity (normal distribution for log ν)
log-likelihood = -7525.62 (no. of iterations =9)

VARIABLE	ESTIMATE	S.E.	MULTIPLIER
COUNT (2)	1.42	0.09	4.1
COUNT (3)	0.64	0.11	1.9
AGE	-0.0015	0.005	
STAGE (2)	1.41	0.14	4.1
STAGE (3)	2.14	0.30	8.5
weib. param.	1.09	0.04	
intercept	-8.90	0.41	

may be due to misspecified models and small sample sizes.

The second problem restricting the use of mixture models is the absence of appropriate software. Researchers have used GLIM macros, as above, and optimisation subroutines (e.g. the NAG [13] subroutine library) but these approaches are difficult to operationalise and tend to be remarkably slow (Davies and Ezzet [5]).

The results below were obtained for the conventional Weibull model (no heterogeneity), the Weibull-normal model, and the Weibull model with nonparametric heterogeneity, respectively, using the research software MIXTURE (Ezzet and Davies [7]) which incorporates faster procedures. MIXTURE uses numerical integration for the normal frailty distribution and, following Laird [10], Heckman and Singer [9] and others, a mass point method for fitting models with nonparametric heterogeneity. Further details are given by Davies [4].

Comparing Table 3 for the conventional Weibull model and Table 4 for the Weibull-normal model shows that there is an appreciable reduction in the log-likelihood when frailty is represented explicitly in the model and, although formal likelihood ratio tests are not appropriate, the nonparametric representation of frailty (Table 5) results in a further notable improvement. Again, there is strong evidence of heterogeneity between individuals in excess of that attributable to the explanatory variables. Moreover, it appears that the normal distribution is unable to account fully for this residual heterogeneity.

Theoretically, ignoring heterogeneity leads to underestimation of the effects of explanatory variables and the standard errors and also to spurious reduction of the Weibull parameter (Lancaster and Nickell [12]). This is consistent with the results for AGE and STAGE in Tables

Table 5: With a nonparametric distribution for the frailty parameter
log-likelihood = -7518.13 (no. of iterations = 49)

VARIABLE	ESTIMATE	S.E.	MULTIPLIER
COUNT (2)	1.33	0.11	3.8
COUNT (3)	0.52	0.12	1.7
AGE	-0.018	0.005	
STAGE (2)	1.34	0.15	3.8
STAGE (3)	1.75	0.29	5.8
weib. param.	1.10	0.04	

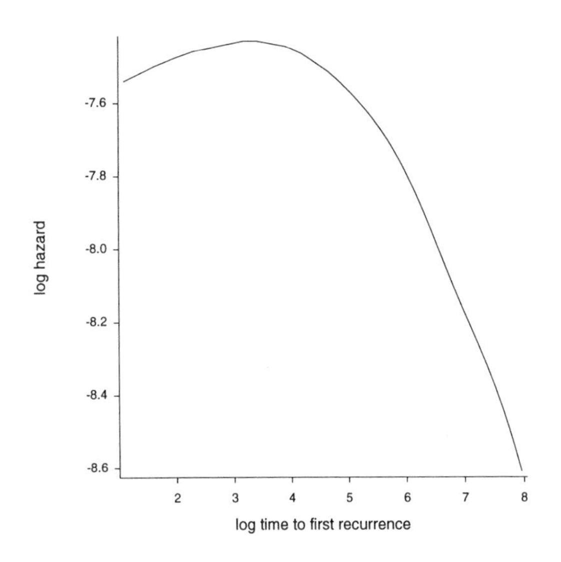

FIGURE 4: Overall hazard estimated from Weibull-normal model

3 to 5 and it appears that the Weibull model without heterogeneity provides a quite misleading picture of the baseline hazard declining with time; the heterogeneous models indicate that the hazard actually increases. The results for COUNT are more difficult to interpret because this is an endogenous rather than exogenous explanatory variable. The nonparametric calibration required a terminal session of several hours; the algorithm becomes slower and there is an increasing risk of "sticking" or converging to a local optimum as the likelihood surface becomes more complex. However, the results are very similar to the more-readily calibrated Weibull-normal model.

Figure 4 shows the overall estimated rate of first recurrence using the Weibull- normal model and is a parametric equivalent of Figure 1. Although the hazard for each woman increases over time in this model, the overall rate of recurrence shows the typical decline after an initial increase. As anticipated in Section 2, this arises through a statistical effect whereby the less frail progressively dominate the survivors. Conversely, these results provide no evidence that the decline in the recurrence rate shown in Figure 1 is attributable to any regenerative process.

31.6 Concluding comments

There is only space in this short paper to provide some indication of the different frailty models which may now be operationalised. However, while it is premature to offer guidance about the relative merits of parametric and nonparametric specifications, it is evident that ignoring a frailty effect risks serious inferential error.

Acknowledgements

This research was supported, in part, by a grant from CAPES - Brazil, proc. 189/90-1 and the ESRC (Research project R000233850).

31.7 References

[1] Breslow, N. (1974) "Covariance Analysis of Censored Survival Data", *Biometrics* **30**, 89-99.

[2] Clayton, D. G. (1988) "The Analysis of Event History Data: A Review of Progress and Outstanding Problems", *Statistics in Medicine* **7**, 819-841.

[3] Cox, D. R. (1975) "Partial Likelihood", *Biometrika* **62**, 2, 269-276.

[4] Davies, R. B. (1993) "Nonparametric Control for Residual Heterogeneity in Modelling Recurrent Behaviour", *Computational Statistics and Data Analysis* **16**, 143-160.

[5] Davies, R. B. and Ezzet, F. L. (1988) "Software for the Statistical Modelling of Recurrent Behaviour", in *Longitudinal Data Analysis: Methods and Applications*, M. D. Uncles, eds., 124-138, London Papers in Regional Science 18. Pion, London.

[6] Elbers, C. and Ridder G. (1982) " True and Spurious Duration Dependence: The Identifiability of the Proportional Hazard Model", *Review of Economic Studies* **49**, 403-409.

[7] Ezzet, F. L. and Davies, R. B. (1988) *A Manual for MIXTURE*, Centre for Applied Statistics, Lancaster University, Lancaster, England.

[8] Francis, B. , Green, M., and Payne, C. (1993) *The GLIM System: Release 4 Manual*, Oxford: OUP.

[9] Heckman, J. J. and Singer, B. (1984) "A Method for Minimizing the Impact of Distributional Assumptions in Econometric Models for Duration Data", *Econometrica* **52**, 271-320.

[10] Laird, N. (1978) "Nonparametric Maximum Likelihood Estimation of a Mixing Distribution", *JASA* **73**, 805-811.

[11] Lancaster, T. (1979) "Econometric Methods for the Duration of Unemployment", *Econometrica* **47**, 939-956.

[12] Lancaster, T. and Nickell. S. (1980) "The Analysis of Re-Employment Probabilities for the Unemployed", *JRSS* **143**, Series A, 141-152.

[13] NAG (1983) *Numerical Algorithms Group Library Manual*, Oxford, England.

32

Optimal design of reflective sensors using probabilistic analysis

Aaron Wallack and Edward Nicolson

(wallack@robotics.eecs.berkeley.edu) and (nicolson@robotics.eecs.berkeley.edu)
Computer Science Division
University of California
Berkeley, CA 94720

ABSTRACT Linear stepper, or Sawyer, motors have become popular in robotic mechanisms because of their high positional accuracy in open loop control mode [1]. Precision and repeatability are prerequisites in manufacturing and assembly. However the motor's actual position becomes uncertain when it is subject to external forces. Position sensors mounted on the motors can solve this problem and provide for force-control [2].
This paper describes a sensor, a technique for determining the robot's position, and an analysis technique for determining the optimal sensor configuration. Reflective optical sensors are used to generate raw data which is scaled and then processed using Bayesian probability methods. We had wanted to analyze different sensor configurations by marginalizing the performance over predicted data. Since marginalizing over the entire state space is infeasible due to its size, Monte Carlo techniques are used to approximate the result of the marginalization. We implemented the positional technique, and measured its performance experimentally; the sensors estimated the robot's position to within $\frac{2}{1000}''$, in line with the probabilistic analysis.

32.1 Introduction

The Sawyer motors are part of a multiple robot workspace, a RobotWorld™ system, at U.C. Berkeley. In RobotWorld, multiple robots cooperate to perform tasks, suspended from a 0.8 by 1.3 meter rectangular platen. The robots possess end-effectors for manipulating objects in the workspace (refer Figure 1). Each robot is driven by two orthogonal Sawyer motors which provide motion in the plane; other dc servo motors provide motion in the vertical and rotational directions. This report is concerned only with the motion in the plane due to Sawyer motors.

The use of Sawyer motors in electro-mechanical assembly tasks is hampered by their open-loop nature. The motors are at their commanded position only when there are no external forces. In chip placement tasks, where there are no external forces, the motors can be positioned to 20 microns. During assembly tasks, forces will be generated during part mating, and the motors will be displaced from their commanded position by an amount equal to the force multiplied by their compliance, which was measured to be 59 microns/N.

Reflective light sensors take advantage of the variance in the reflectivity across the platen. The platen, or stator, of the stepper motors is formed by machining evenly spaced grooves in a steel plate in both the x and y directions. The grooves are then filled with epoxy, resulting in a surface of evenly spaced steel squares. As steel reflects more light than epoxy, light sensors can be used to determine the robot's position. Two sets of sensors are mounted on each robot flush

---

[0]First author supported by Fannie and John Hertz Foundation; second author by NSF Grant IRI-9114446
[1]*Selecting Models from Data: AI and Statistics IV.* Edited by P. Cheeseman and R.W. Oldford. ©1994 Springer-Verlag.
[1]RobotWorld is a registered trademark of Yaskawa Co.

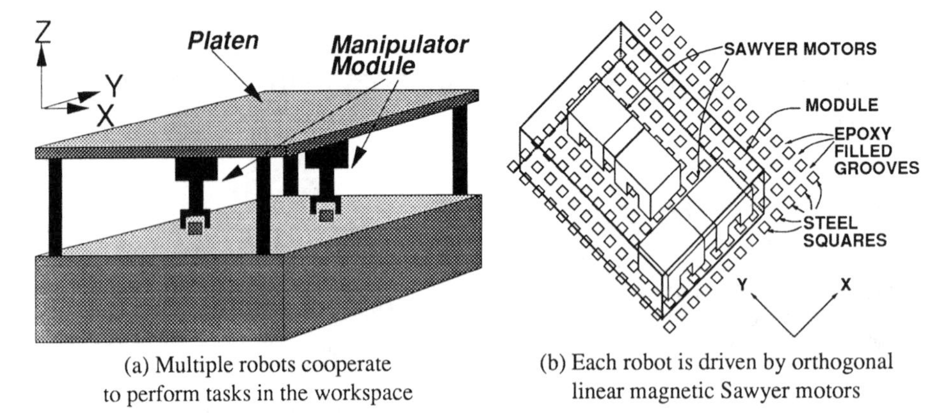

(a) Multiple robots cooperate to perform tasks in the workspace

(b) Each robot is driven by orthogonal linear magnetic Sawyer motors

FIGURE 1: The RobotWorld Environment

with the motor, facing the platen (see Figures 2(a) and 2(b)): one set for determining x position, and one set for determining y position.

Unfortunately, the platen is not uniformly reflective, due to dirt, scratches, and other factors. For these reasons, the periodic signal generated by the sensors varies in peak amplitude. The use of multiple sensors at different orientations improves the position estimate. The accuracy of the position estimate will vary with the number of sensors and with their orientations.

At each position, we assumed that the sensed reflectivities for each sensor were probabilistically independent. The position probabilities function furnish a maximum likelihood estimate, which we call the position, and also a standard deviation of that estimate, which we call the positional accuracy.

We wanted to estimate how many sensors, and what sensor configurations or phase orientations, would be necessary to achieve a sufficient accuracy. We probabilistically analyzed data from a single reflective sensor to answer the following questions:

- how many sensors should be used?
- how should the sensors be spread out?

The positional accuracy of a configuration was measured by the expected deviation of the position estimate. We define the expected deviation to be expected spread of the probability function (position estimate) with respect to the actual position. Normally, the error variance is computed relative to the mean position, but in this case, we computed the estimated variance with respect to the actual position because, in this context, we are interested in the determining the accuracy of the position estimate with respect to the actual position. Normally, the expected deviation is approximately the standard deviation, because, usually, the mean position estimate approximates the actual position.

By marginalizing the expected deviation over all positions and predicted sensor readings, we can predict the expected deviation for a configuration. Unfortunately, marginalization was infeasible due to the size of the state-space of sensor configurations. Therefore, we used Monte Carlo techniques to estimate the results of the marginalizations. We constructed and mounted

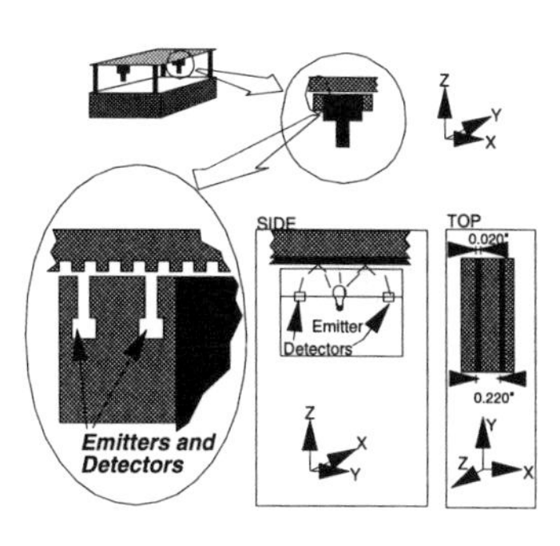

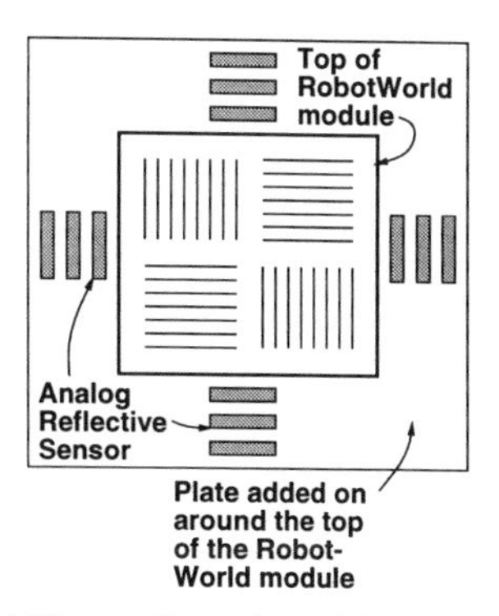

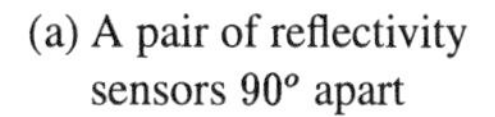

(a) A pair of reflectivity sensors 90° apart

(b) The configuration of the sensors around the top of the module

FIGURE 2: Details of the sensors

four reflectivity sensors and implemented the position algorithm. The technique estimated the robot's position to within $\frac{2}{1000}$", in agreement with the estimated precision.

32.2 Determining robot position from reflectivities

Since the grid pattern repeats with period 0.04 " (≈ 1 mm), the reflective sensors, as shown in Figures 2(a) and 2(b), only detect the robot's position modulo one full step. The term microposition refers to the intrastep position of the robot, and the term microstep denotes $\frac{1}{64}$ of a step.

Assuming that the reflected light varies linearly with the amount of overlap between the sensor and the steel squares, the response should be a periodic sawtooth wave. Figure 3 shows a histogram with respect to reflectivity and robot microposition. The reflectivity, as a function of microposition, is slightly sinusoidal owing to the distance between the sensor and the platen, and the rounded corners on the steel squares. To adapt to variation in the platen's reflectivity, the peak reflectivities were recorded, and the data was normalized accordingly.

A large sample set (10 million trials) was collected by performing many reflectivity experiments. Approximate probability functions were generated using this sample set. Since the probability functions were generated by collecting experimental data, the functions implicitly describe the platen's non-uniform reflectivity and the phase difference between the sensors.

In the following equations, M refers to the robot's microposition in microsteps, S_i refers to an individual sensor reading, and $\vec{S}$ refers to the set of sensor readings $(S_1, S_2, \ldots)$. M and $\{S_i\}$ are treated as discrete variables for two reasons: because the approach simplifies both constructing and utilizing the functions and because the underlying system is imprecise at such small resolution. By the principle of indifference, we assume that the *a priori* probability of the position is distributed uniformly with respect to the microposition M (refer equation (32.1)).

Figure 3 is a histogram showing the frequency of different $Microposition = M, SensorValue = S_i$ observations. The function $Occurrences\ (\ Microposition = M \wedge SensorValue = S_i\ )$ describes to the number of observed occurrences of such a configuration in the sample set. We used the Dirichlet-multinomial model to guarantee non-zero probabilities: α and β are *a priori*

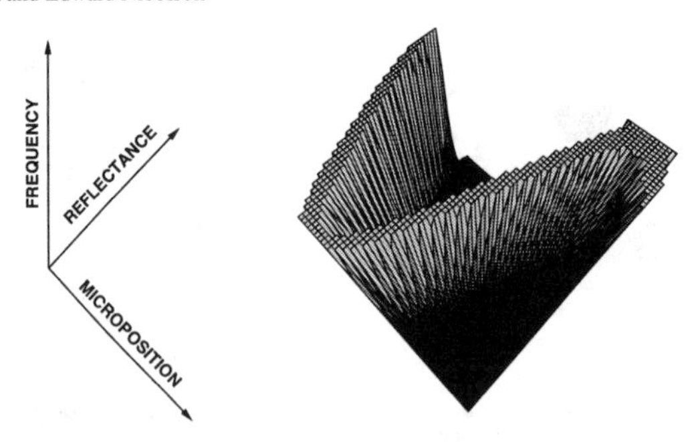

FIGURE 3: A histogram showing the number of observed occurrences of microposition M and sensed scaled reflectivity S_i (the axis of micropositions is *northwest*, and the axis of reflectivities is *northeast*).

initialization values, and $|i|$ refers to the number of different discrete sensed reflectivities).

$$\Pr(Microposition = M) = \frac{1}{|M|} \tag{32.1}$$

$$\Pr(Microposition = M \mid SensorValue = S_i) = \frac{Occurrences(Microposition = M \wedge SensorValue = S_i) + \alpha}{Occurrences SensorValue = S_i + |M|\alpha} \tag{32.2}$$

$$\Pr(SensorValue = S_i \mid Microposition = M) = \frac{Occurrences(SensorValue = S_i \wedge Microposition = M) + \beta}{Occurrences(Microposition = M) + |i|\beta} \tag{32.3}$$

To simplify the analysis, the sensed reflectivities are assumed independent. This assumption disregards some contextual information, such as the local reflectivity of the platen, and, in doing so, produces a conservative estimate.

Combining Bayes theorem (equation (32.6)) and the independence assumption for reflectivities (equation (32.4)) generates an equation for $\Pr(Microposition = M \mid SensorValues = \vec{S})$ (equation (32.7)).

$$\Pr(SensorValues = \vec{S} \mid Microposition = M) = \prod_i \Pr(SensorValue_i = S_i \mid Microposition = M) \tag{32.4}$$

$$\Pr(SensorValues = \vec{S}) = \sum_M Pr(SensorValues = \vec{S} | Microposition = M) Pr(Microposition = M) \tag{32.5}$$

$$\Pr(Microposition = M \mid SensorValues = \vec{S}) = \frac{\Pr(SensorValues = \vec{S} \mid Microposition = M)\Pr(Microposition = M)}{\Pr(SensorValues = \vec{S})} \tag{32.6}$$

$$\Pr(Microposition = M \mid SensorValues = \vec{S}) = \frac{\Pr(SensorValues = \vec{S} \mid Microposition = M)}{\Pr(SensorValues = \vec{S})|M|} \tag{32.7}$$

The probabilities are stored in logarithmic form in order to facilitate computation of equation (32.4). $\Pr(SensorValues = \vec{S})$ is computed by summing over all of the possible micropositions. The total probability function, $\Pr(Microposition = M \mid SensorValues = \vec{S})$, can be computed in $O(\#\ sensors)$ time because each sensor S_i probability function is non-infinitesmal in only two continuous ranges of micropositions M. We can screen out all infinitesmal probabilities by using range operations to determine the intersection of all of the non-infinitesmal regions.

32.3 Estimation of a sensor configuration's accuracy

The accuracy of a sensor configuration was measured as the expected deviation between an actual microposition M and the associated probability function of micropositions inferred from the sensed reflectivities. We assume that the positional error is distributed normally, and we expect errors of approximately 2σ, due to the fact that 95% of a normal gaussian distribution is within two standard deviations of the mean.

We used the term $\hat{\sigma}_{M,\vec{S}}$ (refer equation (32.8)) to describe a variant of the *standard deviation*, which we term the expected deviation. The normal definition of the standard deviation is the square root of the variance, and the normal definition of the variance is the expectation of $\sum_i (x_i - \overline{x})^2$. In this report, we define the expected variance to be the weighted sum square difference between the probability function and the actual position M, as opposed to the average position $\overline{x}$. This expected deviation characterizes the spread of the probability function $\Pr(Microposition = M \mid SensorValues = \vec{S})$. Normally, the expected deviation and the standard deviation will be approximately equal, since the mean position estimates the actual position.

Each sensor configuration's accuracy was measured by marginalizing the expected deviations, $\hat{\sigma}_{M,\vec{S}}$, over all micropositions M and sensed reflectivities $\vec{S}$. The earlier conservative independence assumption implies that this technique overly estimates the expected deviation $\hat{\sigma}$ (accuracy) of each sensor configuration.

To complete the analysis, the standard deviation of the expected variances was also computed (refer equation (32.10)). The standard deviation of the expected variances is used to model the distribution of expected variances. The expected variance was used, rather than the expected deviation, because the expected variance will be approximately Gaussian, and the expected deviation will not be approximately Gaussian.

$$\hat{\sigma}^2_{M,\vec{S}} = \textstyle\sum_i (i - M)^2 \Pr(Microposition = i \mid SensorValues = \vec{S}) \tag{32.8}$$

$$\hat{\sigma}^2 = \textstyle\sum_{M,\vec{S}} \hat{\sigma}^2_{M,\vec{S}} \Pr(SensorValues = \vec{S} \wedge Microposition = M) \tag{32.9}$$

$$\Pr(SensorValues = \vec{S} \wedge Microposition = M) =$$
$$\Pr(SensorValues = \vec{S} \mid Microposition = M) \Pr(Microposition = M)$$
$$\sigma(\hat{\sigma}^2) = \sqrt{E((\hat{\sigma}^2)^2) - (E(\hat{\sigma}^2))^2} \tag{32.10}$$

32.3.1 *Monte carlo approximation*

Estimating $\hat{\sigma}$ by marginalizing over all of the possible positions and reflectivities would produce the desired result, but the computations cannot be performed in a reasonable amount of time. For instance, for a ten sensor configuration, where each sensor can take on 64 discretized values, there are $|S| = 64^{10}$ possible different reflectivities. Marginalizing $\hat{\sigma}_{M,S}$ over the state-space involves $O(|M||S|\#\ sensors)$ time.

Monte Carlo estimation techniques are used to produce a result in a reasonable amount of time. Monte Carlo techniques essentially involve rolling dice to choose random configurations, and then computing values based upon the random configurations. The Monte Carlo assumption is that in the limit, the expected value for a large number of random trials is equal to the true value.

Each trial was performed as follows:

1. Microposition $\mathbf{M}$ was randomly chosen from a uniform distribution.
2. The sensor values $\vec{S}$ were predicted using the $\Pr(SensorValue = S_i \mid Microposition = \mathbf{M})$ distributions.
3. We computed the probabilities $\Pr(SensorValues = \vec{S} \mid Microposition = M)$ for each microposition M (refer equation (32.4)).
4. $\Pr(SensorValues = \vec{S})$ was computed by summing the probabilities for all of the micropositions, M, from the previous step (refer equation (32.5)).
5. We computed $\Pr(Microposition = M \mid SensorValues = \vec{S})$ using Bayes' theorem and the distributions from the previous two steps (refer equation (32.7)).
6. The expected deviations $\hat{\sigma}_{M,S}$ were computed from $\mathbf{M}$ and $\Pr(Microposition = M \mid SensorValues = \vec{S})$ probability distributions (refer equation (32.8)), and these expected deviations, and expected variances were recorded.

Normalizing values usually involves weighting the value for each configuration by the probability of observing the configuration. In the Monte Carlo technique, the expected deviations were not weighted by $\Pr(SensorValues = \vec{S} \wedge Microposition = \mathbf{M})$ because this probability of observing this configuration was implicitly taken into account when the configurations were randomly chosen.

32.4 Results

A 100,000 trial Monte Carlo simulation was run for different sensor configurations. Sensor configurations are characterized by $\#\ sensors$ and $periodicity$. The sensors were placed at phases $\frac{2\pi}{periodicity}$ with respect to the platen.

Figure 6 shows the expected deviations for the scaled data set for each of the simulated sensor configurations $\{\langle \#\ sensors = 3, periodicity = 3\rangle, \langle \#\ sensors = 3, periodicity = 4\rangle, \ldots \langle \#\ sensors = 4, periodicity = 3\rangle, \ldots\}$. The expected deviations are presented in units of the sensor's discretization ($\frac{1}{64}$ microsteps). Figure 4 shows that four reflective sensors placed evenly at intervals of $\frac{2\pi}{9}$ ($periodicity = 9$) have an estimated expected deviation of 0.752 $\frac{1}{64}$ microsteps (peak error of $< 2\frac{1}{64}$ microsteps). Eight sensors are necessary in order to produce peak error accuracy of less than $\frac{1}{64}$ microstep.

The variances of the expected deviations are used to characterize the error distribution. In most robotic applications, the average error is not as critical as the peak error during some operation. The smaller the standard deviation of the variance, the smaller the range of expected errors. These variances would be used to determine the minimal sensor configuration for which 99% of the time the positional error is less than 1 microstep. As the number of sensors increases,

Figure 4 shows that the variances of the deviations decrease at a rate eight times faster than the expected deviation.

$\# sensors$	$periodicity$	$\hat{\sigma}$	$\sigma(\hat{\sigma}^2)$
3	3	0.877	3.130
4	3	0.767	1.747
5	3	0.646	0.360
6	3	0.587	0.327
7	3	0.564	0.301
8	3	0.517	0.284
9	3	0.433	0.257
10	3	0.437	0.255

$\# sensors$	$periodicity$	$\hat{\sigma}$	$\sigma(\hat{\sigma}^2)$
3	6	0.894	1.000
4	9	0.752	0.628
5	3	0.646	0.360
6	3	0.587	0.327
7	14	0.523	0.300
8	9	0.451	0.262
9	15	0.433	0.251
10	6	0.391	0.235

FIGURE 4: In Figure 6, the estimated error for sensor configurations using scaling achieve local minima at $periodicity = 3$ and the periodicity corresponding to the minimum expected variation. The estimated expected deviations, and the standard deviations of the expected variances, for both of these cases are given in the tables above. The values are based upon Monte Carlo estimation sampling of 100,000 cases (from Figure 6).

The probabilistic analysis delivered a non-intuitive result: it showed that sensor spacing at $\frac{2\pi}{3}$ is only slightly suboptimal (within 10%) for all sensor configurations. This result leads us to consider constructing three-sensor unit with 120° phasing, as opposed to two-sensor units. Multiple three sensor units could be installed to improve position estimates.

The fact that sensor spacing at $\frac{2\pi}{3}$ is only slightly suboptimal may be due to the combination of non-redundant information and oversampling. Maximum oversampling would occur at sensor spacing $\frac{2\pi}{2}$, but this spacing is inherently bad because the reflectivity response is slightly symmetric, sampling at opposite angles on the unit circle is effectively redundant sampling. (This *redundancy* does not violate the independence assumption because the sensed reflectivities, as functions of position, are independently chosen from their respective distributions.)

32.4.1 Experimental results

Using four separate reflective sensors to record the x-axis of one of the RobotWorld modules, we experimentally measured the difference between the commanded position and the computed position; a short run covering 3.5 steps is shown in Figure 5. For this sensor configuration, the reflectivity information was evenly uninformative at extreme phases, so we spaced the sensors at orientations of $\frac{\pi}{2}$ with respect to the platen. It was difficult to both position the sensors and achieve the desired performance with the prototype sensors. These results were achieved by utilizing eight bit resolution in the probability tables.

The difference between the sensed and commanded positions is given in Figure 5. The peak errors are approximately twice as large as the expected deviation of 1.028 $\frac{1}{64}$ step. This experiment agreed with the predictions and confirmed the probabilistic analysis.

32.5 Conclusion

In this paper, we presented a technique for estimating the minimum number of reflective sensors necessary for localizing RobotWorld modules to an accuracy of $\frac{1}{64}$ microsteps for position and force-control applications. Probabilistic methods, model simulation, and Monte Carlo estimation

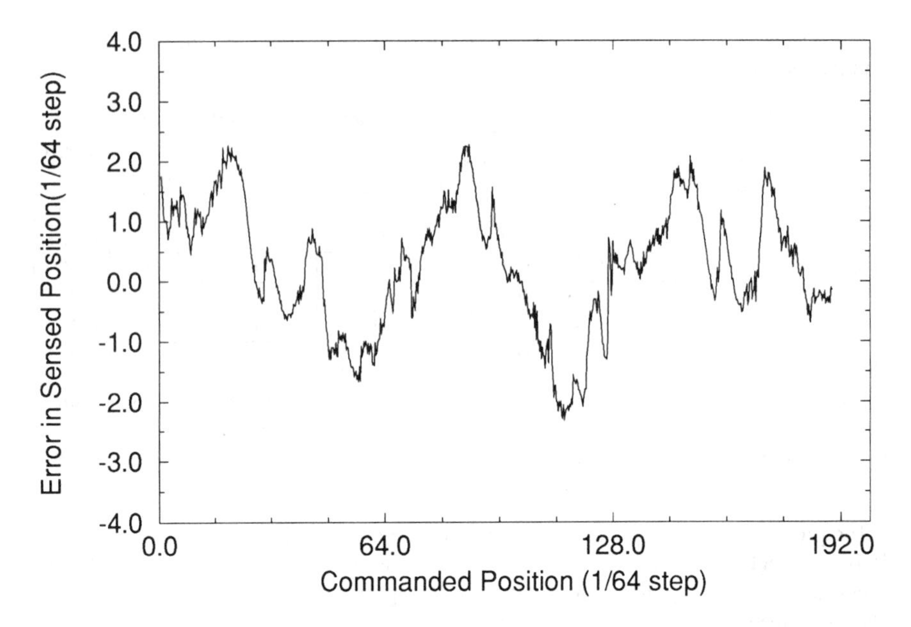

FIGURE 5: Experimentally determined error in sensed position. The standard deviation for the run is $1.028 \frac{1}{64}$ step, and the peak position error is approximately twice the expected deviation ($2.2 \frac{1}{64}$ step).

were utilized to estimate the variance of the error for different sensor configurations. Experimental position error data matched the analytical estimate, validating the technique. Using this analysis, we found that an eight sensor configuration should provide enough "noisy" information to accurately determine the position to within $\frac{1}{64}$ microsteps ($\frac{1}{1000}''$). In addition, sensor phasing of 120° provides nearly optimal performance compared with other sensor phasings.

Acknowledgements

The authors wish to acknowledge: Clark Lampson for a prototype sensor, Barak Yekutiely for engineering, and fabricating the reflective sensors, Steve Burgett, and Peter Cheeseman and Wray Buntine for constructive criticisms on the original report.

32.6 References

[1] "Description of a Robot Workspace Based on a Linear Stepper Motor". *AT&T Technical Journal*, 67(2):6–11, 1967.

[2] Jehuda Ish-Shalom and Dennis G. Manzer. "Electronic Commutation and Control of Step Motors". Technical Report RC 11084, IBM Thomas J. Watson Research Center, 1985.

[3] Edward Nicolson, Steve Burgett, Aaron Wallack, and Barak Yekutiely. Optimal position sensing for closed loop control of linear stepper motors. In *JSME International Conference on Advanced Mechatronics*, pages 322–327, 1993.

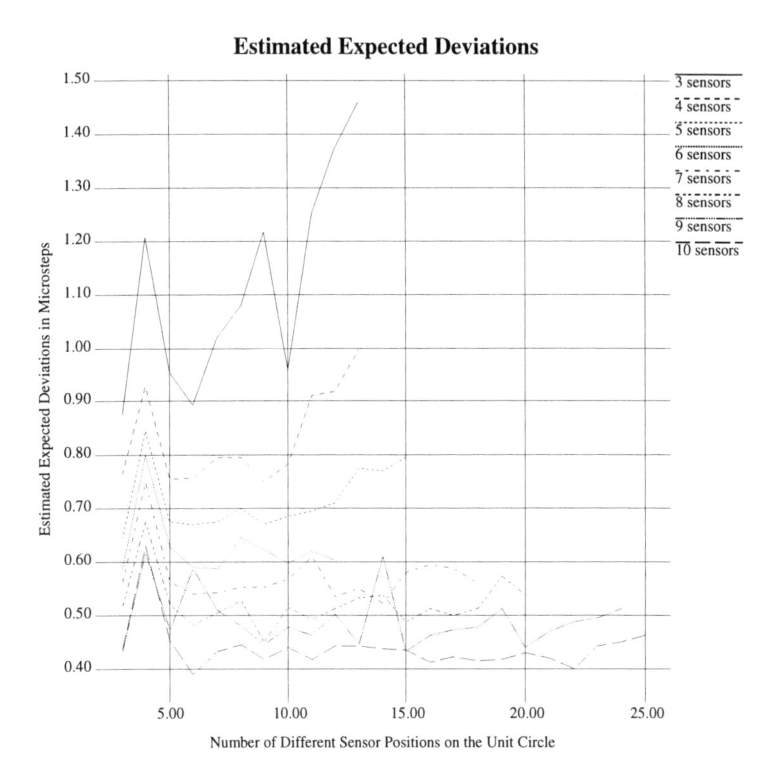

FIGURE 6: The estimated expected deviations of different sensor configurations computed by Monte Carlo estimation with 100,000 trials. The expected deviations are in terms of $\frac{1}{64}$ step. Each line in the graph represents a fixed number of sensors with different sensor configurations. Notice the expected deviations for $\boldsymbol{periodicity} = 3$ is only slightly greater than the minimum expected deviation, and that the expected deviations rise when $\boldsymbol{periodicity} > 2|\# \ \boldsymbol{sensors}|$ (sensors are positioned at angles $< \frac{2\pi}{\boldsymbol{periodicity}}$).

Part V

Similarity-Based Models

33
Learning to Catch: Applying Nearest Neighbor Algorithms to Dynamic Control Tasks

David W. Aha and Steven L. Salzberg

Navy Center for Applied Research in Artificial Intelligence
Naval Research Laboratory, Code 5514
Washington, DC 20375 USA
aha@aic.nrl.navy.mil
and
Department of Computer Science
Johns Hopkins University
Baltimore, MD 21218
salzberg@cs.jhu.edu

ABSTRACT This paper examines the hypothesis that local weighted variants of k-nearest neighbor algorithms can support dynamic control tasks. We evaluated several k-nearest neighbor (k-NN) algorithms on the simulated learning task of catching a flying ball. Previously, local regression algorithms have been advocated for this class of problems. These algorithms, which are variants of k-NN, base their predictions on a (possibly weighted) regression computed from the k nearest neighbors. While they outperform simpler k-NN algorithms on many tasks, they have trouble on this ball-catching task. We hypothesize that the non-linearities in this task are the cause of this behavior, and that local regression algorithms may need to be modified to work well under similar conditions.

33.1 Introduction

Dynamic control problems are the subject of much research in machine learning (e.g., Selfridge, Sutton, & Barto, 1985; Sammut, 1990; Sutton, 1990). Some of these studies investigated the applicability of various k-nearest neighbor (k-NN) methods (Dasarathy, 1991) to solve these tasks by modifying control strategies based on previously gained experience (e.g., Connell & Utgoff, 1987; Atkeson, 1989; Moore, 1990). However, these previous studies did not highlight the fact that small changes in the design of these algorithms may drastically alter their learning behavior. This paper describes a preliminary study that investigates this issue in the context of a difficult dynamic control task: learning to catch a ball moving in a three-dimensional space, an important problem in robotics research (Geng *et al.*, 1991).

Our thesis in this paper is that agents can improve substantially at physical tasks by storing experiences without explicitly modelling the physics of the world. Instead, they model the data itself, using more general learning capabilities. Although it is useful to exploit physical models whenever possible, we believe that by collecting and modelling large numbers of stored experiences, a learning device can be more flexible, since it does not need to understand (explicitly, that is) the physics of the environment. For example, professional tennis players train by repeating their actions thousands of times, rather than by studying the physics of spinning

[1]*Selecting Models from Data: AI and Statistics IV.* Edited by P. Cheeseman and R.W. Oldford. ©1994 Springer-Verlag.

tennis balls. In other words, the skills we wish to study are gained through practice, during which the learner stores experiences, which are retrieved for later use by pattern-matching and data-modelling methods.

To investigate our hypothesis, we conducted experiments using a simulated robot world in which the robot moves freely on a flat surface, while a ball flies through the air above it. The robot does not have a model of the physics of motion that guide the ball, but it does have memory and data-modelling abilities. In particular, we evaluated several k-NN variants on this learning task and addressed the following questions:

1. Can nearest neighbor quickly learn rough approximations for control tasks?
2. Does k-NN ($k > 1$) outperform simple 1-NN, and does either outperform a simple non-learning control strategy?
3. Can nonlinear scaling functions of the instances' attributes increase learning rate?
4. Given k neighbors, how should the agent induce a value to control its own motion?

We conducted experiments to investigate each of these questions. The last question is perhaps the most crucial. The robot must decide at each time step (i.e., every 0.1 seconds) which direction to move. The input to the robot is a four-dimensional vector, and the robot is essentially learning a function that maps this vector to an angle in the range $[0, 2\pi)$. Because it is learning a function rather than performing a classification task, local regression techniques can be used to predict a value from the k neighbors. We investigated two such techniques.

33.2 Related work

One of the most well-known learning tasks involving dynamic control is the cart-and-pole problem (Michie and Chambers, 1968), which requires the agent to balance a pole standing on a cart. Learning involves selecting which of two directions to push the cart along a one-dimensional track of fixed length. This task is difficult because, when the pole falls, it is not obvious which action was responsible for the fall, since the system may have been in trouble for many time steps. Neural net (Selfridge, Sutton, & Barto, 1985) and nearest neighbor approaches (Connell & Utgoff, 1987) have performed reasonably well on this problem.

Another interesting delayed-feedback problem is that of "juggling" a ball so that it falls repeatedly on a paddle held by a robot (Rizzi, Whitcomb, & Koditschek, 1992; Aboaf, Drucker, & Atkeson, 1989). Aboaf *et al.* programmed their robot to learn how to calibrate its motions. That is, the ball-juggling strategy was built into its software, but the vision system and the robot effectors introduced noise and calibration errors. Thus, the robot could not juggle the ball without substantial tuning. Therefore, they taught their robot how to improve its performance by adjusting its aim according to a polynomial fitted to the robot's errors.

More recently, Atkeson and his colleagues investigated a range of techniques, including some of the methods used here, for learning functions such as controlling a robot arm (Moore & Atkeson, 1992). Their results indicate that local weighted regression algorithms can outperform other k-NN variants on several simulated and (hardware) robotics tasks. We examine the performance of similar algorithms in this paper, attempting to extend this work for learning a difficult task in a 3-D space. Although we currently are working with a simulation, we hope to test these methods on hardware in the near future.

33.3 Function learning

During training, the robot tries to catch the ball. One instance is recorded for each time step during the ball's flight. The robot can move at a maximum speed of ten meters per second in any direction in its two-dimensional plane. After the ball lands, then the direction from each instance's location to the ball's landing location is computed and recorded with the instance. This encodes one point for the target function (i.e., all instances are "positive" examples of the function). The learning task is to closely approximate this function across the complete set of possible instances, despite the fact that the instance space is large, and only a small portion of the space is encountered during training. Thus, generating accurate predictions relies on correctly generalizing from the stored instances.

33.3.1 Learning algorithms and experiments

The algorithms we investigated all use as their basis a form of k-nearest neighbor learning. This family of algorithms has been accorded several names in the machine learning and related literatures, including *instance-based* learning (Bradshaw, 1987; Aha, Kibler, & Albert, 1991), *exemplar-based* learning (Salzberg, 1991), *memory-based* reasoning/learning (Stanfill & Waltz, 1986; Atkeson, 1989), and *case-based* learning (Aha, 1991). We include in this family of algorithms more sophisticated methods that first extract a set of neighbors, and then attempt relatively expensive function estimation methods using, for example, local weighted regression.

Initially, the robot has no stored instances to use for predicting which direction to move; it uses the simple default rule "move towards the ball" during the first training flight. One instance is computed for each time step during each flight, but instances of a flight are not stored until the flight completes.

All stored instances include information on which direction the robot should move. This information is obtained by subtracting the robot's current two-dimensional location from the ball's (eventual) landing location. For subsequent flights (i.e., after the first one), the learning algorithms compute similarities between the current instance and all stored instances, select the k most similar instances, and uses a function of their stored directions to determine which direction to move the robot.

The selection of the instance representation is crucial for simple k-NN algorithms. If the instance's dimensions are relatively uninformative, due to the fact that they do not encode important relationships between the ball's flight and the robot's location, then no learning algorithm can be expected to have fast learning rates. Furthermore, because k-NN is intolerant of irrelevant attributes (Aha, 1989; Kelly & Davis, 1991), it is important to minimize the number of irrelevant attribute dimensions used to represent the instance. Our algorithms should have faster learning rates when the instance representations contain informative and relevant attributes, and some of our experiments demonstrate that this is, in fact, the case. Our preliminary experiments led us to represent instances with four attributes: ball-robot distance, the angle θ of the ball's motion with respect to the robot (See Figure 1), the height of the ball, and the vertical velocity of the ball.

We conducted many learning trials with several different k-nearest neighbor algorithms. Each trial consisted of a sequence of training throws that were randomly generated according to a constrained uniform distribution (i.e., the ball flies for 4.5 seconds and lands within 30 meters of the robot's initial position). Each algorithm was tested on a separate but similarly-generated

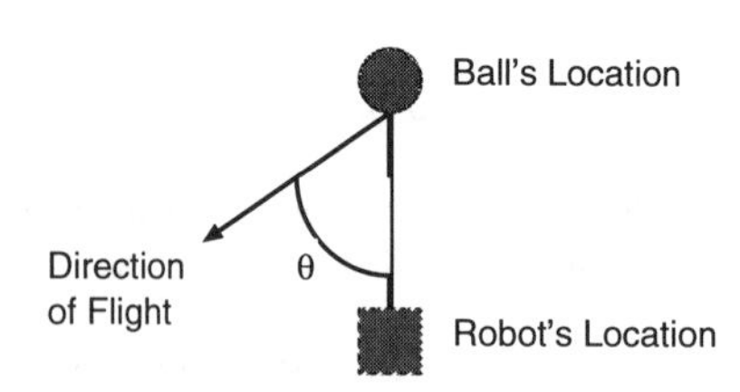

FIGURE 1: Overhead view of the ball-catching task. The angle θ is one of four dimensions used to represent instances. The others are the robot-ball distance, the ball's current height, and its vertical velocity.

Algorithm	Average distance to ball	Percent caught
1-NN	0.45	94.4
5-NN	0.37	98.2
5-NN (weighted)	0.57	91.5
Local regression	0.60	90.3
Local weighted regression	0.60	90.1

TABLE 33.1: Comparison of the k-NN algorithms on two dependent measures.

set of 100 throws. The goal of the robot was to minimize its distance to the ball when the ball hit the ground (i.e., to catch the ball). Since the robot can move at speeds up to ten meters per second, it can theoretically catch every ball.

Figure 2 provides some intuition regarding the robot's movements. This figure displays the flight of the ball, commencing from home plate at the top of the graphs, and the movements chosen by the robot when using the basic 1-NN algorithm. The upper left graph demonstrates that, on the first training instance, the robot simply moves toward the ball's (x, y) position. Thus, the ball eventually flies over the robot's head and lands far away (i.e., 6.14 meters) from the robot's final position. The other three graphs show where the robot moves on the same test flight after 1, 10, and 100 training flights have been processed. After only one training flight, the robot naturally moves in the wrong direction (i.e., towards the right) since all the instances derived from the first training flight tell the robot to move to the right. In this case, the ball landed 37.84 meters from the robot. After 10 training flights, the robot has a much richer set of instances to draw from and it moves much closer to its goal, missing by just 3.69 meters. Finally, the lower-right graph shows that, after 100 training flights, the robot more smoothly tracks and catches the ball, ending up just 0.35 meters from the ball's landing location.

33.3.2 Varying the prediction mechanism

Given a value of $k > 1$, we faced the question of how values should be induced by the direction-selecting function. Table 33.1 shows a small sample of our experimental results using different k-NN methods. Each line in the table shows the performance of an algorithm after 500 training throws. For 1-NN, no interpolation was necessary. For 5-NN, our approach simply averaged the directions stored with the k-nearest neighbors. Weighted 5-NN is a straightforward similarity-weighted averaging algorithm, in which the closer instances' directions are counted

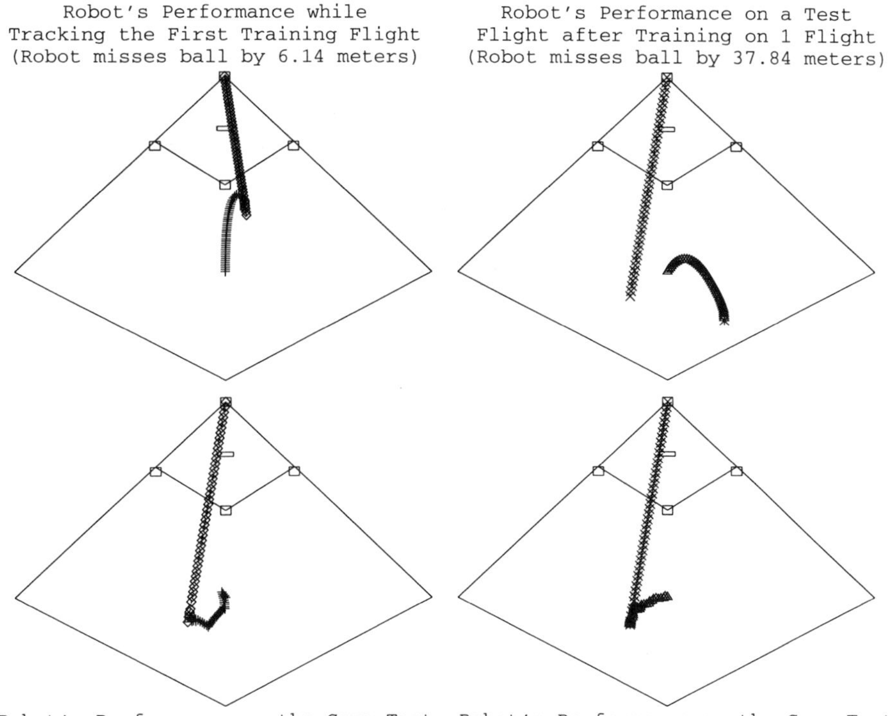

(Ball is Launched from Home Plate (Top); Robot Starts from Center Field)
Robot's Performance while
Tracking the First Training Flight
(Robot misses ball by 6.14 meters)
Robot's Performance on a Test
Flight after Training on 1 Flight
(Robot misses ball by 37.84 meters)
Robot's Performance on the Same Test
Flight after Training on 10 Flights
(Robot misses ball by 3.69 meters)
Robot's Performance on the Same Test
Flight after Training on 100 Flights
(Robot catches ball at 0.35 meters)

FIGURE 2: These graphs show where the robot moves as it follows the ball during the first training flight (upper left) and as it tracks a test flight after processing one, ten, and 100 training flights respectively.

more heavily (i.e., according to their relative similarity). Finally, we also tested two local linear regression approaches (Moore and Atkeson, 1992), which compute a regression equation from the k-nearest neighbors and then use it to make predictions. The weighted regression approach differs in that it gives closer neighbors greater influence in the computation of the regression equation (i.e., based on relative similarity).

Although local regression algorithms work well on other tasks (Moore & Atkeson, 1992), the linear regression methods did not outperform simpler k-NN methods on this task. In the table, we regard a ball as "caught" if the robot's final location is within one meter of the ball's landing location. Although not shown in the table, the robot's final location was always close to the ball's landing location even when the robot failed to catch the ball. This was true for all the methods tested.

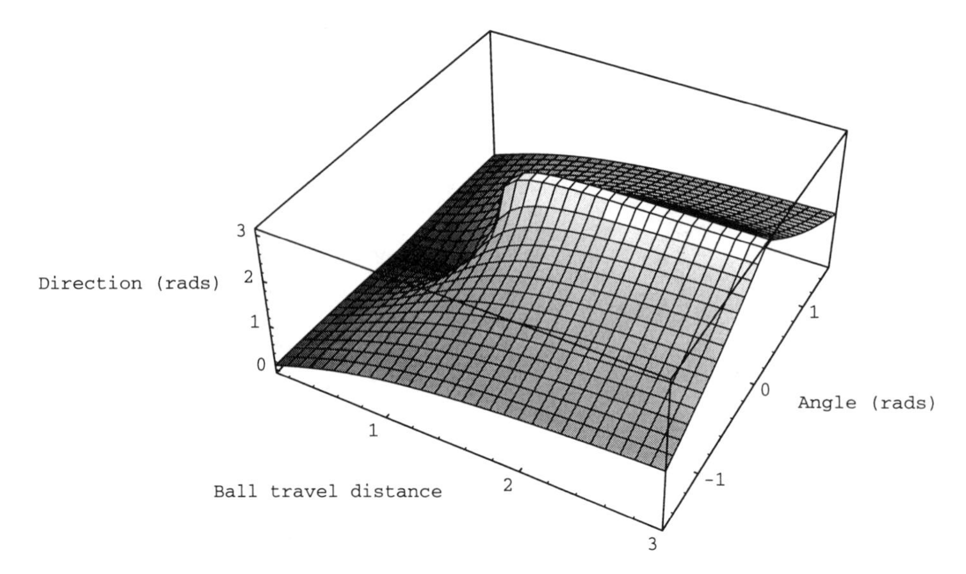

FIGURE 3: Direction to move when the ball is one meter from the robot.

33.4 Discussion and conclusion

The target function is highly non-linear, and in fact contains discontinuities. However, in most regions of the attribute space, this function is relatively smooth. A small piece of it is shown in Figure 3, which helps to illustrate this problem. The vertical axis in this figure measures the direction the ball is to move, in radians. The axis marked "angle" indicates the direction the ball is currently moving with respect to the robot, and the "ball travel distance" refers to how far the ball will travel before it hits the ground, in meters. The distance between the ball and the robot was held constant at one meter for this graph. Note that, when the angle is zero, the function hits a discontinuity; an infinitesimal change in angle at that point will cause a change in the robot's direction from left to right. A more serious problem occurs when the ball travel distance is one meter (i.e., the ball will land exactly where the robot is currently standing). A problem arises because, if the ball travels slightly less than one meter, then the robot should move straight ahead. If the ball instead travels slightly more, then the robot should move straight back, in exactly the opposite direction.

These sharp changes in the function may have decreased the performance of the linear regression methods for the following reasons. First, many of the crucial prediction tasks actually encountered are close to the non-linearities. The crucial decisions are those made when the ball is close to the ground and about to fall. Robot movements made at this time determine whether it will catch the ball or be slightly outside of the one meter radius of the ball's landing location. The robot almost always gets close to the ball's actual landing location before it lands. Thus, predictions must frequently be made for instances lying close to the non-linearities (i.e., ball's relative angle is close to zero and the distance to be traveled by the ball in the remainder of its flight is almost equal to the ball-robot distance). Second, local linear regression is biased towards assuming that the target function is locally linear in the feature space. Therefore it is likely to make inaccurate predictions near non-linearities. Since non-linearities frequently exist near many instances encountered while solving this task, local regression methods are at a slight disadvantage here.

The goal of our learning algorithms is *not* to induce the rules of physics governing the ball's flight, but rather to learn a mapping from instances describing the ball's relative position and velocity to an action that the robot should take. We applied variants of k-nearest neighbor algorithms to solve this ball-catching task. Virtually all the methods we have used work well when the ball is far away, guiding the robot in roughly the correct direction. However, when the robot is near the landing location of the ball, most methods have difficulty due to discontinuities in the target function. We are working on more elaborate control and data-modelling strategies that we hope will fix this problem. In our future work, we plan to improve the simulation by adding sensor noise, air friction, and effector noise. This will allow us to test the ability of various learning algorithms to handle more realistic control problems.

33.5 References

[1] Aboaf, E., Drucker, S., & Atkeson, C. (1989). Task-level robot learning: Juggling a tennis ball more accurately. In *Proceedings of the 1989 IEEE International Conference on Robotics and Automation* (pp. 1290–1295). Scottsdale, AR: IEEE Press.

[2] Aha, D. W. (1989). Incremental, instance-based learning of independent and graded concept descriptions. In *Proceedings of the Sixth International Workshop on Machine Learning* (pp. 387–391). Ithaca, NY: Morgan Kaufmann.

[3] Aha, D. W. (1991). Case-based learning algorithms. In *Proceedings of the DARPA Case-Based Reasoning Workshop* (pp. 147–158). Washington, D.C.: Morgan Kaufmann.

[4] Aha, D. W., Kibler, D., & Albert, M. K. (1991). Instance-based learning algorithms. *Machine Learning*, *6*, 37–66.

[5] Atkeson, C. (1989). Using local models to control movement. In D. S. Touretsky (Ed.), *Advances in Neural Information Processing Systems 2*. San Mateo, CA: Morgan Kaufmann.

[6] Bradshaw, G. (1987). Learning about speech sounds: The NEXUS project. In *Proceedings of the Fourth International Workshop on Machine Learning* (pp. 1–11). Irvine, CA: Morgan Kaufmann.

[7] Connell, M. E., & Utgoff, P. E. (1987). Learning to control a dynamic physical system. In *Proceedings of the Sixth National Conference on Artificial Intelligence* (pp. 456–460). Seattle, WA: Morgan Kaufmann.

[8] Dasarathy, B. V. (Ed.). (1991). *Nearest neighbor(NN) norms: NN pattern classification techniques.* Los Alamitos, CA: IEEE Press.

[9] Geng, Z., Fiala, J., Haynes, L. S., Bukowski, R., Santucci, A., & Coleman, N. (1991). Robotic hand-eye coordination. In *Proceedings of the Fourth World Conference on Robotics Research* (pp. 4–13 to 4–24). Pittsburgh, PA: Society of Manufacturing Engineers.

[10] Kelly, J. D., Jr., & Davis, L. (1991). A hybrid genetic algorithm for classification. In *Proceedings of the Twelfth International Joint Conference on Artificial Intelligence* (pp. 645–650). Sydney, Australia: Morgan Kaufmann.

[11] Michie, D., & Chambers, R. (1968). Boxes: An experiment in adaptive control. In E. Dale & D. Michie (Eds.), *Machine Intelligence 2*. Edinburgh, Scotland: Oliver and Boyd.

[12] Moore, A. W. (1990). Acquisition of dynamic control knowledge for a robotic manipulator. In *Proceedings of the Seventh International Conference on Machine Learning* (pp. 244–252). Austin, TX: Morgan Kaufmann.

[13] Moore, A., & Atkeson, C. (1992). *Memory-based function approximators for learning control.* Manuscript submitted for publication.

[14] Rizzi, A. A., Whitcomb, L. L., & Koditschek, D. E. (1992). Distributed real-time control of a spatial robot juggler. *IEEE Computer, 25(5)*, 12–24.

[15] Salzberg, S. L. (1991). A nearest hyperrectangle learning method. *Machine Learning, 6*, 251–276.

[16] Sammut, C. (1990). Is learning rate a good performance criterion for learning? In *Proceedings of the Seventh International Conference on Machine Learning* (pp. 170–178). Austin, TX: Morgan Kaufmann.

[17] Selfridge, O. G., Sutton, R. S., & Barto, A. G. (1985). Training and tracking in robotics. In *Proceedings of the Ninth International Joint Conference on Artificial Intelligence* (pp. 670–672). Los Angeles, CA: Morgan Kaufmann.

[18] Stanfill, C., & Waltz, D. (1986). Toward memory-based reasoning. *Communications of the Association for Computing Machinery, 29*, 1213–1228.

[19] Sutton, R. (1990). Integrated architectures for learning, planning, and reacting based on approximating dynamic programming. In *Proceedings of the Seventh International Conference on Machine Learning* (pp. 216–224). Austin, TX: Morgan Kaufmann.

34

Dynamic Recursive Model Class Selection for Classifier Construction

Carla E. Brodley and Paul E. Utgoff

Department of Computer Science
University of Massachusetts
Amherst, MA 01003

ABSTRACT Before applying an automated model selection procedure, one must first choose the class (or family) of models from which the model will be selected. If there is no prior knowledge about the data that indicates the best class of models, then the choice is difficult at best. In this chapter, we present an approach to automating this step in classifier construction. In addition to searching for the best model, our approach searches for the best model class using a heuristic search strategy that finds the best model class for each recursive call of a divide-and-conquer tree induction algorithm. The end result is a hybrid tree-structured classifier, which allows different subspaces of a data set to be fit by models from different model classes. During search for the best model, the method considers whether and why a model class is a poor choice, and selects a better class on that basis. We describe an implementation of the approach, the MCS system, and present experimental results illustrating the system's ability to identify the best model (and model class) efficiently.

34.1 The importance of model class selection in statistics and machine learning

In applied statistics, automatic model selection methods search through a class of models and select the one that best fits the data. These methods assume that the user has specified the class of models from which the selection will be made [Linhart & Zucchini, 1986; Box, 1990; Lehmann, 1990]. This specification is of critical importance to the success of the model selection process; if the model class does not contain a model that fits the data well, then even an exhaustive search must fail to find one. For example, if the best model for a particular data set is a second-order linear function, but the specified model class is the set of first-order linear functions, then no model from this class will provide satisfactory results.

An analogous problem exists in the machine learning community: selecting a concept representation language for a classifier construction task. Classifier construction is the process of finding a generalization of a set of training instances, each instance being described by a set of features and labeled with a class name. The success of this process is measured by the resulting classifier's accuracy when it is used to predict the class name for previously unseen instances. The learning performance of any one algorithm, for a particular task, depends primarily on whether the algorithm's concept representation language contains an accurate classifier for the examples. In statistics terminology, a classifier is a model, a concept representation language is a model class and constructing a classifier from the examples corresponds to model selection. For example, the underlying model class of a linear discriminant function is the set of first-order linear combinations. Selecting a model from this class involves finding a set of coefficients (or weights) and choosing which features to include in the function.

[1] *Selecting Models from Data: AI and Statistics IV.* Edited by P. Cheeseman and R.W. Oldford.

In both the statistics and machine learning communities, the last few decades have seen the creation of many methods for automatic empirical model selection. However, the success of any one method depends primarily on whether a model class that contains an accurate model has been chosen. In some cases, the choice of model class can be made on subject-matter considerations. However, in others there may not be any prior knowledge about the correct class of models. In such situations, someone wanting to find a model will be confused by the myriad choices, and will need to try many of them in order to be satisfied that a better model will not be found easily. In this paper we describe an approach to automating model class selection in addition to selecting a particular model from that class.

34.2 Automatic model class selection

A trivial solution would simply automate the process of trying each model class, and then select the classifier that best fits the specified objective (e.g., maximizing accuracy). Our approach differs from this in two fundamental ways. Firstly, our approach to selecting a model class is guided by various kinds of model *pathology* that develop when using a model class that is a poor choice. We define model pathology to be those recognizable symptoms that occur when the model fits the data poorly. When a model class is inappropriate, one needs to find a different model class that is appropriate. The ability to do this depends both on recognizing whether and why the model class is a poor choice, and on using this information to select a better model class. We have encoded this knowledge in a set of heuristic rules. Secondly, our approach is applied recursively in a divide-and-conquer manner, providing a mechanism for mixing the available model classes into a hybrid. This ability is desirable when different subspaces of the instance space are best fit using models from different model classes.

From the data analyst's point of view, it should become possible to solve classification learning tasks in a much more straightforward manner. Automation of the model class selection step will make machine learning techniques accessible to a wider range of scientists. Ultimately, there will be less need to understand the biases of the available learning algorithms, because this step in the analysis of data will now be automated.

34.3 Heuristic model class search

Our approach searches for a model from the available model classes using a best-first search. A set of heuristic rules is used to decide which model classes to try, which model classes should be avoided, and to determine when the best model has been found. The rules are based on knowledge of the strengths and weaknesses of each model class, and on how these characteristics are made manifest for different types of classification data sets. The instance space is partitioned according to the best model found, and the search is applied recursively to each resulting subset.

The heuristic rules are created from practical knowledge about how to detect when a model is a good fit, or whether a better one could be found in a different model class. For example, the model class of univariate decision trees is a poor choice when the features are related numerically. In this case, the features will be tested repeatedly in the decision tree, giving evidence that a series of tests are being used to approximate a non-orthogonal partition of the data that is not easily represented by a series of hyper-rectangles. The following rule detects this situation, and directs the search to the appropriate model class.

IF two or more features are tested repeatedly in a path of the tree
THEN switch from univariate test to linear test
AND fit a linear test of these features to the instances at the top of the path

Our approach addresses the problem of overfitting to the training data, in domains that contain noisy instances, by pruning back a hybrid classifier to maximize the accuracy on a set of instances that is independent from the training data [Quinlan, 1987; Breiman, Friedman, Olshen & Stone, 1984]. There are two types of noise: the class label is incorrect or some number of the attribute values are incorrect. Overfitting occurs when the training data contain noisy instances and an algorithm induces a classifier that classifies all instances in the training set correctly. Such a tree will usually perform poorly for previously unseen instances. This problem is analogous to the problem of outliers in regression analysis.

Our approach to pruning a hybrid classifier differs from the traditional approach to pruning decision trees. Traditionally, each non-leaf subtree is examined to determine the change in the estimated classification error if the subtree were to be replaced by a leaf labeled with the class of the majority of the training examples used to form a test at the root of the subtree. The subtree is replaced with a leaf if it lowers the estimated classification error; otherwise, the subtree is retained. During the construction phase of the hybrid classifier, our approach saves an alternative model at a node if it appears almost as good as the model selected. During pruning, in addition to deciding whether to retain a subtree or replace it with a leaf, our approach examines whether replacing the subtree with the alternative would result in fewer errors than either of the two other choices.

34.4 A recursive approach to combining model classes

Our method builds a decision tree; each test node is created by selecting and applying a primitive learning algorithm, each with a different underlying model class. The test at a node produces a partition of the instance space. For each subset of the partition, a recursive call to our procedure will either terminate or produce a subpartition recursively. The ability to combine model classes in a single tree structure allows a space of models that is strictly richer than the union of its constituent primitive model classes. The option of picking a single model class is not lost, but the approach also permits a hybrid classifier.

34.5 MCS: A model class selection system

We have implemented the approach in a system that we call the *Model Class Selection (MCS) System*. Currently, our system combines three primitive model classes that have been used extensively in both machine learning and statistics algorithms: linear discriminant functions, decision trees and k-nearest-neighbor classifiers.

For each model class there are many different algorithms for searching for the classifier that best fits the data. The *fitting algorithms* for each model class included in MCS are:

Univariate Decision Trees: To build a univariate decision tree, MCS chooses the features to include in the tree based on the information gain-ratio metric [Quinlan, 1986].

Linear Discriminant Functions: For two-class tasks, the system uses a linear threshold unit, and for multiclass tasks it uses a linear machine [Nilsson, 1965]. To find the weights of a linear discriminant function, the system uses the Recursive Least Squares procedure [Young, 1984] for two-class tasks and the Thermal Training rule [Frean, 1990; Brodley & Utgoff, in press] for multiclass tasks. To select the terms to include in a linear discriminant function, one of four search procedures is applied: sequential backward elimination (SBE) [Kittler, 1986], a variation of SBE, DSBE[3], that examines the form of the function to determine which terms to eliminate [Brodley & Utgoff, in press], sequential forward selection (SFS) [Kittler, 1986], and a method that examines a decision tree for replicated tests of a subset of the features, to suggest which terms to include in the function. The choice of which of these methods to apply is determined dynamically during learning, depending on the models that have already been formed and what their pathologies suggested.

K-**Nearest Neighbor:** This classifier stores each instance of the training data. To find k, the system estimates the classifier's accuracy with the following measure: for each instance in the training data, classify that instance using the remaining instances. The system selects the value of k that produces the highest number of correct classifications on the training data.

34.5.1 Rule base

The heuristic rules contain five types of pathology-detection criteria to guide the search for the best model. The following criteria detect the success or failure of a model and provide information about the reasons of failure:

1. **Information Theory:** This measure is used to compare any partition of the instances, whether it be a univariate test, a linear discriminant function, a k-NN classifier, or even an entire subtree. For a k-NN classifier, we employ a leave-one-out strategy for generating the class counts to compute the information-gain ratio of the classifier. Specifically, we classify each of the instances in the classifier using the remaining instances.

2. **Codelength:** Our use of this measure is based on the Minimum Description Length Principle, which states that the best "hypothesis" to induce from a data set is the one that minimizes the length of the hypothesis plus the length of the exceptions [Rissanen, 1989]. The *codelength* of a classifier (the hypothesis) is the number of bits required to encode the classifier plus the number of bits needed to encode the error vector resulting from using the classifier to classify the instances.

3. **Concept Form:** Pagallo and Matheus each demonstrate the utility of examining the form of a classifier to guide the search for the best model [Pagallo, 1990; Matheus, 1990]. For example, during the learning process we can examine the partially formed tree to determine whether a set of univariate tests is tested repeatedly in the tree. The action that is taken depends on the form of the replication. In addition, examination of the coefficients of a linear combination test yields information about the relative importance of the different features to classification.

---

[3] DSBE requires that the input values be normalized; MCS normalizes instances using zero mean and unit standard deviation

TABLE 34.1: A selection of rules from MCS's rule base

1.) IF the number of instances $<$ hyperplane capacity THEN find best univariate test
ELSE form a linear test LT_n based on all n features

2.) IF the best univariate test is 100% accurate THEN select it and stop.
ELSIF the test does not compress the data THEN fit a k-NN (k=2), continue
ELSIF the average difference in info-score of each feature to best features is $< \epsilon$
THEN start with best feature and search for a better test using SFS
ELSE select the univariate test and recurse.

3.) IF the instances are linearly separable THEN continue feature elimination
ELSIF a few weights are larger than the others THEN find the best univariate test
ELSIF many of the weights are close to zero THEN use DSBE
ELSE use SBE to find the best linear test based on the fewest features

4.) IF info-score(LT_{i-1}) $\geq$ info-score(LT_i) AND $i > 1$ THEN cont. feature elimination
ELSIF the best linear test compresses the data THEN select that test and recurse
ELSE find the best univariate test

5.) IF neither a univariate test nor a linear test compress the data THEN
find the best value of k for an instance based classifier AND
select from available tests, the one with the maximum info-score, recurse

6.) IF info-score(best univariate test) $>$ info-score(LT_n) THEN search for LT_i using SFS
ELSE examine form of the linear test to choose SBE or DSBE (see Rule 3)

4. **Accuracy:** Another comparative measure of two models is their classification accuracy. In addition, the accuracy of a linear combination test indicates whether the space is linearly separable.

5. **Instance Set:** We measure two different aspects of the set of training instances: the number of instances and the number of features.

Each rule uses one or more of these sources of information. The knowledge about the situations in which a particular model class is best is encoded explicitly in the rules. In Table 34.1 we show a selection of rules in the current implementation. Rule 1 begins the search and is based on the observation that when there are fewer than $2n$ instances (n is the number of features), the capacity of a hyperplane, a linear test will *underfit* the training instances; when there are too few instances there are many possible correct orientations for the hyperplane, and there is no basis for determining which is the best test.

Rule 2 examines the univariate test to determine if the initial decision was correct. If the test is not 100% accurate and the average squared difference between the information score of each of the other features and the best feature is less than a threshold (10% of the magnitude of the best score in our implementation), then MCS searches for a multivariate test using SFS. This

TABLE 34.2: MCS vs. learning in one model class: Accuracy

Model Space	Breast	Bupa	Heart	Glass	Pixel
k-NN (k=1)	95.58 ± 1.1	61.16 ± 3.5	77.33 ± 3.9	63.85 ± 5.5	94.58 ± 0.8
LT_n	95.63 ± 0.9	65.00 ± 4.8	83.07 ± 4.4	52.50 ± 9.0	83.71 ± 2.8
Univariate DT	94.54 ± 1.8	65.23 ± 5.1	79.87 ± 4.7	66.15 ± 4.3	94.72 ± 1.8
MCS	95.86 ± 1.0	67.67 ± 3.7	83.07 ± 3.9	67.12 ± 4.4	94.53 ± 0.8

rule is motivated by the observation that if many of the features appear equally informative then a combination of these features may yield a better partition of the data. In Rules 2 and 5, MCS checks to see that the best test found thus far compresses the training data, which occurs when the codelength of the test and the error vector resulting from using the test to classify the data is less than the codelength of the error vector of the instances. If the number of bits to code the test is not less than the number of bits to code the errors that the test corrects, then MCS fits a k-NN classifier to the data. For cases in which univariate and linear combination tests do not compresses the data, a k-NN classifier may perform better, as there may not be a generalization of the instances that performs well, whereas using all of the information in the instances may yield better performance.

Rule 3 determines whether the initial selection of a linear test was appropriate. If the instances are linearly separable then MCS tries to eliminate any noisy or irrelevant features from the test; otherwise MCS examines the form of the test to decide where next to direct the search. In a linear combination test, if a few weights are of much larger magnitude than the others, this indicates that only a few features are important to classification. In these cases it may be better to start with the best univariate test and perform a SFS search. However, if many of the weights are close to zero, then this indicates that these features are not relevant and can be eliminated quickly, in which case one can use DSBE, which is a version of SBE that uses the magnitudes of the weights to determine which features to eliminate from the linear combination test. If neither of these two cases is true then the system will use an SBE search to try to eliminate irrelevant features. Rule 4 evaluates the best linear test found in the search; if the test compresses the data then it is selected; otherwise a search for the best univariate test is conducted (if one has not been found yet). Rule 6 starts an SFS search for the best linear test if the information score of a univariate test is higher than that of a linear combination test based on all n features.

34.5.2 MCS versus learning in one model class

We compare our system to a univariate decision tree algorithm, a linear discriminant algorithm (which constructs a linear machine for multiclass tasks) that builds a classifier using all of the input features, and a k-nearest neighbor algorithm ($k = 1$). For each of the tasks, the data were split randomly into training (75%) and test (25%) sets, with the same split used for all algorithms. For each of five data sets, Table 34.2 shows the sample average and standard deviation of the classification accuracy for ten runs.

The Breast data was collected at the Institute of Oncology, Ljubljani. The 699 instances are described by nine numeric and symbolic features. The two classes represent the reoccurrence or non-reoccurrence of breast cancer after an operation. The task for the Bupa data set is to determine whether a patient has a propensity for a liver disorder based on the results of six blood

TABLE 34.3: Model classes in the MCS classifiers

Model Class	Breast	Bupa	Heart	Glass	Pixel
k-NN	0.2	0.1	0.0	0.6	1.1
Linear Test	1.8	3.6	1.1	0.4	3.3
Univariate Test	0.2	1.1	0.0	6.6	2.9
Number of Hybrids	8	10	1	6	9

tests. There are 353 instances. The Heart data set consists of 303 patient diagnoses (presence or absence of heart disease) described by thirteen symbolic and numeric attributes [Detrano, et al.]. For the Glass data set, the task is to identify a glass sample taken from the scene of an accident as one of six types of glass using nine numeric features. The 213 examples were collected by B. German of the Home Office Forensic Science Service at Aldermaston, Reading, UK. In the pixel segmentation domain the task is to learn to segment an image into one of seven classes. Each of the 3210 instances is the average of a 3x3 grid of pixels represented by nineteen low-level, real-valued image features.

For three of the five data sets (Breast, Bupa and Glass) MCS constructed a classifier that is more accurate than each of the other algorithms. For two of these cases (Bupa and Glass) the difference is statistically significant at the 0.05 level using a paired *t*-test. For the Heart data set, the classifier produced by MCS has the same accuracy as the best of the other algorithms and for the Pixel task, MCS produced a classifier that is less accurate than two of the other algorithms; although this difference is not statistically significant.

The problem that no one model class will be best for all tasks is illustrated by the results for the individual model classes: for the Breast and Heart data sets, the linear discriminant model achieved the highest accuracy of the three individual model classes, and for the Bupa, Glass and Pixel data sets, the univariate decision tree achieved the highest accuracy of the three. In contrast, the classifiers found by MCS for each data set were never significantly less accurate than the best of the primitive algorithms and were sometimes significantly more accurate than the best of the primitive algorithms; indicating that a hybrid can provide a significant performance improvement. Note that they were always more accurate than the second and third best of the primitive algorithms. This property of robustness is desirable in an automated model selection system.

These results show that using MCS provides a significant advantage for data sets that are best learned with a hybrid classifier. However, even for data sets for which a hybrid classifier is not advantageous, MCS has the ability to find the best of the primitive algorithms. For some data sets, such as the Pixel data, three of the algorithms performed similarly, and we conjecture that the performance achieved by these algorithms for this data is an upper limit on the accuracy that can be achieved due to noise in the data set.

Table 34.3 shows the average number of nodes of each type of model class in the MCS classifiers and the number of classifiers produced by MCS that were hybrids. Note that for the Heart data set, for nine out of ten runs, MCS found a classifier that was identical to the best of the primitive algorithms (a linear combination test) and in only one case produced a hybrid classifier. For the Breast, Bupa and Glass data set, for which MCS produced classifiers with higher accuracy, the most frequent model found by MCS was a hybrid. For the Glass data set, the four runs for which MCS did not produce a hybrid it produced a univariate decision tree which is the best of the other algorithms. For the Pixel data set, nine of the ten runs produced a hybrid classifier, but this did not increase the accuracy over the other algorithms.

34.6 Conclusion

In this paper we have presented a heuristic recursive approach to model class selection for classifier construction in machine learning. The empirical results illustrate that multiple model classes can be searched effectively to select the best model for a given task. This approach is applicable to statistical methods for automated model selection. An experienced statistician recognizes when the assumptions about the data are incorrect and when the choice of model class is poor. Often this knowledge comes from applying diagnostic tools that identify aspects of the problem that do not conform to the chosen modeling process [Cook & Weisberg, 1982]. There are many different statistics that can be used to assess the fit of a selected model to the data. The difficulty an inexperienced practitioner faces is how to choose which of these diagnostic methods to use and how to utilize the resulting information to alter the modeling process. By encoding this knowledge into a heuristic search strategy, the expert's knowledge becomes available to less experienced practitioners.

Acknowledgments

This material is based upon work supported by the National Aeronautics and Space Administration under Grant No. NCC 2-658, and by the Office of Naval Research through a University Research Initiative Program, under contract number N00014-86-K-0764. The Pixel data set was donated by B. Draper from the Visions Lab at UMASS. The Breast, Bupa and Heart Disease data sets are from the UCI machine learning database.

34.7 REFERENCES

[Box, 1990] Box, D. R. (1990). "Role of models in statistical analysis." *Statistical Science, 5*, 169-174.

[Breiman, Friedman, Olshen & Stone, 1984] Breiman, L., Friedman, J. H., Olshen, R. A., & Stone, C. J. (1984). *Classification and regression trees*. Belmont, CA: Wadsworth International Group.

[Brodley & Utgoff, in press] Brodley, C. E., & Utgoff, P. E. (in press). "Multivariate decision trees." *Machine Learning*. Kluwer Academic Publishers.

[Cook & Weisberg, 1982] Cook, R. D., & Weisberg, S. (1982). *Residuals and influence in regression*. Chapman and Hall.

[Detrano, et al.] Detrano, R., Janosi, A., Steinbrunn, W., Pfisterer, M., Schmid, J., Sandhu, S., Guppy, K., Lee, S., & Froelicher, V. (1989). "International application of a new probability algorithm for the diagnosis of coronary artery disese." *American Journal of Cardiology, 64*, 304-310.

[Frean, 1990] Frean, M. (1990). *Small nets and short paths: Optimising neural computation*. Doctoral dissertation, Center for Cognitive Science, University of Edinburgh.

[Kittler, 1986] Kittler, J. (1986). "Feature selection and extraction." *Handbook of pattern recognition and image processing.*

[Lehmann, 1990] Lehmann, E. L. (1990). "Model specification: The views of Fisher and Neyman, and later developments." *Statistical Science, 5*, 160-168.

[Linhart & Zucchini, 1986] Linhart, H., & Zucchini, W. (1986). *Model selection.* NY: Wiley.

[Matheus, 1990] Matheus, C. J. (1990). *Feature construction: An analytic framework and an application to decision trees.* Doctoral dissertation, Department of Computer Science, University of Illinois, Urbana-Champaign, IL.

[Nilsson, 1965] Nilsson, N. J. (1965). *Learning machines.* McGraw-Hill

[Pagallo, 1990] Pagallo, G. M. (1990). *Adaptive decision tree algorithms for learning from examples.* Doctoral dissertation, University of California at Santa Cruz.

[Quinlan, 1986] Quinlan, J. R. (1986). "Induction of decision trees." *Machine Learning, 1*, 81-106.

[Quinlan, 1987] Quinlan, J. R. (1987). "Simplifying decision trees." *Internation Journal of Man-Machine Studies, 27*, 221-234.

[Rissanen, 1989] Rissanen, J. (1989). *Stochastic complexity in statistical inquiry.* New Jersey: World Scientific.

[Safavian & Langrebe, 1991] Safavian, S. R., & Langrebe, D. (1991). "A survey of decision tree classifier methodology." *IEEE Transactions on Systems, Man and Cybernetics, 21*, 660-674.

[Young, 1984] Young, P. (1984). *Recursive estimation and time-series analysis.* New York: Springer-Verlag.

35

Minimizing the expected costs of classifying patterns by sequential costly inspections

Louis Anthony Cox, Jr. and Yuping Qiu

US WEST Advanced Technologies
4001 Discovery Drive, Boulder, CO 80303

ABSTRACT In many applications, an expert classifier system has access to statistical information about the prior probabilities of the different classes. Such information should in principle reduce the amount of additional information that the system must collect, e.g., from answers to questions, before it can make a correct classification. This paper examines how to make best use of such prior statistical information, sequentially updated by collection of additional costly information, to optimally reduce uncertainty about the class to which a case belongs, thus minimizing the cost or effort required to correctly classify new cases. Two approaches are introduced, one motivated by information theory and the other based on the idea of trying to prove class membership as efficiently as possible. It is shown that, while the general problem of cost-effective classification is NP-hard, both heuristics provide useful approximations on small to moderate sized problems. Moreover, a hybrid heuristic that chooses which approach to apply based on the characteristics of the classification problem (entropy of the class probability distribution and coefficient of variation of information collection costs) appears to give excellent results. The results of initial computational experiments are summarized in support of these conclusions.

Key Words: Classification expert system, pattern classification, recursive partitioning, machine learning

35.1 Introduction

In real-time expert systems and decision systems for which information collection and/or observation processing costs are very high, the ability to quickly guess correct answers based on minimal case-specific information, augmented with substantial statistical experience about similar cases, can potentially lead to dramatic improvements in performance. This paper presents a simple model of an expert classification system seeking to classify objects at minimum expected cost (or as quickly as possible, if the "costs" refer to inspection times). It discusses an exact dynamic programming algorithm and two heuristics for using prior statistical information to guess the correct class of a case as quickly (or, more generally, as cheaply) as possible. Practical applications include the following:

o *Optimal system diagnosis:* If a complex system has a known set of possible failure states with known prior probabilities, then the problem of sequentially inspecting components to minimize the expected cost of determining which failure state the system is in is an instance of the problem solved in this paper.

o *Design of interactive dialogues for diagnostic and advisory systems:* In many expert consultation systems, the system determines which questions to ask based on its knowledge base and on some reasoning about what conjunctions of facts will suffice to justify conclusions. This paper

---

[1] *Selecting Models from Data: AI and Statistics IV.* Edited by P. Cheeseman and R.W. Oldford.

presents heuristics for using statistical information about the prior probabilities that different conclusions are correct to adaptively formulate a sequence of questions that will minimize the expected effort (number of questions to answer) required from the user.

o *Pattern recognition and string-matching:* Applications in telecommunications, molecular biology, C^3I and computer science require a system to identify patterns and/or classify strings of symbols. Our heuristics suggest which symbols to examine first to match a new string to one of a set of known possible ones as cheaply as possible.

o *Optimizing applied scientific research:* Our heuristics can also potentially be used to design cost-effective applied research programs – e.g., to determine whether a chemical is a probable human carcinogen – by adaptively sequencing the tests to be performed.

o *Cost-effective statistical pattern classification:* Parametric statistical models for pattern classification, including linear discriminant analysis and logistic regression models, often lead to decision rules with the generalized linear model form "Assign an object to class j if the score $f(\beta x)$ has a value in interval I_j", where βx represents the inner product of pattern vector x with a weighting vector β and f is a (possibly nonlinear) linking function. This paper assumes that such a β is known and considers the order in which components of x should be examined to minimize the expected cost of identifying the class corresponding to βx.

In all these applications, classification is an expensive operation because it requires information about the object being classified that can only be obtained through costly inspections or experiments. Classification technologies (e.g., standard implementations of neural nets) that require all of the features of an object to be available as input may then be needlessly costly. Instead, we focus on recursive partitioning ("tree-structured") algorithms that lend themselves naturally to economical information-collection. The work reported on here complements earlier work [Cox 1990] that treated patterns as sets of statistically independent attribute values (e.g., "working" or "not working" for each statistically independent component in a multi-component reliability system). This paper assumes that the possible patterns (sets of attribute values) and their prior probabilities are known, so that the observed values of attributes are in general statistically dependent. Thus, where our previous work presented heuristics for determining at minimum expected cost whether a complex system (e.g., a space shuttle) would function correctly when activated, this paper addresses situations such as diagnosing at minimum expected cost which one of several possible failure states has occurred.

To obtain a quick grasp of the type of problem solved in this paper, consider the following two-person game. One player (called "Hider") selects a row of the following matrix:

1	0	0	1	
0	1	0	1	A simple two-person zero-sum game ("Mastermind").
0	1	1	1	Hider chooses a row; Inspector observes the contents
1	0	0	0	of columns for the row that Hider has selected.

The other player ("Inspector") tries to identify the selected row in a minimum number of moves by sequentially inspecting the contents of the columns. If Inspector must choose in advance which columns to inspect, then the minimum number of columns required to guarantes that the row will be identified is 3. By using a sequential (closed-loop) strategy, Inspector can guarantee that no more than 2 inspections are needed. [This "minimax" strategy is to inspect the first (left-most) column first, and then the third column if a 0 is seen and the fourth column otherwise.] Now suppose that, instead of playing this game against an intelligent opponent, Inspector knows that

"Nature" selects rows 1–4 with probabilities 0.1, 0.2, 0.3, and 0.4, respectively. (For example, the rows of the table may represent the possible failure states of a four-component reliability system, where each column represents a component and "0" and "1" represent "failed" and "working".) Is there an inspection strategy that requires fewer than 2 inspections on average to determine the selected row? The algorithms and heuristics presented in the following sections show how to answer this question by constructing a minimum expected cost sequential inspection strategy. (The reader may wish to show that, in the above game against nature, an optimal sequential inspection strategy reduces the average number of inspections required to identify the row to 1.9. Searching for the least-cost solution drives home the combinatorial nature of such problems.)

35.2 Problem formulation

A *logical classification expert system* for assigning cases to classes (where classes may be interpreted as diagnoses, predictions, or prescriptions, for example) consists of a set of rules that determine the unique class to which each case belongs from the truth values of various propositions about the case. ("Unknown" can be included as one of the classes.) Suppose that the propositions that determine class membership have been reduced to disjunctive normal form (DNF); then the classification expert system can be expressed as a set of rules, each of the form

(case z belongs to class j) if [propositions $(X_{j1}, \ldots, X_{jn})$ all hold for case z] (1).

Here, the X_{ij} are atomic propositions that can be asserted about the case (they are literals or negations of literals, so that the rules need not be strictly Horn); the disjunctions of the DNF have been distributed across multiple rules; and the conjunctive terms have become the antecedent conjunctions of the rules. (The dummy variable z, which might be regarded as implicitly universally quantified over the set of all cases, is bound in any application, so that propositional rather than predicate calculus applies.)

An expert classification system whose rules are in the canonical form (1) can be represented by an $M \times N$ matrix of ones, zeros, and blanks and an M-vector of class labels, where M is the number of distinct classes and N is the total number of literals that can be asserted (or denied) about a case. The interpretation is that columns correspond to the literals, rows to the classes, and if the pattern of truth values of the literals for a case matches the pattern of zeros and ones in row i (with the usual coding 0 = false, 1 = true), then the case belongs to the class whose label is associated with row i. The same label could appear more than once in the M-vector of labels, since there may be more than one way to prove membership in a class. However, this paper assumes that there is a one-to-one correspondence between classes and rows. This is a natural restriction in many cases (see below). Blanks in the matrix may be used to represent "don't care" conditions, i.e., values of literals that do not enter into the antecedent of a rule. Such a matrix may be called the *canonical matrix* for a logical classification system. It maps each N-vector of truth values for literals into exactly one corresponding class label. Let $L \leq M$ denote the number of distinct classes: then the classification system may be viewed abstractly as computing a discrete classification function $f : \{0,1\}^N \longrightarrow \{1, 2, \ldots, L\}$. The challenge is to develop algorithms or heuristics for computing the value of this function, for any set of instance data, as quickly (or as inexpensively) as possible.

Let x denote the (initially unknown) vector of truth values for the set of N literals applied to a new (as-yet uninspected) case. (In pattern recognition applications, x could be a binary "feature vector" stating which features are present and which ones absent.) Let A be the canonical matrix of an expert classification system. If x were known, it could be compared to the rows of A until a match is found (i.e., until the truth values of the components of x match the truth values specified in the row). Then the classification of x would be given by the class label of that row. In practice, the following two considerations make classification less straightforward:

o *It may be costly to observe the components of* x. To capture this pragmatic aspect of classification, we will let c denote an N-vector of *inspection costs* for determining the truth values of the N components of x.

o *The answers to some questions may be more or less predictable in advance.* We will assume that prior information is represented by an L-vector of prior *class probabilities*, denoted by p. These induce conditional probabilities for the answers to questions: as some classes are eliminated by questioning, the probabilities of those that remain (that are consistent with all answers received so far) must be updated via conditioning.

In summary, a general classification problem with prior information about class probabilities is represented in canonical form by a triple (A, p, c) and a vector of uncertain but discoverable truth values, x, where A is the canonical matrix; x is one of L possible truth value configurations (typically interpreted as patterns or "states" for a case) having respective probabilities of $p(1), p(2), \ldots, p(L)$ for classes $1, 2, \ldots, L$; and c is the vector of inspection costs, with c_i being the cost of determining the truth value of literal i for $i = 1, 2, \ldots, N$.

A candidate *solution* for a logical classification problem (A, p, c) can be represented as a rooted binary classification tree in which each nonterminal node represents a literal (a property in our simplified model), the two branches descending from such a node correspond to the two possible truth values or test outcomes for that property (left branch = absent, right branch = present), and the terminal nodes contain the names of the classes. Such a tree specifies which property the system should ask about next, given the answers it has received so far, until it has obtained enough information to deduce which class the case belongs to. (A slight generalization would allow it to stop short of establishing the correct class with certainty, instead halting and making a best guess when the expected value of additional information, measured in terms of the expected reduction in the cost of classification error, is less than the expected cost to collect it. The algorithms discussed in this paper can easily be modified to incorporate this refinement.) The *expected cost* of a candidate solution can be defined recursively as follows:

(i) The expected cost of a terminal node is zero.

(ii) The expected cost of a nonterminal node is the inspection cost for the corresponding literal (i.e., for testing whether that property holds) plus the sum of the costs of each of its two child nodes weighted by their probabilities (i.e., by the probability that the case has the property, for the right child node, and weighted by one minus this probability for the left child node.)

An *optimal solution* is a classification tree having minimum expected cost among all possible classification trees. It thus represents a cost-effective strategy for reducing uncertainty about class membership by optimally sequencing questions about a case's properties. Section 3 presents algorithms and heuristics for constructing optimal or nearly optimal classification trees.

As an example, consider an A matrix consisting of the following four pattern vectors (rows): (1, 0, 0, 1), (0, 1, 0, 1), (1, 0, 1, 0), and (0, 1, 1, 1). Let the prior probabilities of these four patterns be 0.4, 0.2, 0.3, and 0.1, respectively; thus, $p = (0.4, 0.2, 0.3, 0.1)$. The inspection cost vector is $c = (3, 1, 4, 1)$. Given a randomly generated pattern instance $x = (x_1, x_2, x_3, x_4)$, what is the best inspection strategy for examining the components of x to determine which pattern it is an instance of? The answer is not obvious to a human even though the problem is small.

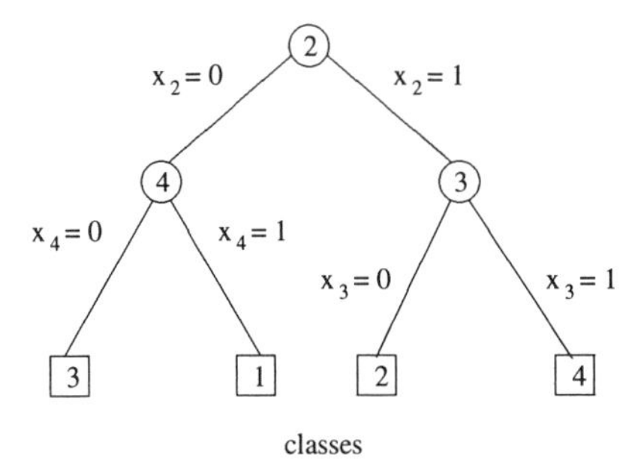

FIGURE 1: An example classification tree.

Using the dynamic programming procedure discussed in Section 3, an optimal answer is found to be as follows (see Figure 1): Inspect component 2 of x first (corresponding to column 2 of the A matrix). If there is a zero in column 2 (which occurs with probability 0.4 + 0.3 = 0.7) then inspect component 4 next; otherwise, inspect component 3 next. The outcome of these two inspections will determine which pattern x is an instance of, and no other inspection strategy has a smaller expected cost. (The expected cost of this strategy is 2.9. There is another optimal strategy that also achieves this expected cost that the reader may try to find.) The CPU time required to find this solution via dynamic programming is less than 0.01 seconds.

35.3 Exact and heuristic procedures

35.3.1 Exact solutions

Constructing an optimal (minimum expected-cost) classification tree is intrinsically difficult. Although exact optimal solutions can be found using dynamic programming, the time-complexity required is in general exponential in the number N of columns of A.

Theorem 1: *Finding minimum expected-cost classification trees for systems (A, p, c) is NP-hard.*

Proof: The proof is based on a polynomial-time reduction of the well-known NP-hard *set covering problem* to an optimal classification problem with equal costs. It is given in the appendix. If an "inspection" is interpreted as a data base retrieval of a stored value, then this result is compatible with the major result in [Greiner 1991].

Optimal classification strategies for small problems (e.g., for A of dimension less than about 20×20) can be found by dynamic programming, as follows. Let $C(A, p, c)$ denote the minimum expected cost of classifying a case using system (A, p, c). Then an optimal classification tree can be constructed by solving the following dynamic programming recursion for j, the index of the first column of A to be inspected:

$$C(A, p, c) = \min_{1 \leq j \leq n} [c_j + w_{j0} C(A_{j0}, p_{j0}, c) + (1 - w_{j0}) C(A_{j1}, p_{j1}, c)] \qquad (2).$$

where A_{jk} denotes the submatrix consisting of those rows of A having a truth value of k in column j, for $k = 0, 1$; p_{jk} = the normalized probability vector for the submatrix A_{jk}, found by rescaling the row probabilities for A_{jk} to sum to 1; and w_{jk} = prior probability that the truth value in column j is k, for $k = 0, 1$ (i.e., w_{jk} = the sum of the prior row probabilities for all rows in A_{jk}). The time complexity of this recursion is exponential in N, the number of columns of A (it is $N!$ in the special case where A is an identity matrix and of order not greater than $n(n-1)\cdots(n-m+2)2^{m-1}$ in general). If the number of rows of A is bounded by a constant, K, however, then the time complexity is $O(N^{K-1})$, i.e., it is polynomial in N for any given number of rows of A.

35.3.2 *Information-theoretic and signature-based heuristics*

When all inspection costs are equal, the optimal classification problem is to minimize the average number of questions (inspections) needed to determine the class of a case. If the rows of A could be designed by the user, then a sharp lower bound on the average number of questions required would be given by the entropy of the class probability vector p. A larger number of questions is typically required when A is exogenously specified, but the criterion of entropy reduction still provides a useful guide for selecting columns to minimize the number of questions that must be asked. If inspections are costly, then this criterion can be generalized to entropy reduction per unit cost.

Information-theoretic (entropy) heuristic: *At any stage of the sequential inspection process, inspect next the column that yields the greatest expected reduction in classification entropy per unit of inspection cost, given the results of all inspections made so far.*

The classification entropy at any stage is a measure of the uncertainty about the true class: it is defined as $H(p) = -\sum_i p_i \log_2 p_i$ where the sum is taken over all classes i that are consistent with the observations made so far and where the prior probabilities p_i for this subset of classes have been normalized (rescaled) to sum to 1. Intuitively, the information-theoretic heuristic always prescribes performing next the test that is expected to be "most informative" per unit cost, where informativeness is defined by expected entropy reduction. Letting $H(A)$ denote the classification entropy of matrix A, this amounts to always picking next the column j that maximizes $\{H(A) - [w_{j0}H(A_{j0}) + (1 - w_{j0})H(A_{j1})]\}/c_j$. Motivation for the information-theoretic heuristic is provided by the following often rediscovered "folk theorem". In the special case of uniform p and c, the maximum-entropy column is the one with the most nearly equal numbers of ones and zeros among the remaining (unexcluded) rows of A, and the information-theoretic heuristic calls for inspecting next this most nearly balanced column. Similar entropy-reduction heuristics (without costs) have been widely applied in *recursive partitioning algorithms* for learning classification rules from noisy data [Breiman 1984,Quinlan 1986]. Our optimal classification problem is different, since (A, p, c) contains an explicit model of the data-generating process (making it unnecessary to learn it from sample data) and since the classification rules are already known. The entropy reduction criterion finds a new use in this setting by suggesting how to use the rules most cost-effectively to make a classification decision.

An alternative approach to cost-effective information collection follows a *guess-and-verify strategy*. Consider first the case of uniform inspection costs. Rather than looking for the most informative column to inspect, the guess-and-verify approach instead identifies the row of A that is most likely to be the true one, given all observations made so far. Then, it tries to prove or disprove efficiently the hypothesis that the case being classified matches this most likely row. To test this hypothesis cost-effectively, a *signature* is created for each row. A *signature* for row i is a subset of column values that occurs only in row i. Thus, if these values are observed for a case, then the case belongs to class i. Given a signature for the currently most likely row, the next column to be inspected is the one corresponding to the element of its signature that is *least* likely to match observations. (If the case being classified does not have the hypothesized signature, then it is desirable to find this out as quickly as possible instead of wasting resources making easy matches, only to eventually discover that the hypothesized signature is wrong after all.) In the general case of unequal inspection costs, the heuristic works as follows.

Signature heuristic:

Step 1 (Signature Generation): Generate a signature for each row using the following greedy heuristic [Liepens 1991]. For each row, i, choose some initial column, j_0, as a seed. The value in column j_0 of row i discriminates this row from all rows with the opposite value in column j_0 (assuming that any informationless column, having all identical entries, has been deleted). Add to this partial signature j_0 the column having the greatest ratio of discrimination probability to inspection cost, where the "discrimination probability" for a column is defined as the sum of the probabilities of the rows that differ from row i in this column. Continue to select columns by this ratio criterion and add them to the partial signature until a complete signature (discriminating row i from all other rows) has been formed. Apply this signature-growing procedure starting with each of the N possible elements of row i as a seed, and choose the signature with the smallest total cost (sum of the inspection costs of its columns) as the one to be used for row i.

Step 2 (Row Selection): At any stage, choose as the next row to test the one with the greatest ratio of success probability to cost. The "success probability" in the numerator is the conditional probability (after eliminating all rows that are inconsistent with previous observations) that the case is an instance of this row. The "cost" in the denominator is the cost of the signature for the row, i.e., it is the sum of the inspection costs of its as-yet unobserved components.

Step 3: (Column Selection): The next column to be inspected is the one that contains the *least* likely element in the signature of the selected row.

Interpretively, Step 2 selects the "easiest" problem to work on (as measured by the probability-to-cost ratio) while Step 3 prescribes working on the "hardest" part of this problem first, so as not to waste effort prematurely on the easier parts. The "problems" considered consist of proving that a case instantiates a particular row .

35.4 Experimental results and a hybrid heuristic

To evaluate the performance of the information-theoretic and signature heuristics, we randomly generated several thousand test problems and solved them exactly, using the dynamic programming recursion in equation (2), and also approximately using the heuristics. Although careful discussion of the experimental design used and the results obtained are beyond the scope of this paper, the highlights of the experiments were as follows.

1. For randomly generated A matrices of a given size (numbers of rows and columns), the comparative performances of the information-theoretic and signature heuristics depend systematically on the entropy of the prior class probability vector, $H(p)$, and on the coefficient of variation of the inspection costs (i.e., on the ratio of the standard deviation of the components of c to their mean value). Therefore, for experiments with given numbers of rows and columns of A, we generated a fixed number (50 in exploratory experiments and up to three hundred in final experiments) of classification problems in each cell of a 5×10 grid of *entropy x coefficient of cost variation* values. The five entropy "bins" used for class probability vectors were chosen to cover the range from zero to the maximum possible entropy (equal to $\log_2 M$ for a problem with

M rows) in equal intervals on a log scale; thus, for example, for a problem with $M = 10$ rows as possible patterns, we would generate an equal number of prior class probability vectors in each of the entropy ranges 0.0 – 0.66, 0.67 – 1.3, 1.3 – 2.0, 2.0 – 2.65, 2.65 – 3.3. The inspection cost vectors were binned into ten intervals based on their coefficients of variation, ranging systematically from 0.0 – 0.1 through 0.9 – 1.0. The distribution of relative errors (compared to the exact solution) was studied for each heuristic in each of the 50 cells in this grid, for a variety of problem sizes.

Tables 1–3 summarize the results of these experiments. (To save space, only five of the ten coefficient-of-variation columns are shown.) These results are for A matrices of dimension 10×10, i.e., ten patterns and ten properties. The numbers shown in each cell are the average relative errors for 50 randomly generated test problems falling in the corresponding entropy bin for p and coefficient of variation bin for c.

TABLE 35.1: Relative Percentage Errors for the Entropy Heuristic (10×10 Problems)

	Coefficients of Variation of Costs				
Entropy	0.1	0.3	0.5	0.7	0.9
0 – 0.67	20.1	17.1	15.4	13.4	8.6
0.67 – 1.3	13.8	16.3	10.8	8.0	7.8
1.31 – 2.0	12.75	10.8	7.3	7.4	5.6
2.1 – 2.65	6.2	5.1	4.5	3.8	3.9
2.66 – 3.3	1.8	1.8	1.5	2.3	1.6

TABLE 35.2: Relative Percentage Errors for the Signature Heuristic

	Coefficients of Variation of Costs				
Entropy	0.1	0.3	0.5	0.7	0.9
0 – 0.67	0.39	0.58	1.5	0.56	0.83
0.67 – 1.3	8.2	7.5	5.0	6.2	4.3
1.31 – 2.0	9.6	9.4	7.9	6.9	7.2
2.1 – 2.65	8.2	10.4	8.6	8.2	5.4
2.66 – 3.3	7.9	10.8	9.2	9.6	5.9

TABLE 35.3: Relative Percentage Errors for the Hybrid (Entropy/Signature) Heuristic

	Coefficients of Variation of Costs				
Entropy	0.1	0.3	0.5	0.7	0.9
0 – 0.67	0.14	0.11	0.81	0.15	0.27
0.67 – 1.3	3.9	2.9	1.5	1.4	1.35
1.31 – 2.0	3.2	2.7	2.1	1.35	1.6
2.1 – 2.65	1.8	1.6	1.3	0.44	1.65
2.66 – 3.3	0.48	0.65	1.08	0.98	0.64

2. From the experimental results in Tables 1–3, it is clear that the information-theoretic and signature classification heuristics have complementary strengths. The signature algorithm does better (by up to 30-fold on 10×10 problems) on relatively uniform problems having low entropy

and low coefficient of cost variation. The information-theoretic algorithm does better (typically by a factor of between 2 and 9 on 10×10 problems) on all problems with sufficiently high entropy of the prior class probability vector; at lower entropy levels, its performance (measured by relative error compared to the true optimum) improves as the coefficient of variation of the cost vector increases. For example, on 10×10 problems, the signature algorithm dominates in the first two (low) entropy bins (i.e., for problems with prior classification entropies of less than 1.4 bits) while the information-theoretic algorithm dominates in the upper two entropy bins (problems with prior classification entropies of greater than 2.6 bits).

3. The average performance of the heuristics deteriorates gradually as problem size increases. We examined performance on problems ranging from 4×4 to 16×16, with a variety of non-square matrices for comparison. (For problems of size 16×16 , the dynamic programming algorithm took on the order of 1 hour on a Symbolics 3630 Lisp machine to find optimal solution trees, and we have not yet considered larger problems.) For example, average percentage errors for the signature algorithm increased from about 1.4% on a set of one hundred 5×5 test problems to about 8.7% on a set of one hundred 14×14 test problems.

4. Because the information-theoretic and signature algorithms have complementary strengths, a hybrid of the two works much better than either of them alone. They can be hybridized by generating subtrees from each node using both heuristics and keeping the one producing the better result (lower expected cost). Using such a hybrid heuristic, we were able to reduce the average error on the test set of 10×10 problems from about 9% to about 1% (see Table 3). Although the theoretical worst-case computational complexity of this hybrid heuristic is exponential, in practice it appears to take less than twice as long to run as the signature heuristic alone.

5. Relative computational efficiency for the hybrid heuristic and both of its components (the information-theoretic and signature heuristics) increases with problem size. For problems smaller than 8×8, the dynamic programming solution takes less than a second (even without correcting for garbage collection and other sources of machine-induced variation) and there is little point in using a heuristic. For 14×14 problems, the dynamic programming solution takes over ten minutes, while the most computationally expensive version of the hybrid (trying both signature and information-theoretic criteria at each node) takes about 2 seconds. Computing time for the dynamic programming algorithm rises sharply with problem size: 13×13 problems take less than six minutes to solve, while 16×16 problems take over an hour. The two heuristics take less than a second for all problem sizes examined, with the hybrid taking slightly longer (about 1.5 seconds on 10×10 problems compared to 0.92 seconds for the signature heuristic and 0.24 seconds for the entropy heuristic). For larger problems, the hybrid heuristic is expected to achieve better than 95% of the optimal solution in less than 1% of the time needed to reach full optimality. It is easily checked that both our heuristics have polynomial-time complexity (bounded above by $O(mn^2 \cdot \min\{m, n\})$ for the signature heuristic and $O(mn \cdot \min\{m, n\})$ for the entropy heuristic based on the numbers of operations that must be performed. The factor $\min\{m, n\}$ arises from the fact that each inspection eliminates at least one row and one column.) Thus, their advantage relative to the exact solution procedure continues to increase as problem sizes grow larger.

35.5 Conclusions

We have presented two heuristics for obtaining approximate solutions to the NP-hard problem of deciding in what order to ask questions so as to minimize the average cost of classifying cases. The preliminary experimental results reported here suggest that these heuristics are effective on small problems. The most exciting discovery, however, is that they can easily be hybridized to obtain substantial improvement in performance. Additional experiments are currently being designed and extensions of these results to arbitrary logical classification systems (when more than one row may represent the same class) are being developed. For practical purposes, it appears that the hybrid of the signature and entropy heuristics achieves near-optimal solutions with at most a few seconds of computational effort in problems with fewer than about 20 classes. (The time required by the heuristics is relatively insensitive to the number of attributes available to be inspected since signature length grows slowly as columns are added.) This range of problem sizes contains many applications of practical interest.

35.6 References

[Cox 1990] Cox, L.A., Jr., "Incorporating statistical information into expert classification systems to reduce classification costs", *Annals of Mathematics and Artificial Intelligence*, **2**, 93-108, 1990.

[Greiner 1991] Greiner, R., "Finding optimal derivation strategies in redundant knowledge bases", *Artificial Intelligence*, **50**, 95-115, 1991.

[Breiman 1984] Breiman, L., J. H. Friedman, J., R. A. Olshen, and C.J. Stone, *CART: Classification and Regression Trees*. Wadsworth, Belmont, CA, 1984.

[Quinlan 1986] Quinlan, R., "The effects of noise on concept learning", in R.S. Michalski et al, *Machine Learning: An Artificial Intelligence Approach*. Morgan Kaufmann, Los Altos, CA, 1986.

[Liepens 1991] Liepens, G., and W.D. Potter, "A genetic algorithm approach to multiple-fault diagnosis", in L. Davis (ed), *The Handbook of Genetic Algorithms*. Von Nostrand Reinhold, New York, 1991.

Appendix: Proof of Theorem 1

Proof of Theorem 1. We present a polynomial-time reduction from the set covering problem to a specially constructed pattern classification problem (A, p, c) with equal costs. The set covering problem is a well-known NP-complete problem and can be stated as follows:

Given a collection F of subsets of a set S and a positive integer k, does F contain a subset F' (called a cover) with $|F'| \leq k$ and such that $\cup_{f \in F'} f = S$?

For an instance of the set covering problem, we construct a corresponding pattern classification instance in polynomial time. Let $m = |S| + 1$ and $n = |F|$. Define the pattern matrix A as follows: $a_{ij} = 1$ if $i < m$ and the ith element of S is in the jth subset in F; $a_{ij} = 0$ otherwise. Finally, set $p_i = (m-1)^{-1}(n+1)^{-1}(1 \leq i \leq m)$, $p_m = n/(n+1)$, $c_i = 1(1 \leq i \leq n)$.

Given this construction, if the true pattern is a_m in the pattern classification problem, then it can be distinguished from the other patterns by inspecting the components corresponding to an arbitrary cover in the set covering problem. Conversely, the components that must be inspected in order to separate a_m from the other patterns always induce a cover in the set covering problem. An inspection policy is represented by a binary decision tree with $m - 1$ nonterminal nodes (decision nodes labeled by components) and m terminal nodes (corresponding to the

m patterns). For any given inspection tree, let E_i denote the subset of distinct components that are on the path from the top node to the terminal node corresponding to pattern i. Then $|E_i| \le n$ and E_m corresponds to a cover in the set covering problem. Let v denote the expected inspection cost associated with the inspection tree: then it must be the case that $p_m|E_m| < v = p_1|E_1| + \cdots + p_m|E_m| \le p_m|E_m| + (1 - p_m)n$.

To complete the proof we show that the answer to the set covering problem is "yes" if and only if the optimal inspection tree in the corresponding classification problem satisfies $|E_m| \le k$. If $|E_m| \le k$, then the subsets corresponding to the components in E_m forms a desired cover in the set covering problem. On the other hand, if $|E_m| \ge k + 1$, then the minimum expected inspection cost satisfies $v^* > p_m(k + 1)$. We now show by contradiction that there is no cover with size less than or equal to k in this case. Suppose that, to the contrary, there is a cover F' composed of k or less subsets. Then the components corresponding to F' form the set E_m for some feasible inspection tree. The expected inspection cost of such a tree satisfies $v \le p_m|E_m| + (1 - p_m)n \le p_m k + (1 - p_m)n$. Therefore, we would have $p_m(k + 1) < v^* \le v \le p_m k + (1 - p_m)n$, which implies that $p_m < n/(n + 1)$. But this contradicts the fact that $p_m = n/(n + 1)$. This completes the proof. $\diamond$

36

Combining a knowledge-based system and a clustering method for a construction of models in ill-structured domains

Karina Gibert and Ulises Cortés

Department of Statistics and Operations Research
karina@eio.upc.es
and
Software Department
ia@lsi.upc.es
Universitat Politècnica de Catalunya
Pau Gargallo, 5. Barcelona 08028. Spain.

ABSTRACT
Standard statistical methods usually ignore the additional information that an expert has about the domain structure. Direct treatment of symbolic information is not a very common characteristic of statistical systems. KLASS is a statistical clustering system that provides the possibility of using either quantitative and qualitative variables in the domain description. The user may also declare part of its knowledge about the domain structure.
The system is especially useful when dealing with ill-structured domains (*i.e.* a domain where the consensus among the experts is weak as mental diseases, sea sponges, books, painters...). That is why it is also useful from the artificial intelligence point of view. The output is a partition of the target domain. Conceptual and extensional descriptions of the classes can also be achieved.

36.1 Introduction

Ascending hierarchical clustering methods mainly consist of generating classes from a set of individuals described in terms of relevant characteristics or variables (see [Roux85], [Everitt81] ...). As known, there are different types of variables which can be used in these descriptions (in general terms: quantitative — continuous or not—, qualitative — ordinal, nominal, binary,...).

In general, after data collection a matrix is built up where the individuals to be classified are in the rows and the variables used to describe them are in the columns. In most cases this constitutes a non homogeneous database where qualitative and quantitative information about these individuals coexists.

Clustering methods are frequently used for dealing with quantitative information, but very often the data matrix to be classified also contains qualitative variables. These methods have two important degrees of freedom: the first one is the criterion used to generate a class (aggregation criterion); the second is the function used to measure the proximity between the elements of the sample (*i.e.* metrics).

Although research has been done in this field [Gower71], the main problem when qualitative variables are present for clustering is the difficulty of measuring distances between individuals

[1] *Selecting Models from Data: AI and Statistics IV.* Edited by P. Cheeseman and R.W. Oldford. ©1994 Springer-Verlag.

partially described by qualitative variables and partially by quantitative ones, specially when a symbolical representation is used for qualitative variables. In this case, several alternatives can be used:

- The codification of quantitative variables to obtain the corresponding qualitative ones, with the subsequent loss of information.
- The application of an analysis technique that deals with qualitative information only. Often correspondence analysis is used, although the objective of these qualitative methods is not exactly that of clustering techniques.

On the other hand, the need to handle nominal data from the Artificial Intelligence point of view has generated other kinds of clustering. Indeed, several well known expert systems, as MYCIN [Shortlife76] or others, are actually classifiers. [Michalski83] initiated *conceptual clustering*. Diday, who introduced in 1972 the "*nuées dinamiques*" algorithm, at present is working on *conceptual pyramidal clustering* [Diday92].

Nowadays, many research is being done on different approaches to clustering. Among others we can mention the works of [Cheeseman90] on *bayesian clustering*; those of [Bezdek87] on *fuzzy clustering* or [Fisher92] on *concept formation*.

All in all, it is highly interesting to obtain a classification of a given group of individuals described in this way. This opens a door to the automated generation of classification rules, extremely useful in knowledge-based environments (especially the diagnostic-oriented ones). In fact, when a knowledge-based system has to be built, the knowledge of the experts has to be formalized. They often use qualitative terms to express their expertise, and sometimes it is really difficult to quantify the concepts.

In this paper, the main characteristics of KLASS are presented in section 36.2, while section 36.3 shows the possibility of introducing the expert's knowledge as a bias of the clustering process.

After that, in section 36.4, a real application on the biological domain is expounded and, finally, some conclusions are presented in section 36.5.

36.2 Measuring distances and aggregation criteria

KLASS is a modular programme, implemented in **COMMONLISP**, that may be integrated into bigger environments in an easy way. The clustering method implemented is an adapted version of the chained reciprocal neighbors algorithm. One of its particularities is that **KLASS** can be used with any metrics or any aggregation criteria, since these two features are treated as parameters of the algorithm. In the following sections, our own proposals referring to the treatment of these parameters are presented.

36.2.1 Distances

The euclidean metrics has been traditionally used to measure distances between individuals in environments with only quantitative information.

Otherwise, when all the variables are qualitative, the data matrix use to be transformed into a *complete incidence table* by splitting each qualitative variable in a set of binary variables, one for each different value contained in the domain of the original variables. When data is presented in such a way, the χ^2 metrics is used to calculate distances between objects.

In **KLASS** a new distance has been defined for a qualitative variable expressed in symbolic terms. It is equivalent to the χ^2 distance which operates on complete incidence tables. Using it, splits of qualitative variables can be definitely avoided, and qualitative variables can be symbolically represented.

The distance between objects i and i' is then $d(i,i') = \sum_{\forall k \in \mathcal{Q}} \frac{1}{K^2} d_k^2(i,i')$, where $\mathcal{Q}$ is the set of qualitative variables. Taking n_k as the number of modalities of variable k; I_{ik} as the number of individuals of the sample that chose the same modality as individual i for variable k; I_{k_s} as the number of individuals of the sample that are in modality s for variable k and $f_i^{k_s} = I_{k_s} / \sum_{\xi=1}^{n_k} I_{k_\xi}$, $d_k^2(i,i')$ can be defined as follows:

and

$$d_k^2(i,i') = \begin{cases} 0, & \text{if } i \text{ and } i' \text{ belong to the same modality} \\ \frac{1}{I_{ik}} + \frac{1}{I_{i'k}}, & \text{otherwise and with ordinary objects} \\ \frac{(f_i^{k_s}-1)^2}{I_{k_s}} + \sum_{j=\neq s}^{n_k} \frac{(f_i^{k_j})^2}{I_{k_j}}, & \text{if } i \text{ is of modality } s \text{ and } i' \text{ is } \textit{extended}^{3} \\ \sum_{s=1}^{n_k} \frac{(f_i^{k_s}-f_{i'}^{k_s})^2}{I_{k_s}}, & \text{in the general case} \end{cases}$$

Given an homogeneous database of qualitative variables, distances between individuals can be calculated as said before. The problem arises when both quantitative and qualitative variables contribute to the description of the individuals. In this case, partial distances using only quantitative (C) or qualitative (Q) information are still computable. A pooled distance, called *mixed distance*, is defined using these two components.

$$d^2(i,i') = \alpha \sum_{\forall j \in C} \frac{(x_{ij} - x_{i'j})^2}{s_j^2} + \beta \frac{1}{n^2} \sum_{\forall k \in \mathcal{Q}} d_1(i,i',k)$$

The weights of each term are robust to the presence of outliers and take into account the number of variables of each type (n_C, and n_Q), as well as the magnitude of the quantitative (d_C) and qualitative (d_Q) components.

$$\alpha = \frac{n_C}{d_C}, \text{ and } \beta = \frac{n_Q}{d_Q}$$

36.2.2 Aggregation criteria

Among all the existing aggregation criteria, the centroid and Ward criteria have been selected. The first one has a geometric basis, while the second is related with the concept of information.

Representing every new class or subclass by its *mass center*, a conceptual description for the class is obtained, and dealing with classes as if it were ordinary objects is allowed. However, more precise matters have to be defined with the components related to the qualitative variables. In some cases vectorial components may appear, thus generating what is called an *extended object*.

Recurrent expressions have been developed to calculate the representative of each class in order to save calculations. The *mass center* of a new class is calculated from the *mass centers*

[3] See section 36.2.2 for a description of extended objects.

of the two aggregate subclasses. Therefore, the complexity of calculating the *mass center* of a new class is always the same, and independent of how many objects the class contains. This represents a great improvement in the last steps of the algorithm, where the subclasses to be aggregate contain many individuals.

36.3 Using the expert knowledge

The application of standard statistical methods to large databases, especially when the domain is ill-structured, is expensive, and *blind.* In most cases, the expert has enough knowledge about the domain as to organize part of it in entities that make sense. Statistical methods hardly use this information. The main idea of this work is to use that knowledge in a declarative form for building rules to subdivide the classification space in coherent *environments*, and then classify each *environment* independently of the rest. This kind of bias was suggested by Martín *et al.* ([Martín91]) to be used in the context of symbolic inductive classifiers.

The partition induced by the rules is composed of classes that contain less elements than the initial set of objects. It has been observed that the separate classification on each subspace requires all in all less time than the classification of the global database.

36.3.1 Knowledge representation

In order to make easy the obtaining of expert's knowledge about the domain, it is expressed in terms of rules in a declarative way. The rules are composed of a right hand side indicating some class K and a left hand side composed by the conjunction of conditions to be satisfied by a given object to belong to class K. Actually, the left hand side specifies the values that some of the variables have to take in class K. The syntax of the rules is as follows:

$$\bigwedge_i (operator\ (values^*)\ attribute_i) \rightarrow K \tag{36.1}$$

Each term of the left hand side expresses the functional relation to be satisfied by the attribute values of the objects that belong to class K. In what concerns the **LISP** representation, each rule is a list of functional relations between an attribute and its possible values. The final element of the list is the right hand side of the rule (*i.e.* the class).

A real-world case has been studied (see section 36.4 for a detailed description of data). The subject is an ill-structured domain: `sea sponges`. Besides the data matrix, the experts provided a set of seven rules to be considered in the classification process (see [Béjar92]). These rules look like the following:

$$\begin{array}{ll} & (=(1)\ \mathit{cortex}) \wedge (=(1)\ \mathit{papill\ae}) \\ \wedge & (=(1)\ \mathit{principal\text{-}spicule\text{-}type\text{-}tylostyle}) \\ \wedge & (=(\mathit{without\text{-}microscleres}\ \mathit{microstyles\text{-}and\text{-}microxeas})\ \mathit{kind\text{-}of\text{-}microsclere}) \\ \wedge & (=(\mathit{massive}\ \mathit{thick\text{-}encrusting})\ \mathit{final\text{-}growth\text{-}shape}) \Longrightarrow k1 \end{array} \tag{36.2}$$

The **LISP** representation of the rule presented in 36.2 is:

$$\begin{array}{ll} (& (=(1)\ \mathit{cortex})\ (=(1)\ \mathit{papillae}) \\ & (=(1)\ \mathit{principal\text{-}spicule\text{-}type\text{-}tylostyle}) \\ & (=(\mathit{without\text{-}microscleres}\ \ \mathit{microstyles\text{-}and\text{-}microxeas})\ \mathit{kind\text{-}of\text{-}microsclere}) \\ & (=(\mathit{massive}\ \ \mathit{thick\text{-}encrusting})\ \mathit{final\text{-}growth\text{-}shape}) -> k1) \end{array}$$

This formalism allows the set of representable rules to be easily extensible. Rules expressing more complex relationships between the attributes and its values can be easily included, in particular the set of forbidden values for a given attribute in an specific class. For example, to express that in class $K1$, the attribute $kind - of - microsclere$ cannot take the values $microstyles - and - microxeas, without - microscleres$, the corresponding rule would be:

((neq (without-microscleres microstyles-and-microxeas) kind-of-microsclere)$- > k1)$

It is important to remark that the shortness of this kind of rules implies certain degree of generality: the less terms a rule has, the more general it is and the more instances satisfy it. A long rule use to be very specific, but more precise; that is, the set of individuals satisfying it is small, but very homogeneous.

Actually, the *quality* of the partition induced by the rules is directly depending on the structure of those rules. That's why it is useful to be careful when selecting the attributes that belong to the rules, choosing the most *relevant*.

There exist some automatic tools that may help the experts to quantify the relevance of the attributes used in the objects description, and to identify the ones having a significant relevance (see [Belanche91] for a proposal on this line).

The technique proposed in [Belanche91] identifies 18 *relevant* attributes, what represents the 40% of the total. Indeed, the rules provided by experts only include these 19 *relevant* attributes.

36.3.2 Classifying with rules

The set of rules $\mathcal{R} = \{r_1 \dots r_\rho\}$ given by the expert induce a partition $\mathcal{P} = \{p_0 \dots p_\rho\}$ on the total set of individuals $\mathcal{I} = \{i_i : i = (1 : n)\}$ to be studied, where there is one class p_j, $(j = 1 \dots \rho)$ directly determined by each rule $r_j \in \mathcal{R}$ besides p_0, which is also called the *residual class*:

$$\mathcal{P} = \left\{ \begin{array}{ll} p_j = \{i \in \mathcal{I} : i \text{ satisfies rule } r_j \in \mathcal{R}\}, & \forall j = 1 : \rho \\ p_0 = \{i \in \mathcal{I} : i \text{ is a non-classified object}\} & \end{array} \right\}$$

A non-classified object satisfies either none rule or several rules that have different right hand side. So, the *residual class* $p_0 \in \mathcal{P}$ can be seen as the class of those objects for which no information is available or no consistent information is provided.

As presented at the begining of this section, a local classification on each element of $\mathcal{P}$ is to be performed. As a result, the conceptual representation of every class p_j, $j = 1 : \rho$ is obtained.

Let O_j, $j = 1 : \rho$ be the representative of the elements of $p_j \in \mathcal{P}$. Taking into account the possibility of treat these conceptual descriptions as ordinary objects, a final classification of $F = p_0 \cup \{O_j : j = 1 : \rho\}$ is performed at the end of the process to obtain an only ascendant hierarchical tree for the total domain ($\mathcal{I}$) that integrates the classes determined by the rules with the residual class in a unique structure.

This allows to take advantage of any partial knowledge on the domain: experts need not to give a set of rules complete to improve the classification, but only those rules they really know, and let the system look for the structure of those parts of the domain for which no knowledge is provided.

On the other hand, the computational cost of the local classifications of p_j, $j = 1 : \rho$ is not significant, since their sizes use to be very little compared to the total sample size. In the application presented here, the 44.74% of the objects are included in some of the classes directly defined by a rule. Furthermore, these local classifications may be executed in parallel, because p_j, $1 : \rho$ are disjuncts sets of individuals.

So, from a computational point of view, classifying F is the single relevant process. Nevertheless, F use to be smaller than $\mathcal{I}$ (in our example F contains one half of the total sample), thus considerably reducing the total computational cost.

36.4 A real application

In this section a biological case is presented. To be precise, O.*Hadromerida* of C.*Demospongiæ* of Porifera are studied. The *Hadromeridæ* are an interesting target domain, due to the lack of a definitive taxonomy accepted by all the experts (see [Domingo90]). At present, several classifications are accepted by different schools. The purpose of this application is to provide a classification of these animals by computational means, presumably more objectives than human criteria. This task is often complex and expensive because of the need of high skilled biologists and the problems presented, in general, by Porifera.

The database is composed of 76 species belonging to 27 *genera* of O.*Hadromerida*, described by means of 45 variables, three of which are quantitative, 27 are qualitative and 15 are binaries.

In the general classification, executed with mixed metrics and Ward criterion, the main families distinguished by the experts can be identified, but although the results are better than those obtained using other metrics, some species still remain bad classified.

As shown in section 36.3, it is possible to use the partial expert knowledge to improve the quality of the classification, by expressing it as rules. In our example, the partition induced by the rules contains 8 proper elements in addition to the residual class. These classes have species of the following *genera*:

C1: Only Cliona
C2: Only Diplastrella and Timea
C3: Oxycordyla y Stylocordyla
C4: Only Polymastia
C5: Rhizaxinella y Suberites
C6: Only Spirastrella
C7: Tethya

As said before, the local classification of every class generates a conceptual description of each one, which may be incorporated to the residual class as new *extended objects*. In fact, these conceptual descriptions represent ficticious *intermediate* objects of the initial classes. For example, class **C1** includes the species *Cliona carteri, Cliona schmidti* and *Cliona viridis*. Its conceptual description is the gravity center of these three elements, having, for the three first variables, the following aspect:

$$Cortex - layers = ((1\_layer, \frac{1}{3})(2\_layers, 0)(3\_layers, 0)(no\_cortex, \frac{2}{3}))$$

$$Internal - cortex - layer = no - internal - cortex - layer$$

$$Cortex = 0.33333$$

These three components may be interpreted in the following way: none of the objects of class **C1** has `internal cortex layer`; the 33% of them have `cortex` and, from the third variable, it is known that those cortex have `one only layer`.

Having the seven extended objects resulting from the local processes included in the residual class, a new classification may be performed. The classes coming from an α-cut at level 14.5 are the following:

Figure 1 presents the ascendant hierarchical tree of the residual class and the representative of the expert-determined classes.

C1: Rhizaxinella_pyriphera, Rhizaxinella_biseta, Rhizaxinella_elongata, Rhizaxinella_uniseta, Suberites_carnosus_v.incrustans, Suberites_carnosus_v.ramosus, Suberites- carnosus_v.typicus, Suberites_domuncula, Suberites_ficus, Suberites_gibbosiceps.

C2: Laxosuberites_ectyonimus, Laxosuberites_rugosus, Prosuberites_epiphytum, Prosuberites_rugosus, Prosuberites_longispina, Terpios_fugax, Diplastrella_bistellata, Diplastrella_-ornata, Timea_hallezi, Timea_mixta, Timea_stellata, Timea_unistellata, Spirastrella_cunctatrix, Spirastrella_minax, Pseudosuberites_sulfureus, Pseudosuberites_hyalinus, Laxosuberites_ferrerhernandez.

C3: Cliona_carteri, Cliona_celata, Cliona_labyrinthica, Cliona_schmidti, Cliona_viridis.

C4: Alectona_millari.

C5: Timea_chondrilloides.

C6: Oxycordyla_pellita, Stylocordyla_borealis.

C7: Tethya_aurantium, Tethya_citrina.

C8: Quasilina_intermedia, Ridleya_oviformis, Polymastia_radiosa, Trachyteleia stephensi, Tylexocladus_joubini, Tentorium_papillatus, Tentorium_semisuberites, Suberites_caminatus, Spinularia_spinularia, Polymastia_invaginata, Weberella_bursa, Weberella_verrucosa.

C9: Sphærotylus_antarcticus, Sphærotylus_capitatus, Polymastia_conigera, Polymastia fusca, Polymastia_grimaldi, Polymastia_hirsuta, Polymastia_infrapilosa, Polymastia_mammillaris, Polymas-tia_martæ, Polymastia_polytylota, Polymastia_robusta, Polymastia_tenax, Polymastia_ectofibrosa, Polymastia_tissieri.

C10: Quasilina_richardii,Quasilina_brevis.

C11: Polymastia_corticata, Polymastia_agglutinaris, Polymastia_littoralis, Proteleia_sollasi.

C12: Trichostema_hemisphæricum, Trichostema_sarsi.

C13: Polymastia_inflata, Polymastia_uberrima, Polymastia_spinula, Aaptos_aaptos.

The analysis of this classification shows that the main groups found by classifying without rules are maintained, but the bad classified species are almost disappeared (see [Gibert92] for a detailed description of these results). We can also say that classification with rules presents more structured classes for those *genera* that use to appear scattered to the other groups.

Finally, it is also interesting to observe that in the classification of the residual class, the structure presented for those elements which do not satisfy any rule is exactly the same obtained in the general classification without rules, as expected.

36.5 Conclusions

First of all, using rules is a good bias of the classification process since it reduces the computational cost as well as it improves the quality of the final classification. On the one hand, the classes induced by the declarative rules are a partition. So, it is possible the execution of parallel classification processes, one for each class, with the corresponding time saving. On the other hand, the reciprocal neighbors algorithm allows the treatment of a conceptual description of the classes and subclasses. It is then possible to include the prototypical element of each $p_j \in \mathcal{P}$ in the residual one. This operation integrates all the objects in an only classification tree. In the case presented here, the classification time using the rules represents the 37% of the time needed to perform the classification directly on the initial set of objects. Hence, there is a considerable computational saving when declarative information can be acquired from the experts, and the achieved performance regarding to the classification itself does not decrease.

Michalski, Fischer and other authors have been making efforts in this direction, but always working with very little and well-structured domains. We introduce here an example of combining statistical tools with inductive learning methods traditionally used in artificial intelligence to analyze ill-structured domains.

On the other hand, manipulations of the classes induced by the rules enable the exploration of the data base, and, in consequence, generation of new models for the domain may be performed.

Finally, some effort have been made to *correct* and enhance the expert's rules by adding new attributes to the original ones when there is a common set of objects that satisfy two or more rules. This work more related with the automatic rule-generation is actually in progress. The observation of those objects that simultaneously satisfy two or more rules, leads on a major knowledge about the domain by the expert, who can realize why his/her initial rule was too general. It will then be possible a rule-specification process, and a formulation of more accurate rules. In fact, the expert will increase his capacity of formalizing or expliciting his expertise.

36.6 References

[Béjar92] Béjar, J. and Cortés, U. (1992) "LINNEO+: Herramienta para la adquisición de conocimiento y generación de reglas de clasificación en dominios poco estructurados", *Proc. IBERAMIA-92*, 471-481, Noriega Eds, México.

[Belanche91] Belanche, L. and Cortés, U. (1991) "The nought attributes in knowledge-based systems". *Proc. EUROVAV-91*, 77-102, Logica Cambridge Ltd., England.

[Bezdek87] Bezdek, J. (1987) "Some non-standard clustering algorithms", Developments in Numerical Ecology, Springer-Verlag, Berlin.

[Cheeseman90] Cheeseman, P. (1990) "On finding the most probable model", Comp. Models of Discovery and Theory Formation, Morgan Kauffmann, San Mateo, CA.

[Diday92] Diday, E., Brito, P., Mfoumoune, E. (1992) "Modelling probabilistic data by conceptual pyramidal clustering", *Proc. AI and Stats.*, 213-219, Ft. Laud., FLO.

[Domingo90] Domingo, M. (1990)" Aplicació de tècniques d'IA (LINNEO) a la classificació sistemàtica. O. HADROMERIDA (DEMOSPONGIÆ- PORIFERA)", Master thesis, Ecology Dep., Univ. of Barcelona, SP.

[Everitt81] Everitt, B. (1981) "*Cluster analysis*", Heinemann, London.

[Fisher92] Fisher, D. H., Langley P. (1986) "Conceptual clustering and its relation to numerical taxonomy", *AI and Stats.*, Addison-Wesley, Reading, MA.

[Gibert92] Gibert, K. and Cortés, U. (1992) "KLASS: Una herramienta estadística para la creación de prototipos en dominios poco estructurados", *Proc. IBERAMIA-92*, 483-497, Noriega Eds, México.

[Gibert91] Gibert, K. (1991) " KLASS: Estudi d'un sistema d'ajuda al tractament estadístic de grans bases de dades,", Master thesis, *Rep. LSI–91–7*, UPC, Barcelona, SP.

[Gower71] Gower, J. C. (1971) "A general coefficient for similarity and some of its properties", *Biometrics* **27**, 857-872.

[Martín91] Martín, M. , Sangüesa, R. and Cortés, U. (1990) " Knowledge acquisition combining analytical and empirical techniques", *ML* **91**, 657–661, Evanston, Illinois.

[Michalski83] Michalski, R. S., Stepp R. (1983) "Automated construction of classifications: conceptual clustering *vs* numerical taxonomy", *IEEE Trans. PAMI*, **5**, 219-243.

[Roux85] Roux, M., (1985) "*Algorithmes de classification*", Masson, Paris.

[Shortlife76] Shortlife, E. H. (1976) "MYCIN: A rule-based computer program for advising physicians regarding antimicrobial therapy selection" Ph. D. Thesis Stand. Univ.

FIGURE 1: Ascendant hierarchical tree for sea sponges, with a α-cut = 14.5.

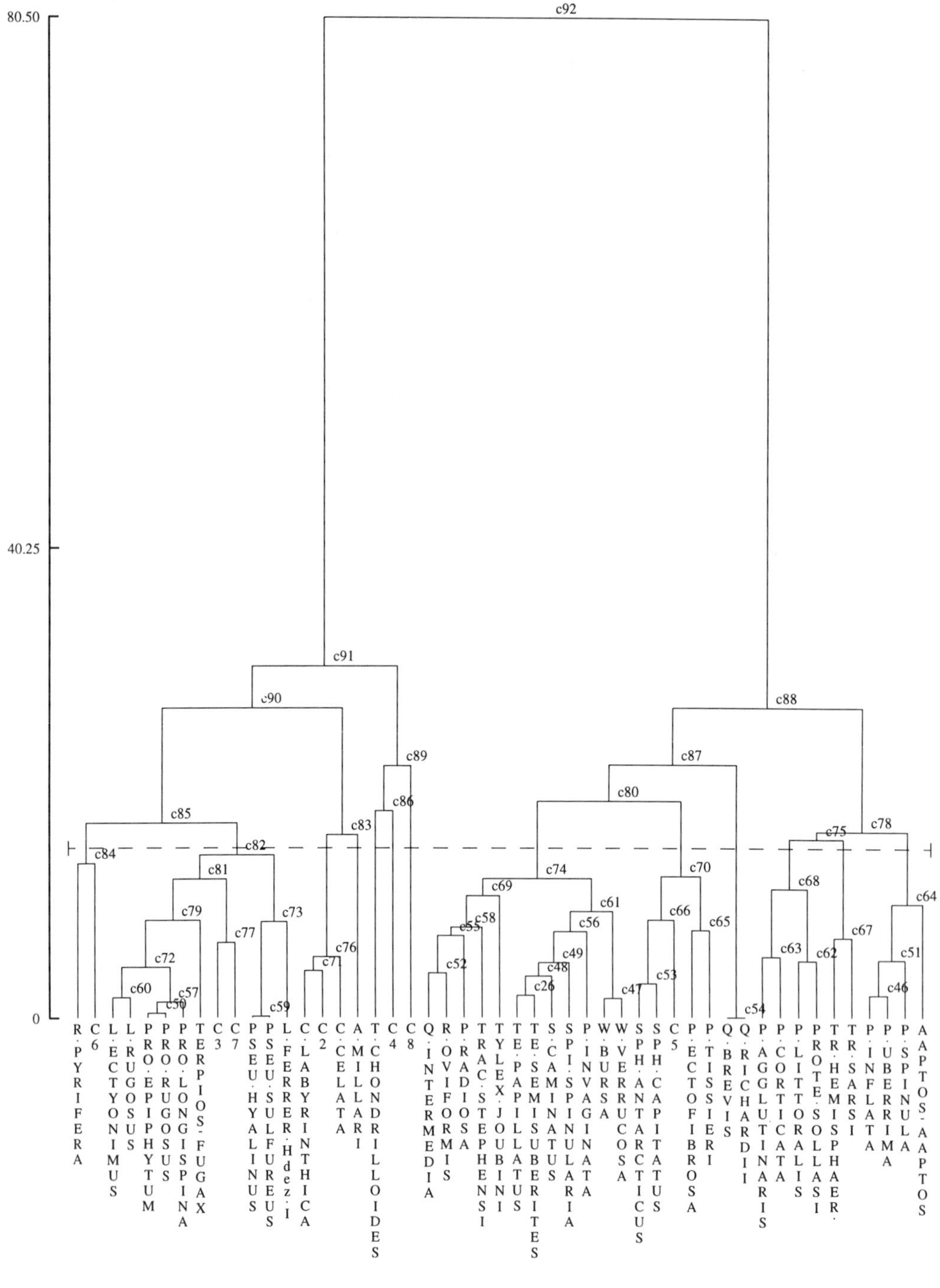

37
Clustering of Symbolically Described Events for Prediction of Numeric Attributes

Bradley L. Whitehall and David J. Sirag, Jr.

United Technologies Research Center
East Hartford, CT 06108

ABSTRACT This chapter describes a new conceptual clustering system capable of constructing classes for the prediction of a single numeric attribute. The clustering for single numeric attribute prediction (CSNAP) system clusters data described by a variety of symbolic attributes. The system trades off the accuracy of the predicted values against the clarity of the descriptions produced to produce classes which are both predictive and meaningful to a human observer. CSNAP has been used to develop classes for time based events and has demonstrated an ability to learn complex cyclic patterns.

37.1 Introduction

Clustering systems are used to partition a set of events into classes. These classes are intended to provide useful insight into differences and similarities between events in that set. There are a wide variety of clustering systems including Numeric, Conceptual, General, and Goal-directed clusterers. Numerical clustering systems partition events based upon a "distance" metric placing events that are "close" to each other in the same class [Anderberg 73]. Conceptual clustering systems partition events into classes based upon the conceptual cohesiveness of the resulting classes [MichStep 83]. The cohesiveness of a class is a function of the complexity of the class description and the ability of the class to predict attribute values. General systems cluster events to provide more information about all or many of the attributes describing events in the universe of discourse. Goal-directed systems assume that the clustering process will take place with respect to a particular feature. This paper describes a new goal-directed approach to conceptual clustering, where the goal of the system is to enable accurate predictions about a particular event attribute.

The clustering for single numeric attribute prediction (CSNAP) system is a goal-directed approach to conceptual clustering. CSNAP limits its predictive claims to a single attribute in order to maximize the cohesiveness of the resulting classes with respect to that attribute. CSNAP produces classes which are both predicative of a single attribute and easy to describe using the other provided attributes.

The goal of CSNAP is to produce easily understood classes of known events suitable for predicting the value of a designated numeric attribute for events where the value of that attribute is unknown. This is accomplished by finding clusters of events that can be distinguished from other events by an understandable description and have a low variance on the designated numeric attribute. The next section describes the type of problems that CSNAP is designed to solve. The third section explains why previous clustering systems are not designed to meet the specific goals of CSNAP. Also described in the third section is how CSNAP systems relate to learning

[1]*Selecting Models from Data: AI and Statistics IV.* Edited by P. Cheeseman and R.W. Oldford.

from example systems. The fourth section presents the implementation details of this approach. Section five contains empirical results for the CSNAP system. The final section contains our concluding remarks.

37.2 Class of CSNAP problems

There are four important characteristics of the problems CSNAP was designed to handle:

1. Predictions are wanted for one numeric attribute of the examples (the dependent variable).
2. The value of the dependent variable is probabilistic in nature. In other words, two events with identical descriptions will not (necessarily) have the same value for the dependent variable.
3. The event description attributes are not completely independent. For example, attributes of time-based events such as days, weeks, and holidays, are related in complex ways.
4. The user of the system wants accurate predictions and meaningful descriptions of the found classes.

Many everyday phenomena possess these characteristics, such as predicting interstate congestion, determining the best time to go to the bank, and understanding job queues in a machining shop. CSNAP is appropriate when each event has a large number of attributes describing a situation or state and another specially designated numeric attribute exists, upon which the statistical properties of the found classes are to be based. A sample problem for this would be learning the length of the queue for a particular bank. The system would be presented with a number of examples describing the time the sample was taken, the state of the bank, and observed queue length. The goal of the CSNAP system is to build classes of events so that for future situations, one can predict what the queue length will be. Each event of this problem contains the following attributes:

Attribute	Example Values
queue length	2
time minutes	45
time hour	3pm
day of week	Thursday
near holiday	no
drive up open	yes

Provided with a number of events from various times, a CSNAP system could produce classes such as:

- (day of week = Friday) and (time hour = 3pm, 4pm, or 5pm)
  mean queue length = 15.7, variance = 3.2
- (drive up open = yes) and (number of tellers = 3)
  mean queue length = 2.3, variance = 4.5

The system constructs a classification tree from these classes. The tree is organized with the root covering all event descriptions and its branches covering more specific subsets. This insures that all possible event descriptions are covered by some node. Now, before going to the bank, the classification tree can be queried to determine how long the lines are likely to be.

The classification trees produced by CSNAP provide the following capabilities:

1. Answer a query about the dependent variable for a specific state (time).

2. Determine the state with predicted optimal dependent attribute value.

3. Incorporate user's knowledge into the classification tree.

The first capability, answering queries such as "Is now a good time to go to the bank?", was explained in the above scenario. The second capability just requires a search through the tree for a node matching the conditions of optimality. For example, find the best time (time with shortest queue length) to go to the bank between 10am and 3pm. The third capability is possible because the classification trees are easy to understand and manipulate. A user could add nodes to the tree by hand (or through a simple interface) to extend knowledge to the structure that might not have been observed in the examples. One might add a node indicating that the bank is closed on Monday because it is President's Day.

37.3 How CSNAP is different

CSNAP differs from traditional clustering systems in a number of ways. First, CSNAP has a specific goal to accomplish with the resulting classification of events. Standard clustering algorithms try to discover good clusters of events based upon distance metrics or conceptual cohesiveness. Standard clustering systems are able to discover new and interesting relationships among the attribute values in order to define new classes. The goal of such systems is the formation of generally useful classes, where exactly what is meant by "useful" refers to the metric of the system and not necessarily useful to the user. CSNAP on the other hand, does not cluster events in what might be the most interesting fashion from a general view point. Instead, the events are divided into classes so that each class has specific statistical properties for the values of one attribute of the events in the class e.g., a mean and variance. However, CSNAP is not a pure statistical classification system. The descriptions of the classes must be useful and meaningful to the users of the system. Humans want to gain insight from the resulting classes. Thus, CSNAP must tradeoff the accuracy of a complex class verses the less accurate, but conceptually simpler classes it discovers.

CSNAP is a goal-directed clustering system [StepMich 86]. Goal-directed clustering systems use examples differently than standard clustering systems. For CSNAP, the dependent attribute is used to guide the clustering process because clusters with small variances in the dependent attribute are sought. Thus the examples help determine which class they belong in based upon the value of this attribute. In some respects, CSNAP is more closely related to learning from example systems than many clustering systems. Like learning from example systems [Quinlan 86,Michalski 75,BrFrOlSt 84], each event has a specific attribute that is identified as being important to predict. Unlike learning from example systems though, the dependent variable does not strictly determine the appropriate class. CSNAP will place events with different dependent variable values into the same class if the class description and variance for events covered by that description are good. It is the cohesiveness of the whole class that is important, not the specific classification of any single example.

To help explain some of the unique aspects of CSNAP, it will be contrasted with two other clustering systems AUTOCLASS [ChKeSSTF 88] and COBWEB [Fisher 87]. The clusters produced by AUTOCLASS and COBWEB are intended for different purposes than the classes produced by CSNAP, so it is not surprising that they produce clusterings with different characteristics. Unlike AUTOCLASS (and many numeric classification systems), CSNAP systems use a hierarchical refinement of concepts to provide complete coverage of the event space. AUTOCLASS provides a midpoint definition for its classes. COBWEB maintains the events contained in each cluster and produces predictions based upon the number of occurrences of each attribute value. Another distinction between these approaches is that CSNAP clusters events based upon a single numeric attribute. The other attributes are used to explain the classes, but are not the primary concern in clustering. AUTOCLASS uses several attributes in conjunction to define differences between classes. Taken together, several attributes with small differences can describe very different classes. This gives AUTOCLASS the capability to discover interesting classes which differ only slightly in several different attributes. Similarly, COBWEB does not focus on any single attribute and so is able to learn predictions for all attributes for each cluster. A third distinction between CSNAP and AUTOCLASS is that the attributes describing CSNAP events do not have to be independent as required by AUTOCLASS. In many domains one of the most important attributes is time. When learning time based descriptions it is natural to use attributes such as minute, hour, day, month, etc. These attributes are all cyclic and are very interdependent. CSNAP is able to create meaningful class descriptions using these attributes without unusual encoding or special purpose logic. The classes produced by CSNAP do not reflect the rates of co-occurrence of attribute values like COBWEB.

37.4 CSNAP Implementation

The algorithm for CSNAP combines a statistical approach with a requirement that good descriptions be produced. The basic algorithm is given in Figure 1. The essence of the approach is in steps 2 and 3, in which a kernel of events are selected and an initial description is formed. First the system splits the events to find clusters of events which minimize the variance of the dependent attribute. The procedure to build the descriptions then uses these initial clusters as a guide to constructing the classes. Events are moved into and out of the initial cluster in order to construct a cohesive class. As points are moved, the variance of the clusters tends to increase, but the moves are selected to make the classes easier to describe. There is a tradeoff between concise, clear descriptions and low variance. The user can adjust weights that are used to balance these conflicting goals.

The outlined approach is similar to ID3 [Quinlan 86] from the standpoint of splitting off points, but similar to AQ [Michalski 75] in the way that descriptions are constructed. It is a clustering system because the class memberships are determined dynamically as the clusters are built.

CSNAP (points, current-node)

1. Sort points on dependent variable
2. Find the cluster of points with smallest variance
3. Build a description of the cluster
   (Move points to/from cluster as needed to maintain a simple description)
4. Set new-points = points covered by description
5. Create new-node with new-points as child of current-node
6. Call CSNAP (new-points, new-node)
7. Set points = points - new-points
8. Call CSNAP (points, current-node)

FIGURE 1: CSNAP Algorithm

Attributes	**Type**	**Values**
rate	real	≥ 0
seconds	circular	0-60 (integer)
minutes	circular	0-60 (integer)
hour	circular	midnight, 1am, 2am, . . ., 11pm
day-or-night	nominal	day, night
day-of-week	circular	mon, tue, . . ., sun
day-type	nominal	weekend, weekday
day	circular	1-31 (integer)
week	circular	first, second, third, fourth, fifth
month	circular	jan, feb, . . ., dec
year	linear	≥ 1980
season	circular	winter, spring, summer, autumn

FIGURE 2: Building Model Attributes

37.5 Empirical Tests

In this section, results of running CSNAP on a model of building traffic is presented. This model is not used to help form classes, only to generate the training events. The dependent variable is the traffic rate of a building, the number of people entering and exiting the building. The examples from the model are described with eleven attributes (shown in Figure 2) and the dependent variable rate. The model takes into account a number of factors when determining the rate including season and time of day. When the model creates a data point, it produces the typical traffic rate for the specified time period. The actual number of passengers observed is

	Real Data				Model			
	AAE	Var	Max	Ave SSE	AAE	Var	Max	Ave SSE
CSNAP	1.04	3.47	30.55	2.13	0.57	1.89	24.95	1.49
ID3	1.46	14.55	47.00	4.08	1.31	14.56	47.17	4.03
Mov Ave.	1.64	13.12	35.60	3.97	1.51	12.74	29.60	3.87

FIGURE 3: Results on Traffic Data

generated by sampling a distribution determined by the traffic rate supplied by the model. It is this non-deterministic number of passengers that is used by CSNAP for learning. This captures the idea that in real buildings the rate will not be the same everyday at the same time; there exist natural fluctuations in the data.

Two months of data, sampling the traffic every five minutes, produced over 17,000 examples for training. The testing set was produced by running the model over a one week period (five minute sampling) immediately following the two month training period. Results are shown in Figure 3 comparing CSNAP, ID3, and a 10-point moving average output for both the actual data point, and the true model value. In the figure, AAE is the average absolute error of the value from the learned model, Var is the variance of those values, Max is the maximum value difference, and Ave SSE is the average sum of squared error. For the ID3 results, the rates were rounded to the nearest integer, and each integer rate defines a unique class. As can be seen, CSNAP performs much better than the other approaches. Figure 4 graphically shows the data for a single day with the true model, the actual data, and the traffic predicted by CSNAP. This figure illustrates that CSNAP has learned a pattern very similar to the original model, based only on noisy samples drawn from that model. CSNAP did not require a manually developed model of the environment, a description of numbers or types of classes to be found, or any theoretical assurances that the attributes are independent. Even so, CSNAP was able to create an accurate and fairly easy to understand/modify empirical model of the underlying process.

Figure 5 shows how the average error of the predictions decreases as the system constructs the classification tree. The solid line is the average error over the one week testing period in passengers per minute. The dashed line is the average error for the tree plus one standard deviation of the error. CSNAP was designed to construct classes that have small variance and this graph indicates CSNAP is accomplishing that goal. By taking the difference between the two lines, it can be seen that the standard deviation of error decreases from 4.2 to 1.3 passengers per minute. This graph also indicates that the CSNAP system does not over learn from the training data. If it had, the error would increase as the tree grew. The completed tree has 470 nodes. While this might not be as small as desired, the ID3 tree for this problem had over 67,000 nodes[3].

In the future, it would be interesting to compare CSNAP predictions with AUTOCLASS results, but currently it is not obvious how to generate real-valued predictions based on the classes formed by AUTOCLASS. Other new work on CSNAP will likely include development of an incremental form and methods for automatic "optimization" of its performance parameters. This "optimization" process would attempt to automatically determine reasonable tradeoffs between accuracy and description complexity.

[3]A standard version of ID3 with CHI squared pruning was used.

37.6 Conclusion

CSNAP is closely related to many existing clustering algorithms, but its differences give it additional power for the solution of many practical problems. The large volume of research on related approaches such as AUTOCLASS, ID3, and COBWEB has provided a wide range of techniques for use in new algorithms as well as valuable information on their various strengths and weaknesses. Combining techniques selected in response to real world requirements has resulted in a practical approach with many of the best attributes of the original algorithms.

CSNAP is able to solve problems that other systems are not able to handle. It can accept examples described by attributes which are related in complex ways. It can make use of knowledge of the attribute type - Are the attribute's values nominal, ordered, real, or circular? CSNAP is able to construct simple models capturing the essential patterns in non-deterministic environments. It provides its results in a structure that allows easy human modification and understanding. CSNAP is able to provide good predictions for events.

37.7 References

[Anderberg 73] Anderberg, M., (1973) *Cluster Analysis for Applications.* Academic Press, New York.

[BrFrOlSt 84] Breiman, L., Friedman, J.H., Olshen, R., and C.J. Stone (1984) *Classification and Regression Trees* Wadsworth International Group, Belmont, CA.

[ChKeSSTF 88] Cheeseman, P., Kelly, J., Self, M., Stutz, J., Taylor, W. and Freeman, D. (1988) "AutoClass: A Bayesian Classification System", *Proceedings of the Fifth International Machine Learning Conference*, 54-64.

[Fisher 87] Fisher, D. H. (1987) *Knowledge Acquisition via Incremental Conceptual Clustering*, PhD Thesis, Department of Information and Computer Science, University of California at Irvine.

[Michalski 75] Michalski, R. S. (1975) "Synthesis of Optimal and Quasi-optimal Variable Valued Logic Formulas", *Proceedings of the 1975 International Symposium on Multiple-Valued Logic*, 76-87.

[MichStep 83] Michalski, R.S. and Stepp, R.E. (1983) "Learning from Observation: Conceptual Clustering", in *Machine Learning: An AI Approach*, R.S. Michalski, J.G. Carbonell and T.M. Mitchell eds., 331-364, Tioga Publishing Company.

[Quinlan 86] Quinlan, J.R. (1986) "Induction of Decision Trees", *Machine Learning*, **1**, 81-105.

[StepMich 86] Stepp, R.E. and Michalski, R.S. (1986) "Conceptual Clustering: Inventing Goal-Oriented Classifications of Structured Objects", in *Machine Learning: An AI Approach, Volume II*, R.S. Michalski, J.G. Carbonell and T.M. Mitchell eds., 471-498, Morgan Kaufmann Publishers, Inc.

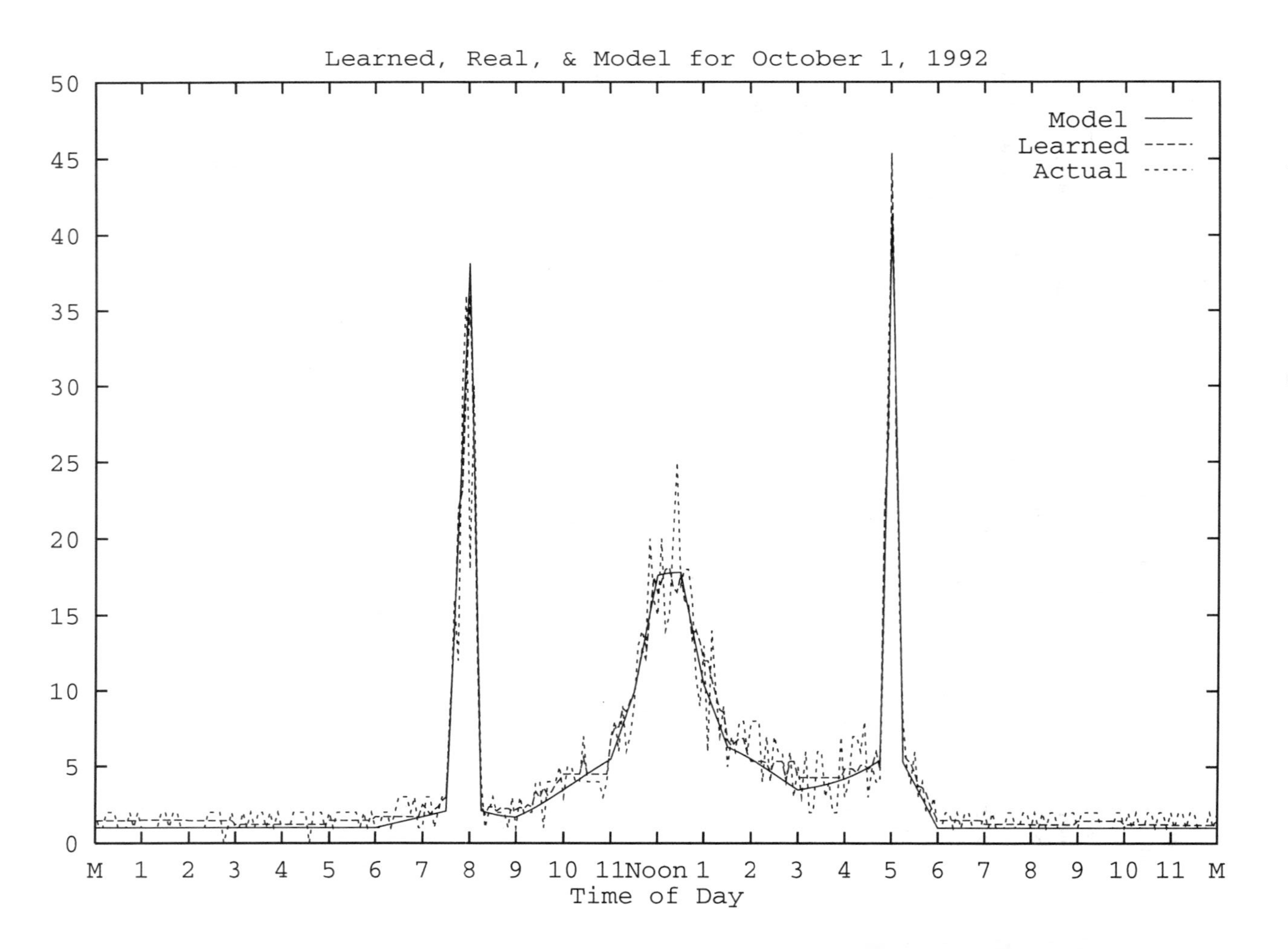
Learned, Real, & Model for October 1, 1992
Model
Learned
Actual
50
45
40
35
30
25
20
15
10
5
0
M 1 2 3 4 5 6 7 8 9 10 11Noon 1 2 3 4 5 6 7 8 9 10 11 M
Time of Day
Rate in passengers per minute

FIGURE 4

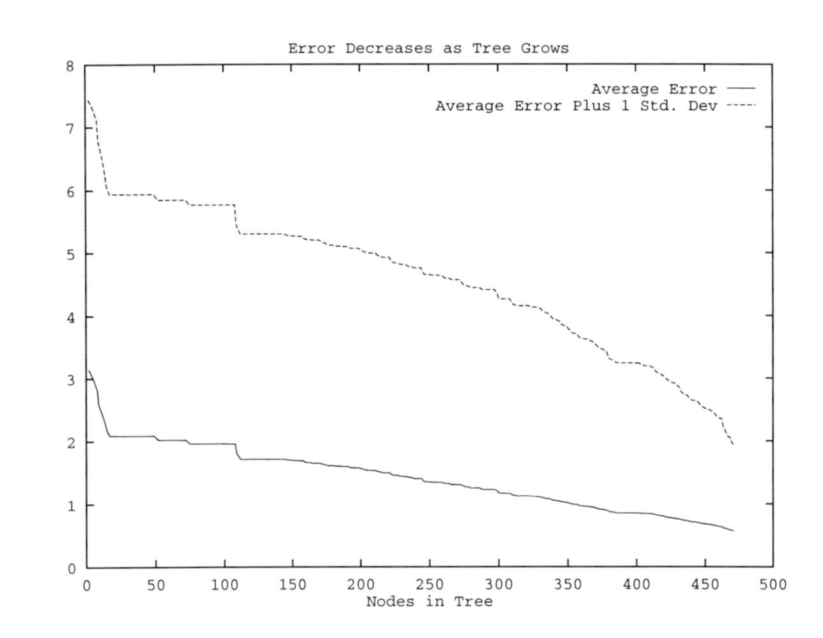

FIGURE 5: Error and uncertainty decrease as tree grows

38
Symbolic Classifiers: Conditions to Have Good Accuracy Performance

C. Feng, R. King, A. Sutherland, S. Muggleton, and R. Henery

The Turing Institute Ltd., Glasgow, U.K., and
Dept. Computer Science
150 Louis Pasteur/Priv.
University of Ottawa
Ottawa, Ontario, Canada K1N 6N5
cfeng@csi.uottawa.ca,

Department of Statistics, Livingstone Tower
Strathclyde University
Glasgow, UK.,

Strathclyde University,

University of Ottawa,

and

Strathclyde University.

ABSTRACT Symbolic classifiers from Artificial Intelligence compete with those from the established and emerging fields of statistics and neural networks. Traditional view is that symbolic classifiers are good in that they are easier to use, are faster, and produce human understandable rules. However, as this paper shows, through a comparison of fourteen established state-of-the-art symbolic, statistical, and neural classifiers on eight large real-world problems, symbolic classifiers also have superior, or at least comparable, accuracy performance when the characteristics of the data suit them. These data characteristics are measured using a set of statistical and qualitative descriptors, first proposed in the present (and the related) work. This has implications for algorithm users and method designers, in that the strength of various algorithms can be exploited in application, and in that superior features of other algorithms can be incorporated into existing algorithms.

38.1 Introduction

Classification algorithms from Artificial Intelligence, statistics, and neural networks are applied to solve practically similar problems, though they use different data modeling methods and produce different classification functions/rules. The proliferation of various classification methods in recent years has made it more complex to decide when, and where, to apply a particular method. Intuitively, an experienced user would choose an algorithm depending on the particular problem and the strength of that algorithm. Indeed, this is the approach taken by Bayesian statistics practitioners and has been supported by in-depth theoretical analysis. In the lack of a coherent, complete theory, symbolic learning methods are commonly accepted as being superior, when human understandability takes the highest priority for a particular problem. They

---

[1] *Selecting Models from Data: AI and Statistics IV.* Edited by P. Cheeseman and R.W. Oldford.

are also known to be faster and easier to use. However, their strength (or weakness) in accuracy performance has remained unclear. To investigate this, as part of the STATLOG[3] project, we have conducted a series of comparisons of symbolic (CART, INDCART, C4.5, NewID, CN2), statistical (linear discrimination, quadratic discrimination, logistic regression, projection pursuit, k-nearest neighbor, kernel density estimation, Bayes network classifier), and neural net (back-propagation, radial basis functions) classifiers on eight large real-world problems.

The algorithms are chosen because they are representatives of the *established* state-of-the-art, and are generally considered to be good algorithms even though sometimes there are variations that improve some aspects of the methods. The datasets were selected because they have already gained strong interest in *application*, and are from real-world problems. They include three datasets from image processing (vehicle silhouette recognition, segment image recognition, and satellite image recognition), two from engineering (space shuttle control, Belgian power control), and one each from handwritten character recognition (KL digits), finance (U.K. credit assignment), and medicine (diabetes).

The algorithms are tested on centrally pre-processed datasets that are measured using a set of statistical and qualitative descriptors that we developed in order to investigate what characteristics of a dataset make it suitable for a particular algorithm. The results are produced by a group of distributed partners in the STATLOG project who are experts in the particular algorithms that they are testing, and are selectively validated by novice users who rerun the algorithms using the reports filled by the expert users. The three criteria that we use to measure algorithm performance are accuracy, training and classification/testing time. These criteria can, to some extent, reliably measure the performance of algorithms (we differentiate performance based on around two percentage points).

38.2 The problem

Empirical comparisons have frequently been used to evaluate newer and more advanced classification algorithms in Artificial Intelligence, statistics, and neural networks. Most have been within one field or another, but there are still many, but fewer, comparisons between two or three of these areas. Generally, particular methods were found to do well in some specific problems and for some performance measures, but not in all applications. For example, k-nearest-neighbor performed very well in accuracy in recognizing handwritten characters [Aha 92], but not as well on the sonar-target task [GormSej 88]. Most comparisons were uncoordinated and suffered from several handicaps:

- few algorithms were compared, and different algorithms were involved so that the results are difficult to gather or the collected results may be meaningless;
- frequently the algorithms that were compared to were not good representatives, or were not up-to-date in their respective fields;
- comparisons were sometimes biased to demonstrate the capability of the authors' algorithm, e.g., by the selective choice of datasets;

[3] It is initiated by the European Commission and is part of ESPRIT. The full title is "The Comparative Testing of Statistical and Logical Learning Algorithms on Large-Scale Applications to Classification, Prediction and Control". In this paper, we are only interested in classification.

- datasets were small and their selection was not sufficiently broad, or sometimes variations of them were used, which makes the results difficult to generalize;
- the characteristics of datasets were under-studied to provide meaningful explanation for why particular algorithms performed well on particular problems.

These handicaps reduce the strength of the conclusions that can be drawn from comparison studies.

A few attempts have been made to overcome these limitations. [ThBaBlBr 91] conducted a co-ordinated comparison of many symbolic and neural net algorithms on two related, and artificially created problem. They found that symbolic rule-based classifiers (e.g., AQ11) outperformed symbolic decision-tree (ID3) and neural net (e.g., back-propagation) classifiers. Weiss and colleagues compared symbolic (CART [BrFrOlSt 84] and PVM) and statistical (e.g., linear discrimination, logistic regression) algorithms [WeisKul 91]. They concluded that PVM and CART were more accurate (on most datasets). Shavlik, Mooney, and their colleagues compared ID3, a perceptron, and back-propagation algorithms on five datasets [ShavMoTo 91], and found that back-propagation were more accurate than, or at least equal to, ID3. Similar comparisons support this conclusion. There is also a large body of comparisons supporting that certain neural net algorithms were better than, or at least as accurate as, some statistical and symbolic algorithms. [Ripley 92] compared a diverse set of statistical and neural net algorithms, together with a decision-tree classifier. He reported that nearest neighbor and conjugate gradient (a neural net) were more accurate on two related datasets. A similar view on k-nearest neighbor is supported by [Aha 92].

Despite efforts by these authors, it is unclear when, and where, a particular algorithm will perform better. These results are confusing, to say the least, when studied from an application practitioner's viewpoint. In the context of this paper, what makes a symbolic algorithm (not) suitable for particular problems, and (perhaps) why?. This question remains unanswered. One common limitation of these comparisons is that there has been no measurable characteristics of data proposed to explain the findings.

38.3 Organization of comparisons

The algorithms that we have tested include:

- Symbolic classifiers: three decision-tree based, CART [BrFrOlSt 84], INDCART (a variant of CART using 0-Standard Error for pruning), C4.5 [Quinlan 87], and NewID[4], one rule-based, CN2 [ClarBos 91].
- Statistical classifiers: k-nearest neighbor[5], linear discriminant (Discrim), quadratic discriminant (Quadra), logistic regression (LogReg), projection pursuit (SMART) [FrieStu 81], kernel density estimate (Alloc80) [HerHaBu 74], and a Bayes network classifier (CASTLE) [AcCaGoMo 91].
- Neural networks: back-propagation (Backprop), and radial-basis functions (RBF) [PoggGir 90].

[4]This is a post-pruning version of ID3; it is available from The Turing Institute Ltd., UK.

[5]This version is based on the geometrical distance function. Note its ties with instance-based learning.

Datasets	Satellite	KL Digits	Vehicle	Credit	Shuttle	Belgian	Segment	Diabetes
Ex.No.	6435	18000	846	8900	58000	1250	2310	768
Att.No.	36	40	18	16	9	21	11	8
Cat.Att.	0	0	0	8	0	0	0	0
Classes	6	10	4	2	7	2	7	2
$cancor_1$	0.94	0.92	0.84	0.76	0.97	0.89	0.98	0.55
$cancor_2$	0.93	0.91	0.82	0.00	0.70	0.00	0.96	0.00
$cancor_3$	0.79	0.84	0.36	NA	0.22	NA	0.83	NA
$cancor_4$	0.24	0.78	0.00	NA	0.15	NA	0.59	NA
$fract_1$	0.36	0.17	0.47	1.00	0.63	1.00	0.31	1.00
$fract_2$	0.71	0.34	0.91	1.00	0.95	1.00	0.61	1.00
$fract_3$	0.97	0.48	1.00	NA	0.98	NA	0.84	NA
$fract_4$	0.99	0.61	1.00	NA	1.00	NA	0.95	NA
skew	0.73	0.18	0.83	1.21	4.44	0.43	2.96	1.06
kurtosis	4.17	2.92	5.18	4.40	160.31	2.66	24.48	5.83
scale	yes	yes	no	no	no	no	no	no

TABLE 38.1: Statistical characteristics of datasets

[WeisKul 91] presents descriptions of a few algorithms. [Ripley 92] discusses neural networks and the competing methods from statistics viewpoints.

The most important characteristics of the datasets are that they are large real-world problems and have already gained strong interest in *application*, especially in Europe[6]. All these datasets are available from Department of Statistics and Modeling Sciences, University of Strathclyde, UK (e-mail: bob@stams.strath.ac.uk). Some have already been placed in the standard dataset repository of U.C. Irvine, CA, USA (e-mail: pmurphy@ics.uci.edu). Table 38.1 shows details about the datasets. The table contains the number of examples of the datasets (Ex.No.), the number of attributes (Att.No.), the number of categorical attributes (Cat.Att.), the number of classes (Classes), and a set of statistical characteristics that we propose as data descriptors.

Canonical discriminant correlations measure the number of discriminant classes sufficient to classify the examples. $cancor_1$ measures the success of the first canonical discriminant which gives the best single linear combination of attributes that discriminates between the classes. $cancor_2$ measures that of the second canonical discriminant which is the best single linear combination orthogonal to the first, and so on. $fract_1$ is the percentage of the successful discrimination of the examples that can be explained by the first discriminant. $fract_2$ is that of the second discriminant, and so on. The figures for the above two measures can be used to illustrate the degree of difficulty involved in the discrimination task.

Univariate skewness, $skew$, measures the non-normality of the attributes when considered separately (0 for normal distribution). Similarly, the univariate kurtosis, $kurtosis$, should be 3 for normal distribution. The scaling factor of the data is measured by a binary property $scale$, meaning that the attributes of the data are equally scaled and equally relevant to classification.

To ensure a fair comparison, care has been taken to reduce bias towards one category of algorithms or another. A two stage schedule is designed to minimize the possibility of "cheating". In the first step, an expert user carries out the evaluation of the algorithm and fills what has been done in a report. In the second step, an independent (novice) user selectively tries to repeat the

[6] It seems that real-world data tends to contain a higher proportion of image processing and the majority also contain fewer categorical attributes.

work based on the record and compares how close the proposed use of the algorithm is to the actual use of the algorithm. In all cases, the dataset has been centrally processed with missing values replaced and training/testing sets randomly partitioned. This measure guarantees that all datasets received by the users have the same properties as described in Table 38.1. We discouraged the further processing of data by the users (except discretization and numericalization of attributes). When there is any departure from the centrally issued data we would report what had been done. (The main reason for further processing, e.g., dropping attributes or using part of the training data, is that the algorithm was not able to process the data in full.)

Depending on the sizes of datasets, two different partitioning methods were used to split the dataset into a set to train the algorithm and another to test the function/rule produced. Table 38.2 shows the method used for each dataset. A train/test split was used when the dataset is large. For

Name	Split method	Atts. No.	Class No.	Training size	Test size
Satellite	train/test	36	6	4435	2000
KL Digits	traing/test	40	10	9000	9000
Vehicle	9×cross-validation	18	4	752	94
Segment	10×cross-validation	11	7	2079	231
Diabetes	12×cross-validation	8	2	704	64
Credit	train/test	39	2	6230	2670
Shuttle	train/test	9	7	43500	14500
Belgian	train/test	9	2	1250	1250

TABLE 38.2: Dataset training/testing split

large dataset, the estimated testing result on the test data is normally reliable [WeisKul 91]. For smaller datasets, n-fold cross-validation was used, which gives more accurate estimation of the accuracy (error rate). The results are measured by classification accuracy (together with standard deviation), and training and testing time, which is measured in CPU seconds and corrected to the standard machine of SUN SPARCstation IPC.

38.4 Results and discussion

The algorithm users (expert and novice) all confirmed that symbolic algorithms are easy to use, and the figures also show that they are much faster (so are the three discriminants). In accuracy, comparing symbolic methods (C4.5, CART, INDCART, NewID, CN2) with algorithms from statistics and neural networks, they performed well on the Shuttle, Segment, and Credit datasets. Examining Table 38.1, we note that Shuttle and Segment have the highest values for kurtosis and skew, which means that they are the furthest from (multivariate) normal (a separate study indicates a tentative splitting point is $skew > 1$ and $kurtosis > 7$, but these are not clear-cut). Symbolic methods are nonparametric, i.e., they do not make assumptions on the underlying distributions. Theoretically, this should make them robust to large kurtosis and skew. A few other algorithms that we have tested are nonparametric (e.g., Alloc80, SMART, and Backprop), but symbolic algorithms seem to be more robust to extreme distributions than these other nonparametric methods.

Both Segment and Shuttle also have very high $cancor_1$ (> 0.97), indicating that it is easier to obtain high accuracy (with the first discriminant). Shuttle exhibits multimodality (analyzed using S-plus, but we have no good statistic to measure this). We analyzed the trees produced for

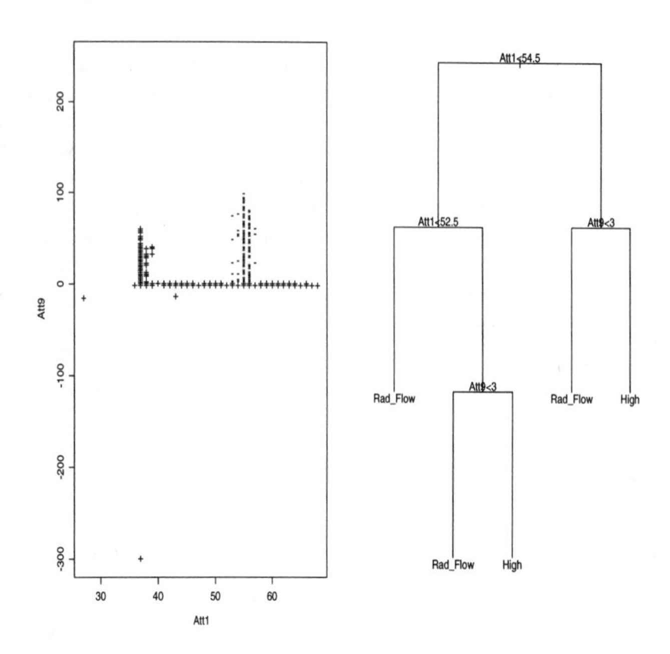

FIGURE 1: Shuttle data analysis

Shuttle, and found that only a few attributes are needed to classify the data. For example, a tree with five leaves using two attributes would perfectly classify two classes "High" and "Rad_flow" (the right of Figure 1). This seems to indicate that attributes selected in this data are very well suited to the symbolic algorithms. The data seem to consist of isolated islands or clusters of points, each of which is pure (belongs to only one class), with one class comprising several such islands. However, neighboring islands may be very close and yet come from different classes. Their boundaries seem to be parallel with the coordinate axes. This suggest that it may be possible to classify the combined dataset with arbitrarily near 100% accuracy using a decision tree.

The characteristic that stands out about Credit is that it has a large number (50%) of binary/categorical attributes (obviously some must be relevant to classification). It is possible to get 87% accuracy using just one categorical attribute (many algorithms do worse than this). Most symbolic algorithms agree that this attribute produces the best rule (NewID, CART, and CN2), and this is also the conclusion of CASTLE. On the other hand, SMART and Backprop achieve about the same accuracy, but the simple structure of the problem is masked. The three discriminants also fail to find the simple rule. In separate studies, symbolic algorithms performed at least as well as back-propagation [ShavMoTo 91] on datasets with high percentage of categorical attributes.

Symbolic algorithms achieve lower accuracy on satellite image, KL digits, vehicle, and diabetes datasets. In the Satellite dataset, they are surpassed by k-nearest neighbor (and perhaps its related methods Alloc80 and radial basis functions), on the KL Digits data as well as by back-propagation and SMART, and by the three discrimination algorithms. The main characteristic of Satellite and KL Digits is that they both have equally scaled and equally important attributes, and KL Digits also have very near (multivariate) normal skew (0.18) and kurtosis (2.92) – against 0 and 3 respectively. Theoretically, both datasets suit k-nearest neighbor, similarly, KL Digits data suits linear discriminant, quadratic discriminant, and logistic regression; these are also confirmed by the empirical data.

On the satellite dataset, the first three discrimination functions would be sufficient to separate all six class means ($fract_3 = 0.9691$). This points to the seriation of three remaining classes. So it is hard to improve accuracy further; the result of k-nearest neighbor would not be too far from the optimum. Equally, this result may also indicate that the original four attributes may be successfully reduced to three with no (or little) loss of information. KL Digits is processed by Karhunen-Loeve method from the originally 16×16 pixelated data using 40 main eigenvectors. So it is artificially near normal.

The Vehicle data is also an image processing dataset. But its attributes have been processed so they are not equally scaled and equally important. This has a great impact on k-nearest neighbor and radial basis functions, but the more sophisticated Alloc80 seems to resist such disruptions. The departure from normality of this data is not severe (skew $= 0.83$, kurtosis $= 5.2$), so linear discriminant, quadratic discriminant, and logistic regression perform relatively well. This similar pattern appears on the diabetes data, where (skew$= 1.06$ and kurtosis$= 5.83$). Back-propagation and SMART are robust compared to k-nearest neighbor.

On the Belgian dataset, most algorithms achieve very high accuracy. The symbolic algorithms perform similarly to the discriminants (results for logistic regression is not reliable as it used only part of the attributes). This dataset is also not too far from normal, but the attributes are not equally scaled. The conclusion would be that no particular methods have a significant edge (except for SMART, perhaps).

38.5 Conclusion

In this paper we presented the results of comparing fourteen symbolic, statistical, and neural net classifiers on eight large real-world problems. We analyze the results using a set of statistical and qualitative data characteristics to discover when symbolic algorithms from Artificial Intelligence can be expected to produce better, or at least comparable, accuracy. If the data has extreme distribution, i.e., it is far from normality as measured by its skew and kurtosis (the tentative division is $skew > 1$ and $kurtosis > 7$), if the data exhibits multimodality, i.e., it contains data centers in a scattered space, and if the data has a very high proportion of relevant categorical attributes, then symbolic algorithms can be expect to be good. This is in addition to the traditional view of that symbolic methods are easy to use, are faster, and produce human understandable rules.

But in certain situations, symbolic algorithms are definitely poor methods. These include that the data has equally scaled and equally relevant attributes where k-nearest neighbor is better, that the data has near normal distribution (tentatively, $skew < 1$ and $kurtosis < 5.5$) where the three discriminants are better. In other situations symbolic algorithms may be expected to produce comparable accuracy as either k-nearest neighbor or the three discriminants. In such situations, back-propagation, kernel density estimation, and projection pursuit algorithms can be good compromise methods (especially when no clearly better algorithms can be identified).

Acknowledgements

We thank the European Commission and all the partners in STATLOG, in particular the Turing Institute and Strathclyde University. Special acknowledgements are due to Ashwin Srinivasan, David Aha and Wray Buntine for contributing datasets or algorithms.

38.6 REFERENCES

[AcCaGoMo 91] S. Acid, S. L. de Campos, A. González, R. Molina, and Pérez de la Blanca. *CASTLE : A Tool for Bayesian Learning*. In *Esprit Conference 1991*.

[Aha 92] D. Aha. Generalizing case studies: a case study. In *9th Int. Conf. on Machine Learning*, pages 1–10, San Mateo, Cal., 1992. Morgan Kaufmann.

[BrFrOlSt 84] L. Breiman, J. H. Friedman, R. A. Olshen, and C. J. Stone. *Classification and Regression Trees*. Wadsworth, Belmont, 1984.

[ClarBos 91] P. Clark and R. Boswell. Rule induction with CN2: some recent improvements. In *EWSL '91*, pages 151–163, Porto, Portugal, 1991. Berlin: Springer-Verlag.

[FrieStu 81] J. H. Friedman and W. Stuetzle. Projection pursuit regression. *Journal of American Statistics Association*, 76:817–823, 1981.

[GormSej 88] R. P. Gorman and T. J. Sejnowski. Analysis of hidden units in a layered network trained to classify sonar targets. *Neural networks*, 1 (Part 1):75 – 89, 1988.

[HerHaBu 74] J. Hermans, J. D. F. Habbema, and A. T. Van der Burght. Cases of doubt in allocation problems, k populations. *Bulletin of International Statistics Institute*, 45:523–529, 1974.

[PoggGir 90] T. Poggio and F. Girosi. Networks for approximation and learning. *Proceedings of the IEEE*, 78(9, September):1481 – 1497, 1990.

[Quinlan 87] J. R. Quinlan. Simplifying decision trees. *International journal of man-machine studies*, 27((3) September):221 – 234, 1987.

[Ripley 92] B. D. Ripley. Statistical aspects of neural networks. In *Invited talk at SemSat, Sandbjerg, Denmark, April 1992*. Chapman and Hall, 1992.

[ShavMoTo 91] J. W. Shavlik, R. J. Mooney, and G. G. Towell. Symbolic and neural learning algorithms: an experimental comparison. *Journal of Machine learning*, 6(2, March):111 – 143, 1991.

[ThBaBlBr 91] S. Thrun, J. Bala, E. Bloedorn, and I. Bratko, editors, *The MONK's problems - a performance comparison of different learning algorithms*. Carnegie Mellon University, Computer Science Department, 1991.

[WeisKul 91] S. M. Weiss and C. A. Kulikowski. *Computer systems that learn: classification and prediction methods from statistics, neural networks, machine learning, and expert systems*. Morgan Kaufmann, San Mateo, CA, 1991.

Algorithm	*Accuracy(%)*		*Time(sec.)*	
	Train	Test	Train	Test
INDCART	92.0	92.0±.53	206	193
C4.5	94.3	91.2±.53	1086	11
SMART	89.5	89.1±.60	2151	5
CASTLE	88.3	88.1±.63	81	33
CART	87.2	87.0±.65	19	19
NewID	100	87.0±.65	380	4
Discrim	87.2	87.0±.65	71	16
RBF	87.5	87.0±.65	837	54
LogReg	87.3	86.9±.65	251	30
Backprop	88.2	86.9±.65	28819	19
CN2	100	86.7±.66	2309	13
Quadra	86.3	86.0±.67	78	20
Alloc80	97.7	83.0±.72	24	738
k-N-N	100.0	80.6±.77	0	1851

TABLE 38.3: Credit results.

Algorithm	*Accuracy(%)*		*Time(sec.)*	
	Train	Test	Train	Test
Alloc80	96.8	97.0±1.12	1248	279
NewID	100.0	96.6±1.19	64	69
CART	99.5	96.0±1.29	139	4
C4.5	98.7	96.0±1.29	142	93
CN2	99.7	95.8±1.32	114	3
INDCART	98.8	95.5±1.36	248	234
SMART	96.1	94.8±1.46	16362	1
k-N-N	100.0	92.3±1.75	5	28
LogReg	90.3	89.1±2.05	302	8
CASTLE	98.9	88.8±2.07	377	31
Discrim	88.8	88.4±2.11	74	7
Quadra	84.5	84.3±2.39	50	16

TABLE 38.5: Segment data results.

Algorithm	*Accuracy(%)*		*Time(sec.)*	
	Train	Test	Train	Test
NewID	100	99.99±.008	6180*	NA
CN2	100.0	99.97±.014	11160*	NA
C4.5@	99.90	99.96±.017	11131	11
IND-CART@	99.96	99.92±.023	1152	16
k-N-N	99.61	99.56±.055	65270	21698
SMART	99.39	99.41±.064	110010	93
Alloc80	99.05	99.17±.075	55215	18333
CASTLE	96.34	96.23±.158	819	263
LogReg@	96.06	96.17±.159	6946	106
Discrim@	95.02	95.17±.178	508	102
Backprop	95.10	95.10±.179	28800	75
Quadra@	93.65	93.28±.208	709	177

TABLE 38.4: Shuttle control data results: * combines training and testing time, "@" indicates that part of the training data was used because they could not use the full set with given machine resources — 32760 for C4.5 (full training data); 32625 for INDCART; and 20000 for Discrim, Quadra, and LogReg.

Algorithm	*Accuracy(%)*		*Time(sec.)*	
	Train	Test	Train	Test
k-N-N	91.1	90.6±.65	2105	944
RBF	88.9	87.9±.73	723	74
Alloc80	96.4	86.8±.76	63840	28757
INDCART	98.9	86.3±.77	2109	9
CART	NA	86.2±.77	348	14
Backprop@	88.8	86.1±.77	54371	39
NewID	93.3	85.0±.80	296	53
C4.5	95.7	84.9±.80	449	11
CN2	98.6	84.8±.80	1718	16
Quadra	89.4	84.7±.80	276	93
SMART	87.7	84.1±.82	83068	20
LogReg	88.1	83.1±.84	4414	41
Discrim	85.1	82.9±.84	68	12
CASTLE	81.4	80.6±.88	75	80

TABLE 38.6: Satellite image result: @ was terminated (perhaps prematurely) after 20 hours of CPU time.

Algorithm	*Accuracy(%)*		*Time(sec.)*	
	Train	Test	Train	Test
k-N-N	100.0	98.0±.15	0	13881
Alloc80	100.0	97.6±.16	48106	48188
Quadra	98.4	97.5±.16	1990	1648
Backprop	95.9	95.1±.23	129840	240
LogReg	96.8	94.9±.23	3538	1713
RBF	95.2	94.5±.24	2280	580
SMART	95.7	94.3±.24	389448	58
Discrim	93.0	92.5±.28	141	54
CASTLE	87.4	86.5±.36	49162	45403
NewID	100.0	83.8±.39	785	109
INDCART	99.7	83.2±.39	3550	53
CN2	96.4	82.0±.40	9183	103
C4.5	95.0	82.0±.40	1437	35

TABLE 38.7: KL Digits results.

Algorithm	*Accuracy(%)*		*Time(sec.)*	
	Train	Test	Train	Test
SMART	99.7	99.4±.22	5853	12
LogReg#	99.8	99.3±.24	103	27
Discrim	97.8	97.5±.44	46	28
CN2	100.0	96.8±.50	272	17
NewID	100.0	96.7±.51	32	19
CART	99.1	96.6±.51	280	18
INDCART	99.3	96.6±.51	151	150
Alloc80	97.4	95.6±.58	3676*	0
C4.5	98.8	95.5±.59	66	12
CASTLE	97.1	95.3±.60	477	199
Quadra	96.4	94.8±.63	85	41
k-N-N	100.0	94.1±.67	0	137

TABLE 38.9: Belgian data results: * combines training and testing time, # seven of the original attributes are linear functions of the others but they were removed in the application of logistic regression.

Algorithm	*Accuracy(%)*		*Time(sec.)*	
	Train	Test	Train	Test
Quadra	91.5	85.0±3.68	251	29
Alloc80	100.0	82.7±3.90	30	10
LogReg	83.3	80.9±4.05	758	8
Backprop	83.2	79.3±4.18	14411	4
Discrim	79.8	78.4±4.24	16	3
SMART	93.8	78.3±4.25	3017	1
C4.5	93.5	73.4±4.56	153	1
k-N-N	100.0	72.5±4.61	164	23
CART	NA	71.6±4.65	25	1
NewID	97.0	70.2±4.72	18	1
INDCART	95.3	70.2±4.72	85	1
RBF	90.28	69.3±4.76	1736	12
CN2	98.2	68.6±4.79	100	1
CASTLE	49.5	45.5±5.14	23	3

TABLE 38.8: Vehicle silhouettes results.

Algorithm	*Accuracy(%)*		*Time(sec.)*	
	Train	Test	Train	Test
LogReg	78.1	77.73±5.20	31	7
Discrim	78.0	77.47±5.22	27	6
SMART	82.3	76.82±5.28	314*	NA
CART	77.3	74.48±5.45	61	2
CASTLE	74.0	74.22±5.47	29	4
Quadra	76.3	73.83±5.49	24	6
C4.5	86.9	73.05±5.55	12	1
INDCART	92.1	72.92±5.55	18	17
CN2	99.0	71.09±5.67	38	3
NewID	100.0	71.09±5.67	10	10
Alloc80@	71.2	69.92±5.73	115*	NA
k-N-N	100.0	67.58±5.85	1	2

TABLE 38.10: Diabetes data results: @ achieves its best results using only attributes 2, 8, 6, and 7, * combines training and testing time.

Part VI

Regression and Other Statistical Models

39

Statistical and neural network techniques for nonparametric regression

Vladimir Cherkassky and Filip Mulier

EE Department
University of Minnesota
Minneapolis, Minnesota 55455, U.S.A.
email: cherkass@ee.umn.edu

ABSTRACT The problem of estimating an unknown function from a finite number of noisy data points has fundamental importance for many applications. Recently, several computational techniques for non-parametric regression have been proposed by statisticians and by researchers in artificial neural networks, but there is little interaction between the two communities. This paper presents a common taxonomy of statistical and neural network methods for nonparametric regression. Performance of many methods critically depends on the strategy for positioning knots along the regression surface. A novel method for adaptive positioning of knots called Constrained Topological Mapping(CTM) is discussed in some detail. This method achieves adaptive placement of knots of the regression surface by using a neural network model of self-organization.

39.1 Neural networks and nonparametric statistics

In the last decade, neural networks have given rise to high expectations for model-free statistical estimation from a finite number of samples(examples). However, several recent studies clearly point out that Artificial Neural Networks(ANNs) represent inherently statistical techniques subject to well-known statistical limitations [BarrBarr 88,GemBieDou 92,White 89]. Moreover, many learning procedures can be readily recognized as a special case of stochastic approximation techniques proposed in the early 50's. Whereas early ANN studies have been mostly empirical, recent research successfully applies statistical notions (such as overfitting, resampling, bias-variance dilemma, the curse of dimensionality etc.) to improve neural network performance. Statisticians can also gain from analyzing ANN models as a new tool for data analysis.

Here we consider regression problems, i.e., we seek to estimate a function f of $N-1$ predictor variables (denoted by vector X) from a given set of n training data points, or measurements, $Z_i = (X_i, y_i)$ $(i = 1, \ldots, n)$ in N-dimensional sample space:

$$y = f(X) + error \tag{39.1}$$

where $error$ is unknown (but zero mean) and its distribution may depend on X. The distribution of training data in X is also unknown and can be arbitrary.

Nonparametric methods make no or very few general assumptions about the unknown function $f(X)$. Nonparametric estimation from finite training set is an ill-posed problem, and meaningful predictions are possible only for sufficiently smooth functions [PoggGiro 90]. Additional complications arise due to inherent sparseness of high-dimensional data sets (this is known as *the*

---

[1] *Selecting Models from Data: AI and Statistics IV.* Edited by P. Cheeseman and R.W. Oldford.

curse of dimensionality) and due to the difficulty in distinguishing between signal and additive noise in expression (39.1).

There is a range of conflicting opinions on the utility of ANNs for statistical inference. Whereas high expectations of neural network enthusiasts are usually caused by their ignorance in statistics, negative attitude towards empirical neural network research on the part of some statisticians can be traced to the view that algorithmic approaches are inferior to analytical proofs. It may be interesting to note that many recently proposed statistical methods have their neural network counterparts. For example, there is a close similarity between MARS [Friedman 90] and a tree-structured network proposed by [Sanger 91]; the popular backpropagation networks use the same functional representation as Projection Pursuit Regression [FrieStue 81]; and Breiman's PI-method [Breiman 91] is related to sigma- pi networks [RumMcCPDP 86] as both seek to represent an output(function) in the sum-of-products form. A recent meeting between statisticians and ANN researchers helped to clarify the relationship between the two fields [CheFriWec 93]. The following differences have been noted:

1. *Problem/model complexity*. Usually ANNs deal with large amount of training data (i.e. thousands of samples), whereas statistical methods use much smaller training samples. Hence, ANN models usually have higher complexity than statistical methods.

2. *Goals of modeling*. In Statistics, the usual goal is *interpretability*, which favors structured models (i.e. CART and linear regression). In ANN research, the main objective is *generalization/prediction*. Hence, the ANN models usually have little, if any, interpretability. However, for many high-dimensional problems even structured methods are difficult to interpret, due to large model size (i.e. large CART tree).

3. *Comparisons and search for the best method*. Most statistical and ANN methods are asymptotically "good", i.e. can guarantee faithful estimates when the number of training samples grows very large. Unfortunately, real-world provides finite and usually sparse data sets. For such ill-posed problems, asymptotic performance is irrelevant, and the best method should conform to the properties of data at hand. Hence, no single method dominates all others for all possible data sets. The real research goal is not to find the best method, but to characterise the class of functions/mappings (along with assumptions about the noise, smoothness etc.) for which a given method works best. An example of such an approach is recent work of Barron [Barron 93] on characterization of functions that can be approximated with feedforward networks. Another related relevant problem is characterization of real data sets from important application domains.

4. *Batch vs flow-through processing*. Most statistical methods utilize the whole training set (batch mode), whereas ANNs favor iterative processing (one sample at a time) known as flow-through methods in Statistics. Iterative ANN processing requires many presentations of training data and uses slow computational paradigm (gradient descent). Statistical methods are usually much faster.

5. *Computational complexity and Usability*. As a result of (4), statistical methods tend to be more computationally complex and difficult to use by non-statisticians. ANN methods are computationally simple, and can be easily applied by novice users. Also, hardware implementations favor simpler ANN methods.

6. *Robustness.* ANN methods appear more robust than statistical ones with respect to parameter tuning. Even suboptimal choice of parameters (network size, learning rate etc.) usually gives reasonable results. Another important aspect of robustness is providing confidence intervals for prediction estimates. Confidence intervals are routinely provided in statistics, but usually lack in ANN application studies.

Many ANN application studies succesfully tackle problems with tens and hundreds variables, in spite of the fact that nonparametric methods face the problem of sparse data in high dimensions. The reasons for this disagreement between theory and practical applications of ANNs are not very clear. A possible explanation is that application studies employ clever preprocessing or data encoding techniques, so that the network has to learn very simple functions or decision surfaces[CherkNaja 92]. Another possibility is that in high-dimensional applications many input variables are interdependent, so that the *effective* problem dimensionality is small. Yet another plausible explanation is that high-dimensional real data typically has non-uniform distribution (i.e. appears in clusters) - this may explain the success of *local* empirical methods, such as Radial Basis Functions and Fuzzy Logic. Another view is that introducing the right bias into the model [GemBieDou 92] is much more important than "learning from examples". This point of view is based on an observation that much of hard human learning(e.g. in perception and language) is "one-shot". Therefore, humans must have prewired powerful representations formed during thousands of years of evolution, so that nonparametric statistical theory becomes somewhat irrelevant.

39.2 Taxonomy of techniques for regression

Due to empirical nature of ANN research, it is important to provide common taxonomy of ANN methods for nonparametric regression. Cherkassky [Cherkassky 92] provides such a common taxonomy loosely following statistical framework [Friedman 90], as briefly outlined next. Computational methods for regression fall into three major groups:

- *Global parametric methods*, which assume that an unknown function in (1) has a fixed (parametric) form, or can be represented as a combination of basis functions globally defined in X. Examples include Linear Regression and fixed-topology networks, i.e. feedforward nets with fixed number of hidden units.

- *Piecewise parametric or locally parametric methods* try to approximate an unknown function by several simple parametric functions each defined over a different subregion of domain of independent variables. These methods are also equivalent (at least in low dimensions) to local or kernel smoothers [FrieSilv 89]. In these methods the training data enters explicitly into a parametric model. Examples are piecewise-linear regression, splines [de Boor 78] and statistical kernel methods.

- *Adaptive computation methods* which dynamically adjust their strategy to the training data, in order to take into account the behavior of the unknown function [Friedman 90]. The motivation for adaptive methods is due to the fact that for high-dimensional problems locally parametric techniques fail to estimate the underlying function faithfully because there is usually not enough training samples. Since global parametric methods inevitably introduce bias and local parametric methods are applicable only to low-dimensional problems (given a limited-size training set), *adaptive* methods are the only practical alternative in high-dimensions.

We further distinguish several types of adaptive methods:

- *Global Adaptive Methods* that seek regression estimate within a rich class of functions defined globally in the domain of predictor variables X. Global adaptive methods are based on existence theorems from Function Approximation stating that any "reasonable" function can be represented as superposition of simpler functions of lower dimensionality [BarrBarr 88,Breiman 91]. Frequently, such a superposition of parametric and/or non-parametric functions(components) can be regarded as a network of components. In ANN research, these methods are known as *adaptive* networks, where the network topology (i.e. the number of hidden units) is not fixed in advance. Examples include: Projection Pursuit [FrieStue 81]; variable-topology feedforward networks, polynomial networks[BarrBarr 88] etc;

- *Local Adaptive Methods* are generalization of kernel smoothers where the kernel functions and kernel centers are determined from the training data by some adaptive algorithm. Examples of local adaptive methods include radial basis function (RBF) networks, regularization networks, networks with locally tuned units etc [BroLow 88,MoodDark 89,PoggGiro 90,Specht 91];

- *Recursive Partitioning Methods* seek to partition domain of X into disjoint subregions, so that each subregion is using its own (simple) parametric regression function, using recursive partitioning strategy. Examples include CART and MARS [BreFriOlsSto 84,Friedman 90]. A closely related to MARS tree-structured adaptive network for function approximation has been proposed by Sanger [Sanger 91]. The main difference is that MARS uses complete training set to specify tree partitioning(batch method), whereas in Sanger's method data points are presented one at a time, and computations are performed locally at the nodes of a tree;

- *Adaptive Function Quantization Methods* provide discrete approximation of the regression surface using (non-recursive) partitioning methods borrowed from Pattern Recognition and ANN research. The goal is to provide piecewise-constant approximation of the unknown function by specifying (X, y) locations of the finite set of knots or network units. The domain of independent variables is partitioned into many (unequal) subregions, assuming constant unknown response value for each subregion. The center of each subregion corresponds to the knot location, and the goal of learning is to estimate (X, y) coordinates of knot locations from the training data.

39.3 Adaptive knot location

39.3.1 Motivation: the problem of knot location

One of the most challenging problems in practical implementations of adaptive methods for regression is adaptive positioning of knots along the regression surface. Typically, knot positions in the domain of X are chosen as a subset of the training data set, or knots are uniformly distributed in X. Once X-locations are fixed, commonly used data- driven methods can be applied to determine the number of knots. However, de Boor [de Boor 78] showed that a polynomial spline with unequally spaced knots can approximate an arbitrary function much better than a spline with equally spaced knots. Unfortunately, the minimization problem involved in determination of the optimal placement of knots is highly nonlinear and the solution space is not convex [FrieSilv 89]. Hence, the performance of many recent algorithms that include adaptive knot placement (e.g. MARS) is difficult to evaluate analytically. There are additional factors that contribute to the complexity of knot positioning problem:

- *High-dimensional problems.* For a single-variable function, it can be shown that the optimal local knot density is roughly proportional to the squared second derivative of the function and the local density of the training data, and inversely proportional to the local noise variance [Friedman 91]. No analytic results are available for higher dimensions or data-dependent noise, but similar results probably hold true.

- *Estimating the second derivative* of a true function is necessary for optimal knot placement. Yet, the function itself is unknown and its estimation depends on the good placement of knots. This suggests the need for some iterative procedure that alternates between function estimation (smoothing) and knot positioning steps.

- *Finite-sample variance.* Given a (sparse) training sample, it may not be a good strategy to place knots at a subset of data set, as done in many statistical methods. A sparse sample may be a poor approximation of an underlying distribution. Hence, knot location can be determined better by applying some smoothing to the scatterplot of X- coordinates of training data, in order to find better estimate of the unknown distribution.

39.3.2 Self-organizing maps for adaptive knot location

In statistical methods knot locations are typically viewed as free parameters of the model, and hence the number of knots directly controls the model complexity. Alternatively, one can introduce topological ordering between the knots, and then impose *local regularization* constraints on adjacent knot locations, so that neighboring knots cannot move independently. Such an approach is effectively implemented in the model known as Kohonen's Self-Organizing Maps (SOM) [Kohonen 84]. This model uses a set of units ("knots") with neighborhood relations between units defined according to a fixed topological structure (typically 1D or 2D grid). During training or self- organization, data points are presented to the map iteratively, one at a time, and the unit closest to the data moves towards it, also pulling along its topological neighbors. The goals of the original SOM algorithm are vector quantization (density estimation) and dimensionality reduction, when a high-dimensional distribution is approximated using a low-dimensional map.

An M-dimensional map consists of units (knots, neurons) arranged in M-dimensional grid topology. The neighborhood relation between the units i and j on the map is specified as the *distance* between these units, or simply the number of hops in the shortest path from i to j. Cherkassky and Lari-Najafi [CherkLari 91] proposed to interpret the units of a topological map as the dynamically movable knots for regression problems. Correspondingly, the problem of finding regression estimate can be stated as the problem of forming an M-dimensional topological map using a set of samples from N-dimensional sample space (where $M \leq N - 1$). Unfortunately, straightforward application of the Kohonen Algorithm to regression problem does not work well. The reason is that the original Kohonen's algorithm is intended for density estimation/ vector quantization and it does not preserve the functionality of the regression surface. In other words, the presence of noise in the training data can fool the algorithm to produce a map that is a multiple-valued function of predictor variables in the regression problem (39.1). This problem is overcome in the Constrained Topological Mapping (CTM) algorithm, where the best-matching unit is found in the space of predictor variables, rather than in the input(sample) space [CherkLari 91]. Hence, regression fitting is achieved by training a map to approximate training samples $Z_i = (X_i, y_i)$ in N-dimensional input space(using notation adopted in section 1). Note that in the context of CTM method the knots (units) *cannot* be viewed as free parameters.

CTM algorithm forms an M-dimensional map in N-dimensional space as follows:

Step 0 Initialize the M-dimensional topological map in N- dimensional sample space.

Step 1 Given a training input Z in N-dimensional sample space, find the closest (best matching) unit i in the subspace of independent variables:

$$\| Z^*(k) - W_i^*(k) \| = \min_j (\| Z^*(k) - W_j^*(k) \|) \tag{39.2}$$

where $Z^*(k)$ is the projection of the input vector onto X-space, $W_j^*(k)$ is the projection of the weight vector of a unit j, and k is the discrete time step. The Euclidean distance is used as a metric in (39.2).

Step 2 Define symmetric neighborhood of units j surrounding the "winning" unit i, $C(i,j)$, and adjust the weights of the winner and all nodes in its neighborhood

$$W_j(k+1) = W_j(k) + b(j,k)(Z(k) - W_j(k)), j \in C(i,j) \tag{39.3}$$

where $b(j,k)$ is the scalar learning factor (monotonically decreasing with k) and

$$W_j(k+1) = W_j(k), j \notin C(i,j) \tag{39.4}$$

Step 3 Reduce the learning factor and the neighborhood, increase iteration number k and return to step 1.

We also need specify the learning factor and the neighborhood function, as follows.
Learning Factor for the winning unit:

$$b(k) = b_0 (\frac{b_f}{b_0})^{k/k_{max}} \tag{39.5}$$

Where b_0 and b_f are initial and final values of the learning rate(user defined), k is the iteration number, corresponding to k-th presentation of the training data to the map, k_{max} is the max number of iterations, defined as the product of the training set size and the number of times this set was repeatedly presented to the network during training.
Neighborhood Function:

$$C(i,j) = exp(\frac{-(i-j)^2}{(b(k)S_0)^2}) \tag{39.6}$$

Where S_0 is the number of knots per dimension.
Learning rate for unit j in the neighborhood of unit i :

$$b(j,k) = b(k)C(i,j) \tag{39.7}$$

Note that both the learning rate and the neighborhood are gradually reduced as the training continues. Gradual decrease in the learning rate is typical for neural network training viewed as a special case of stochastic approximation [White 89]. However, decrease of the neighborhood function does not have an obvious statistical interpretation/explanation.

Based on empirical evidence, CTM algorithm appears competitive against other methods for low-to-medium- dimensional problems. Unfortunately, theoretical properties of the CTM algorithm are difficult to derive. Hence we provide some qualitative insights into the properties of CTM and its relationship to other methods. CTM algorithm interpretation can be greatly simplified by considering training(self-organization) process in X-space of predictor variables *separately* from y-dimension (response variable). It can be readily seen, that X-coordinates of CTM units during training follow standard Kohonen's SOM algorithm. Therefore, CTM units effectively approximate distribution of the training data in the space of predictor variables. This capability is related to other methods, such as Sammon's method for low-dimensional projections [Sammon 69] and Principal Curves/Manifolds [HastStue 89]. On the other hand, in y-dimension CTM method is similar to the flow-through version of kernel methods known as the Method of Potential Functions [MendelFu 70]. The main distinct feature of CTM is the concept of the topological map and neighborhood, that can be interpreted as *local regularization* constraint.

39.4 CTM comparisons

Maechler et al [MaMaScCsHw 90] compare prediction performance of a backpropagation network vs Projection Pursuit(PP) for several 2- variable regression problems of various complexity, i.e. simple interaction, radial, harmonic, additive and complicated interaction functions. They conclude that both methods achieve similar prediction accuracy, but that PP is computationally much faster(by two orders of magnitude). Later, Cherkassky et al [CherkLeeLari 91] applied CTM method to the same data set, and achieved overall improvement in the prediction accuracy at a somewhat lower computational cost. These results are particularly encouraging, since CTM results are degraded by the quantization error peculiar to the current CTM implementation (piecewise-constant). Friedman [Friedman 91] applied MARS to the same data, and obtained further improvement over CTM for all functions, except the harmonic function, where CTM still performed best.

Cherkassky and Mulier [CherkMuli 92] presented another comparison between CTM, PP and MARS for 2-variable regression problems. Their results indicate that PP is inferior to MARS or CTM, for all data sets studied. Comparisons between MARS and CTM are not conclusive: whereas MARS gives somewhat better prediction for simple interaction functions, CTM outperforms MARS for a "difficult"(rapidly varying) harmonic function. Results of the latter comparison are not surprising, since CTM is similar to kernel methods which are known to work best for harmonic functions. Also, MARS is designed for high-dimensional problems. In terms of computational speed, MARS is at least one order of magnitude faster than CTM.

In order to compare performance of different methods on regression problems with collinear variables, consider artificial data sets in table 1. For example, in the case of function 1, the variable y is dependent on only two variables a and b, however, the data set consists of three predictor variables x_1, x_2 and x_3. It is important that the modeling method be able to take advantage of the implicit constraints in order to reduce the effective problem dimensionality.

TABLE 39.1: Functions used to create the artificial data sets

number	*function*	*range*	*noise*
1	$x_1 = a^2, x_2 = b^2, x_3 = \cos(a^2 + b^2), y = a + b$	$a, b \in [0, 1]$	$\sigma = 0.1$
2	$x_1 = a^2, x_2 = \sin(b), x_3 = \cos(a^2 + b^2), y = ab$	$a, b \in [0, 1]$	$\sigma = 0.1$
3	$x_1 = a^2, x_2 = 4(a^2 - 0.5)^2, y = a$	$a \in [0, 1]$	$\sigma = 0.1$
4	$x_1 = \sin(2\pi a), x_2 = \cos(2\pi a), y = a$	$a \in [0, 1]$	$\sigma = 0.1$
5	$x_1 = a, x_2 = \sin(a), y = \cos(a)$	$a \in [-1, 1]$	$\sigma = 0.05$
6	$x_1 = a, x_2 = a^2, y = \sqrt{1 - 0.5(a^2 + a^4)}$	$a \in [-1, 1]$	$\sigma = 0.05$
7	$x_1 = a, x_2 = a, x_3 = a, y = \sqrt{1 - a^2}$	$a \in [-1, 1]$	$\sigma = 0.05$
8	$x_1 = a, x_2 = \cos(a), x_3 = a, x_4 = a^2, x_5 = a^3, y = \sin(a)$	$a \in [-1, 1]$	$\sigma = 0.05$
9	$x_1 = a, x_2 = sin(2\pi a), y = \cos(2\pi a)$	$a \in [-1, 1]$	$\sigma = 0.05$

TABLE 39.2: Best normalized RMS error of the test set for the 9 functions with noise

method	*1*	2	3	4	5	6	7	8	9
k nearest neighbors	.2947	.4967	.4368	.4748	.4014	.2803	.3723	.1201	.1080
GMBL	.2826	.5343	.3691	.4058	.3564	.2339	.2680	.1081	.0916
MARS	.2750	.4696	.3634	.4547	.3367	.2254	.2662	.1089	.1031
CTM	.2598	.4881	.3698	:3606	.3494	.2697	.3052	.1091	.1463

The training and test sets each consisted of 100 samples with normally distributed noise (with standard deviation σ given in Table 1). The normalized RMS error for the test set was used as a performance index. Four methods were compared: k-nearest neighbors, CTM, MARS, and GMBL. Generalized memory-based learning (GMBL)[Atkeson 90] is an advanced version of locally weighted regression, i.e., it applies weighted nearest-neighbor or locally weighted linear interpolation with parameters selected by cross-validation. For each method, the smoothing parameters were varied to achieve the best results. Results in Table 2 indicate that the modeling difficulty is not related to the dimensionality of the presented data set. This can be seen by comparing functions 5 and 8. Both of these functions have a one dimensional constraint and have similar sinusoidal dependence, but function 8 has 5 predictor variables while function 5 has 2. The modeling results for function 8 are much better than those of function 5. A possible explanation for this is that some of the predictor variables of function 8 are powers of a which are the terms in the Taylor approximation for the sinusoid. The variable y depends on some of the predictor variables linearly, so this is in fact a simple modeling task. This example shows that it is misleading to judge the difficulty of a problem based solely on the dimensionality of the data set. The above results show that CTM can be successfully applied to regression problems with dependent variables. Similarly, CTM is capable of handling regression problems with *statistical* dependencies between variables as well.

Acknowledgement: This work was supported, in part, by 3M Corporation.

39.5 REFERENCES

[Atkeson 90] Atkeson,C.G.,(1990) "Memory-Based Approaches to Approximating Continuous Functions", in *Proc. Workshop on Nonlinear Modelling and Forecasting.*

[BarrBarr 88] Barron, A.R. and R.L.Barron, (1988) "Statistical Learning Networks: a Unifying View", in Computing Science and Statistics, *Proc of the 20th Symp. Interface*, E. Wegman, ed, *American Stat. Assn*, 192-203.

[Barron 93] Barron, A.R.,(1993), "Universal Approximation Bounds for Superposition of a Sigmoidal Function", *IEEE Trans. on Information Theory*, 1993 (to appear).

[de Boor 78] de Boor, C., (1978) , *A Practical Guide to Splines*, N.Y., Springer Verlag.

[BreFriOlsSto 84] Breiman, L., Friedman, J.H., Olshen, R.A., and C.J. Stone, (1984), *Classification and Regression Trees*, Wadsworth, Belmont, CA.

[Breiman 91] Breiman, L., (1991), "The PI Method for Estimating Multivariate Functions From Noisy Data", *Technometrics*, v.3, no.2, 125-160,.

[BroLow 88] Broomhead, D.S. and D. Lowe, (1988), "Multivariable Functional Interpolation and Adaptive Networks", *Complex Systems*, v2, 321- 355.

[Cherkassky 92] Cherkassky, V., (1992), "Neural Networks and Nonparametric Regression", in *Neural Networks for Signal Proc.*, S.Y. Kung et al (Eds), IEEE Press, 511-521

[CherkLari 91] Cherkassky, V. and H. Lari-Najafi, (1991), "Constrained Topological Mapping for Nonparametric Regression Analysis", *Neural Networks*, Pergamon, v.4, 27-40.

[CherkLeeLari 91] Cherkassky, V., Lee, Y. and H. Lari-Najafi, (1991), "Self-Organizing Network for Regression: Efficient Implementation and Comparative Evaluation", *Proc IJCNN*, v.1, 79-84.

[CherkNaja 92] Cherkassky,V.and H. Najafi, (1992), "Data Representation for Diagnostic Neural Nets", *IEEE Expert*,7,5,43-53.

[CherkMuli 92] Cherkassky,V. and F. Mulier, (1992), "Conventional and Neural Network Approaches to Regression", *Proc SPIE Conf on Applic of Artificial Neural Nets*, SPIE vol 1709, 840-848.

[CheFriWec 93] Cherkassky, V., J.H. Friedman and H. Wechsler (Eds), (1993), *From Statistics to Neural Networks*, NATO ASI, Les Arcs, France. Proceedings to be published by Springer Verlag, 1994.

[Friedman 90] Friedman, J.H., (1990), "Multivariate Adaptive Regression Splines", *Technical Report no.102* (Revised with discussion), Dept of Statistics, Stanford Univ.

[Friedman 91] Friedman, J.H., Personal communication, 1991.

[FrieSilv 89] Friedman, J.H. and B.W. Silverman, (1989), "Flexible Parsimonious Smoothing and Additive Modeling", *Technometrics*, v.31, 1, 3-21.

[FrieStue 81] Friedman, J.H. and W. Stuetzle, (1981), "Projection Pursuit Regression", *JASA*, v.76, 817-823.

[GemBieDou 92] Geman,S., E.Bienenstock and R.Doursat, (1992), "Neural Networks and the Bias/Variance Dilemma", *Neural Computation*,4,1-58.

[HastStue 89] Hastie,T, and W.Stuetzle, (1989), "Principal Curves", *JASA*,v.84,no.406,502-516.

[Kohonen 84] Kohonen, T., (1984), *Self-Organization and Associative Memory*, Berlin, Springer-Verlag.

[MaMaScCsHw 90] Maechler, M., Martin, D., Schimert, J., Csoppensky, M. and J.N. Hwang, (1990), "Projection Pursuit Learning Networks for Regression", *Proc. Int'l Conf. on Tools for AI*, IEEE Press, 350-358.

[MoodDark 89] Moody, J. and C.J. Darken, (1989), "Fast Learning In Networks of Locally Tuned Processing Units", *Neural Computation 1*, 281.

[MendelFu 70] Mendel, J.M. and K.S. Fu, Eds.,(1970) , *Adaptive Learning and Pattern Recognition Systems: Theory and Applications*, Academic Press, NY.

[PoggGiro 90] Poggio, T. and F. Girosi , (1990) , "Networks for Approximation and Learning", *Proc. of the IEEE*, v.78, no.9, 1481-1497.

[RumMcCPDP 86] Rumelhart, D.E., McClelland, J.L. and the PDP Research group (Eds.), (1986), *Parallel Distributed Processing*, MIT Press.

[Sammon 69] Sammon,J.W., (1969), "A Non-Linear Mapping for Data Structure Analysis", *IEEE Trans. Comp.*, C-18, 401-409.

[Sanger 91] Sanger, T.D., (1991), "A Tree-Structured Additive Network for Function Approximation in High-Dimensional Spaces", *IEEE Trans. on Neural Networks*, v.2, no.2, 285-293.

[Specht 91] Specht, D.F., (1991), "A General Regression Neural Network", *IEEE Trans. Neural Nets*, 2,6,568-576.

[White 89] White, H., (1989), "Learning in Artificial Neural Networks: a Statistical Perspective", *Neural Computation*, 1, 425-464.

40

Multicollinearity: A tale of two nonparametric regressions

Richard D. De Veaux and Lyle H. Ungar

Princeton University and University of Pennsylvania

ABSTRACT
The most popular form of artificial neural network, feedforward networks with sigmoidal activation functions, and a new statistical technique, multivariate adaptive regression splines (MARS) can both be classified as nonlinear, nonparametric function estimation techniques, and both show great promise for fitting general nonlinear multivariate functions.
In comparing the two methods on a variety of test problems, we find that MARS is in many cases both more accurate and much faster than neural networks. In addition, MARS is interpretable due to the choice of basic functions which make up the final predictive equation. This suggests that MARS could be used on many of the applications where neural networks are currently being used.
However, MARS exhibits problems in choosing among predictor variables when multicollinearity is present. Due to their redundant architecture, neural networks, however, do not share this problem, and are better able to predict in this situation. To improve the ability of MARS to deal with multicollinearity, we first use principal components to reduce the dimensionality of the input variables before invoking MARS. Using data from a polymer production run, we find that the resulting model retains the interpretability and improves the accuracy of MARS in the multicollinear setting.

40.1 Introduction

In many estimation problems, the functional form governing the relationship between predictors and response is not known. Examples abound everywhere and include such diverse applications as modeling chemical plants, time series modeling, and relationships between spectrographic data and chemical concentrations. When many predictor variables are present, choosing the best subset of predictors is a formidable task, even if one assumes that the response is a linear function of the predictors. When the linearity assumption fails, as it often does, the problem becomes daunting.

Historically, many different equation forms have been suggested, some based on experience, others by mathematical convenience. Stepwise procedures of systematically adding and removing terms have often led to reasonable, if not optimal models. However, when the predictors exhibit a high degree of multicollinearity, great instability in the selection process can be present. In the linear case, a great deal can be learned by examining the correlation structure of the predictor variables. Solutions such as principal component regression (PCR), ridge regression or partial least squares (PLS) may then be appropriate. Such projection methods take linear combinations of the original variables to create new variables, of which a subset often gives an accurate model.

Neural networks, in many forms, are also being widely used to learn nonlinear mappings (see Rumelhart *et al.*, 1986 or Ripley, 1992 for an introduction to neural networks). These networks have very large numbers of parameters (called weights) estimated by minimizing an error over

---

[1] *Selecting Models from Data: AI and Statistics IV.* Edited by P. Cheeseman and R.W. Oldford.

a set of training patterns. Because they are nonlinear projection methods, and because of their tendency to overparameterize, neural networks tend to be fairly insensitive to problems of multicollinearity. However, precisely because they are overparameterized, they are typically not used for interpretation of the system, but only for prediction. Neural networks offer a nonlinear method which uses projection; the input variables are linearly combined before being passed through their nonlinear transformations.

Multivariate adaptive regression splines (MARS, Friedman 1991), use forward selection to adaptively build a set of basis functions for the function approximation. Unlike projection methods, MARS works in the original coordinate system and finds linear and nonlinear combinations of these coordinates. Recently, De Veaux *et al.* (1992) evaluated and compared neural networks and MARS via their performance on several benchmark problems. In most cases, MARS was able to predict as well or better as the neural network. However, this superiority was not uniform over the test problems.

It is the goal of this paper to better understand when different types of methods work or fail, particularly on nonlinear problems. We will show that whether projection or selection methods perform better is situation dependent. We furthermore show how a nonlinear selection-based method, MARS, can be combined with linear projection methods such as principal components to give models which are as accurate as neural networks, but are simpler and more interpretable.

The paper is structured as follows. The next section recalls certain properties of MARS and neural networks. Section 3 presents the problem of multicollinearity and gives a solution to the collinearity problem which combines principal components and MARS. In section 4 the method is applied on a data set from a polymer production run and compared to a neural network. Section 5 summarizes our results.

40.2 MARS and neural networks

We assume that most readers are familiar with principal component analysis (PCA) and with neural networks, and so describe them only very briefly. MARS is less widely used, but due to space limitations we must refer the reader to descriptions in Friedman (1991), De Veaux *et al.* (1992) or Sekulic and Kowalski (1992a).

40.2.1 MARS

The goal of MARS is to produce a simple, understandable model of a response to an arbitrary function of a set of predictor variables. MARS adds basis functions by forward selection. For each predictor variable, x_i, and every possible value, t of x_i, MARS divides the data into two parts, one on either side of the "knot", t. MARS keeps the knot and variable pair which gives the best fit. To each part, it fits the response using a pair of linear functions, each non-zero on one side of the knot.

After one variable has been selected, a split on a subsequent variable can occur in two different ways. The split can either depend on the previous split (splitting the input space only on one side of the previous knot), or it can ignore the previous split, splitting the entire input space on the new knot. In this former case, the associated basis functions of the original knot are said to be parents of the new basis functions.

MARS adds to the set of basis functions using a penalized residual sum of squares, or generalized cross validation criterion. Like any forward selection procedure, when confronted with two highly correlated input variables, it must choose between them. This may be reflected in the final set of basis functions, where only one of the two may be represented. In many applications, this may not be the optimal choice. For example, when multiple sensors are present, a weighted average of the sensors, may be preferable to the single one with the highest correlation with the response.

In general, the set of basis functions selected will be too large, and must be pruned back using backward elimination. The final model is obtained by performing backward selection using the generalized cross validation criterion. The linear basis functions are then replaced by cubic splines to make the approximation smoother.

40.2.2 Neural networks

The most common form of neural networks, feedforward networks with sigmoidal activation functions or, as they are sometimes called, "backpropagation" networks can, in the simplest case of a single layer of N hidden nodes, be written in the form

$$y_i = \sum_{j=1}^{N} w_{ij}\sigma_j(x) \tag{40.1}$$

where

$$\sigma_j(x) = 1/(1 + \exp - \sum_k w_{jk}x_k) \tag{40.2}$$

and the weights w_{ij} and w_{jk} are selected by a nonlinear optimization method to minimize the mean squared error over the training set.

The simplest algorithm to use is gradient descent ("backpropagation") in which each weight is iteratively changed proportionately to its effect on the error, but more advanced methods such as conjugate gradient methods and sequential quadratic programming are being used increasingly.

Neural networks are attractive as automatic model-building tools because they can be proven, given enough nodes, to be able to represent any well-behaved function, with arbitrary nonlinear interactions between the inputs. They are also, besides being suitable for implementation on massively parallel computers, relatively robust to outliers and poor data. In short, they have been seen as a way of doing nonlinear model-building without the pain of learning statistics. They have one major disadvantage: due to the high degree of interaction and collinearity between the variables and basis functions, it is almost impossible to interpret the models.

40.2.3 Comparisons

MARS and neural networks have some obvious similarities and differences. Both are methods of deriving nonlinear models from data. Both require significant amounts of data to build a reasonable model, and do so by using models with (potentially) large numbers of parameters - sufficiently large numbers of parameters that the models may be considered to be nonparametric. The methods differ in that MARS is based on subset selection, while neural networks offer a nonlinear projection method.

MARS and neural networks also share a subtle but important similarity. Both can be viewed as methods which select basis functions adaptively, based on the data. When there are multiple independent variables, methods which construct basis functions based on the data can be shown to give more accurate models (for a given amount of data) than methods which use a fixed set of basis functions (Barron, 1992). Thus neural networks and MARS are more efficient than Fourier series expansions or higher order NARMA models.

To understand further how the two methods work, it is useful to look at a their performance on an example problem. MARS was compared with neural networks on modeling a nonlinear model of a chemical plant (a non-isothermal continuously stirred tank reactor) which has been used to test different nonlinear control schemes (Psichogios *et al.*, 1992). The reactor temperature, the concentrations of the two chemical species in the reactor, and the temperature of the feedstream to the reactor were used to predict the reactor temperature and concentrations one sample time in the future.

When trained with sufficient data (over 100 data points), both MARS and neural networks were found to be have substantially lower predictive error than linear regression. (Testing must, of course, be done on a separate validation data set since comparisons on the training data are not meaningful.) MARS was found to be much faster than neural networks and, for sufficiently large data sets, it also produced more accurate models.

MARS is, however, not universally superior to neural nets. When not all the state variables are accessible, for example if only the inlet temperature and one of the concentrations on the above reactor problem are available, an autoregressive moving average (ARMA) model must be constructed by using past measurements. For this time series problem, the accuracies for the neural network and for MARS were not significantly different. As will be explained below, we believe that this "failure" of MARS is due to the collinearity of the input data. MARS was also found to be more sensitive to outliers and high leverage points in the predictor space.

40.3 Multicollinearity

In linear regression,

$$y = \beta_0 + \beta_1 x_1 + \ldots + \beta_p x_p + \epsilon, \tag{40.3}$$

severe estimation problems arise if the $x_1, \ldots, x_k$ are highly collinear. The problem is manifested through increased variance of the estimated parameters. Prediction, *per se*, of the response is unaffected, however, as the least squares fit will ensure that the residuals are still as small as possible. However, prediction of points not contained in the original data set may be affected due to the instability of the coefficients. The variance inflation factor (VIF) for each predictor is defined as

$$VIF = \frac{1}{1 - R_i^2} \tag{40.4}$$

where R_i^2 is the multiple correlation between x_i and the other predictors, $x_j (j \neq i)$. The VIF indicates how much the variance of the predicted coefficient β_i of equation 40.3 is inflated compared to orthogonal predictors. In highly collinear data sets, the VIF may be hundreds of times higher than an orthogonal set, rendering the interpretation of the coefficient meaningless.

A general approach to the problem is to consider the principal components of the predictor variables as predictors rather than the original variables. Usually, the predictor variables are taken to be standardized to have mean 0 and variance 1, so that principal components can be derived from the correlation rather than the covariance matrix. We shall assume that this is the case here. These new variables are orthogonal linear combinations of the original (standardized) predictors in decreasing order of variance. The idea of principal component regression is to use only the first k ($k < p$) principal components in a regression equation of the form:

$$y = \alpha_0 + \alpha_1 z_1 + \ldots + \alpha_k z_k + \epsilon, \tag{40.5}$$

where z_i is the ith principal component. It is then hoped that this will still retain good predictive behavior and that the z_i will be interpretable.

For an example of a case where principal components are natural, consider the temperature of a flow measured by six different devices at various places in a production process. Even though the inputs are highly correlated, a better prediction of the response may be gained by using a weighted combination of all six predictors rather than choosing the single best measurement. Variable subset selection attempts to circumvent the problem by choosing a small subset of the original p predictors which are not collinear, but still able to predict y to a reasonable degree of accuracy. However, as in the case of the "redundant" temperature measurements, there are cases when dropping a predictor out of the equation completely is not desired.

Neural networks and MARS deal with the problem of collinearity in very different ways. Due to its overparameterization, the coefficients or weights of a neural network are inherently difficult to interpret. However, it is this very redundancy that makes the individual weights unimportant. That is, at each level of the network, the inputs are linear combinations of the inputs of the previous level. The final output is a functions of very many combinations of sigmoidal functions involving high order interactions of the original predictors. Thus neural networks guard against the "problems" of multicollinearity at the expense of interpretability.

MARS, on the other hand, builds its set of basis functions via forward selection of the predictors, and at the end performs a backward selection to prune the number of selected basis functions. It is the forward selection procedure that makes MARS vulnerable to multicollinearity. If two predictor variables are both are correlated with the response, at some stage of the forward selection procedure MARS may be forced to choose between placing a knot on one of these predictors. The choice may be somewhat arbitrary if both result in roughly the same penalized residual sum of squares or gcv criterion value. Unfortunately, the choice has potentially profound impact on the choice of all further variable and knot selections and thus on the final model as well. This is a result of the tree structure in MARS and the forward method by which it selects. In an extreme case, it may happen that the choice of one variable may be slightly better at the current step, but that a much better model would result if the other predictor had been chosen. As in any subset selection procedure, the interpretability of the final model when one correlated predictor is chosen over another is degraded as well.

40.3.1 Principal component MARS

In the case of linear regression, the use of principal components stabilizes the coefficients since the predictor variables are orthogonal. The problems of using a selection method are ameliorated by using orthogonal predictors. Multicollinearity in the predictors may make selection procedures for nonlinear methods such as MARS even worse than in the linear case. Friedman (1991) was aware of this problem and proposed two strategies for dealing with multicollinearity. First, he suggested fitting a series of increasing interaction order models, comparing their gcv scores and then selecting the model with the lowest interaction order with an acceptable fit. His second strategy was to invoke a penalty on introducing new variables into the model, so that the change of entering two highly collinear inputs would be lessened.

Unfortunately, neither of these strategies alleviates the problem of choosing between two inputs in a collinear setting. Merely blocking out one does not ensure optimality or even reasonableness of fit. Moreover, increasing the interaction order says nothing directly about the problem of choosing a sub-optimal predictor at an early stage.

We propose to first transform the predictor variables using principal components before using MARS. This preprocessing of the variables into orthogonal variables will alleviate the problem of selection among highly collinear inputs. While this is certainly not the only method for reducing collinearity, it serves as a generic way to relieve the multicollinearity of tree based subset selection procedures. We will discuss other possible strategies in section 5, but first show how the method can dramatically improve the resulting fit from MARS on data from a chemical processing production run.

40.4 Example

To illustrate the problem with MARS in dealing with multicollinearity, we will use a data set consisting of measurements taken every 4 hours from a polymer production run. The predictors consists of 18 "control" variables which the engineers have some ability to adjust during the run, and 5 "material" variables which describe properties of the raw material being introduced into the system. The response y is the viscosity of the material as measured by a surrogate of molecular weight. A great deal of multicollinearity is present as is evident in Table 40.1. The variance inflation factors range from a low of about 9 to a high of over 2000.

Control Predictors								
Number	Name	VIF	Number	Name	VIF	Number	Name	VIF
x_1	Upper feed flow	239	x_2	Lower feed flow	139	x_3	Pyro pressure	181
x_4	C02 flow	1119	x_5	Pyro pot temp	11	x_6	Mix flow rate	168
x_7	Brine temp	33	x_8	Hot purge rate	31	x_9	Poly pressure	49
x_{10}	Poly level	16	x_{11}	Spray flow rate	45	x_{12}	Spray temp	258
x_{13}	Spray cat feed	550	x_{14}	Suction leg level	31	x_{15}	MEOH inj rate	80
x_{16}	Swirl temp	166	x_{17}	Swirl pump amps	40	x_{18}	Swirl cat feed	2156
Material Predictors								
Number	Name	VIF	Number	Name	VIF	Number	Name	VIF
x_{19}	H20	31	x_{20}	HCOOH	25	x_{21}	SALT	11
x_{22}	MEOH	23	x_{23}	CH2O	9			

TABLE 40.1: Predictors and their variance inflation factors

If we consider only the 18 control variables as our predictors, restrict the interaction order to 2, and use the default parameters settings of MARS, we find two curves (x_6, x_9), and two surfaces ($x_8 x_{17}$ and $x_{12} x_{17}$) using 11.5 degrees of freedom with a resulting R^2 value of 0.597. In an analogy with linear regression, one might expect that adding the 5 additional material variables will result in a model which may increase the number of degrees of freedom used, with the benefit of an increase in R^2. However, the introduction of these new predictors, results in a model using only one surface ($x_8 x_{17}$) with 4.0 degrees of freedom and an R^2 of 0.359. It seems that not only have the 5 new input variables not appeared in the new model, but their introduction has changed the stepwise procedure significantly.

The output from the MARS model shows what has happened. The first output shows the knot placement with only the control variables:

```
forward stepwise knot placement:
basfn(s)    gcv        variable        knot   parent
   0       313.3
   1       335.7          17.          92.41     0.
  3 2      363.2           8.          3.149     1.
  5 4      472.7           9.          6.848     0.
  7 6      749.5           6.          17.24     0.
  9 8      1278.          12.          50.84     1.
 11 10     6603.           9.          6.813     7.
```

$R^2 = 0.597.$

The second shows the placement for the full model with all 23 predictors:

```
forward stepwise knot placement:
basfn(s)    gcv        variable        knot    parent
   0       313.3
   1       335.7          17.          92.41      0.
  3 2      377.6           8.          3.155      1.
  5 4      522.2          10.          42.80      0.
  7 6      767.5          22.          0.1500     0.
   8       1409.          14.          77.26      5.
   9       3365.           9.          5.742      5.
```

$R^2 = 0.359.$

At the third knot placement, MARS has to choose between two relatively close alternatives (in terms of the gcv), placing a knot on variable x_9 or x_{10}. These result in very different future choices. In the second case, this knot generates two daughters which are eventually pruned out, as is the knot itself, resulting in a model with only x_8 and x_{17}. However, if variable x_9 is chosen, the future choices remain after the pruning, and result in a four variable model with a much better fit, as measured by R^2.

To help alleviate the problem of knot placement in the presence of multicollinearity, we use the linear principal components of the predictors rather than the original variables. Rather than choose the first k components, we allow MARS to select among all 23 principal components. The knot placement is shown below:

```
forward stepwise knot placement:
basfn(s)    gcv        variable          knot          parent
   0       313.3
  2 1      343.3           8.           0.3229           0.
   3       354.2           2.          -3.074            1.
   4       358.9           5.          -3.210            2.
   5       365.6          23.        -0.2505E-01         1.
  7 6      595.2          21.        -0.1507E-01         0.
  9 8      1948.           1.          -1.559            0.
```

$R^2 = 0.784.$

The last two pairs of basis functions are pruned, resulting in model with three surfaces: $(x_2 x_8)$, $(x_5 x_8)$ and $(x_{23} x_8)$. The loading vectors (or weights) for the 4 principal components chosen are shown in Table 40.2.

The predictor involved in all three surfaces, PC_8 is made up (essentially) of three material variables (HCOOH, SALT and CH_2O), Brine temperature (related to SALT) and the suction

Predictor	PC2	PC5	PC8	PC23	Predictor	PC2	PC5	PC8	PC23
Upper feed flow	-0.18	0.34	0.20	-0.04	Lower feed flow	-0.02	0.10	-0.05	0.07
Pyro press	-0.30	0.07	-0.07	0.00	C02 flow	0.05	0.14	-0.03	-0.52
Pyro pot temp	-0.22	-0.08	-0.09	0.01	Mix flow rate	0.16	-0.29	-0.16	-0.04
Brine temp	-0.30	-0.14	-0.35	-0.04	Hot purge rate	-0.18	-0.16	0.11	0.02
Poly press	-0.07	0.08	-0.08	0.02	Poly level	0.07	0.27	-0.20	-0.04
Spray flow rate	-0.13	0.03	0.15	-0.03	Spray temp	-0.15	-0.05	0.07	0.12
Spray cat feed	-0.17	0.13	0.08	-0.36	Suction leg level	-0.17	-0.33	0.35	-0.02
MEOH inj rate	0.24	0.09	-0.08	-0.04	Swirl temp	-0.38	0.15	0.10	-0.10
Swirl pump amps	-0.38	0.11	-0.08	0.02	Swirl cat feed	-0.06	0.17	0.01	0.75
H20	-0.10	-0.09	0.03	0.02	HCOOH	-0.02	-0.59	-0.31	-0.01
SALT	-0.39	0.04	-0.45	0.00	MEOH	-0.13	-0.13	0.09	-0.02
CH2O	-0.19	-0.24	0.50	-0.01					

TABLE 40.2: Principal component loadings for chosen directions

leg level. The next most important variable, PC_{23} is essentially three control variables, the swirl catalyst feed rate, the spray catalyst feed rate and the CO_2 flow. PC_2 involves SALT, two variables involving swirl, brine temperature and pyro pressure. Finally, PC_5 is made up of HCOOH, two feed flows and suction leg level.

We have used PCA and MARS to select a small number of linear combinations of variables for non parametric fitting. The technique is not specific to principal components or MARS, but to any dimension reduction technique followed by subset selection and then nonparametric fitting. As an example, we could fit a loess (Chambers and Hastie, 1992) surface to the viscosity using the three combinations (PC_2,PC_8), (PC_5,PC_8) and (PC_8,PC_{23}). This model gave an R^2 value of 0.92. Undoubtedly, MARS could be fine-tuned to produce a better model than the one illustrated above. The point, however, is that MARS automatically chooses the surfaces or curves to fit using subset selection. Trying all single and pairwise combinations in a generalized additive model setting is computationally infeasible. The resulting fitting is a relatively minor matter after the relevant variables have been selected.

40.5 Discussion and future work

The results presented above show that in cases where there is high collinearity among the predictors, it may be important to linearly project them with a method such as PCA before using a nonlinear model selection method such as MARS. Since nonlinear projection techniques such as neural networks include such linear combination, the same preprocessing will have no effect on their performance.

For linear problems where the original coordinate system is a meaningful one, linear subset selection methods such as stepwise regression work well. If there is a high degree of correlation between the predictor variables, and significant noise in them projection methods such as PCA or PLS may be warranted. If the problem is nonlinear, and a high degree of correlation between predictors exists, neural networks or MARS with a PCA front end will tend to outperform MARS alone.

For nonlinear situations, knowing whether to preprocess need not be obvious. Spectrographic data from IR, NMR and similar methods have very highly correlated values, and PCA and PLS methods have proven fruitful for linear analysis, but a combined PLS/MARS method works less well than MARS on the original variables (Sekulic and Kokalski, 1992b). Why is this? Interactions which occur between individual predictor variables (*e.g.* the strength of given spectrographic peaks) may be masked when interactions between the principal components are chosen, if these are chosen on a linear basis.

The proceeding section of this paper has shown that there are important cases where nonlinear subset selection (*e.g.* MARS) does work well on the principal components. In this case, the results may be more interpretable than those provided by methods such as neural networks which do not involve subset selection. The required computation is also much lower with MARS, and extensive cross validation is not required to avoid overfitting or excessively complex (and hence high variance) models.

In the examples presented in this paper, PCA was used as the projection method to be combined with MARS as the nonlinear model/selection method. It is also possible to use other projection or nonlinear modeling methods. An attractive projection method is PLS, which differs from PCA in that outputs as well as inputs are used in picking the principal directions. (see also Wold *et al.* 1989, Holcomb and Morari, 1992 and Qin and McAvoy, 1992). This complicates interpretation of the model, but has the advantage that one need only keep the dominant "principal components" as (unlike the PCA) they account contain combinations of the inputs which have the largest effect on the outputs. Since PLS uses the output variables and needs an assumed form of relationship between the inputs and outputs (generally, but not necessarily linear), using PLS complicates the algorithm for combining the projection method with the nonlinear model. An optimal method, therefore, requires an iterative scheme in which one first picks a projection for a given form of inner relationship (initially linear) between the projection weights and the outputs and then finds an inner relationship for the projection weights as found in the first step. The new inner relationship requires that the first step be performed again, so the method iterates until convergence.

Many of the methods discussed in this paper are relatively new, and the collective wisdom on their overall attributes and performance in a variety of settings is thin. We have attempted to provide a framework for the possibility of combining some of the features of these tools, and have indicated that some dramatic improvement is possible. Much work remains to be done.

40.6 References

[1] Barron, A. R. "Approximation and Estimation Bounds for Artificial Neural Networks.", in press, *Machine Learning* (1992).

[2] Chambers, J. M. and Hastie, T. J., "Statistical Models in S", Wadsworth & Brooks/Cole, Pacific Grove, California, 1992.

[3] De Veaux, R. D., Psichogios, D. C., and Ungar, L. H. "A Comparison of Two Non-Parametric Estimation Schemes: MARS and Neural Networks." *Comp. Chem. Engng.* **17:8** (1993) 819-837.

[4] Friedman, J. H. "Multivariate adaptive regression splines." *The Annals of Statistics* **19:1** (1991) 1-141.

[5] Holcomb, T. R., and Morari, M. "PLS/Neural Networks." *Comp. Chem. Engng.* **16:4**, (1992) 393-411.

[6] Psichogios, D. C., De Veaux, R. D., and Ungar, L. H. "Non-Parametric System Identification: A Comparison of MARS and Neural Networks." *ACC* **TA4** (1992) 1436-1440.

[7] Qin, S. J., and McAvoy, T. J. "Nonlinear PLS Modeling Using Neural Networks." *Comp. Chem. Engng.* **16:4**, (1992) 379-391.

[8] Ripley, B. D. "Statistical Aspects of Neural Networks." Invited lectures for SemStat, Sandbjerg, Denmark, 25-30 April 1992.

[9] Rumelhart, D., Hinton, G., and Williams, R. "Learning Internal Representations by Error Propagation." *Parallel Distributed Processing: Explorations in the Microstructures of Cognition, Vol 1: Foundations* Cambridge: MIT Press (1986), 318-362.

[10] Sekulic, S., and Kowalski, B. R. "MARS: A Tutorial." *J. Chemometrics* **6**, (1992).

[11] Sekulic, S., and Kowalski, B. R. "Nonlinear Multivariate Calibration Methods Combined with Dimensionality Reduction." submitted to *J. Chemometrics* (1992).

[12] Wold, S., Kettaneh-Wold, N., and Skagerberg, B. "Nonlinear PLS Modeling." *Chemometrics and Intelligent Laboratory Systems*, **7** (1989) 53-65.

41
Choice of Order in Regression Strategy

Julian J. Faraway

Department of Statistics, University of Michigan

ABSTRACT Regression analysis is viewed as a search through model space using data analytic functions. The desired models should satisfy several requirements, unimportant variables should be excluded, outliers identified, etc. The methods of regression data analysis such as variable selection, transformation and outlier detection, that address these concerns are characterized as functions acting on regression models and returning regression models. A model that is unchanged by the application of any of these methods is considered acceptable. A method for the generation of all acceptable models supported by all possible orderings of the choice of regression data analysis methods is described with a view to determining if two statisticians may reasonably hold differing views on the same data. The consideration of all possible orders of analysis generates a directed graph in which the vertices are regression models and the arcs are data-analytic methods. The structure of the graph is of statistical interest. The ideas are demonstrated using a LISP-based analysis package. The methods described are not intended for the entirely automatic analysis of data, rather to assist the statistician in examining regression data at a strategic level.

41.1 Introduction

Textbooks on linear regression have several chapters, each devoted to one particular aspect of building a regression model and checking its adequacy. One chapter may study variable selection and another diagnostics for the detection of outliers. If these may be viewed as tactics in the pursuit of a regression model, what of the strategy? Very little is said about how these various techniques fit together, other than that it is a skill gained by experience. [Daniel 80] is a notable exception in this respect. There are outstanding questions concerning the interaction between these tactics and the order in which they should be carried out. In this article, we will look at, not a particular specific of regression analysis, but the process as a whole.

Regression strategy is usually discussed within the context of expert systems for Statistics, but production of such systems for the automatic analysis of data has been stalled primarily by the difficulty of integrating the real-world context of the data. Nevertheless, this does not preclude worthwhile study of regression strategy. This article does not describe an expert system and the tools discussed are not intended for the completely automatic analysis of regression data. These methods should be regarded in the same way as the usual tools of the regression analyst, such as Box-Cox transformations. The difference is that they are designed for strategic, not tactical, application. Just as the tactical tools require the supervision of a statistician for appropriate use, these strategic tools are not meant to be blindly or automatically applied. [Gale 86] and [Phelps 87] contain several articles in which expert systems and statistical strategies are discussed.

We view the pursuit of an appropriate model by the statistician in the following way: The desired model (or models) should satisfy several requirements: unimportant variables should be excluded, outliers identified, variables appropriately transformed, etc. An initial model is proposed and these requirements are checked sequentially by numerical or graphical methods

[1] *Selecting Models from Data: AI and Statistics IV.* Edited by P. Cheeseman and R.W. Oldford.

and if necessary the model is changed in a way suggested by the particular method. For example, variable selection methods can detect redundant variables and propose their elimination. When the current best model satisfies all the requirements the analysis ends. The choice of requirements and methods used to test them and make appropriate model changes is made by the statistician. We take particular interest in the order of application of the methods. We must automate the methods used due to the amount of repetitive analysis required, which leads to the two main difficulties with the tools we propose. Graphically-based methods are difficult to automate because they rely on human perception and because the automated methods are context-free, the statistician must examine the results to determine the sense or lack of it. These tools are not intended to replace the standard analysis - they are an additional aid and the results should be interpreted with due care.

Regression analysis is influenced by the taste of the statistician. The choice and ordering of methods are not universally agreed upon and so it is possible that two experienced analysts will construct different valid models and come to different conclusions from the same data. If one was aware of this, then one would hesitate to seize upon one conclusion and discard the other, rather one might say that the data do not support any strong conclusion. However, it is quite possible that the first two statisticians may agree and a third disagree, or a fourth or a fifth. Given that the number of reasonable analyses is likely to be large, one is unlikely to have the resources to collect all these opinions. Often, one will be the sole analyst so that it will be difficult to know if the data support many or only one conclusion.

[Lubinsky 87] discuss the search for "good" regression models although their methodology and motivation differ from ours. See also [Brownstone 88] and [Adams 90]

We show a way, given a particular choice of methods, of generating all of the acceptable models arrived at by different orderings of the methods chosen. Thus the statistician may discover if there are several competing models for the data which support different conclusions, in which case suitable doubt may be expressed, or that only one model is indicated, whence the conclusion may be infused with greater confidence.

Regression analytic methods are sometimes quite inexact, perhaps depending on the statistician's interpretation of a plot, but since a large number of possible regression analyses need to be considered, these methods need to be exactly specified. This we do in section 2, so that they may be programmed. In section 3, we discuss the generation of acceptable models derived by the various orderings of the methods. We also address the question of which of the generated models is best in section 4.

41.2 Characterizing regression data analysis

The process of building a regression model may consist of several stages such as outlier detection, variable selection and transformation which we shall call regression analytic procedures (RAPs). At each stage there is a candidate model that may be supplanted by another model according to the result of the RAP. A regression model might be specified by the original data, the functions specifying the link between the predictors and the response and weights on the observations. Of course, a richer formulation of regression models is possible, if not desirable, but for simplicity we will proceed with this.

We wish to characterize RAP's as functions acting on regression models and returning regression models. For some procedures such as variable selection and transformation, this is relatively straightforward, but for other diagnostic based methods depending on graphics, it is not so easy. Also the physical context of the data can be included to a limited extent by restricting the RAP's from transforming or eliminating certain variables or points but it is difficult to include knowledge concerning which functional forms are more appropriate than others. Thus, the RAP's provide only an approximation to a regression data analysis by a human but the information provided by this approach is additional to that given by a standard analysis, not a replacement, so nothing is lost and much may be gained.

When constructing a regression model, we have certain requirements for what is an acceptable final choice - for example, that there be no redundant predictors, that the expected response should be linear in the predictors, that there be no outliers included in the model etc. The RAP's are a response to these requirements in that they examine a candidate model and if necessary change that model to make it acceptable with respect to that particular requirement. Thus, a minimal list of RAP's is determined by the requirements we wish our model of choice to satisfy. An acceptable final model would be one which is not changed by the application of any of the RAP in the list.

The following RAP's have been programmed. The names for the methods are given brackets for reference. Check and remove outliers (outlier-test), check and remove influential points (test-influence), check for non-constant variance and reweight if necessary (hetero-test), check for and apply a Box-Cox transform on the response(box-cox-test), check for transformations of the predictors (tran-predictors), perform variable selection using the backward elimination method (bw-elim), perform variable selection using the forward selection method (fw-sel) and restore points previously eliminated that are not now outliers, but may be influential (restore-points). See [Faraway 92] for more details on these procedures.

This list is obviously not exhaustive, but is representative of the sort of data-analytic actions that may occur in practice and is appropriate for some common requirements for acceptable regression models. I certainly do *not* claim that these are the best methods to use all the time only that the ideas that follow are not restricted by these particular choices.

These functions have been programmed to take any regression model as input, and output a (possibly changed) regression model. The flexibility of Lisp and the object-oriented programming system that comes with LISP-STAT ([Tierney 90]) makes it easier to program these functions in full generality to keep track of the numeric (the data and weights) and non-numeric (the transformations and variable names) components of the model.

41.3 Generation of acceptable models

In this section, we consider the generation of acceptable models by changing the order in which RAP's are applied. A natural, although arbitrary, choice of initial model is the regression of all possible predictors on the response with unit weights on the observations. Clearly, there are situations when there will be several reasonable initial choices, but for now we will use only that one initial model. Our ideas are not affected by this restriction.

First we should select a list of RAP's considered appropriate for the particular problem. Starting with the initial model, we apply each of the RAP's individually. Some will cause no change, but others may return different models. To each newly found model, we apply each of the RAP's until no new models are generated. The total analysis may be viewed as a directed graph, where the vertices are models and the arcs linking the vertices are RAP's that resulted in a change. Loops indicating RAP's which had no effect on a particular model could be explicitly drawn, but can be implicitly assumed. The absorbing vertices (having out-degree zero) in the graph are acceptable models. This approach to generating the acceptable models is much more efficient and comprehensive than simply generating all combinations of the list of RAP's and then applying them sequentially because much redundant calculation is eliminated, particularly since an RAP may occur more than once in a given path.

I have written some experimental software in LISP-STAT to construct these regression analyses. The user inputs the data and makes a selection from the available RAP's. The program then constructs the digraph from this information and displays it graphically. The user may interact with the graph by clicking on the vertices to obtain a regression summary of the model used at the selected vertex. The software is obtainable from Statlib. (Anonymous FTP to `lib.stat.cmu.edu`, login as `statlib`, file is `xlispstat/regstrat` or by e-mail to `statlib@lib.stat.cmu.edu` containing the message `send regstrat from xlispstat`).

The work of [Lubinsky 87] differs from ours in that their graph is strictly a tree where the nodes of the tree are *features* of regression models and the arcs indicate the result of tests for the presence of such features. The strategy (ordering of the tests) is determined by calibrating the order of analysis on known data, and is then fixed. The user then chooses a model (or models) from the tree trading the simplicity of the model against its accuracy. We produce all the acceptable models determined by a particular fixed choice of RAP's and leave the user to choose amongst them. We do not fix the order.

Now, certainly some of these analyses might consist of sequences of RAP's that no statistician would ever try in practice, but this of little consequence since the final model is the ultimate subject of interest. The statistician should examine these models to ensure that they are physically sensible and discard those that are not. The difficulty is not that unreasonable models will be included amongst those considered, since the statisticians can easily screen these out, rather that important models will not be discovered due to the inflexibility of the RAP's. Thus we do not claim that this method will produce all reasonable models, but it may well find some that otherwise might have been missed.

We shall use several datasets in the following discussion:

The Galapagos dataset: 29 cases being islands, 5 geographic predictors (area, elevation, distance to nearest island, distance from Santa Cruz island and area of adjacent island) and number of species as the response, described in detail in [Andrews 85].

The Chicago dataset: 47 cases being zip codes in Chicago, 5 socio-economic predictors (% minority composition, fire rate, theft rate, age of housing and income) and no. homeowner insurance policies is the response, described in detail in [Andrews 85].

The Swiss dataset: 47 cases being provinces in 1888 Switzerland, 5 socio-economic predictors and a standardized fertility measure is the response. Described in [Mosteller 77].

Some data will generate several possible models, others only one. For example, using the RAP's outlier-test, test-influence, box-cox-test, bw-elim, fw-sel and restore-points, and forcing the inclusion of the % minority composition variable in Chicago dataset produces the digraph shown in Figure 1. The initial model chosen was the one with all cases and predictors included and all variables untransformed. The Chicago dataset analysis considers 17 models in all, of which 2 are acceptable. In contrast analyzing the Galapagos dataset using the RAP's outlier-test, restore-points, box-cox-test and tran-predictors produces the digraph shown in Figure 2 where 24 models are considered and 5 found to be acceptable. In contrast the Swiss dataset produces

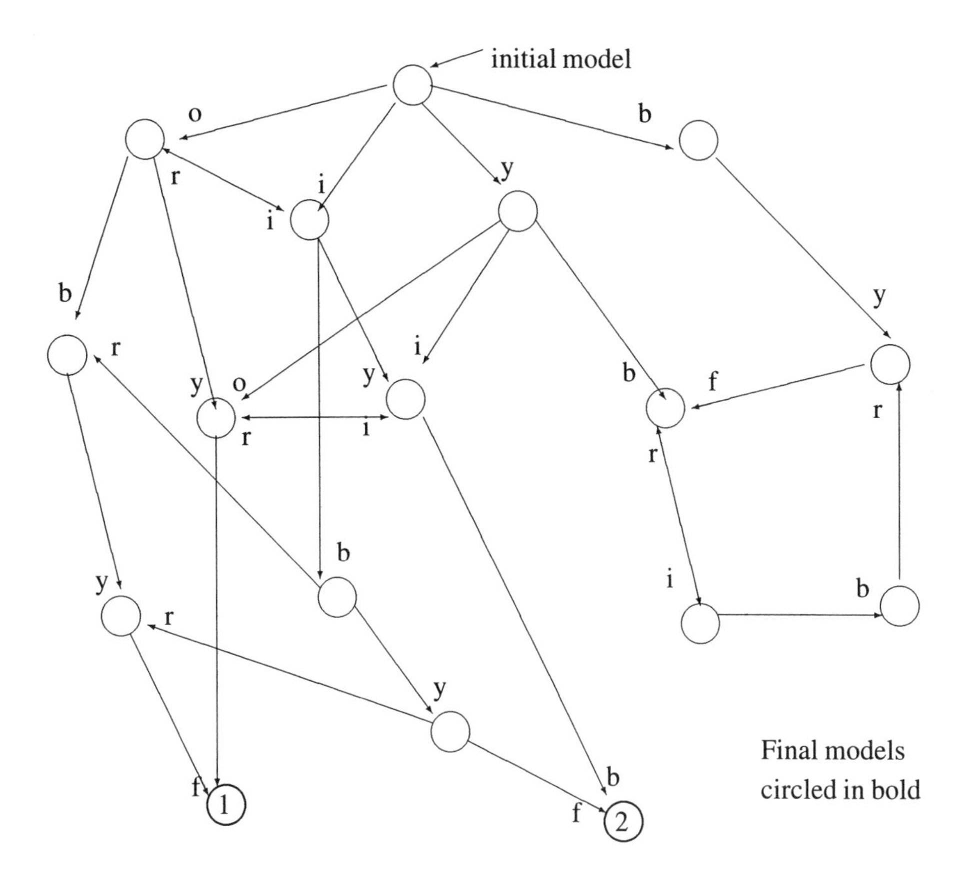

Key: o - outlier test, i - influential test, r - restore points,
f - forward selection, b - backward elimination.

FIGURE 1: Analysis of the Chicago data

only one acceptable model, as the result of applying backward elimination to the initial model with no other RAP having any effect.

The Galapagos analysis required 4x24=96 regression actions - presumably beyond the patience of many human regression analysts. For some choices of RAPs the graph can become much larger. For instance, including an RAP that can transform predictors to the Chicago analysis results in digraph containing 99 vertices, although again only two acceptable models are found. This is too unwieldy for profitable graphical display and some restriction on the application of the RAPs may be desirable to cut down on the size of the graph. The nature of the RAP's used here mean that the graph must be finite because variables and cases can either be included or excluded and the choice of transformations is finite, but the potential for infinite graphs exists if

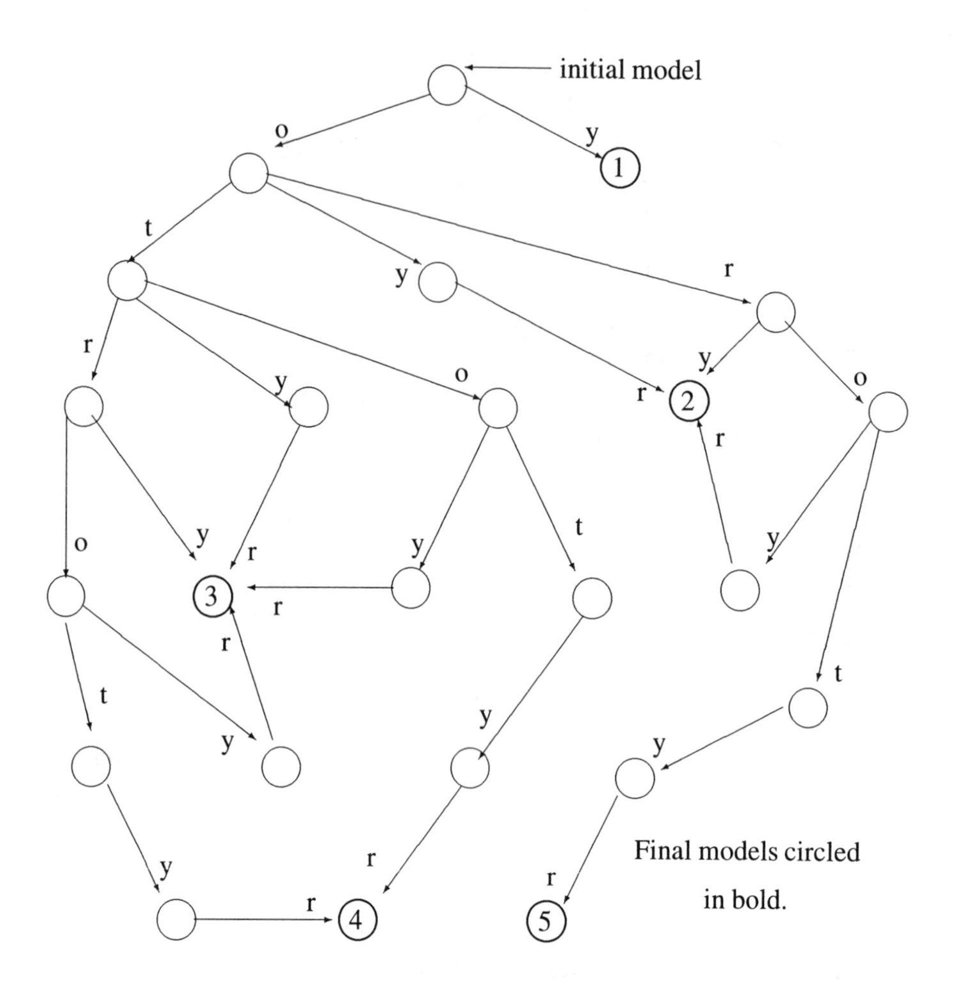

FIGURE 2: Analysis of the Galapagos Data

the RAP's are not constructed carefully.

Notice how cycles may occur in the graphs, for example in the Chicago data there is a loop on the right of the graph so that any search method that applies RAP's sequentially without memorizing past models would be vulnerable to being caught in such a cycle.

Different vertices in the graph may be reached by different paths indicating different orders in the analysis. It is simple to see the effect of eliminating an RAP from the analysis by removing the appropriate arcs and vertices from the graph. If information on the models represented by the vertices is stored, RAP's may be added to the graph, without redoing what has already been computed. It is also possible to identify crucial stages in the analysis that led to the choice of one model or another. The models and RAP's where the branch occurs could be more closely investigated by the statistician. The software will any vertex to be selected to display that model and allow further analysis.

41.4 Model Selection

Having generated a list of acceptable models, can we choose which one is best? Expert knowledge of the particular area may allow one to choose one model with confidence or at least eliminate some of the competitors. Failing that we might be tempted to use criterion based model selection methods such as the adjusted R^2 or the Akaike information criterion. Given that the response may be transformed in different ways in the competing models, the criterion may have to allow for this, so Mallow's C_p may be inappropriate. [Lubinsky 87] discuss some ways of choosing from models that differ in structure, allowing for the simplicity and accuracy of the model.

We could simply pick the model that maximizes the chosen criterion, but this may be precipitous. Suppose, prediction is our goal then the predictions from the acceptable models may vary greatly even if the criterion does not. For example, consider the regression analysis depicted in Figure 2. Suppose we wish to use one of the five models indicated to make a prediction for the number of species for an island with predictors (order as above): (0.01 25 0.2 0 0.03). This produces five predictions ranging from 0.78 to 17.1 - a range of variation that would not be anticipated given the standard errors. Yet, the adjusted R^2's for these models differ hardly at all. It would be rash to simply pick the one that has the maximum adjusted R^2 and discard the rest. Small perturbations in the data or model might change the selection as might the use of a different criterion.

If we refuse to choose one model, then how should we combine the predictions made by the five models? One possibility is to take the five predictive distributions and form a weighted sum. The weights need not be equal and could be motivated by contextual knowledge of which models are likely to be more appropriate than others. In any case, the conclusions should be checked for robustness by varying the weights. The software can form these weighted predictive distributions given user-supplied weights.

Another objective in regression analysis is to assess the dependence of the response on a particular predictor. This dependence can be quantified by the appropriate regression parameter. If different transformations are used in the acceptable models, say a log transform on the response in one model and a square root in another, it will be difficult to directly compare the relevant parameter estimates. One possibility for a consistent method of comparison is to assess the change in the response as the relevant predictor is changed (both in the original scale) at a specific point in the range of X, X_0, that should be chosen with respect to the context of the data, i.e at points of particular interest in the predictor space. Since the effect may differ over this space, several X_0's might be considered. Another concern in interpreting regression coefficients is collinearity, which is not specifically addressed here.

In the Chicago dataset regression analysis depicted in Figure 1 there are two acceptable models. There is also a loop of four models on the right of the graph. Closer study reveals that the loop is caused by the inclusion of one variable depending on one case which is influential. At least of one of these four models should also be considered in addition to the two other acceptable models. Fortunately, a square root transformation was used on the response in all potential final models so there is no difficulty in comparing the parameter estimates across models. We find that, although the % minority composition variable is significant in all the models, the magnitude of the effect can differ by a factor of two.

Given that the analyst is aware that there are several possible candidate models giving different interpretations, selecting one of them capriciously and discarding the rest would ignore the real uncertainty in the estimate. It would be far better to report the full range of acceptable models and the predictions they make. If, however, there is only one acceptable model generated, then the analyst can be a lot more confident in the estimate. This is the advantage of having the list of acceptable models available.

Another concern in simply selecting a "best" model is in the reliability of inference from that model. If one accepts that allowance for the data analysis that precedes model selection should be made in the inference that follows, then since the amount of data analysis here has been increased substantially over the usual amount, naive inference from a "best model" is likely to be even more optimistic with regard to estimated standard errors.

It should be emphasized here that it would imprudent to rely on the generated models alone. We advise that the statistician perform their usual analysis without using the RAP's and paying particular attention to graphical methods and physical context. A weakness of the RAP's is they lack the human perception of graphical displays and thus may miss important features. The generated models should be regarded as additional information not as a replacement for a standard analysis.

41.5 Discussion

Some work has been done on finding an optimal regression strategy that may be applied to all datasets. However, examination of Figures 1 and 2 indicate that the paths leading to acceptable models can be quite different. A fixed strategy would find only one model which as we have seen may not tell the whole story. This suggests that searching for an optimal strategy is both unrealistic and inappropriate.

One might be tempted to turn to all inclusive methods such as MARS [Friedman 91] to avoid the difficulty in choosing a regression strategy and a final model, but the advantage of more traditional methods is that much can be learned about the data in the process of the analysis that is not seen in black box method. There is also the potential to find much simpler models.

We wish to emphasise that without the full incorporation of physical context into the RAP's, which is a quantum leap beyond what we have here, and without a much more comprehensive set of RAP's, the methods we have discussed here are only appropriate for careful use by statisticians and not for unguided application by the uninitiated.

41.6 References

[Adams 90] Adams J. (1990) *American Statistical Association Proceedings of the Statistical Computing Section* 55- 62

[Andrews 85] Andrews D. & Herzberg A (1985) "Data : a collection of problems from many fields for the student and research worker" *New York, Springer-Verlag.*

[Brownstone 88] Brownstone D. (1988) "Regression strategies" *Proceedings on the 20th Symposium on the Interface, Ed. Wegman E. et al.* 74-79

[Daniel 80] Daniel C. & Wood F. (1980) "Fitting Equations to Data, 2nd Ed." *New York, John Wiley.*

[Faraway 92] Faraway J. (1992) "On the Cost of Data Analysis" *Journal of Computational and Graphical Statistics* **1** 215-231

[Friedman 91] Friedman J. (1991) "Multivariate Adaptive Regression Splines" *Annals of Statistics* 1-141

[Gale 86] Gale W. (Editor) (1986) "Artificial intelligence and statistics" *Addison-Wesley, Reading Mass.*

[Lubinsky 87] Lubinsky D. & Pregibon D. (1987) "Data Analysis as Search" in "Interactions in Artificial Intelligence and Statistical Methods" edited by Phelps B. *Gower Technical Press, Aldershot, Hants*

[Mosteller 77] Mosteller F. & Tukey J. (1977) "Data Analysis and Regression" *Addison-Wesley, Reading Mass.*

[Phelps 87] Phelps B. (Editor) (1987) "Interactions in Artificial Intelligence and Statistical Methods" *Gower Technical Press, Aldershot, Hants*

[Tierney 90] Tierney L. (1990) "Lisp-Stat: An object-oriented environment for statistical computing and dynamic graphics." *Wiley, New York*

42 Modelling response models in software

D.G. Anglin and R.W. Oldford

Department of Statistics and Actuarial Science
University of Waterloo
Waterloo ON, N2L 3G1
CANADA.

ABSTRACT We describe our software design and implementation of a wide variety of response models, which model the values of a response variable as an interpretable function of explanatory variables. A distinguishing characteristic of our approach is the attention given to building software abstractions which closely mimic their statistical counterparts.

42.1 Introduction

One of the great workhorses of statistical science is the response model. This model supposes that there are $p+1$ variables $x_1, \ldots, x_p$ and y whose values are observed for each of n independent realizations. Interest lies in modelling the values of the *response* variable y as an interpretable function of the *explanatory* variables $x_1, \ldots, x_p$. The explanatory variables are taken to be fixed at their observed values – either because they are under our control and the response is observed after their setting, or because we choose to condition the response on the observed values of the explanatory variables. Once fitted to data, the model is an interpretable summary of the dependence of the response on the explanatory variables as well as a predictor of values of the response for any values of the explanatory variables.

The most widely used response model is the linear model, which expresses the response as a function of the explanatory variables and a random disturbance ϵ. This function is linear in its unknown parameters α and $\beta_1, \ldots, \beta_p$ and is typically written as

$$y = \alpha + \beta_1 x_1 + \cdots \beta_p x_p + \epsilon. \tag{42.1}$$

In its simplest form, the model takes each ϵ to be an independent realization from some distribution having zero mean and finite variance σ^2. When the distribution of ϵ is also assumed to be the Gaussian distribution, we will call this the *Gaussian linear model.*

Software for this model has long been the mainstay of statistical packages. The fitting algorithm is simply least-squares (or equivalently maximum likelihood when ϵ is Gaussian) and requires only data corresponding to the response and explanatory variables. Summary results run the gamut from printed tabular output to interactive graphics for model criticism and exploration.

As computational resources have grown, so too has the response model. The relatively simple linear model has been generalized to response models which are at once more computationally intensive and more flexible. Even so they retain much of the interpretability of the linear model.

[0]Research supported by NSERC in the form of a Postgraduate Scholarship for the first author and an operating grant for the second.

[1]*Selecting Models from Data: AI and Statistics IV.* Edited by P. Cheeseman and R.W. Oldford.

We review some of these models in Section 2.

Good statistical modelling is a largely iterative process. Many models might be selected, examined, and discarded before the analyst settles on some hopefully small set of competing models worth reporting (if any). Statistical computing environments must be designed to support this process of data analysis and modelling.

In what follows we describe our software design and implementation of a wide variety of response models. A distinguishing characteristic of our approach is the attention given to building software abstractions which closely mimic statistical counterparts (see *eg.* [Oldford 87]). That is, we intend to build software models of response models. The response models we consider are reviewed in Section 2 and our software is described in Section 3. Section 4 relates a very short example of the use of the software. In Section 5 we describe our approach in relation to that of others. Section 6 contains some discussion of implementing other response models not covered in previous sections and of implementing statistical models in general.

42.2 Response Models

As described earlier, by a response model we mean one which designates some variables (typically one) as response variables and the remainder as fixed explanatory variables. We suppose further that values of all response variables are independent from one realization to the next. Of interest then, is inference about the conditional distribution of the response variables given the explanatory variables.

The linear model equation (42.1) is such a model. To emphasize this we rewrite the Gaussian linear model (42.1) as

$$y|x_1, \ldots, x_p \sim N(\mu, \sigma^2) \quad (42.2)$$

$$\mu = \alpha + \beta_1 x_1 + \cdots \beta_p x_p. \quad (42.3)$$

Here, $N(\mu, \sigma^2)$ denotes the Gaussian distribution with mean μ and variance σ^2. σ^2 is the unknown conditional variance of y and the explanatory variables enter the model only through the conditional mean μ of y.

An important natural generalization is to replace the Gaussian distribution by one that is more appropriate for the response. For example, if each realization of the response is a proportion of people who respond to some medical treatment, we may wish to use a Binomial distribution with mean proportion π. However, modelling the mean π as a linear combination of the explanatory variables may lead to estimated values of π outside of $[0, 1]$. The logit, or log-odds, defined as $\log(\pi/(1-\pi)) = \alpha + \beta_1 x_1 + \cdots \beta_p x_p$ is one possible alternative. This is the logistic regression model and is a special case of an extension due to Nelder and Wedderburn [NeldWedd 72] called a *generalized linear model* (see also [McCuNeld 89]).

A generalized linear model is determined by three relations:

$$y|\boldsymbol{x} \sim f(\mu, \phi) \quad (42.4)$$

$$\eta = \boldsymbol{x}^T \boldsymbol{\beta} \quad (42.5)$$

$$g(\mu) = \eta \quad (42.6)$$

for explanatory variables $\boldsymbol{x} = (1, x_1, \ldots, x_p)$ and parameters $\boldsymbol{\beta} = (\alpha, \beta_1, \ldots, \beta_p)$. Now the response is a random variable whose (conditional) distribution $f(\mu, \phi)$ is known to be a member of a restricted exponential family. Instead of σ^2, this more general setup has a dispersion parameter ϕ (which may be known; see [McCuNeld 89] for further detail). The *family* $f(\cdot)$ induces a function $V(\cdot)$ such that the variance of y is $\phi V(\mu)$. However, it is no longer the conditional mean μ which is a linear function but rather the function η. The two are related through a *link function* $g(\cdot)$. The class can be extended to so-called quasi-likelihood models by not specifying $f(\cdot)$ completely but by only asserting its mean and its variance (as a function of the mean).

In the binomial proportion example, y is a binomial proportion with conditional mean $\mu = \pi$ and link function $g(\mu) = \log(\mu/(1-\mu))$. This latter quantity is modelled by some linear combination of the explanatory variables, η.

We note that η and consequently μ are functions of the explanatory variables $\boldsymbol{x}$. To emphasize this point we will sometimes write these as $\eta(\boldsymbol{x})$ and $\mu(\boldsymbol{x})$. Typically all functions $f(\cdot)$, $\eta(\boldsymbol{x})$, and $g(\mu(\boldsymbol{x}))$ are specified; interest lies in the choice of values for $\boldsymbol{\beta}$.

A different way in which the linear model (42.1) has been generalized is to write y as a sum of smooth functions $s_j(\cdot)$ affected by random disturbance

$$y = \alpha + \sum_{j=1}^{p} s_j(x_i) + \epsilon. \tag{42.7}$$

This model is called the *additive model* [HastTibs 90]. It retains the advantage of easy interpretation of the linear model, but adds the increased flexibility of some nonlinear impact of changes in the x_j's upon the value of y, with the unknown $s_j(\cdot)$'s now playing the role previously belonging to the β_j's.

The additive model and the generalized linear model are brought together in *generalized additive models* [HastTibs 90]. These use the additive term

$$\eta(\boldsymbol{x}) = \alpha + \sum_{j=1}^{p} s_j(x_j) \tag{42.8}$$

of the additive model and otherwise have identical structure to the generalized linear model.

42.3 Implementation

Our software implementation of the wide class of response models consists of software components which mimic very closely the statistical concepts they represent. The object-oriented facilities of Common Lisp [Steele 90] are used to create distinct *classes* of objects which represent the statistical concepts discussed in Section 42.2. The strength of this approach is that the statistical meaning and relationships between these different response models is preserved and enforced through the class structures and their accompanying functions. This clarifies the software for both the user and the developer.

As can be seen from the above discussion, response models are built from interpretable components. Consequently we use a variety of *model object* classes whose components mimic those of the response model. In particular, the class of a response model object is determined by its systematic component as represented by the model formula $y \sim \eta(\boldsymbol{x})$, by the specific functional forms of $\eta(\boldsymbol{x})$, by its stochastic component describing the nature of the random variation in y, and by the link function $g(\cdot)$ which joins them.

The variables involved in a response model and the functional form of $\eta(\cdot)$ are represented by *formula objects*. This describes the systematic part of the model. The formula $\eta(\boldsymbol{x})$ is directly related to a function $\mu(\boldsymbol{x})$, where $\mu(\boldsymbol{x})$ describes the property of the response which we are modelling. Stochastic variation in y for a given $\boldsymbol{x}$ is described in terms of $\mu(\boldsymbol{x})$, though perhaps differently for different classes of response model. This random component of the model, corresponding to (42.4), is represented by a *family object*. Interpretation of the model formula amounts to specifying the relationship between the systematic component, described by $\eta(\boldsymbol{x})$, and the random component, described in terms of $\mu(\boldsymbol{x})$ [McCuNeld 89]. This relationship $g(\mu(\boldsymbol{x})) = \eta(\boldsymbol{x})$ is the link function (42.6), and is represented by *link objects*.

Using these various components we create a *hierarchy* of model objects. For example, we represent in software the fact that linear models are a special case of generalized linear models by making the class linear-model a *subclass* of the class generalized-linear-model. Subclasses *inherit* properties of their superclasses, since any particular instance of the subclass is also an instance of the superclass. For example, both generalized-linear-model and linear-model have a link component (42.6), but in the case of the linear model (42.3), the link $g(\cdot)$ is restricted to be the identity.

Parallel to the hierarchy of models will be *model fits* — as distinct from *models*. Model fits, represented by *fit objects*, result from fitting a model object to a *data object* using a certain *fitting procedure*.

42.3.1 Formula objects

Response models have the characteristic that one variable, the *response* y, is separable from the remaining variables, the *explanatory variables* $\boldsymbol{x} = (x_1, \ldots, x_p)$. We take to be common among statistical models the existence for each model of a *model structure* which identifies the variables in the model and at least to some extent describes the *systematic* relationship between them. The structure of the response model, in this sense, is provided by a *model formula* $y \sim \eta(\boldsymbol{x})$.

This structure is represented by class response-formula, which has slots identifying the variables involved in the model, identifying the response variable from amongst these, and specifying a function $\eta(\cdot)$ of the explanatory variables.

It is clear from Section 42.2 that different types of function $\eta(\cdot)$ will play an important role in specifying the structure of a response model. To represent the restrictions on the response-formula for the generalized additive model, we define the subclass additive-formula, instances of which enforce the requirement that $\eta(\boldsymbol{x})$ is of form (42.8). The further subclass linear-formula represents the more specific case (42.5) used in generalized linear models, linear models, and Gaussian linear models.

42.3.2 Family objects

A class family-object represents in the abstract the *stochastic* component of the model. In order to facilitate dispatching to appropriate methods, specific distributions $f(\cdot)$ each have their own subclass of family-object with the following properties:

- The name of the family
- A variance function which produces the variance in this family for a single observation with mean μ

- A deviance function which produces the contribution to the deviance of an observation y with expected mean μ.

A top-level variable with the same name as the family for f is defined which is the sole instance of these subclasses. For example, for errors binomially distributed about the mean $\mu(\boldsymbol{x})$, there exists a subclass of family called binomial-family with name "Binomial", variance function $V(\mu) = \mu(1-\mu)$ and deviance function

$$D(\mu, y) = -2(y \log \mu + (1-y)\log(1-\mu)),$$

and a variable binomial-family bound to an instance of class binomial-family.

42.3.3 *Link objects*

Link objects follow the same strategy as family objects. Subclasses of link-object correspond to specific link functions $g(\cdot)$, have slots containing the name of the link, the link itself, and the inverse $g^{-1}(\cdot)$ of the link. The link and the link-inverse are represented by a class of function objects which stores information on derivatives and other function properties; in particular, first derivatives of the link and link-inverse, which are useful in fitting the model, are available. In the same way as for specific distributional families, there is defined a top-level variable bound to an instance of the specific subclass of link of the same name. For example, there is a link-object subclass called logit-link with name "Logit" and link function object $g(\mu) = \log(\mu/(1-\mu))$, and a variable logit-link bound to an instance of this subclass.

42.3.4 *Model objects*

The software representation of a particular response model is an object of class response-model, which possesses the slot structure. This slot represents the relationship between the explanatory variables and the response, and is restricted to contain an object which is a response-formula.

The specific examples of response models which we described earlier have more structure yet, and these will be represented in software by subclasses of response-model. Using the objects we have just outlined, the hierarchy below response-model is straightforward to describe (see Figure 1). Class generalized-additive-model has slots structure (inherited from response-model), family, and link representing the systematic, the random, and the link components of the model. For generalized-additive-model the structure slot must contain an instance of class additive-formula. Slot family must have a specific family-object subclass instance as its value; similarly for slot link.

The characteristic feature of additive models as a subclass of generalized additive models is that the link function is the identity function. The link subclass identity-link represents link $g(\mu) = \mu$, and the link slot for class additive-model always has the value identity-link. An alternative specialization of generalized additive models is to further restrict the class of formulae. By restricting the contents of slot structure to instances of class linear-formula, we create the class generalized-linear-model.

The class which has both additive-model and generalized-linear-model as superclasses is the classical linear-model, since this class inherits both the property that the link is the identity, *and* the property that $\eta(\cdot)$ is linear. The further special case of a Gaussian family gives rise to the common gaussian-linear-model.

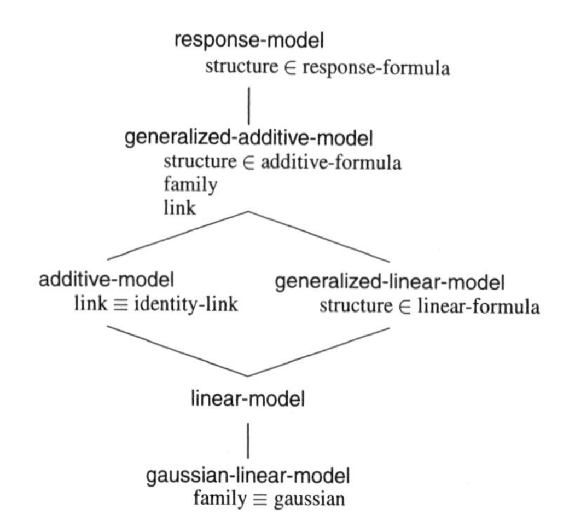

FIGURE 1: Class hierarchy of response models

The class of a model object can be used by generic functions to dispatch to the most computationally efficient methods for a particular model class. For example, the fitting procedure for a gaussian-linear-model will be more computationally efficient than for more general models. Certainly we would write a method which applies to this special case for the generic function that performs the fit. This behavior is accentuated by careful creation of instances: if a program requests an instance of a linear-model which has the gaussian-family, the appropriate gaussian-linear-model is returned; similarly, a request for a generalized-linear-model instance with the identity-link will produce a linear-model instance.

42.3.5 Fit objects

The choice of a model class for a certain problem involves careful consideration on the part of the analyst of the type and structure of the data. Utilizing a suitable class is important to valid conclusions, and in many cases the data themselves may indicate that the selected model is inappropriate. Accordingly not only model summaries but also model assessment techniques are vital tools of the data analyst.

A major benefit of interactive statistical programming environments is they simplify, and can even encourage, the iterative data analysis procedure of choosing a model class, selecting and fitting a model from within that class, assessing the model, and when necessary, selecting a different model or even a different model class.

To this end, we define in parallel to our model hierarchy a hierarchy of *model fits*, which represent the fit of a model object to a *data* object by some *fitting procedure*. The fit object contains fundamental quantities important to interpretation and assessment of the fit, and there is collection of standard mathematical and graphical devices available for this purpose which accept fit objects as input.

Residuals, and quantities derived from them, are central players in model assessment. For the response models we've discussed above, there are a number of different residual quantities

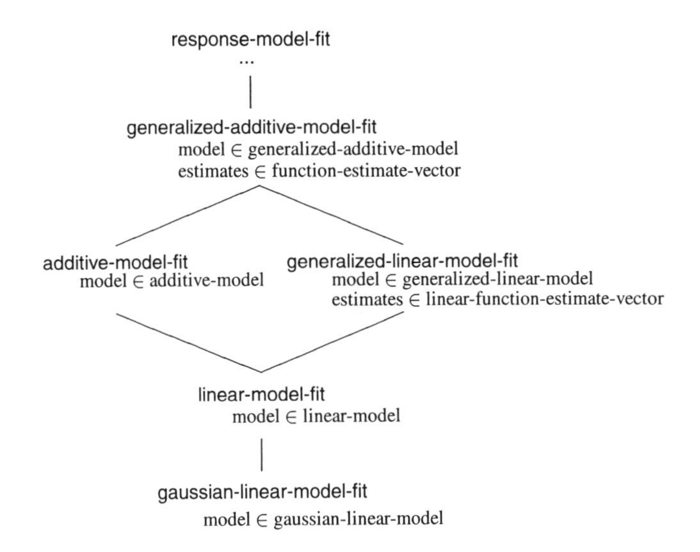

FIGURE 2: Class hierarchy of response model fits

which have been used [McCuNeld 89,PierScha 86]. The generic function residuals is capable of providing any of these from a fitted model object.

Models are often summarized by numerical quantities such as the fitted coefficients and standard errors, and residual sums of squares. Some useful numerical summaries for a given model fit are provided by the summary generic function, which returns an object of the appropriate subclass of summary-object.

There are also a variety of useful plots for model assessment within the models we've been discussing, and plots appropriate for a model under consideration can be produced from an object representing a fit of that model.

42.4 An example

Landwehr, Pregibon, and Shoemakér [LanPreSh 84] present a generalized linear model analysis of long-term survival of 306 breast cancer patients after surgery [Haberman 76]. The data consist of

- Survival, a binary variable which is 1 if the patient survived 5 or more years after surgery, 0 otherwise
- Age, the age of the patient at the time of surgery
- Year, year of the patient's surgery (minus 1900)
- Nodes, number of positive axillary nodes detected in the patient.

The response variable y_i is Survival for patient i, and we take $\boldsymbol{x}_i = (x_{1i}, x_{2i}, x_{3i})$ to be Age, Year, and Nodes for patient i, respectively. Since the response is binary, $\mu(\boldsymbol{x}) = E(y|\boldsymbol{x}) =$

probability that a patient with explanatory variable $\boldsymbol{x}$ will survive five or more years ($y = 1$). As a first step in analysis, we might consider a generalized linear model with

$$\eta(\boldsymbol{x}) = \alpha + \beta_1 x_1 + \beta_2 x_2 + \beta_3 x_3$$

with the logit link $g(\mu) = \log(\mu/(1-\mu))$. The appropriate family is binomial. Supposing that Cancer is a data structure to which our variables Survival, *etc.*, are meaningful, we can fit this as a generalized linear model by

```
(setf cancer-glm-1 (glm "Survival ~ Age + Year + Nodes"
                        Cancer
                        :family binomial-family
                        :link logit-link))
```

which returns a generalized-linear-model-fit object. Note that the keyboard expression of $\eta(\boldsymbol{x})$ omits the parameters. This is in keeping with the Wilkinson and Rogers model specification notation [WilkRoge 73]. The notation and its extensions (see [ChamHast 92]) are a convenient (and common) way to specify response models.

The above command structure hides much of the implementation. Closer examination would reveal that the function glm requests that the generic function model return a generalized-linear-model object, which object has a structure slot containing a linear-formula object appropriate to the string provided. In particular, for our example, the linear-formula object has slots identifying Survival as the response variable, and Age, Year, and Nodes as explanatory variables.

Once glm has a model-object in its possession, it passes this object, the data, and a fitting procedure (maximum likelihood, by default) to a generic function fit. Based on the formula from the model, the fit procedure requests the data it needs by name from the data object, performs the fitting procedure, and returns the appropriate fit-object. This object is subsequently returned by glm. In our example, the returned object cancer-glm-1 is of class generalized-linear-model-fit, and can be queried for the estimates $\hat{\alpha}, \hat{\beta}_1, \hat{\beta}_2, \hat{\beta}_3$ of the parameters, for values $\hat{\eta}(x)$ and $\hat{\mu}(x)$, and for the total deviance of the fitted model, the latter obtained by summing $D(\hat{\mu}(x_i), y_i)$ over all patients i.

We can use fitted model objects like cancer-glm-1 as input to summaries and plots. If we do partial residual plots for cancer-glm-1, for example, we observe that the relationship between Nodes and the partial residuals for Nodes is quite nonlinear, suggesting a transformation may be appropriate. Exploration of Nodes and the other explanatory variables yields a more complex model involving a logarithmic transformation of Nodes ([LanPreSh 84]). We could alternatively have tried using generalized additive models via the function gam.

42.5 Related work

In traditional statistical systems models have had no explicit software representation. They are defined at each call for a fit. Sometimes there is an implicit current fitted model which can be examined and changed by adding and deleting terms from its structure (*eg.* GLIM [BakeNeld 78]). In other systems, many fitted models may exist simultaneously but the model is again defined implicitly as part of the fit (*eg.* [AbraRizz 88], [BecChaWi 88]). A notable early exception to this approach is the econometric modelling system TROLL ([TROLL 82]) which in fact predates the common interactive systems mentioned above. In TROLL, models are separate data structures involving many equations which relate variables (endogenous and exogenous distinguished), parameters, and random quantities. They can be fitted by various procedures and, because the random structure is specified, once fitted these models are used to simulate future outcomes. Our approach to model representation is closer in spirit to TROLL than to the more common statistical systems.

In distinguishing model fits from the models themselves, we depart from other authors. Consider the approach taken in the book *Statistical Models in S* (*SMS*) [ChamHast 92]. In *SMS*, there is a hierarchy of model fits corresponding to ours. A fit of the appropriate class is constructed and returned by a model-specific fitting procedure, such as glm, based on arguments specifying the formula, the data, and family and link objects. However, at no point is there created a model object, *per se*, and consequently there is no model hierarchy. This of course precludes the development of procedures which operate exclusively on models (*eg.* nesting operations, combination operations, comparison operators, *etc.*). The formulae of *SMS* have class formula, but this class specializes no further. By contrast a hierarchy of formula classes plays a major role in the definition of our model hierarchy. Section 6 below suggests extensions of formula to cover other statistical models.

Various flavours of object-oriented programming have been used to build software representations for statistical concepts (*eg.* see [Oldford 87], [Pedersen 91], [Tierney 91], [ChamHast 92], [Oldford 90], and the references therein). *SMS* is the first comprehensive treatment of software representations of statistical response models and has certainly influenced our work (particularly on summaries of model-fits). Perhaps the major distinction between our implementation and that in *SMS* or [Tierney 91] is a different interpretation of object-oriented programming. In our implementation, all hierarchies of objects begin with the most general of objects, and grow downwards through progressively more specific classes. 'Downward' in this context refers to the direction of inheritance — from superclass 'down' to subclass. This is the classic approach described for example in SmallTalk-80 [GoldRobs 83]. In *SMS* and [Tierney 91], the hierarchies are reversed: conceptually more general classes inherit from the more specific ones. In particular, generalized-linear-model is a subclass of linear-model. A strong argument against this nonstandard subtyping when designing a class hierarchy can be found in [HalbO'Br 87]. Because linear models are conceptually special cases of generalized linear models, we have chosen to have linear-model appear as a dependent of generalized-linear-model. Then any instance of a linear-model behaves exactly as a generalized-linear-model should the user wish it.

42.6 Other statistical models

We think of a statistical model as a relationship between variables that involves some *stochastic* (or *random*) component(s). To capture this intentionally vague description, we define a class called model-object to be the top of our model class hierarchy. The class response-model previously discussed is a direct subclass of model-object.

The model-object class has a single slot called structure whose contents represent the known relationship between the variables. Often the defining characteristic of more specific model classes will be a more restricted class for the contents of structure. In particular, the class response-model is a subclass of model-object for which the slot structure is a formula that separates the response variable from the explanatory variables (*i.e.* a **response-formula**). But for other models **structure** might be something more complex: a graph representing the joint distribution of the variables (*eg.* [Whittake 89]); a system of differential equations for structural equations models (*eg.* [TROLL 82]); and so on.

Other statistical models fit naturally into this framework. The class of tree-based models (*eg.* [BrFrOlSt 84]), for example, would be a subclass of **response-model**. Other examples include non-linear regression [BateWatt 88], locally-weighted regression [ClevDevl 88], and multivariate adaptive regression splines [Friedman 91]. Implementation of subclasses of **model-object** which are not response models remains to be explored.

42.7 Concluding remarks

Constructing software models which match the corresponding statistical models in organization makes the software easier to understand. In particular, a coherent hierarchy of model classes naturally representing their conceptual counterparts can be used easily by other code from a variety of levels. While an interface at the level of the **glm** command of Section 42.4 is possible even when the software is not structured this way, a less superficial implementation encourages evolution of more sophisticated interaction [OldfPete 88]. One can write clear and simple procedures which operate on objects in the way in which we usually think of them. The class of individual objects in this environment in large part defines what operations may be performed on them.

Formal software representations of abstract concepts like data, model, estimation procedure, and fit permit their manipulation in novel ways. Some examples illustrate the point. A TROLL model can be estimated, decomposed, linearized, simulated, and bootstrapped [TROLL 82]. Bates and Chambers [BateCham 87] also describe a bootstrap resampling procedure in this context: given a model instance, an estimation procedure, and a set of sample data sets, the bootstrap function estimates the model for each data set in the set of sample data sets. In *SMS* there are tools for model searching which use formula objects to bound the search.

With others we have developed a statistical system called Quail which is based on matching software constructs to statistical analysis concepts. We anticipate that the opportunity for further research in this area is substantial. Quail is available via anonymous ftp from setosa.uwaterloo.ca.

42.8 References

[AbraRizz 88] Abrahams, D.M. and F. Rizzardi (1988) *BLSS: the Berkeley interactive statistical system.* W. W. Norton & Company, New York, NY.

[BakeNeld 78] Baker, R.J. and J.A. Nelder (1978) *The GLIM System, Release 3, Generalized Linear Interactive Modeling.* Numerical Algorithms Group, Oxford.

[BateCham 87] Bates, D.M. and J.M. Chambers (1987) *Statistical Models as Data Structures* AT&T Bell Labs Statistical Research Report. 6 pages.

[BateWatt 88] Bates, D.M. and D.G. Watts (1988) *Nonlinear Regression Analysis and its Applications.* John Wiley & Sons, New York, NY.

[BecChaWi 88] Becker, R. A., Chambers, J. M., and Wilks, A. R. (1988) *The New S Language: A Programming Environment for Data Analysis and Graphics.* Wadsworth & Brooks/Cole,

Pacific Grove, CA.

[BrFrOlSt 84] Breiman, L., Friedman, J.H., Olshen, R., and C.J. Stone (1984) *Classification and Regression Trees* Wadsworth International Group, Belmont, CA.

[ChamHast 92] Chambers, J.M. and T.J. Hastie (1992) *Statistical Models in S.* Wadsworth & BrooksCole, Pacific Grove, CA.

[ClevDevl 88] Cleveland, W.S. and S.J. Devlin (1988) Locally-weighted Regression: An Approach to Regression Analysis by Local Fitting. *JASA* **83**, 596-610.

[Friedman 91] Friedman, J.H. (1991) "Multivariate Adaptive Regression Splines", *Ann. Stat.* **19**, 1-141.

[GoldRobs 83] Goldberg, A. and Robson, D. (1983) *Smalltalk-80. The Language and Its Implementation*, Addison-Wesley, Reading, MA.

[Haberman 76] Haberman, S.J. (1976) "Generalized Residuals for Log-Linear Models", *Proc. 9th Intl. Biometrics Conf., Boston*, 104-122.

[HalbO'Br 87] Halbert, D. C. and O'Brien, P. D. (1987) "Using Types and Inheritance in Object-Oriented Programming," *IEEE Software*, Sept. 1987, 71-79.

[HastTibs 90] Hastie, T.J. and R.J. Tibshirani (1990) *Generalized Additive Models.* Chapman and Hall, London.

[HurlOldf 89] Hurley, C.B. and R.W. Oldford (1989) "A Software Model for Statistical Graphics," Technical Report STAT-89-13 (University of Waterloo, Department of Statistics and Actuarial Science, Waterloo, ON). Also appears in: *Statistical Computing and Graphics*, A. Buja and P.A. Tukey, eds., 77-94, Institute for Mathematics and its Applications, University of Minnesota (1991).

[LanPreSh 84] Landwehr, J.M., Pregibon, D., and A.C. Shoemaker (1984) "Graphical Methods for Assessing Logistic Regression Models", *JASA* **79**, 61-71.

[McCuNeld 89] McCullagh, P. and J.A. Nelder (1989) *Generalized Linear Models* (Second Edition). Chapman and Hall, London.

[NeldWedd 72] Nelder, J.A. and R.W.M. Wedderburn (1972) "Generalized Linear Models" *JRSS (A)* **135**, 370-384.

[Oldford 87] Oldford, R. W. (1987) "Abstract Statistical Computing," *Bulletin of the International Statistical Institute: Proceedings of the* 46th *session*, **52**, Book 4, 387-398.

[Oldford 90] Oldford, R. W. (1990) "Software Abstraction of Elements of Statistical Strategy," *Annals of Mathematics and Artificial Intelligence*, **2**, 291-308.

[OldfPete 88] Oldford, R.W. and S.C. Peters (1988) "DINDE: Towards more sophisticated software environments for statistics," *SIAM Journal on Scientific and Statistical Computing*, **9**, 191-211.

[Pedersen 91] Pedersen, J. (1991) "Situations, Summaries, and Model Objects," in *Statistical Computing and Graphics*, A. Buja and P.A. Tukey, eds., 139-185, Institute for Mathematics and its Applications, University of Minnesota (1991).

[PierScha 86] Pierce, D.A. and D.W. Schafer (1986) "Residuals in Generalized Linear Models" *JASA* **81**, 977-986.

[Steele 90] Steele, G. (1990) *Common LISP: The Language* (Second Edition). Digital Press.

[Tierney 91] Tierney, L. (1991) "Generalized Linear Models in Lisp Stat". *Technical Report No. 557*, School of Statistics, University of Minnesota.

[TROLL 82] *TROLL Documentation*. Technical Report from the Center for Computational Research in Economics and Management Science, MIT, Cambridge, Massacusetts.

[Whittake 89] Whittaker, J. (1989) *Graphical Models in Applied Multivariate Statistics*. John Wiley & Sons, Chichester, England.

[WilkRoge 73] Wilkinson, G.N. and C.E. Rogers (1973) "Symbolic Description of Factorial Models for Analysis of Variance" *Applied Statistics* **22**, 392-399.

43
Principal components and model selection

Beat E. Neuenschwander and Bernard D. Flury

Department of Mathematics,
Indiana University, Rawles Hall,
Bloomington, IN 47405, USA

ABSTRACT Let the kp-variate random vector $\mathbf{X}$ be partitioned into k subvectors $\mathbf{X}_i$ of dimension p each, and let the covariance matrix $\mathbf{\Psi}$ of $\mathbf{X}$ be partitioned analogously into submatrices $\mathbf{\Psi}_{ij}$. Based on principal component analysis we suggest a hierarchy of models, where the lowest level assumes independence of all $\mathbf{X}_i$, with identical covariance matrices, and the highest level makes no assumptions about $\mathbf{\Psi}$ beyond positive definiteness. The intermediate levels are characterized by a common orthogonal matrix $\boldsymbol{\beta}$ which diagonalizes $\mathbf{\Psi}_{ij}$ for all pairs (i, j), i.e., $\mathbf{\Psi}_{ij} = \boldsymbol{\beta}\mathbf{\Lambda}_{ij}\boldsymbol{\beta}'$, where $\mathbf{\Lambda}_{ij}$ is diagonal. The hierarchy is motivated by both a practical example and theoretical arguments. Model selection based on likelihood ratio tests and information criteria is discussed.

43.1 Introduction

A main theme of statistics is the description of real world phenomena in terms of stochastic models, which both provide good approximations and sufficient stability. An insufficient fit ("bias") may be improved by a more complex model, but this typically leads to unstable estimation ("noise") of the underlying parameters.

This trade-off between bias and noise is particularly important in multivariate statistics, since this area usually involves models with large numbers of parameters. There are well known methods in statistics, such as regression analysis, where sophisticated strategies for model selection are available.

The problem of parsimonious modeling becomes very clear in estimating covariance matrices (Dempster [6]). In fact, estimating the variances and covariances of a p-dimensional random vector $\mathbf{X}$ involves $p(p+1)/2$ parameters. This can be very troublesome if p is large. Instead of estimating all of these parameters, statisticians search for models which reflect "patterns" in the data.

Patterned covariance matrices (Szatrowski [19]) arise from multivariate data with imposed structure. A classical example is the complete symmetry model of Wilks [20], which assumes the covariance matrix of the p-dimensional random vector $\mathbf{X}$ to be of the form

$$\text{Cov}[\mathbf{X}] = \begin{pmatrix} a & b & \dots & b \\ b & a & \dots & b \\ \vdots & \vdots & \ddots & \vdots \\ b & b & \dots & a \end{pmatrix}$$

Thus, the variances are supposed to be equal, and the covariances among the p variables as well. This covariance matrix is invariant under all permutations of elements of $\mathbf{X}$.

---

[0] First author supported by a grant of the Swiss National Science Foundation

[1] *Selecting Models from Data: AI and Statistics IV.* Edited by P. Cheeseman and R.W. Oldford. ©1994 Springer-Verlag.

Whereas the complete symmetry model and many others rely on group theoretical considerations (Andersson [2]), there exist also other approaches to impose structure on covariance matrices. One of them uses the well known concept of principal components which is based on the spectral decomposition of a symmetric, positive definite matrix.

43.2 Principal components

Let $\mathbf{X}$ be a p-dimensional random vector with covariance matrix $\mathbf{\Psi}$, let $\lambda_1 \geq \lambda_2 \geq \ldots \geq \lambda_p$ be the eigenvalues, and $\boldsymbol{\beta}_1, \ldots \boldsymbol{\beta}_p$ be the corresponding eigenvectors of $\mathbf{\Psi}$. Thus, $\mathbf{\Psi}=\boldsymbol{\beta}\mathbf{\Lambda}\boldsymbol{\beta}'$, where $\mathbf{\Lambda}$ is the diagonal matrix of eigenvalues, and $\boldsymbol{\beta} = [\boldsymbol{\beta}_1, \ldots, \boldsymbol{\beta}_p]$. Traditionally, principal component analysis (PCA) has been used in three ways.

- PCA is a method of transforming correlated variables into uncorrelated ones. In fact, the principal components $U_1, \ldots, U_p$ in $\mathbf{U}=\boldsymbol{\beta}'\mathbf{X}$ are uncorrelated.
- PCA is used for finding linear combinations with relatively large or small variability. Finding the linear combination $U_1 = \mathbf{a}_1'\mathbf{X}$ with maximum variance among all vectors $\mathbf{a}_1$ of unit length leads to $\mathbf{a}_1 = \boldsymbol{\beta}_1$, and the variance is λ_1. Next, the linear combination $U_2 = \mathbf{a}_2'\mathbf{X}$ which has maximum variance and is uncorrelated with U_1 is $U_2 = \boldsymbol{\beta}_2'\mathbf{X}$, and the variance is λ_2. This process can be continued up to the last component U_p which is uncorrelated with all the others and has variance λ_p.
- PCA is a commonly used data reduction technique. If the first q $(q < p)$ principal components $U_1, \ldots, U_q$ account for most of the variability (i.e., if the last $p-q$ eigenvalues are relatively small), then the statistical analysis can be carried out on the first q principal components without losing too much information contained in the data.

Extensive introductions to principal components are given in Flury [9], Jackson [12] and Jolliffe [13].

Example:

Frets [10] reports measurements on the first and second adult son in a sample of N=25 families. The variables are head length (X_1) and head breadth (X_2). The sample covariance matrix is

$$\mathbf{S} = \begin{bmatrix} \mathbf{S}_{11} & \mathbf{S}_{12} \\ \mathbf{S}_{21} & \mathbf{S}_{22} \end{bmatrix} = \begin{pmatrix} 102.83 & 59.62 & 70.33 & 52.68 \\ 59.62 & 51.86 & 44.25 & 40.21 \\ 70.33 & 44.25 & 97.98 & 51.71 \\ 52.68 & 40.21 & 51.71 & 46.24 \end{pmatrix}$$

where $\mathbf{S}_{11}$ and $\mathbf{S}_{22}$ refer to the covariance matrix of the first and second sons respectively, and $\mathbf{S}_{12}$ denotes the covariance between the sons. Consider only the first son. Using the roman analogs $\mathbf{B}$ and $\mathbf{L}$ instead of $\boldsymbol{\beta}$ and $\mathbf{\Lambda}$ to indicate that we are dealing with observed data instead of model parameters, we obtain

$$\mathbf{S}_{11} = \mathbf{B}\mathbf{L}_{11}\mathbf{B}' = \begin{pmatrix} 0.835 & -0.551 \\ 0.551 & 0.835 \end{pmatrix} \begin{pmatrix} 142.18 & 0.00 \\ 0.00 & 12.51 \end{pmatrix} \begin{pmatrix} 0.835 & 0.551 \\ -0.551 & 0.835 \end{pmatrix}$$

Thus,

$$\begin{aligned} U_1 &= 0.835X_1 + 0.551X_2 \\ U_2 &= -0.551X_1 + 0.835X_2 \end{aligned}$$

with $Var(U_1) = 142.18, Var(U_2) = 12.51$. We see that the first principal component (measuring essentially the size of the head) accounts for most of the variability in the data, which is typical for biometrical examples. □

During the last few years, the relationship between principal components and patterned covariance matrices has been given considerable attention. The first approach using PCA for modeling covariance matrices was the common principal component (CPC) model suggested by Flury [8], [9]. The CPC model assumes equality of the eigenvectors of the covariance matrices of k multivariate, independent populations. It is a useful tool for reducing the number of parameters, whenever appropriate, and it therefore helps to improve the stability of parameter estimation.

Example (cont.):

The spectral decomposition of $\mathbf{S}_{22}$ is

$$\mathbf{S}_{22} = \mathbf{B}\mathbf{L}_{22}\mathbf{B}' = \begin{pmatrix} 0.851 & -0.526 \\ 0.526 & 0.851 \end{pmatrix} \begin{pmatrix} 129.93 & 0.00 \\ 0.00 & 14.29 \end{pmatrix} \begin{pmatrix} 0.851 & 0.526 \\ -0.526 & 0.851 \end{pmatrix}$$

Note that the orthogonal matrix $\mathbf{B}$ is almost identical with the one found in the decomposition of $\mathbf{S}_{11}$. It seems therefore appropriate to assume equality of the eigenvectors for the first and second sons. □

Since the classical CPC setup assumes independence between the populations (which is clearly violated in our example), we extend it in the following way.

Definition:

Let the pk-variate random vector $\mathbf{X}$ be partionened into k vectors of dimension p each, such that

$$\mathbf{X} = \begin{bmatrix} \mathbf{X}_1 \\ \mathbf{X}_2 \\ \vdots \\ \mathbf{X}_k \end{bmatrix}$$

and let its covariance matrix be

$$\text{Cov}[\mathbf{X}] = \mathbf{\Psi} = \begin{bmatrix} \mathbf{\Psi}_{11} & \mathbf{\Psi}_{12} & \dots & \mathbf{\Psi}_{1k} \\ \mathbf{\Psi}_{21} & \mathbf{\Psi}_{22} & \dots & \mathbf{\Psi}_{2k} \\ \vdots & \vdots & \ddots & \vdots \\ \mathbf{\Psi}_{k1} & \mathbf{\Psi}_{k2} & \dots & \mathbf{\Psi}_{kk} \end{bmatrix}$$

where each $\mathbf{\Psi}_{ij}$ has dimension $p \times p$. Then the random vector $\mathbf{X}$ fulfills the *common principal component model for dependent random vectors* if there exists an orthogonal $p \times p$ matrix $\boldsymbol{\beta}$ such that

$$\mathbf{\Psi}_{ij} = \boldsymbol{\beta}\mathbf{\Lambda}_{ij}\boldsymbol{\beta}'$$

for all pairs (i, j), where all $\mathbf{\Lambda}_{ij}$ are diagonal. □

The above example served as a motivation to study a model in which all diagonal blocks $\mathbf{\Psi}_{ii}$ of $\mathbf{\Psi}$ have common principal component structure. Why then assume that the same holds for the off-diagonal blocks $\mathbf{\Psi}_{ij}$ $(i \neq j)$? We can argue for this in two ways, using either a latent variable approach or a principle from information theory.

(i) Motivation by latent variables:

Assuming common principal components for $\mathbf{\Psi}_{11}$ and $\mathbf{\Psi}_{22}$, the CPC model for dependent random vectors can be motivated in terms of a latent variable approach (Bartholomew [3], Everitt [7]). Suppose $F_1, \ldots, F_p$ are p independent random variables (so called latent variables, not necessarily observable), and suppose the pairs of random variables (U_i, V_i) are functions of F_i,

$$U_i = h_{i1}(F_i), \quad V_i = h_{i2}(F_i), \qquad i = 1, \ldots, p$$

By the independence of the F_i, the covariance matrix of $\mathbf{U} = (U_1, \ldots, U_p)'$ and $\mathbf{V} = (V_1, \ldots, V_p)'$ is

$$\text{Cov}\begin{bmatrix} \mathbf{U} \\ \mathbf{V} \end{bmatrix} = \begin{bmatrix} \mathbf{\Lambda}_{11} & \mathbf{\Lambda}_{12} \\ \mathbf{\Lambda}_{21} & \mathbf{\Lambda}_{22} \end{bmatrix} =: \mathbf{\Lambda}$$

where the matrices $\mathbf{\Lambda}_{ij}$ are all diagonal. Assuming common principal components for $\mathbf{U}$ and $\mathbf{V}$ we finally get

$$\text{Cov}\begin{bmatrix} \boldsymbol{\beta}\mathbf{U} \\ \boldsymbol{\beta}\mathbf{V} \end{bmatrix} = \text{Cov}\left[(\mathbf{I}_2 \otimes \boldsymbol{\beta})\begin{pmatrix} \mathbf{U} \\ \mathbf{V} \end{pmatrix}\right] = (\mathbf{I}_2 \otimes \boldsymbol{\beta})\mathbf{\Lambda}(\mathbf{I}_2 \otimes \boldsymbol{\beta}') = \begin{bmatrix} \boldsymbol{\beta}\mathbf{\Lambda}_{11}\boldsymbol{\beta}' & \boldsymbol{\beta}\mathbf{\Lambda}_{12}\boldsymbol{\beta}' \\ \boldsymbol{\beta}\mathbf{\Lambda}_{21}\boldsymbol{\beta}' & \boldsymbol{\beta}\mathbf{\Lambda}_{22}\boldsymbol{\beta}' \end{bmatrix}$$

Here $\otimes$ denotes the Kronecker product. Hence, the latent variable approach suggests that $\mathbf{\Psi}_{11}$, $\mathbf{\Psi}_{22}$ and $\mathbf{\Psi}_{12}$ can be diagonalized by the same orthogonal matrix $\boldsymbol{\beta}$.

(ii) Motivation by the principle of maximum entropy:

Assume again that $\mathbf{\Psi}_{11}$ and $\mathbf{\Psi}_{22}$ have common principal components, and assume that the underlying distribution is multivariate normal. Let the singular value decomposition of $\mathbf{\Psi}_{12}$ be

$$\mathbf{\Psi}_{12} = \mathbf{\Gamma}_1\mathbf{\Lambda}_{12}\mathbf{\Gamma}_2$$

Then it can be shown that maximizing the entropy over all orthogonal matrices $\mathbf{\Gamma}_1$ and $\mathbf{\Gamma}_2$ leads to

$$\mathbf{\Gamma}_1 = \mathbf{\Gamma}_2 = \boldsymbol{\beta}$$

For a discussion of the principle of maximum entropy, see Good [11].

43.3 A hierarchy of models

We now suggest a hierarchy of models based on the CPC model defined in the previous section. The different models arise by imposing additional structure on the eigenvalues.

Level 1: The model of k independent random vectors with equal covariance matrices *(IEC)*

$$\mathbf{\Psi} = \mathbf{I}_k \otimes \mathbf{\Psi}_{11} = \begin{bmatrix} \mathbf{\Psi}_{11} & \mathbf{0} & \ldots & \mathbf{0} \\ \mathbf{0} & \mathbf{\Psi}_{11} & \ldots & \mathbf{0} \\ \vdots & \vdots & \ddots & \vdots \\ \mathbf{0} & \mathbf{0} & \ldots & \mathbf{\Psi}_{11} \end{bmatrix}$$

Level 2: the equicorrelation CPC model *(EQUI)*

$$\mathbf{\Psi} = \mathbf{E}_k(\rho) \otimes \mathbf{\Psi}_{11} = \begin{bmatrix} \mathbf{\Psi}_{11} & \rho\mathbf{\Psi}_{11} & \ldots & \rho\mathbf{\Psi}_{11} \\ \rho\mathbf{\Psi}_{11} & \mathbf{\Psi}_{11} & \ldots & \rho\mathbf{\Psi}_{11} \\ \vdots & \vdots & \ddots & \vdots \\ \rho\mathbf{\Psi}_{11} & \rho\mathbf{\Psi}_{11} & \ldots & \mathbf{\Psi}_{11} \end{bmatrix}$$

where $\mathbf{E}_k(\rho)$ is the $k \times k$ equicorrelation matrix, i.e.,

$$\mathbf{E}_k(\rho) = \begin{pmatrix} 1 & \rho & \dots & \rho \\ \rho & 1 & \dots & \rho \\ \vdots & \vdots & \ddots & \vdots \\ \rho & \rho & \dots & 1 \end{pmatrix}, \qquad \rho \in (-\frac{1}{k-1}, 1)$$

Level 3: the proportional CPC model *(PROP)*

$$\boldsymbol{\Psi} = \mathbf{R} \otimes \boldsymbol{\Psi}_{11} = \begin{bmatrix} \rho_{11}\boldsymbol{\Psi}_{11} & \rho_{12}\boldsymbol{\Psi}_{11} & \dots & \rho_{1k}\boldsymbol{\Psi}_{11} \\ \rho_{21}\boldsymbol{\Psi}_{11} & \rho_{22}\boldsymbol{\Psi}_{11} & \dots & \rho_{2k}\boldsymbol{\Psi}_{11} \\ \vdots & \vdots & \ddots & \vdots \\ \rho_{k1}\boldsymbol{\Psi}_{11} & \rho_{k2}\boldsymbol{\Psi}_{11} & \dots & \rho_{kk}\boldsymbol{\Psi}_{11} \end{bmatrix}$$

where $\mathbf{R} = (\rho_{ij})$ is a positive definite matrix of proportionality constants, and $\rho_{11} = 1$.

Level 4: the CPC model for dependent random vectors *(CPC)*, given by

$$\boldsymbol{\Psi} = (\mathbf{I}_k \otimes \boldsymbol{\beta})\boldsymbol{\Lambda}(\mathbf{I}_k \otimes \boldsymbol{\beta}') =$$

$$\begin{bmatrix} \boldsymbol{\beta} & 0 & \dots & 0 \\ 0 & \boldsymbol{\beta} & \dots & 0 \\ \vdots & \vdots & \ddots & \vdots \\ 0 & 0 & \dots & \boldsymbol{\beta} \end{bmatrix} \begin{bmatrix} \boldsymbol{\Lambda}_{11} & \boldsymbol{\Lambda}_{12} & \dots & \boldsymbol{\Lambda}_{1k} \\ \boldsymbol{\Lambda}_{21} & \boldsymbol{\Lambda}_{22} & \dots & \boldsymbol{\Lambda}_{2k} \\ \vdots & \vdots & \ddots & \vdots \\ \boldsymbol{\Lambda}_{k1} & \boldsymbol{\Lambda}_{k2} & \dots & \boldsymbol{\Lambda}_{kk} \end{bmatrix} \begin{bmatrix} \boldsymbol{\beta}' & 0 & \dots & 0 \\ 0 & \boldsymbol{\beta}' & \dots & 0 \\ \vdots & \vdots & \ddots & \vdots \\ 0 & 0 & \dots & \boldsymbol{\beta}' \end{bmatrix}$$

where all $\boldsymbol{\Lambda}_{ij}$ are diagonal. This is the model given in the above definition.

Level 5: *the full model (FULL)*, which assumes no constraints (other than positive definiteness) on $\boldsymbol{\Psi}$.

Under the assumption that $\mathbf{X}$ is kp-variate normal with mean vector $\boldsymbol{\mu}$ and covariance matrix $\boldsymbol{\Psi}$, the maximum likelihood estimators and their asymptotic properties are given in Neuenschwander [14].

43.4 Model selection

In order to select an appropriate member from a set of several competing models (like those given in the hierarchy of the previous section), usually one of the procedures described below are used.

43.4.1 Likelihood ratio tests

For testing two specific models from a hierarchy against each other, likelihood ratio tests can be constructed. The log–likelihood ratio statistic for testing level i against level j $(i < j)$ is

$$\chi^2(i|j) = 2(L_j - L_i)$$

where L_k denotes the maximum of the log-likelihood function under model k. This statistic is asymptotically distributed as chi square on $m_j - m_i$ degrees of freedom, where m_k is the number of parameters of model k (Cox and Hinkley [5]). As an immediate consequence we obtain the decomposition of the "total chi square"

$$\chi^2(1|5) = \chi^2(1|2) + \chi^2(2|3) + \chi^2(3|4) + \chi^2(4|5)$$

Model	IEC	EQUI	PROP	CPC	FULL
level (i)	1	2	3	4	5
m_i	3	4	5	7	10
L_i	-228.24	-217.33	-217.33	-215.25	-212.19
AIC	462.48	**442.66**	444.66	444.50	444.38
BIC	466.14	**447.54**	450.75	453.03	456.57

TABLE 43.1: Frets data: model selection

43.4.2 *Information criteria*

When more than two models are under consideration, model selection by information criteria is often preferred over tests of hypothesis because of the shortcomings in multiple testing. A group of information criteria based on the Kullback-Leibler information is given by

$$IC(i) = -2L_i + c_n m_i$$

where the c_n are constants depending on the sample size n only (Nishi [15]). The model which minimizes IC is considered to be the best one. Suggestions for c_n are $c_n \equiv 2$ (AIC) (Akaike [1], Sakamoto et. al [17]) and $c_n = \log n$ (BIC) (Rissanen [16], Schwarz [18]).

Example (cont.):

Table 1 displays numerical results for the Frets data, including the maximum of the log–likelihood function (L_i) and the two information criteria for all five models. For instance, if we are interested in testing whether the assumption of common principal components is reasonable, we compute the log–likelihood ratio test statistic for level 4 vs. level 5,

$$\chi^2(4|5) = 2(-212.19 - (-215.25)) = 6.12$$

with 3 degrees of freedom and a corresponding p-value of 0.106. The CPC model seems to be a reasonable model for this data set. In fact, the decomposition of the total chi square is

$$\begin{array}{ccccccccc} 32.10 & = & 21.82 & + & 0.00 & + & 4.16 & + & 6.12 \\ (7) & & (1) & & (1) & & (2) & & (3) \end{array}$$

Degrees of freedom are given in parentheses. The dominant part in the decomposition is the first one, showing that the IEC model clearly fails. The information criteria AIC and BIC both take their minimum value for level 2, the equicorrelation CPC model.

43.5 Conclusions

Good statistical modeling attempts to describe the real world in terms of models that are as simple as possible (in terms of numbers of parameters), yet describe observed data well. Finding a proper compromise between a parsimonious but possibly wrong model and a more complicated but possibly overparameterized one is particularly important in multivariate statistics, where the "curse of dimensionality" (Bellman [4]) is apparent. In normal theory models like those presented in this paper, where first and second moments determine a multivariate distribution completely, the number of parameters grows with the square of the dimension p. In other situations, like multivariate contingency tables, the number of parameters may even grow as an exponential function of the dimension p. Hence, the larger the number of variables, the more important it becomes to reduce the number of parameters by imposing proper constraints, as illustrated in the example of this article. Too often the decision which model to fit is based on convenience instead of scientific reasoning (Dempster [6]): for instance, in our example it is much easier to estimate the full model (with 10 parameters) than any of the intermediate models EQUI, PROP, or CPC, which require interative computations (Neuenschwander [14]). However, in the late 20th century computational simplicity is no longer as important as it was some thirty or fifty years ago when most of the now popular statistical techniques were developed. Traditionally, statisticians have been subject to the straightjacket of computationally simple methods, but this is no longer true. Hierarchies like the one presented in section 3 allow the statistician to choose among a variety of competing models without worrying about the computational difficulty. If expert systems are to aid human decision making, it is particularly important that compromises between parsimonious models (with few parameters) and complicated models (with many parameters) be available, such as the intermediate levels 2, 3, and 4 of our hierarchy.

43.6 References

[1] Akaike, H. (1973). "Information theory and an extension of the maximum likelihood principle", in *2nd International Symposium on Information Theory*, Petrov, B.N. and Csaki, F. , eds., Akademiai Kiado, Budapest, 267–281.

[2] Andersson, S.A. (1975). "Invariant normal models", *Ann. Stat.*, **3**, 132-154.

[3] Bartholomew, D.J. (1987). *Latent Variable Models and Factor Analysis*, Oxford University Press, Oxford.

[4] Bellman, R.A. (1961). *Adaptive Control Processes*, Princeton University Press, Princeton NJ.

[5] Cox, D.R., and Hinkley, D.V. (1974). *Theoretical Statistics*, Chapman & Hall, London.

[6] Dempster, A.P. (1972). "Covariance selection", *Biometrics*, **28**, 157-175.

[7] Everitt, B.S. (1984). *An Introduction to Latent Variable Models*, Chapman & Hall, London.

[8] Flury, B. (1984). "Common principal components in k groups", *JASA*, **79**, 892-898.

[9] Flury, B. (1988). *Common Principal Components and Related Multivariate Models*, John Wiley & Sons, New York.

[10] Frets, G.P. (1921). "Heredity of head form in man", *Genetica*, **3**, 193-384.

[11] Good, I.J. (1963). "Maximum entropy for hypothesis formulation, especially for multidimensional contingency tables", *Ann. Math. Stat.*, **34**, 911-934.

[12] Jackson, J.E. (1991). *A User's Guide to Principal Components*, John Wiley & Sons, New York.

[13] Jolliffe, J.T. (1986). *Principal Component Analysis*, Springer, New York.

[14] Neuenschwander, B.E. (1991). *Common Principal Components for Dependent Random Vectors*, Ph.D. Thesis, University of Bern, Department of Statistics.

[15] Nishi, R. (1988). "Maximum likelihood principle and model selection when the true model is unspecified", *J. Multivariate Anal.*, **27**, 392-403.

[16] Rissanen, J. (1978). "Modeling by shortest data description", *Automatica*, **14**, 465-471.

[17] Sakamoto, Y., Ishiguro, M., and Kitagawa, G. (1986). *Akaike Information Criterion Statistics*, D.Reidel Publishing Company, Dordrecht.

[18] Schwarz, G. (1978). "Estimating the dimension of a model", *Ann. Stat.*, **6**, 461-464.

[19] Szatrowski, T.H. (1985). "Patterned covariances", in *Encyclopedia of Statistical Sciences*, Kotz, S., and Johnson, N.L., eds., vol. 6., 638-641, John Wiley & Sons, New York.

[20] Wilks, S.S. (1946). "Sample criteria for testing equality of means, equality of variances, and equality of covariances in a normal multivariate distribution", *Ann. Math. Stat.*, **17**, 257-281.

Part VII

Algorithms and Tools

44
Algorithmic speedups in growing classification trees by using an additive split criterion

David Lubinsky

Department of Computer Science
University of the Witwatersrand
Johannesburg, South Africa
david@concave.cs.wits.ac.za

ABSTRACT
We propose a new split criterion to be used in building classification trees. This criterion called *weighted accuracy* or $wacc$ has the advantage that it allows the use of divide-and-conquer algorithms when minimizing the split criterion. This is useful when more complex split families, such as intervals corners and rectangles, are considered. The split criterion is derived to imitate the Gini function as closely as possible by comparing preference regions for the two functions. The $wacc$ function is evaluated in a large empirical comparison and is found to be competitive with the traditionally used functions.

Tree-based classification methods attempt to process data into rules that are useful in predicting the behavior of the system for future, as-yet-unseen cases. The data necessary consists of a set of cases, and each case has a number of measured attributes and an assigned class. An example would be a set of records of heart disease patients, where each record contains the results of a number of tests, and a diagnosis. The test results information are the measured attributes in this example, while the diagnosis is the assigned class. The goal of classification schemes is to find rules, based on such data, that can predict the class for new and unseen cases. In our example, the goal is to find a diagnosis based only on the attributes.

Traditional tree-classification methods build trees by recursive partitioning. The recursive-partitioning method divides up the attribute space into mutually-disjoint regions, each assigned a single class. The division is based on the values of the attributes and is usually found by a greedy algorithm that minimizes a split criterion at each step. Each division splits the cases into two sets, and the algorithm proceeds recursively. The leaves of the resulting tree are labeled with classes, and the tree can be used to classify new cases by following a path from the root to a leaf according to the values of the attributes of the new case.

The standard tree-growing algorithm has the following steps:

1. Set the current group to the full data set.

2. If the current group contains only cases from one class, or contains fewer than a specified number of cases, label this node with the class which has most representatives at the node and return.

3. For each variable, find the best split, using an appropriate split criterion. For numeric variables (taking values in $\Re$), the split is of the form $x > c$, and for categorical variables

[0]I would like to thank AT&T Bell Laboratories where I was working when much of this research was done.

[1]*Selecting Models from Data: AI and Statistics IV.* Edited by P. Cheeseman and R.W. Oldford. ©1994 Springer-Verlag.

```
c is the vector of classes of the n cases
sort the classes on the values of the attribute
nla = 0
nlb = 0
Best = ∞
for i from 1 to n
  if c[i] = 'a' then nla = nla + 1
    else nlb = nlb+1
  CurrentVal = splitfun(nla,nlb,na-nla,nb-nlb)
  if CurrentVal < Best then Best = CurrentVal
```

FIGURE 1: A simple sorting and scanning algorithm for finding the best cut on a single numeric attribute

(taking one of a finite set of values) the split is of the form $x \in S$.

4. Find the best split s, where s is the split that minimizes the criterion over all splits considered.

5. Use s to split the current group into two sub-groups, and recursively apply the algorithm from Step 2 to each of the sub-groups.

6. Find sub-trees of the final tree that do not appear to contribute to the performance of the tree and prune them[3].

Selecting the next split (Step 3 in the above algorithm) is the most important step in the tree-growing algorithm since this determines the final structure and predictive ability of the resulting tree. There are two components to this: the set of splits considered, and the split criterion used to pick the best among these possible splits. It is these two topics that occupy us in this paper.

44.1 The Need for an Additive Selection Criterion

In traditional tree-growing procedures, finding the best split on a numeric variables at a node involves finding a single optimal split point on each variable and selecting the best one. If there n cases, then at most $n - 1$ split points need be considered for each variable. This can be done in $O(n \log n)$ time by sorting and scanning as is shown in Figure 1. However, when more complex split families are considered, exhaustively testing each possible split may not be a viable approach with large datasets. For example, to find the split interval that minimizes the Gini criterion (see Table 44.2 for definitions of the split functions), the best method we know of is exhaustive enumeration. This is an $O(n^2)$ operation since there are $\binom{n}{2}$ possible intervals.

One approach to finding the optimal split of this and other complex split families more efficiently is to use divide-and-conquer algorithmic techniques. To do this, the split criterion we use must be *additive* in the following sense:

> Given two disjoint sets of cases, C_1 and C_2, each with associated splits S_1 and S_2, where S_1 splits the cases of C_1 into the two sets L_1 and R_1 and S_2 splits the cases

[3]This final step is needed since the complete tree will have many splits that only fit random structure in the data and hence will actually degrade the performance of the tree if not removed.

Class	No. of cases going left	No. of cases going right	Total no. of cases
a	n_{la}	$n_{ra} = n_a - n_{la}$	n_a
b	n_{lb}	$n_{rb} = n_b - n_{lb}$	n_b
Total	n_l	$n_r = n - n_l$	n

TABLE 44.1: Parameters of a split in terms of counts

of C_2 into L_2 and R_2, then the split criterion f is *additive* if

$$f(L_1 \cup L_2, R_1 \cup R_2) = f(L_1, R_1) + f(L_2, R_2).$$

With an additive split criterion, effort spent in finding the optimal split in one set is not wasted when a larger set is examined.

Assuming we have two classes, an example of an additive split criterion is a modified version of inaccuracy which we call *signed inaccuracy*. There are two, individually additive, versions of signed inaccuracy defined as follows:

$$\begin{aligned} inacc_1 &= n_{la} + n_{rb}, \text{ and} \\ inacc_2 &= n_{ra} + n_{lb}. \end{aligned}$$

The notation used in these and later definitions of split criteria is introduced in Table 44.1.

Minimizing $inacc_1$ assumes the majority of cases that go left are of class b, and the majority of cases going right are of class a. The number of misclassified cases is therefore $n_{la} + n_{rb}$. The converse case is taken care of by $inacc_2$.

Now, inaccuracy, which is defined as

$$inacc = min(n_{la}, n_{lb}) + min(n_{ra}, n_{rb})$$

also allows for the two other cases, when one class dominates on both sides of the split. This leads to the alternate form for inaccuracy

$$inacc = min(inacc_1, inacc_2, n_a, n_b).$$

We can therefore minimize $inacc$ by first minimizing $inacc_1$, then minimizing $inacc_2$ and comparing these values to the total number of cases in each class.

If, by using an additive split criterion, we can reduce the time complexity of finding the optimal split from $O(n^2)$ to $O(n \log n)$, then the double work of optimizing separately for $inacc_1$ and $inacc_2$ is well justified.

44.1.1 *Finding the Optimal Interval*

As an example of how an additive split criterion can be exploited in speeding up an algorithm, we show how $inacc_1$ and $inacc_2$ can be used to find the optimal split interval in $O(n \log n)$ time.

The method for finding the optimal interval is based on the algorithm for finding the maximal consecutive sum of a sequence, i.e. given a sequence of values $v_1 \ldots v_n$, find l, r such that $\sum_{i=l}^{r} v_i$ is maximized. This problem is discussed in [Ben80] and has a $O(n)$ solution. A slightly modified version of this algorithm, which finds the best split interval, is given in Figure 2.

To map the problem of finding the optimal interval of a numeric attribute x to maximal consecutive sum, we first order all cases by the value of x, which is an $O(n \log n)$ operation, and

```
MaxSoFar = 0
MaxEndingHere = 0
CurrentLeft = 1
for i from 1 to n
  MaxEndingHere = MaxEndingHere + Y[i]
  if MaxEndingHere < 0 then
    CurrentLeft = i + 1
    MaxEndingHere = 0
  if MaxSoFar < MaxEndingHere then
    Left = CurrentLeft
    Right = i
    MaxSoFar = MaxEndingHere
```

FIGURE 2: Algorithm for interval of greatest sum

then replace cases of class a by 1 and those of class b by -1, giving a vector y_i. The sequence of maximal sum of y_i corresponds to the interval which minimizes $inacc_2$.

This is true since if $s = \sum_{i=l}^{r} y_i$ then $s = n_{la} - n_{lb}$, and $n_{la} = n_a - n_{ra}$, so $s = -(n_{ra} + n_{lb}) + n_a$. At each node, n_a is a constant so maximizing s is equivalent to minimizing $inacc_2$. By substituting -1 for a's and 1 for b's, we can minimize $inacc_1$.

Other problems such as finding the optimal corner and rectangle splits also benefit from large speedups when using an additive split criterion [Lub94].

Finding the optimal corner cut is an $O(n^2 \log^2 n)$ operation under traditional split criteria, whereas [Lub94] contains an optimal $O(n \log n)$ algorithm using the additive split criterion. This is particularly useful, since finding the optimal corner cut is equivalent to a one level lookahead. Using the algorithm based on the additive criterion allows the lookahead to be feasibly implemented for large datasets.

44.2 Extending to Weighted Inaccuracy

In the previous section, we showed how finding the optimal interval could be reduced from an $O(n^2)$ to an $O(n \log n)$ operation by using the inaccuracy criterion rather than one of the traditional non-additive criteria. But, while minimizing inaccuracy is locally the best criterion for improving accuracy, it is well known that inaccuracy is not a good selection criterion for use in generating splits for growing classification trees. This limitation was first mentioned in [MM72].

The problem with inaccuracy arises when one class is much more prevalent than the other. For example, in Figure 3, there are eight x's and three o's indicating the points from the two classes. Any cut has inaccuracy of three, misclassifying all the o's. But, one of the cuts at S_1 or S_2 would be the best, since this could then be followed by the other leading to perfect classification. The cut at S_1 does in fact minimize both Gini and entropy measures.

What we would like to find is an additive criterion which still selects cuts such as S_1. Assume we have a split in which $n_{la} > n_{lb}$ and $n_{rb} > n_{ra}$, we can get some insight into why the Gini

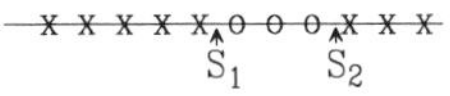

x x x x x o o o x x x
S_1 S_2

FIGURE 3: A case in which the value of the inaccuracy criterion is constant for each cut

function selects cuts such as S_1 by arranging the terms of the function as follows:

$$gini = \frac{n_{la}}{n_{la}+n_{lb}} n_{lb} + \frac{n_{rb}}{n_{ra}+n_{rb}} n_{ra}.$$

Writing the function this way shows that the Gini function is a type of weighted inaccuracy function, since n_{lb} and n_{ra} are the two inaccuracies for this split, and the weights are equal to the proportion of the dominant class in the set. If there is a large proportion of the dominant class then misclassifications tend to get weighted more heavily. The smallest weight of $1/2$ is achieved when each class is equally represented. This accounts for Gini's propensity for choosing large pure[4] splits which makes it such a successful split criterion. The reason that Gini is not additive is that with each considered split, the proportions $n_{la}/(n_{la}+n_{lb})$ and $n_{ra}/(n_{ra}+n_{rb})$ change.

To attempt to mimic the behavior of the Gini function, we can define a weighted inaccuracy $wacc$, as

$$wacc = min(kn_{la}, n_{lb}) + min(kn_{ra}, n_{rb}),$$

for any constant $k > 0$, where k can be viewed as the ratio of cost of misclassification of type a to type b. The two corresponding additive sub-functions, $wacc_1$ and $wacc_2$, are defined as

$$wacc_1 = kn_{la} + n_{rb}$$

and

$$wacc_2 = kn_{ra} + n_{rb}.$$

So $wacc$ can be written as

$$wacc = min(wacc_1, wacc_2, kn_a, n_b).$$

In order to maintain additivity, k must be fixed when selecting among splits at a node. The next section discusses a method of finding a good value for k by imitating $Gini$ as closely as possible.

44.3 Definition of Slope

Given $p_a = n_a/n$, the proportion of cases from class A, and $p_b = 1 - p_a$, the proportion from class B, let $l_a = n_{la}/n$ and $l_b = n_{lb}/n$ be the proportion of cases of each class which *"go left"*. A split can be defined by three of these parameters, since $p_a + p_b = 1$. The choice of which three is arbitrary, but we use p_a, l_a, l_b.

44.3.1 Comparison of Gini, Accuracy and Entropy

Table 44.2 defines the three selection criteria we are interested in, in terms of the above parameters. Since a selection criterion is just a way of choosing among alternate splits, we can compare

[4]A split is pure if one of its branches is composed of cases from only one class.

$$
\begin{aligned}
gini(l_a, l_b, p_a) &= \frac{l_a l_b}{l_a + l_b} + \frac{(p_a - l_a)((1 - p_a) - b1)}{1 - l_a - l_b} \\
wacc(l_a, l_b, p_a, k) &= min(kl_a, l_b) + min(k(p_a - l_a), (1 - p_a - l_b)) \\
entropy(l_a, l_b, p_a) &= -[(E(l_a) + E(l_b) + E(p_a - l_a) + E(1 - p_a - l_b)) - \\
&\quad (E(l_a + l_b) + E(1 - l_a - l_b)) - (E(p_a) + E(1 - p_a))] \\
\text{where } E(x) &= x \log(x)
\end{aligned}
$$

TABLE 44.2: Definition of selection criteria

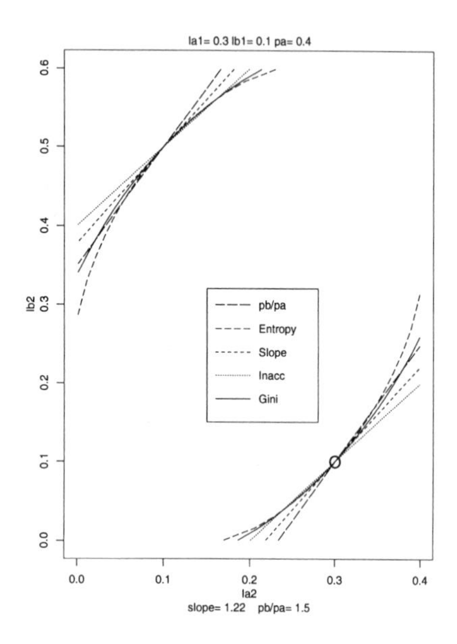

FIGURE 4: The equipreference curves under the 5 selection criteria.

selection criteria by examining the regions of equal preference.

Given p_a and two splits, S_1 and S_2, where S_i is defined by the parameters, l_{ai}, l_{b2}. If we hold l_{a1} and l_{b1} constant, we can find the curve of points for which

$$gini(l_{a1}, l_{b1}, p_a) = gini(l_{a2}, l_{b2}, p_a),$$

by solving for l_{b2} in terms of l_{a2}. There are two solutions each of which is symmetric about the line $y = x$. Plotting these two functions in the plane defined by l_{b1} and l_{b2}, we find that they define three regions. The region between the two curves consists of all splits which are not preferred to S_1, while the two regions outside the curves define splits preferred to S_1.

Figure 4 shows an example where $p_a = 0.4, l_{a1} = 0.3, l_{b1} = 0.1$. The curves are shown for $Gini$, $entropy$, $inaccuracy$ and $wacc$ for a particular value of k that is defined below.

The point labeled 'O', corresponds to $l_{a2} = l_{a1}$ and $l_{b2} = l_{b1}$. Naturally, all equipreference curves pass through this point.

It is interesting to observe how similar the curves for Gini and Entropy are. This explains the empirical result [Min89] that their performance is very similar.

The equi-preference curve for weighted inaccuracy is a straight line. By varying the weight in the definition of $wacc$, we can get a range of lines. To get an additive split criterion which most closely mimics the choices made by Gini, we would like to pick the value of the slope which mostly closely follows the Gini curve. To do this, we define the *slope* of the Gini curve as the slope of the line joining the point where the lower Gini curve meets the x axis, and the point where the curve meets the $x = p_a$ line. This approximates the slope of the lower curve. The slope of the upper curve is the same as the slope of the lower curve.

The function of the lower curve is:

$$sol(l_{a2}) = \frac{-g_1 + 2\,l_{a2}\,g_1 - p_a p_b + \sqrt{g_1 + p_a p_b}\,\sqrt{4\,{l_{a2}}^2 + g_1 - 4\,l_{a2}\,p_a + p_a p_b}}{2\,(-g_1 + p_a)}$$

where g_1 is the value of the Gini criterion for the split S_1.

The function $sol(l_{a2})$ intersect the $l_{b2} = 0$ axis at

$$\alpha = \frac{g_1 - p_a + p_a^2}{g_1 + p_a - 1}$$

and intersects the line $l_{a2} = p_a$ at

$$\beta = \frac{g_1 p_a}{p_a - g_1}.$$

So the *slope* defined as $\beta/(p_a - \alpha)$ is

$$slope = \frac{p_a(g_1 + p_a - 1)}{(p_a - 1)(p_a - g_1)}.$$

This definition of *slope* depends, through g_1, on l_{a1} and l_{b1} as well as on p_a, so it cannot be used in an additive split criterion. In the next section, we show how to pick a value of k based only on p_a in order to recover the additivity of the split function.

44.4 Approximating slope by worst case

The slope defined above is different for each set of values $< l_{a1}, l_{b1}, p_a >$, all of which were held constant in the above analysis. Surface plots of the slope for different values of p_a reveal that each surface is a smoothly varying sail with a ridge down the center. For $p_a < 0.5$, the sail bends up, and for $p_a > 0.5$, the sail bends down, for $p_a = 0.5$ the surface is a constant.

Since we must choose a value of the slope based on p_a alone, we examine those cases where inaccuracy differs most from Gini. This is along the ridge of the sail which corresponds to splits of highest inaccuracy. The condition for being on the ridge is that $l_{a1} = p_a/2$ and $l_{b1} = p_b/2$. Substituting these values in the definition of *slope* above, we find that, the selected slope is

$$k = p_b/p_a.$$

This is a surprisingly simple and natural choice, weighting the a's by their proportion. So, for example, using the values from Figure 3, $k = 3/8$ since there are 8 a's and 3 b's. The weighted inaccuracy measure in this case is:

$$wacc = min(3/8 n_{la}, n_{lb}) + min(3/8 n_{ra}, n_{rb}).$$

The *wacc* score for split S_1 in Figure 3 is $min(15/8, 0) + min(9/8, 3) = 9/8$ and for split S_2 it is $min(15/8, 3) + min(9/8, 0) = 15/8$, so split S_1 is preferred to S_2, and is in fact the optimal split under *wacc*.

44.5 Empirical Evaluation of $Wacc$ in growing trees

We evaluate *wacc* by comparing it to several other split criteria in a reimplimentation of a traditional tree growing algorithm with pessimistic pruning [Qui86]. The performance of the split criteria – in terms of predictive accuracy – was tested in a series of cross validation runs, on seven datasets taken from the UCI repository (see [Sta,For90,RDF,Ger,Ces,ZS,Nat90]), and compared by analysis of variance.

The means in Figure 5 show the percentage accuracy when using seven split criteria in growing trees for seven datasets. Each mean is the mean of ten tenfold cross-validation runs. Each cross-validation run uses a different random seed in creating the partition.

The two functions *acc.gini* and *wacc.gini* are split functions which minimize *inaccuracy* and *wacc* respectively, but use *Gini* as a tie breaker among equally good splits, since these functions will often have many splits which achieve the same minimum. The random splits are included to show how random partitioning performs.

In each column, the split functions are ordered in decreasing accuracy. The letters to the left of each column are the results of a Duncan multiple-range test for each data set. Split functions which have the same letter are not significantly different. So, for example, in the Bupa data, *wacc*, *wacc.gini*, *gini* and *entropy* form a group among which there are no significant differences. There are also no significant differences among *inacc*, *random* and *acc.gini*, but all of this second group are significantly worse (lower accuracy) than those in the first group.

The default accuracy shown at the last line of the table is the proportion of the majority class in each dataset. The difference between this and the predictive accuracies shown in the table is an indication of how much *classification signal* there is in each data set. Note that there is very little signal in the hepatitis dataset. Where the maximum improvement over default is about 3%, while some split criteria actually do worse than the default accuracy. In each of the other cases there is an improvement of at least 10%. Of theses six data sets, either *wacc* or *wacc.gini* is the best criterion in five cases. In the diabetes data, *wacc* and *wacc.gini* actually form a group which significantly better than the rest of the criteria. While, *wacc* and *wacc.gini* do not dominate the other criteria, the experiment does indicate that these functions are certainly competitive with others used in the literature and *wacc.gini* should perhaps be considered as the function of choice due to its overall superior performance.

44.6 REFERENCES

[Ben80] John Bentley. *Programming Pearls*. ACM, 1980.

[Ces] Bojan Cestnik. Hepatitis data. Jozef Stefan Institute, Jamova 39, 61000 Ljubljana, Yugoslavia. From the UCI Machine Learning repository.

[For90] Richard S. Forsyth. Bupa liver disorders. 8 Grosvenor Avenue, Mapperley Park, Nottingham NG3 5DX, 0602-621676, 1990. From the UCI Machine Learning repository.

[Ger] B German. Glass data. Central Research Establishment, Home Office Forensic Science

	Biomed			Bupa			Cleve			Glass	
	Mean	SPLITFUN		Mean	SPLITFUN		Mean	SPLITFUN		Mean	SPLITFUN
A	86.053	Inacc	A	67.181	Wacc	A	78.330	Wacc.gini	A	81.949	Wacc.gini
A			A			A			A		
A	85.585	Inacc.gini	A	66.767	Wacc.gini	A	77.483	Entropy	A	81.740	Wacc
A			A			A			A		
A	85.251	Gini	A	65.787	Gini	BA	76.847	Gini	A	81.107	Entropy
A			A			B			A		
BA	85.056	Entropy	A	65.256	Entropy	BC	75.257	Inacc.gini	BA	79.812	Gini
BA						C			BA		
BA	84.585	Wacc.gini	B	62.289	Inacc	DC	74.230	Inacc	BA	79.444	Inacc.gini
BA			B			D			B		
BA	84.529	Wacc	B	62.092	Random	D	72.830	Wacc	B	77.757	Inacc
B			B								
B	82.997	Random	B	61.878	Inacc.gini	E	66.084	Random	C	69.714	Random
Default Accuracy	65.5			58			54			53	

	Hepatitis			Lymphography			Diabetes	
	Mean	SPLITFUN		Mean	SPLITFUN		Mean	SPLITFUN
A	81.829	Entropy	A	79.667	Wacc	A	75.3769	Wacc.gini
A			A			A		
BA	81.463	Wacc	A	79.667	Gini	A	75.3192	Wacc
BA			A					
BAC	80.162	Inacc	A	79.519	Inacc.gini	B	73.8718	Gini
B C			A			B		
B C	78.801	Inacc.gini	A	79.365	Wacc.gini	B	73.8705	Entropy
B C			A					
B C	78.769	Wacc.gini	A	78.868	Inacc	C	72.4567	Inacc
C			A					
C	78.301	Random	A	78.593	Entropy	D	70.8478	Inacc.gini
C						D		
C	78.292	Gini	B	71.466	Random	D	69.7162	Random
Default Accuracy	79			57			65	

Key

Inacc - Inaccuracy.
Inacc.Gini - Inacuracy with Gini as tie-breaker = Inacc + Gini/n
Wacc - Weighted accuracy
Wacc.gini - Wacc + Gini/(n*max(pa/pb,pb/pa))
Entropy - As defined in text.
Gini - As defined in text
Random - Split chosen at random.

FIGURE 5: Predictive accuracy for various split functions, each mean is the mean of ten, tenfold cross-validation runs.

Service, Aldermaston, Reading, Berkshire RG7 4PN. From the UCI Machine Learning repository.

[Lub94] David J. Lubinsky. Bivariate splits and consistent split criteria in dichotomous classification trees. PhD Thesis, Department of Computer Science, Rutgers University, 1994.

[Min89] John Mingers. An empirical comparison of selection measures for decision-tree induction. *Machine Learning*, 3:319–342, 1989.

[MM72] Robert Messenger and Lewis Mandell. A modal search technique for predictive nominal scale multivariate analysis. *Journal of the American Statistical Association*, 67:768–772, 1972.

[Nat90] National Institute of Diabetes and Digestive and Kidney Diseases. Pima indians diabetes data. From the UCI Machine Learning repository, 1990.

[Qui86] J.R. Quinlan. Induction of decision trees. *Machine Learning*, 1(1):81–106, 1986.

[RDF] Long Beach Robert Detrano, V.A. Medical Center and Cleveland Clinic Foundation. Heart disease database. From the UCI Machine Learning repository.

[Sta] Statlib. Liver disease diagnosis. From CMU statistics library.

[ZS] M. Zwitter and M. Soklic. Lymphography data. University Medical Centre, Institute of Oncology, Ljubljana, Yugoslavia. From the UCI Machine Learning repository.

45
Markov Chain Monte Carlo Methods for Hierarchical Bayesian Expert Systems

Jeremy C. York and David Madigan

Dept of Statistics
Carnegie Mellon University
and
Dept of Statistics
University of Washington

ABSTRACT In a hierarchical Bayesian expert system, the probabilities relating the variables are not known precisely; rather, imprecise knowledge of these probabilities is described by placing prior distributions on them. After obtaining data, one would like to update those distributions to reflect the new information gained; however, this can prove difficult computationally if the observed data are incomplete. This paper describes a way around these difficulties – use of Markov chain Monte Carlo methods.

45.1 Introduction

In a probabilistic expert system, as described in [Pearl 88] and [LaurSpie 88], relationships between variables are quantified via probabilities elicited from an expert. However, in that framework, those probabilities are fixed quantities, and the system is unable to update itself to include empirical information from observed cases. Models for fully Bayesian expert systems were introduced in [SpieLaur 90] (hereafter referred to as SL) and [DawiLaur 89,DawiLaur 93], in which the probabilities that relate the variables are themselves distributed according to some prior law. These frameworks allow for updating of the system to include observed data via Bayes rule. Another advantage of the approach is that the uncertainty associated with any quantity of interest can be assessed via manipulation of the posterior distribution.

However, computational difficulties arise when many of the variables are not observed in the data. In expert systems applications, such missing dimensions are the rule and not the exception. SL and [SpieCowe 92] (hereafter referred to as SC) consider some approximations to circumvent these difficulties. This paper will present an alternate Monte Carlo approach, and discuss its advantages and weaknesses. Since this alternate approach can yield very precise estimates, it allows us to check the accuracy of the approximations used by SC.

[York 92a] reviews the use of the Gibbs sampler in standard, non-hierarchical expert systems; more ambitious discussions of the Gibbs sampler and related Markov chain Monte Carlo (MCMC) methods can be found in [Geyer 92,SmitRobe 93,Tierney 91a,Tierney 91b]. This paper will discuss how MCMC methods can be applied to the hierarchical models of SL. Although it isn't explicitly discussed, under slight changes in notation, all of the following discussion could be applied to the hyper-Dirichlet models of [DawiLaur 93] as well. Examples of the use

[1] *Selecting Models from Data: AI and Statistics IV.* Edited by P. Cheeseman and R.W. Oldford.

of MCMC in a similar context can be found in [MadiYork 93,YorMadHeuLie 93,York 92b].

45.2 Bayesian Expert Systems Models with Complete Data

SL introduce a class of models where a set of variables X is Markov with respect to a directed graph (see [LauDawLarLei 90]); the collection of conditional probabilities θ which defines the probability distribution of X is itself given a probability distribution (to avoid confusion, a probability distribution over θ will be called a *law*). The prior law of θ is defined by putting independent beta or Dirichlet distributions on the conditional probability of each node given a configuration of its parents; $\theta_{x_i|x_{pa(i)}}$ gives the probability that $X_i = x_i$ given that $X_{pa(i)} = x_{pa(i)}$ (here, x_i is a possible value of the node X_i, and $x_{pa(i)}$ is a possible configuration of the set of parents of X_i).

There are two different levels of independence assumptions in the model of SL. The first, called *local independence*, is the assumption that $\theta_{x_i|x_{pa(i)}}$ is independent of $\theta_{x_i|x'_{pa(i)}}$, where $x_{pa(i)}$ and $x'_{pa(i)}$ are different configurations of the parents of X_i. If we write

$$\theta_i = \{\theta x_i | x_{pa(i)} : \text{ all values of } x_i, x_{pa(i)}\}$$

for the set of probabilities of X_i given any state of its parents, the *global independence* assumption is that for any $j \neq i$, all members of θ_i are independent of all members of θ_j.

SL demonstrate that if the data Y are complete, the posterior law of θ is of the same form as the prior law, with the prior parameters updated by data counts. Also, if a prediction is to be made about a future case X, one can write

$$P(X = x \mid Y) = \mathbf{E}(P(X = x \mid \theta)|Y) = \prod_i \mathbf{E}(\theta_{x_i|x_{pa(i)}} \mid Y). \tag{45.1}$$

that is, the probability of a future case is a product of marginal posterior expectations. This happens because the both the quantity $P(X \mid \theta)$ and the posterior law of θ factor in the same way. The global and local independence assumptions are thus crucial for this ease of calculation.

We note, however, that applying algorithms to shunt probabilities around a graph will not necessarily provide the correct results if the marginal expectations are used. Consider two events concerning a future case X; call them A and B, and assume that B contains more than one possible configuration of X. If we knew θ, the conditional probability that X falls in A, given that it falls in B, is $P(A \cap B|\theta)/P(B|\theta)$. If we take the expectation of this quantity with respect to the posterior law of θ, the expectation of a ratio of dependent random variables is not the same as the ratio of the marginal expectations.

However, features of the posterior law of a function of θ can be readily obtained via a standard Monte Carlo simulation. If iid samples of θ are drawn from its posterior law, the function can be evaluated for each simulated θ, and the resulting sample can be used to estimate features of the posterior of the function.

45.3 Bayesian Expert System Models with Incomplete Data

If Z represents the unobserved or missing data, the posterior of θ is given by

$$P(\theta|Y) = \sum_Z P(\theta|Y, Z)P(Z|Y).$$

If there are nodes in Y whose parents where unobserved, then the posterior (which is a mixture distribution) will not usually satisfy the independence assumptions above.

SL approximate this posterior with a law which does obey the global and local independence assumptions. They do so by looking at the posterior law of one particular parameter, which can be written

$$P(\theta_{x(i)|x_{pa(i)}}|Y) = \sum_{Z_{i,pa(i)}} P(\theta_{x(i)|x_{pa(i)}}|Y_{i,pa(i)}, Z_{i,pa(i)})P(Z_{i,pa(i)}|Y_{i,pa(i)})$$

since any element of θ_i will be independent of data on variables other than those in $X_{i,pa(i)} = \{X_i \cup X_{pa(i)}\}$. Note that each marginal posterior of $\theta_{x(i)|x_{pa(i)}}$ given complete data will be beta or Dirichlet, so that marginal posterior given the observed data is a mixture of such laws. SC investigate approximating this marginal posterior law with a beta or Dirichlet distribution with parameters chosen to give the same mean and variance as the mixture. The approximate posterior of θ is given by performing this approximation for all components of θ, and then making the local and global independence assumptions. Predictions about future observations can then be made as discussed in the previous section, and new data can be used to further update the posterior law.

It should be noted that SC only discuss a sequential updating scheme, where data are processed one case at a time. SL note that more elaborate schemes may be possible, where one would keep the current posterior law of θ a mixture with a manageable number of components. New data cases with unobserved values will increase the number of components in the mixture, and so it would be periodically collapsed onto a distribution with a smaller number of components. We only consider the former scheme here, as the coding for the latter approach is fairly difficult.

45.4 Markov Chain Monte Carlo Methods

MCMC methods have been successfully used to overcome problems caused by missing data when using small networks for conventional statistical problems in [MadiYork 93, York 92b, YorMadHeuLie 93]. This section will briefly summarize the approach.

All MCMC methods are ways to produce a stochastic process which has a desired distribution as its stationary distribution. The theory of stochastic processes tells us that the empirical average of a function of the stochastic process will converge to the expectation of that function under the desired distribution; see for example [Tierney 91a, Tierney 91b]. The Gibbs sampler is a particular kind of MCMC scheme; more general schemes are given by [Hastings 70].

In order to approximate $P(X \mid Y)$, or some other function of the system, we need to take an expectation with respect to the unobserved data Z and the parameters θ. The Gibbs sampler does this as follows. First, initialize $Z(0)$ and $\theta(0)$ to some values which are not inconsistent with the observed data Y. For each iteration t, simulate $Z(t)$ from the distribution $P(Z \mid Y, \theta(t-1))$, and simulate $\theta(t)$ from $P(\theta \mid Y, Z(t))$. Continue in this manner, for some large number of iterations N. Specific comments on performing the simulations are provided in the next section.

The following is an estimate of the posterior expectation of any function $f(Z, \theta)$:

$$\hat{f} = \frac{1}{N} \sum_{t=1}^{N} f(Z(t), \theta(t)).$$

Under mild conditions, $\hat{f}$ will converge to $\mathbf{E}(f(Z, \theta) \mid Y))$ with probability one as the Monte Carlo sample size N approaches infinity.

A function f that could be used to estimate $P(X)$ for some future case X is

$$\begin{aligned} f(Z,\theta) &= P(X \mid Y, Z) \\ &= \prod_i \mathbf{E}(\theta_{x_i|x_{pa(i)}} \mid Y, Z) \end{aligned} \tag{45.2}$$

Note that the simulated values of θ are not used in expression (45.2); if the sampling were independent, this would reduce the Monte Carlo variability of the estimate [GelfSmit 91]; this is likely to also be the case in the Markov chain sampling scheme used here. Such an estimate takes advantage of the result of SL that, in the complete data case, predictions can be made using the expected value of θ. The same caution as before applies, however; if the quantity of interest depends upon multiple components of θ and cannot be expressed as a product of functions of those components, using the product of the marginal expectations of those components is not guaranteed to yield correct results.

Other quantities, such as marginal probabilities, posterior variances, cumulative distributions (and thus percentiles) and so on can be estimated in a similar fashion. All that needs to be done is identify a function $f(Z,\theta)$ which has the desired expectation under the posterior.

Additionally, one can derive the posterior distribution of some component $\theta_{i|pa(i)}$ by averaging

$$P(\theta_{i|pa(i)} \mid Y, Z(t))$$

for any specified value of $\theta_{i|pa(i)}$ [WeiTann 90,GeyeTier 92].

45.5 Applying MCMC

In order to implement MCMC, one must be able to simulate from the distribution $P(Z|\theta, Y)$ and from $P(\theta|Y, Z)$. We will first address details of simulating the unobservable data.

If all of the cases in Y have the same pattern of missing data (as is the case in [YorMadHeuLie 93]), simulation of $Z(t)$ can be done by simulating from a multinomial, with probabilities derived from $\theta(t-1)$. This approach will still be useful if a number of cases have common patterns of missing data. Data cases which share no pattern of missing variables with other cases must be handled individually; this situation will not be considered further here.

We first identify a group of neighboring missing nodes U and a collection of observed nodes S which separate the missing nodes from the rest of the graph (we define adjacency, separation, etc to be with respect to the moralized, undirected version of the network). Conditional upon a given configuration of the variables in S, the joint probability distribution $\phi(\theta)$ of the unobserved nodes can be derived (either by brute force or by more efficient algorithms), and the missing data will follow a multinomial distribution with cell probabilities given by $\phi(\theta)$ and number of trials given by the number of data cases with the given configuration of the variables in S; this must be repeated for all such configurations. At every iteration of the simulation, Z will be simulated conditional upon a different value of θ, so it is most efficient to derive the form of $\phi(\theta)$ symbolically.

Together, Y and $Z(t)$ comprise complete data, and so the components of θ given Y and $Z(t)$ have independent beta or Dirichlet laws, with prior parameters incremented by the data counts. Routines for simulating from these distributions can be found in [Devroye 86,Ripley 87]. See [YorkMadi 92] for comments on how one might simulate from the hyper-Dirichlet laws used as priors in the undirected case.

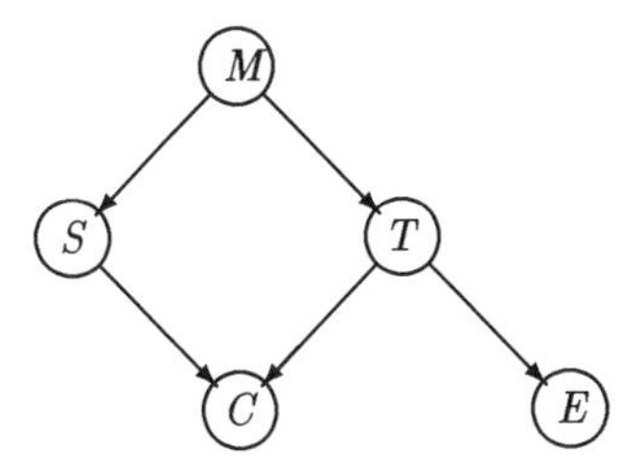

FIGURE 1: A simple Bayesian network

If the system will have to compute estimates of certain quantities, the nature of which are not specified in advance, complete records of $\theta(t)$ and $Z(t)$ could be kept, and estimates could be calculated from those records. This may not be feasible in real time; if, however, the quantities of interest were known in advance and a $f_j(Z(t), \theta(t))$ specified for each one as in the previous section, values of $f_j(Z(t), \theta(t))$ could be calculated as the simulation runs, eliminating the need to store complete records.

45.6 Example

We illustrate the competing approximations with an example loosely borrowed from [Cooper 84], which has been modified and used as an illustration in a number of papers since then. The variables are :

- M : presence of metastatic cancer
- S : positive result on a serum calcium test
- T : presence of brain tumor
- C : subject is in a coma
- E : presence of Papilledema, an abnormal eye behavior

Other authors have used chronic headaches in place of the eye behavior, but the original author has suggested that this is a preferable illustration because chronic headaches and comas cannot co-exist [Cooper 92]. The graphical model is shown in figure 1.

We have considered the case where both M and T are missing from most of the data cases. For a prior, we use either one with marginals centered at values used in [Pearl 88] (namely, $P(M) = .2$, $P(S|M) = .8$, $P(S|\neg M) = .2$, $P(T|M) = .2$, $P(T|\neg M) = .05$, $P(C|S,T) = P(C|S,\neg T) = P(C|\neg S,T) = .8$, $P(C|\neg S,\neg T) = .05$, $P(E|T) = .8$, $P(E|T) = .6$), or a non-informative prior where each component of θ has mean 0.5. Data were simulated from a distribution that was slightly different from that given by Pearl (see Table 45.1); then, the approximation techniques of SC and the MCMC approach were applied.

Since the MCMC method produces a precise estimate of the posterior, we are able to examine both the behavior of the posterior and the accuracy of the SC approximation. Several things can

θ_M	$\theta_{S\|\neg M}$	$\theta_{S\|M}$	$\theta_{T\|\neg M}$	$\theta_{T\|M}$	$\theta_{E\|\neg T}$	$\theta_{E\|T}$	$P(M\|E)$	$P(T\|E)$
0.500	**0.100**	**0.750**	**0.100**	**0.600**	**0.750**	**0.850**	**0.516**	**0.379**

TABLE 45.1: True probabilities used in simulating data. The last two quantities are functions of the θ values. Because of space constraints, components of θ_C are not given above; the values used to produce the data are .1, .8, .8, and .9 respectively when conditioning on $(\neg S, \neg T)$, $(\neg S, T)$, $(S, \neg T)$ and (S, T).

		θ_M	$\theta_{S\|\neg M}$	$\theta_{S\|M}$	$\theta_{T\|\neg M}$	$\theta_{T\|M}$	$\theta_{E\|\neg T}$	$\theta_{E\|T}$	$P(M\|E)$	$P(T\|E)$
MC	**mn**	**0.44**	**0.13**	**0.77**	**0.15**	**0.58**	**0.74**	**0.81**	**0.45**	**0.36**
	s.d.	0.04	0.04	0.05	0.03	0.07	0.02	0.03	0.04	0.04
SC1	**mn**	**0.44**	**0.11**	**0.75**	**0.15**	**0.58**	**0.74**	**0.82**	**0.45**	**0.36**
	s.d.	0.02	0.02	0.03	0.02	0.06	0.02	0.03	0.02	0.03
SC2	**mn**	**0.38**	**0.17**	**0.83**	**0.21**	**0.60**	**0.74**	**0.81**	**0.39**	**0.38**
	s.d.	0.02	0.02	0.03	0.02	0.06	0.02	0.03	0.02	0.03
SC3	**mn**	**0.39**	**0.17**	**0.77**	**0.20**	**0.54**	**0.76**	**0.80**	**0.40**	**0.34**
	s.d.	0.02	0.02	0.04	0.03	0.06	0.02	0.03	0.02	0.03

TABLE 45.2: Posterior means and standard deviations for simulated 1000 simulated data cases, of which only 100 were complete. Prior precision (the sum of the parameters of each beta distribution) was 10; prior means match those of Pearl above. The MCMC estimates are based on a simulation of 25,000 iterations. SC 1, 2 and 3 are estimates generated using the method of SC, with the difference among them being the order in which the data are processed. In the first, the complete cases were processed first. In the second and third, the data cases were processed in random order. The last two columns give posterior means and standard deviations for two quantities of interest; the SC estimates for these were given by 5000 conventional Monte Carlo simulations from their approximation to the posterior.

be seen from this simulation. First is that the SC method is fairly unstable with respect to the order in which the data are processed. Also, the SC method underestimates the posterior variability of some components by approximately 30-50%, particularly θ_M and θ_S. Both methods are able to get reasonable estimates of the true underlying probabilities, which is somewhat surprising in the case of θ_T, for which very little information is available.

In the above calculations, the MCMC approach can estimate correlations between various components of θ. We note that neither the local independence assumption holds, even approximately (the correlation between $\theta_{C|S,\neg T}$ and $\theta_{C|S,T}$ is -.64; other such pairs had correlations

		θ_M	$\theta_{S\|\neg M}$	$\theta_{S\|M}$	$\theta_{T\|\neg M}$	$\theta_{T\|M}$	$\theta_{E\|\neg T}$	$\theta_{E\|T}$	$P(M\|E)$	$P(T\|E)$
MC	**mn**	**0.48**	**0.12**	**0.72**	**0.11**	**0.59**	**0.74**	**0.83**	**0.49**	**0.37**
	s.d.	0.04	0.04	0.05	0.04	0.07	0.02	0.03	0.04	0.04
SC1	**mn**	**0.47**	**0.11**	**0.71**	**0.11**	**0.61**	**0.74**	**0.83**	**0.49**	**0.38**
	s.d.	0.02	0.02	0.03	0.03	0.06	0.02	0.03	0.02	0.03
SC2	**mn**	**0.47**	**0.15**	**0.73**	**0.18**	**0.60**	**0.72**	**0.83**	**0.49**	**0.41**
	s.d.	0.02	0.03	0.03	0.04	0.06	0.02	0.03	0.03	0.04
SC3	**mn**	**0.50**	**0.20**	**0.61**	**0.15**	**0.59**	**0.75**	**0.82**	**0.51**	**0.39**
	s.d.	0.03	0.03	0.03	0.04	0.06	0.02	0.03	0.03	0.04

TABLE 45.3: Posterior means and standard deviations for same data as in Table 45.2. The priors here were all uniform distributions (betas centered at 0.5 with precision 2). All four estimation techniques are the same as described for Table 45.2.

in the range of -.3); nor the global independence assumption (the correlation between θ_M and $\theta_{S|\neg M}$ is -.54).

Other simulations have duplicated the result observed by SC, that if no complete data is available, the method is unable to gain any information about the relationships between the unobserved variables.

45.7 Conclusion

The primary advantage of the MCMC methods in this context is that they provide not only a means for estimating the value of the probability $P(X \mid Y) = \mathbf{E}(P(X \mid \theta) \mid Y)$ but also for estimating the posterior standard deviation, percentiles of the posterior, and so on. In contrast, SL and SC only attempt to approximate $P(X \mid Y)$ and quantities derivable from it. Additionally, the MCMC method makes no untenable independence assumptions, and does not depend on the order in which the data are processed.

The disadvantage is that if one does not know in advance what quantities will be needed, the entire simulation process must be stored so that estimates may be calculated from it later.

Both methods actually require comparable effort in order to implement; the probabilities of unobserved values given observed that we derived symbolically are also used in the SC updating scheme. The main difference between the effort required to implement the two methods is that the MCMC method must do more book-keeping in order to present estimates from the simulated values. In actual running time, the SC method is very fast (under 5 seconds), whereas 25000 iterations of the MCMC routine took approximately 1.5 minutes on a DECstation 5000/120. Crude estimates can be got much more quickly with a smaller number of iterations.

45.8 References

[Cooper 84] Cooper, G. F. (1984) *NESTOR: A computer-based medical diagnostic aid that integrates causal and probabilistic knowledge*, PhD Thesis, Dept of Computer Science, Stanford University.

[Cooper 92] Cooper, G. F. (1992) Personal communication.

[DawiLaur 89] Dawid, A.P. and Lauritzen, S.L. (1989) " Markov Distributions, Hyper-Markov Laws and Meta-Markov Models on Decomposable Graphs, with Applications to Bayesian Learning in Expert Systems," *Technical Report R-89-31*. Institute for Electronic Systems, Aalborg University.

[DawiLaur 93] Dawid, A.P. and Lauritzen, S.L. (1993) "Hyper Markov Laws in the Statistical Analysis of Decomposable Graphical Models," *Ann. Stat.* , in press

[Devroye 86] Devroye, Luc (1986) *Non-uniform Random Variate Generation.* Springer-Verlag, New York.

[GelfSmit 91] Gelfand, A. and Smith, A.F.M. (1991) "Gibbs sampling for marginal posterior expectations," *Communs. Statist. Theory Meth.* **20**, 1747-1766.

[Geyer 92] Geyer, Charles J. (1992) "Practical Markov Chain Monte Carlo," *Stat. Sci.* **7**, 473-483.

[GeyeTier 92] Geyer, Charles J. and Tierney, Luke (1992) "On the Convergence of Monte Carlo Approximations to the Posterior Density," *Technical Report 579*. School of Statistics, University of Minnesota.

[Hastings 70] Hastings, W.K. (1970) "Monte Carlo Sampling Methods Using Markov Chains and Their Applications," *Biometrika* **57**, 97-109.

[LauDawLarLei 90] Lauritzen, S.L., Dawid, A.P., Larsen, B. N. and Leimer, H.G. (1990) "Independence Properties of Directed Markov Random Fields," *Networks* **20**, 491-505.

[LaurSpie 88] Lauritzen, S.L. and Spiegelhalter, D. (1988) "Local Computations with Probabilities on Graphical Structures and their Application to Expert Systems (with Discussion)," *J. Roy. Statist. Soc. ser. B* **50**, 157-224.

[MadiYork 93] Madigan, D. and York, Jeremy C. (1993) "Bayesian Graphical Models," *Technical Report 259*. Department of Statistics, University of Washington.

[Pearl 88] Pearl, J. (1988) *Probabilistic Reasoning in Intelligent Systems*. Morgan Kaufmann, San Mateo.

[Ripley 87] Ripley, B. (1987) *Stochastic Simulation*. John Wiley and Sons, New York.

[SmitRobe 93] Smith, A.F.M. and Roberts, G. O. (1993) "Bayesian Computation via the Gibbs Sampler and Related Markov Chain Monte Carlo Methods," *J. R. Statist. Soc. B* **55**, 3-23.

[SpieCowe 92] Spiegelhalter, D.J. and Cowell, R.G. (1992) "Learning in Probabilistic Expert Systems," in : *Bayesian Statistics 4*, Bernardo, J.M., Berger, J.O., Dawid, A.P. and Smith, A.F.M., eds, Oxford University Press, Oxford.

[SpieLaur 90] Spiegelhalter, D.J. and Lauritzen, S.L. (1990) "Sequential Updating of Conditional Probabilities on Directed Graphical Structures," *Networks* **20**, 579-605.

[Tierney 91a] Tierney, Luke (1991a) "Markov Chains for Exploring Posterior Distributions." *Technical Report 560*, School of Statistics, University of Minnesota.

[Tierney 91b] Tierney, Luke (1991b) "Exploring Posterior Distributions Using Markov Chains," in: *Computing Science and Statistics, Proceedings of the 23rd Symposium on the Interface*, 563-570.

[WeiTann 90] Wei, G.C.G. and Tanner, M. A. (1990) "Calculating the Content and the Boundary of the Highest Posterior Density Region via Data Augmentation, *Biometrika* **77**, 649-652.

[York 92a] York, Jeremy C. (1992a) "Use of the Gibbs sampler in Expert Systems," *Artificial Intelligence* **56**, 115–130.

[York 92b] York, Jeremy C. (1992b) *Bayesian Methods for the Analysis of Misclassified or Incomplete Multivariate Discrete Data*, PhD Thesis, Dept of Statistics, University of Washington.

[YorkMadi 92] "Bayesian methods for estimating the size of a closed population." *Technical report 234*, Department of Statistics, University of Washington.

[YorMadHeuLie 93] York, Jeremy C., Madigan, David, Heuch, Ivar and Lie, Rolv Terje (1993) "Birth Defects Registered by Double Sampling: A Bayesian Approach Incorporating Covariates and Model Uncertainty," submitted for publication.

46

Simulated annealing in the construction of near-optimal decision trees

James F. Lutsko and Bart Kuijpers

Expert Systems Applications Development Group
Katholieke Universiteit Leuven
de Croylaan 46
B-3001 Heverlee, Belgium

ABSTRACT The application of simulated annealing to the optimization of decision trees is investigated. An efficient perturbation procedure is described and used as the basis of the Simulated Annealing Classifier System or SACS algorithm. We show that the algorithm is asymptotically convergent for any choice of global cost function. The algorithm is then illustrated, using the Minimum Description Length Principle as cost function, by applying it to several problems involving both noisy and noise-free data.

46.1 Introduction

One approach to the problem of the automated induction of rules from a database of examples is the construction of decision trees (see [Safavian 91] for a survey of the field). Algorithms for constructing knowledge-based systems by inductive inference from examples have been widely used for some decades in various areas as e.g. pattern recognition, decision table programming, machine learning, medical diagnosis and speech recognition, [Safavian 91], [Payne 77], [Breiman 84].

Many different approaches to the problem of decision tree construction have been proposed [Safavian 91]. One commonly used method is the top-down approach, which is based on a heuristic feature selection criterion. A criticism of this approach is that it is *local* in nature : each choice of attribute may be optimized but there is no guarantee that the resulting tree is in any sense "optimal". The limitations in the top-down approach lead naturally to the idea of searching decision tree-space to find an optimal tree. Such a method is used, e.g., in the well-known CART algorithm [Breiman 84] by which the method of searching tree space has been shown to be both efficient and successful. Methods such as look-ahead and the incremental improvement of decision trees, [Utgoff 89] and especially [Van de Velde 90], provide unconstrained searches of tree space. However, these methods become computationally expensive as the size of tree-space increases. Indeed, a result of Hyafil and Rivest [Hyafil 76], [Garey 79] states that the problem of finding a binary decision tree with the minimum expected number of tests belongs to the class of NP-complete problems. Furthermore, they conjecture that the induction problem with a general cost function is NP-complete [Hyafil 76], [Safavian 91]. This suggests that many global optimization procedures over decision trees (e.g. minimizing the number of nodes or the

[0]The above text presents research results of the Belgian Incentive Program "Information Technology" - Computer Science of the future, initiated by the Belgian State - Prime Minister's Service - Science Policy Office. The scientific responsibility is assumed by its authors.

number of leaves) will be computationally intractable.

Recently, it has become clear that, while truly optimal solutions to such problems are in general infeasible, *near*-optimal solutions can be obtained in polynomial time by means of stochastic methods [van Laarhoven 87]. Here, we propose that one such method, simulated annealing (SA), can be successfully applied to the problem of determining near-optimal decision trees. Simulated annealing has, for practical purposes, successfully "solved" other NP-complete problems, e.g. the traveling salesman problem, in the sense that it produces near optimal solutions [van Laarhoven 87], with a computational effort which scales polynomially with the size of the problem.

In this paper, we first describe the SACS algorithm and we discuss its polynomial time-scaling. We describe a mathematical model, based on the theory of finite Markov chains, which will serve as a framework to study the asymptotic convergence of SACS for any choice of global cost function. Finally, we apply the SACS algorithm, using the Minimum Description Length Principle of Quinlan and Rivest [Quinlan 89] as cost function, to a variety of problems, both noisy and noise free.

46.2 SACS

Simulated annealing, based on the technique of Metropolis sampling [Metropolis 53], is a non-traditional method for minimizing combinatorial functions. Its motivation arises from an analogy with thermodynamics, specifically with the way that liquids freeze and crystallize or metals cool and anneal. Here, we shall limit our discussion of simulated annealing to its use in the SACS algorithm.

46.2.1 General description

In this and the following sections, we follow as much as possible the standard notation in both the description of decision trees [Breiman 84] and of simulated annealing [van Laarhoven 87], [Aarts 90]. Thus, decision trees will generally be denoted as T, nodes of T as $t(T)$ or just t and the subtree of tree T with root t as T_t. The set of leaves of the tree T will be denoted as $\mid \tilde{T}$. The set of all possible attributes will be denoted as $\mathcal{X} = \{X_1, X_2, \ldots\}$ and the attribute assigned to the node t as $X(t)$, while the set of all possible classes will be denoted as $Y = \{y_1, y_2, \ldots\}$. In the following, each attribute will be assumed to take on one of two possible values, 0 and 1. We shall also have occasion to refer to the set of "free" attributes at the node t, $\mathcal{X}(t)$, by which is meant those attributes which could, in a top-down construction of the decision tree, be assigned to node t. This set is also defined for leaves.

SACS begins by generating an initial tree, T_0, from the data set using the top-down growth ID3 algorithm of [Quinlan 86]. ID3 uses the information-theoretic entropy as a heuristic for attribute selection. T_0 becomes the current tree which we denote simply as T. The algorithm then involves the following steps. First, the tree T is perturbed, as described in section 2.2, to produce a new tree T'. The "goodness" of a tree is assumed to be calculable via a cost function (or, in analogy with the physical motivation, an energy function) $E(T)$. To determine whether tree T or tree T' is kept, the other being discarded, we first choose a random number, s, from the unit interval; the new tree is kept if and only if $s < e^{-dE/c}$ where c is a global control parameter. Notice that this implies that if $dE < 0$, then tree T' is kept. Whichever tree is kept then becomes the current tree and the process is repeated. This method of sampling tree-space, known as Metropolis sampling, can be formalized by noting that the probability, $A_{TT'}(c)$, of accepting a tree T' generated from the current tree T at control parameter c is the minimum of 1 and $e^{-dE/c}$. The idea behind this procedure is that dE is, in analogy with thermodynamics, the difference in "energy" of the two configurations and c is the "temperature" of the system. When c is gradually reduced the system will "freeze" into a state which is near the global minimum of the energy function. The two limiting cases, $c = 0$ and $c = \infty$, correspond respectively to greedy algorithms and purely random search algorithms. The details of the initial value of the control parameter, its rate of decrease and the stopping criterion together constitute the *annealing schedule*. The annealing schedule will be discussed in section 2.3.

46.2.2 Perturbing the tree

The most important problem-specific elements of an annealing algorithm are the method of perturbing the current configuration so as to explore configuration space as efficiently as possible and the specification of the cost function. We will call the following method of perturbation *attribute interchange*. Attribute interchange produces a small change in the structure of the tree, while still allowing for the generation of all possible trees.

The first step in the interchange algorithm is to choose a node t in the current tree T. The perturbation will then be confined to the subtree T_t. To choose a node, we assign to every node t in T, upon its creation, a weight $w(t)$ and then select nodes according to the weighted distribution so defined. Clearly, $w(t) = 1$ corresponds to choosing a node randomly. The actual perturbation consists of randomly choosing a new attribute for t from the set of free attributes of t excluding the attribute already assigned to t. Let the attribute selected be X'. Then, in the new tree, T', we have $X(t(T')) = X'$. Now, some nodes in the subtree T_t may already be assigned the attribute X' : let such a node be $\tilde{t}$ so that $\tilde{t} \in T_t$ and $X(\tilde{t}(T)) = X'$. In the tree T' these nodes are assigned the attribute $X(t(T))$ so that $X(\tilde{t}(T')) = X(t(T))$. Thus the origin of the name "interchange" perturbation.

Two special cases occur near the bottom of the tree. First, t may be a leaf. In this case, an attribute is selected as described above and the branches corresponding to the values that the attribute may take on are terminated by new leaves. The second special case is when the selected node t is a terminal node, defined to be a node for which all branches terminate in leaves at the next level. In this case, the list of available attributes is expanded to include the "leaf" attribute. If this attribute is chosen, the children of t are deleted and t is replaced by a leaf.

The so-called one-step generation probability, $G_{TT'}$, that starting with tree T the interchange perturbation generates tree T' is easily developed. If the transition $T \to T'$ is possible, then there is a unique node $\tilde{t} \in T$ which must be chosen, as the first step of the interchange perturbation, so as to generate tree T'. So, if $G_{TT'} \neq 0$, $G_{TT'}$ equals the normalized weight of the node $t(T)$ divided by the number of free attributes at that node. These operations can be carried out very rapidly and allow the transition from any tree T to any other tree $\tilde{T}$ in a finite number of perturbations. It will also be important in section 3 that a one-step perturbation is reversible (in one step).

Before concluding this description, we discuss one difference between the procedure just described and that used in the simulations presented below. There is a considerable inefficiency in the algorithm caused by the fact that leaves containing data points all in one class are nevertheless candidates for the introduction of a new attribute and the consequent growth of the tree. We set the weights of all such leaves to zero so that they are not candidates for perturbation.

In practice, we find no difference in results when this is done except for a savings of as much as a factor of 2 or more in the CPU time due to the smaller size of the neighborhood of the current tree. This practical difference is ignored in section 3 because it would complicate the convergence proof considerably (by the introduction of irreversible one-step-perturbations).

46.2.3 *The annealing schedule*

In SACS we use the annealing schedule proposed by van Laarhoven and Aarts [Aarts 90], [van Laarhoven 87]. The initial temperature, c_0, is determined via an iterative procedure such that the initial acceptance rate, χ, is equal to a user-supplied value (in our case 0.8). The general scheme is then that the temperature is held fixed at its current value until some number, L, of configurations have been generated (although not necessarily accepted). The temperature is then decreased to give a new temperature c_1 and the process repeated. Following [van Laarhoven 87], we take L to be roughly the size of the one-step neighborhood (the neighborhood of T is the set of all trees which can be generated from T by a single perturbation). So, $L = \sum_{t \in T}(\mid \mathcal{X}(t) \mid -1)$. In SACS, after each perturbation, we compare the value of L evaluated with the current tree with the number of configurations since the last temperature decrease : when the number of configurations equals or exceeds L, the temperature is decreased.

The amount by which the temperature is decreased is again given by [van Laarhoven 87]. Specifically, we have $c_{k+1} = c_k / \left(1 + \frac{c_k}{3\sigma(c_k)\log(1+\delta)}\right)$ where $\sigma(c)$ is the standard deviation of the energy function (as determined at the temperature c) and δ is an input parameter which controls the rate of temperature decrease. Typically, one has $\log_{10}(\delta) \in [-1, 1]$.

Finally, each time the temperature is decreased, it is also determined whether the stopping condition $\frac{c_k}{<E>_\infty} \frac{\partial <E>_c}{\partial c} \mid_{c=c_k} < \epsilon$ where $< E >_c$ is the average energy at temperature c, is satisfied. We have taken $\epsilon = 0.01$.

46.2.4 *The energy function*

We have thus far described SACS without explicitly detailing the cost function. Clearly, the quality of the results depends on whether or not we have chosen to optimize the correct quantity. First, we take "quality" to mean, as is usual [Breiman 84], that one decision tree is better than another if it gives a lower misclassification rate when presented with data other than that used to construct it.

In choosing an energy function we wish to find a quantity which is a measure of the accuracy of the current decision tree. Ideally, of course, we would like the fact that a data point ends up in a particular leaf of the tree to uniquely imply its membership in a class. Here, we use the Minimum Description Length Principle (MDLP) of Quinlan and Rivest [Quinlan 89]. The idea is to minimize the amount of information necessary to encode the classification of all of the data points. Generally, it is cheaper to encode a decision tree and the exceptions, i.e. the points which are misclassified, than to transmit the class of each point individually. A weakness of this criterion is that the actual cost function depends on the details of the coding. Here, since we are primarily interested in demonstrating the use of the annealing, we simply follow the coding of [Quinlan 89].

46.3 Convergence and scaling of SACS

The key quantity in a theoretical description of any annealing algorithm is the stationary distribution of configurations at fixed control parameter c, $q_T(c)$. Asymptotic convergence is proven if we can show that, as $c \to 0$, this probability becomes infinitesimal for all trees but those in the optimal subspace.

Given the evident fact that any tree T' can be generated from any other, T, in a finite number of interchange perturbations, a theorem due to Feller [van Laarhoven 87], [Aarts 90], assures the existence of the stationary distribution and shows a way to calculate it. For SACS, the following equation holds for any two trees T and T' : $q_{T'}(c)P_{T'T}(c) - q_T(c)P_{TT'}(c) = 0$. This equation plays a fundamental role in the convergence proof and is known as the condition of *detailed balance*. The concrete form of $G_{TT'}$ and $A_{TT'}(c)$ and the detailed balance equation lead to the following form of $q_T(c)$: $q_T(c) = A_{TT_{opt}}(c)q_T$, where T_{opt} is an arbitrary (but fixed) globally optimal tree. With this information it is easily verified [van Laarhoven 87], [Aarts 90], [Lundy 86] that the distributions converge (for c decreasing to 0) to a (non-uniform) distribution over the set of globally optimal trees. We note that the weights $w(t)$, used in $G_{TT'}$, must satisfy $w(t(T)) = w(t(T'))$ and can not depend on any global property of the tree. We can e.g. take $w(t) = \mid \mathcal{X}(t) \mid -1$ which gives $G_{TT'} = 1/ \mid \mathcal{S}(T) \mid$ where $\mathcal{S}(T)$ is the one-step neighborhood of the tree T. So, very much in the spirit of simulated annealing, the probability of converging to any particular optimal tree is in this case simply proportional to the size of its neighborhood.

In order to properly compare SACS with other methods of searching tree-space, consider the scaling of the computational requirements of the algorithm with problem size. First, we note the trivial point that the storage requirements of the algorithm are no greater than those of many other decision-tree generation algorithms. Thus, the only consideration is CPU time. If N denotes the number of data points available, and $\mid T_m \mid$ denotes the size of the maximal tree, we find after estimating the size of the search space that CPU $\sim N^2 \mid T_m \mid\mid \mathcal{X} \mid \log N$. Thus, we see that SACS scales polynomially with the size of the maximal tree and/or the number of attributes. Since $\mid T_m \mid$ is $\min(N, 2^{|\mathcal{X}|} - 1)$, and, for realistic problems we always have $N \ll 2^{|\mathcal{X}|} - 1$, we conclude that for practical problems, the CPU requirements of SACS indeed scale polynomially in both N, which is usually the number of data points available, and in $\mid \mathcal{X} \mid$. [3]

46.4 Empirical tests

Here, we will describe several problems on which we have tested the SACS algorithm. In the next subsections, we describe first the problems used to study the algorithm: its speed and the dependence of the results on the parameters used.

46.4.1 Description of the test problems

As its name indicates, the Boolean Multiplexer involves binary attributes and classes. Here, in general, for an integer n an object is a string of the form $(a_0, a_1, \ldots, a_{n-1}, b_0,\ b_1, \ldots, b_{2^n-1})$, where the a's and b's are resp. called *addresses* and *channels*. The class of an object is b_s where $s = \sum_{i=0}^{n-1} a_i 2^i$. So, if we consider the $(n + 2^n)$-bit multiplexer and if we take the complete set of

[3] Other annealing schedules [van Laarhoven 87] would give, roughly, the log of this estimate and, thus, true polynomial scaling. However, this formal distinction seems to be of little practical relevance (see [van Laarhoven 87]).

$2^{(n+2^n)}$ objects as training set, we can easily see that knowing the value of a channel-bit, b_i, of a particular data point allows us to predict the class with a probability of $\frac{1}{2} + \frac{1}{2^{n+1}}$ but knowing the value of any of the address bits gives us no information about the class (the probability of predicting it on the basis of this information alone is $\frac{1}{2}$). In the following, we consider the 6-, 11-, 20-, and 37-bit multiplexers which we will abbreviate as the $6bm$, $11bm$, $20bm$ and $37bm$ respectively. We will also examine the $11bm$ with noise (abbreviated as $11bmx$). A given noise level, say x percent noise, is introduced by setting the class bit of x percent of the data set randomly. This means that $x/2$ percent of the classes are incorrect.

The other problem studied here is a pattern recognition problem described by Breiman et al. [Breiman 84]. On electronic calculators and watches with LED or LCD the ten digits are displayed using three horizontal and four vertical lights in a rectangular 8-shaped form in on-off combinations [Breiman 84], [Zhou 91]. So, the digit i $(i = 1, 2, \ldots, 9, 0)$ can be described using a seven-dimensional boolean vector $(X_{i1}, \ldots, X_{i7})$, where $X_{ij} = 1$ (resp. 0) if the j-th light is on (resp. off). The training set for this problem was generated as in [Breiman 84]. We generated vectors $(X_1, \ldots, X_7, Y)$, where the class label Y assumes the values $0, 1, \ldots, 9$ and where an X_i has the correct value with probability 0.9. We refer to this problem below as $digits$. Finally, also following [Breiman 84], we consider this same problem with additional, superfluous, attributes added. Thus, in what is called $digits/17$ below, the data points are initially generated as in $digits$ above, but to each point are added 17 additional binary attributes, giving a total of 24, the values of which are randomly set and completely uncorrelated with the classification of the point.

46.4.2 Basic results

Our procedure in studying these problems was to generate in every case 10 independent sets of data consisting of 200 ($6bm$), 400 ($11bm$), 800 ($20bm$), 2000 ($37bm$), 1000 ($11bm$ with noise) and 200 ($digits$ and $digits/17$) points. The points in each set were generated randomly so that in the smaller data sets some duplication was possible. For every problem, a verification data set consisting of 5000 points was also generated using the same level of noise as in the training set. This was used to obtain a fair estimate of the misclassification rate of the final trees. A second verification set containing 5000 points, but this time without noise, was used to test the "accuracy" of the trees (i.e., the ability of the trees to correctly classify data and thus to filter out the noise). Unless otherwise stated, the results were obtained with the following "standard" values : $\chi = 0.8$, $\epsilon = 0.01$ and $\delta = 1.00$. In all cases, the results given are the "best of run".

The performance of SACS should really be measured by its ability to minimize the cost function, although it is the misclassification rate which is of greatest practical interest. With this in mind, we first observe that the over-all performance of SACS is very good. In the pure multiplexer problems, near optimality in the number of nodes is obtained despite the fact that the deceptiveness of this problem, discussed above, leads to large initial trees. (The optimal number of nodes for the multiplexers are, respectively, 7, 15, 31 and 63). The small misclassification rates, in the absence of noise, are due to the fact that SACS treats the problem as if it were noisy, often generating impure leaves in the course of the simulations. When noise is added to the $11bm$, the computational requirements are greater. This is because the initial trees become larger as the tree-building algorithm attempts to fit the noise. Nevertheless, we see that, up to a level of 50% noise, or 25% of the data corrupted, SACS is still locating near optimal trees for the multiplexer problem leading to near optimal misclassification rates and very high accuracies. In the *digits* problem the performance is similar. This problem has been studied using a wide variety

Problem	Configs.	CPU	E	$\mid T \mid$	R	A
$6bm$	0.6	0.9	77.6 (144.5)	8.1 (20.7)	0 .006	0.994
$11bm$	4.9	8.953	164.0 (580.7)	16.2 (80.9)	0.000	1.000
$20bm$	37.6	96.51	363.02 (1901.7)	33.0 (236.9)	0.001	0.998
$37bm$	266.3	811.2	785.1 (3290.0)	66.3 (319.6)	0.001	0.999
$11bm0$	6.5	17.0	187.4 (691.0)	16.4 (87.0)	0.000	1.000
$11bm25$	27.0	77.46	676.6 (2205.6)	16.5 (380.6)	0.127	0.998
$11bm40$	32.4	101.6	833.73 (2606.33)	16.0 (463.1)	0.200	1.000
$11bm50$	37.6	124.1	937.1 (2892.7)	16.0 (522.7)	0.279	0.945
$digits$	2.5	4.066	419.15 (711.06)	12.2 (66.0)	0.317	1.000
$digits17$	37.4	57.3	401.4 (778.2)	12.0 (69.3)	0.313	0.998

TABLE 46.1: Summary of the results (averages over 10 runs) using the standard values of the input parameters. Configs. is the number of configurations/1000 generated during the run, the CPU-time on a SUN SPARC-2 is given in seconds, E and $\mid T \mid$ (the number of nodes) are shown for the best of run and, in parentheses, the initial tree, R is the misclassification rate, and A the "accuracy" rate.

of methods [Mingers 89]. Generally, misclassification rates are in the range $0.28 < R < 0.32$, since many tree-building heuristics do a good job, in this problem, of recognizing the most important attributes. We see that SACS is performing near the upper boundary of this range even though the results, as discussed above, are only near-optimal. The size of the final tree is also comparable to the results reported in [Mingers 89]. This provides some evidence that the ability of SACS to deal with problems, such as the multiplexer, where most tree building heuristics fail, is not obtained at the price of (substantially) inferior performance in problems which other methods deal with efficiently. When 17 additional variables are added to the *digits* problem, we see that, as expected, considerably more computational time is required due to the much greater size of the search space. However, the final results are comparable to those obtained in the original *digits* problem. This problem was also studied by Breiman et al. [Breiman 84] with similar results.

It is interesting to ask whether SACS is, in the case of the multiplexers, generating better, i.e. more accurate (see A), trees than other methods or just smaller trees. (Generally, the deceptiveness of the multiplexer problem means that any procedure based on a splitting criteria will generate large trees for both noisy or noise-free data.) In order to answer this question, we have implemented the proposal of Gelfand, Ravishankar and Delp [Gelfand 91] (the algorithm will be referred to hereafter as GRD). GRD is similar in spirit to CART [Breiman 84] in that it is primarily a pruning algorithm which is directly controlled by an estimate of the misclassification rate. It is claimed in [Gelfand 91] that GRD is more accurate than CART. Table 2 shows the result of applying GRD to the $11bm$ with noise together with the results from SACS. As expected, SACS generates smaller trees. However, SACS also generates considerably more accurate trees. Indeed, even without noise, GRD has a non-zero misclassification rate. The reason is that when presented with a subset of all possible data points (2048 distinct points are possible in this case), it is possible to generate rules which, e.g., do not test all channels but yield correct classification of the available data. Such rules are false generalizations. It is therefore not surprising that as the noise level is increased, the GRD results worsen relative to SACS. We conclude, on this basis, that SACS is indeed capable of producing more accurate trees than other methods and not just smaller trees.

	GRD		SACS		Optimal	
Noise	$\mid T \mid$	R	$\mid T \mid$	R	$\mid T \mid$	R
$11bm0$	83.4	0.075	16.4	0.000	15	0.0
$11bm25$	82.3	0.228	16.5	0.127	15	0.125
$11bm40$	80.9	0.330	16.0	0.200	15	0.20
$11bm50$	124.1	0.376	16.0	0.279	15	0.25

TABLE 46.2: Comparison of the average number of nodes and the average misclassification rate for the GRD algorithm, SACS and the optimal tree.

δ	Configs.	CPU	E	$\mid T \mid$	R	A
0.1	385.6	795.9	927.5	15.9	0.273	0.956
1.0	37.6	124.1	937.1	16.0	0.279	0.945
10.0	34.0	126.0	955.1	19.8	0.291	0.915
100.0	27.3	110.4	964.5	21.3	0.303	0.894

TABLE 46.3: The result of varying δ using the $11bm50$ problem as a test case.

46.4.3 *Variation of the algorithm*

We now use the $11bm50$ problem to study the dependence of the results on the parameters controlling the annealing and the choice of weights. In order to distinguish relatively small effects, all results in this section were obtained by averaging over 100 simulations.

As stated in section 2, there are three parameters controlling the annealing. They are the initial acceptance rate, χ, the parameter controlling the temperature decrease, δ, and ϵ which sets the stopping criterion. van Laarhoven and Aarts [van Laarhoven 87] show that of the two remaining parameters δ is by far the most important. We have therefore varied δ, holding $\chi = 0.8$ and $\epsilon = 0.01$, in the $11mb50$ problem.

Obviously, only the case $\delta = 0.1$ differs considerably from the three latter cases. The best compromise between CPU and accuracy is obtained in the case $\delta = 1$.

46.5 Conclusion

In this paper, we have investigated the application of simulated annealing to the determination of near-optimal decision trees. Using a particular method of generating trees, the interchange perturbation, and the MDLP cost function we have shown that SA can indeed generate decision trees of comparable or superior accuracy to other methods. The primary advantage of SA is illustrated best by the multiplexer problems : namely that the method is not constrained by a tree-building heuristic. This is in contrast to some other methods, such as CART [Breiman 84] and that of GRD [Gelfand 91], which are primarily concerned with optimizing the pruning of a given, and more or less fixed, tree structure. Also, unlike the iterative methods purposed by [Utgoff 89] and [Van de Velde 90], SACS has been demonstrated to be asymptotically convergent to the optimal subspace and to scale, in CPU requirements, polynomially with the problem size. Compared to the dynamic programming approach of Payne and Meisel [Payne 77], the memory requirements are modest allowing for the consideration of much larger problems.

SACS does not utilize either a tree-building heuristic or a method of pruning, however, it does require a cost function. While we have used here the MDLP, many other possibilities could be used and compared. Tree-building heuristics, such as the asymmetric-τ proposed in [Zhou 91], and pruning procedures, such as the Pessimistic Error Pruning of Quinlan (see e.g. [Mingers 89]) could be adapted to provide global cost functions. This dependence on a (heuristic) cost function means that SACS itself must be considered a heuristic method. It is not clear to us, at this time, how a more precise measure of the misclassification rate of a given decision tree, such as cross validation [Breiman 84], might be incorporated into SACS, although something along the lines of GRD might be possible.

While the CPU requirements of SACS scale polynomially with problem size, it can still be quite expensive for large problems (e.g. problems involving real-valued attributes). This problem might be circumvented by limiting the length of the Markov chains, but it is not clear at this point that this would be a profitable procedure. Alternatively, other tree-building heuristics might be used to construct the initial tree on which SACS can operate to optimize the tree-structure using only those splits which occur in the initial tree. These and other methods are currently under investigation.

Finally, we have stressed that the results presented here are only near-optimal. We believe that for practical applications, some post-processing of the SACS tree should be performed. A particularly promising approach would be to use the SACS tree as the input to the iterative procedure proposed in [Gelfand 91]. Results on more realistic problems using this procedure will be reported at a later time.

46.6 References

[Aarts 90] Aarts, E. H. L. and Korst, J. (1990) *Simulated Annealing and Boltzmann Machines. A Stochastic Approach to Combinatorial Optimization and Neural Computing*. John Wiley & Sons, Chichester.

[Breiman 84] Breiman, L., Friedman, J. H., Olshen, R. A. and Stone, C. J. (1984) *Classification and regression trees*. Wadsworth International Group, Belmont, CA.

[Garey 79] Garey, M. R. and Johnson, D. S. (1979) *Computers and Intractability : A Guide to the Theory of NP-Completeness*. W. H. Freeman and Company, San Francisco.

[Gelfand 91] Gelfand, S. B. , Ravishankar, C. S., and Delp, E. J. (1991) "An Iterative Growing and Pruning Algorithm for Classification Tree Design", *IEEE Transactions on Pattern Analysis and Machine Intelligence*, **13**, 163-174.

[Hyafil 76] Hyafil, L. and Rivest, R. L. (1976) "Constructing optimal binary decision trees is NP-complete", *Inform. Processing Lett.*, **5**, 15-17.

[Lundy 86] Lundy, M. and Mees, A. (1986) "Convergence of an annealing algorithm", *Mathematical Programming*, **34**, 111-124.

[Metropolis 53] Metropolis, N., Rosenbluth, A., Rosenbluth, M., Teller, A., and Teller, E. (1953) "Equation of state calculations by fast computing machines", *J. Chem. Phys.*, **21**, 1087-1092.

[Mingers 89] Mingers, J. (1989) "An empirical comparison of pruning methods for decision-tree induction", *Machine Learning*, **4**, 227-243.

[Payne 77] Payne, H. J. and Meisel, W. S. (1977) "An algorithm for constructing optimal binary decision trees", *IEEE Trans. Comput.*, **26**, 905-916.

[Quinlan 86] Quinlan, J. R. (1986) "Induction of decision trees", *Machine Learning*, **1**, 81-106.

[Quinlan 89] Quinlan, J. R. and Rivest, R. L. (1989) "Inferring Decision Trees Using the Minimum Description Length Principle", *Information and Computation*, **80**, 227-248.

[Safavian 91] Safavian, S. R. and Landgrebe, D. (1991) "A Survey of Decision Tree Classifier Methodology", *IEEE Transactions on Systems, Man, and Cybernetics*, **21**, 660-674.

[Utgoff 89] Utgoff, P. E. (1989) "Incremental induction of decision trees", *Machine Learning*, **4**, 161-186.

[Van de Velde 90] Van de Velde, W. (1990) "Incremental Induction of Topologically Minimal Trees", *Machine Learning: Proceedings of the Seventh International Conference on Machine Learning*, Porter, B. W. and Mooney, R. J., eds., Austin, Texas.

[van Laarhoven 87] van Laarhoven, P. J. M. and Aarts, E. H. L. (1987) *Simulated Annealing: Theory and Applications*. D. Reidel Publishing Company, Dordrecht.

[Zhou 91] Zhou, X. J. and Dillon, T. S.(1991) "A Statistical-Heuristic Feature Selection Criterion for Decision Tree Induction", *IEEE Transactions on Pattern Analysis and Machine Intelligence*, **13**, 834-841.

47
SA/GA : Survival of the Fittest in Alaska

Kris Dockx and James F. Lutsko

Expert Systems Applications Development Group
Katholieke Universiteit Leuven
de Croylaan 46
B-3001 Heverlee, Belgium

ABSTRACT SA/GA is a genetic-algorithm-based approach to combinatorial optimization. It differs from ordinary genetic algorithms in that the acceptance of offspring is subject to the Metropolis sampling criterion [Metropolis 53] which is the basis of simulated annealing. The idea is that of a population subject to a climate that is getting colder. On one hand SA/GA can be viewed as a genetic algorithm with a crossover probability depending on the quality of the children relative to a decreasing temperature. On the other hand SA/GA can be modeled as a Markov chain with temperature-dependent transition probabilities [van Laarhoven 87]. When the population is set to one, the algorithm degenerates to a normal simulated annealing algorithm. We have tested the algorithm by applying it to a set of so-called genetic-algorithm-deceptive problems which are specially constructed to cause genetic algorithms to converge to sub-optimal results. SA/GA is not fooled by the deception and consistently finds the optimal solution. SA/GA also outperforms the pure simulated annealing algorithms while being less sensitive to parameter tuning.

47.1 Introduction

Genetic algorithms (GAs) [Goldberg 89a] are a class of combinatorial optimization techniques. Although many different implementations exist, they share in common the same basic idea. The problem takes the form of a function to be optimized. The starting point for the algorithm is a population of some number of possible solutions to an optimization problem. These possible solutions are usually encoded as strings so that each string represents a possible solution to the optimization problem. Associated with each individual, i.e. each string, is a fitness value which, in the simplest case, is the value of the function for that particular string. New strings are generated by a simulated mating process. A new string becomes a member of the new population which eventually replaces the original population. When the strings encode rules and the evaluation function is a measure of the accuracy of the rules, genetic algorithms provide a very elegant approach to machine learning. Another well-known example of combinatorial optimization is Simulated Annealing (SA). Simulated annealing begins with a single guess at the optimal solution and proceeds as follows. A new solution is created by making a random change in the first solution. The criterion for accepting or rejecting changes is called Metropolis sampling and forms the basis of the so-called Monte Carlo method in statistical physics [Metropolis 53]. This criterion is a function of an external parameter T, called the temperature. Initially, the temperature is "large" and almost all changes are accepted; as the simulation progresses, the temperature is slowly lowered, decreasing the likelihood of acceptance of poor solutions, until

[0]The above text presents research results of the Belgian Incentive Program "Information Technology" - Computer Science of the future, initiated by the Belgian State - Prime Minister's Service - Science Policy Office. The scientific responsibility is assumed by its authors.

the system "freezes" into a particular configuration which is the best guess at the optimum. Simulated annealing has been used successfully in the solution of the Traveling Salesman problem, the design of complex electrical circuits, the training of neural networks and the optimization of decision trees [van Laarhoven 87], [Aarts 90], [Lutsko 91].

Here, we propose that these two approaches to combinatorial optimization be combined to give a hybrid algorithm in the following way. Beginning with the GA described above, the replacement of old strings by new strings, is defined subject to the Metropolis criterion. Specifically, a quantity dE is taken to be the difference in the maximum fitness value of the two parents and the maximum fitness value of the two offspring. Then, a random number from the unit interval is chosen and if it is smaller than $e^{(-dE/T)}$, the two children are added to the new population; otherwise the parents are added to the new population and the children discarded. Again, the temperature is initially sufficiently large that no children are rejected and is slowly lowered throughout the run.

The advantage of this procedure is made clear by interpreting the algorithm from the stand point of both simulated annealing and genetic algorithms. On the one hand, we can view SA/GA as a kind of simulated annealing with a complex mechanism for generating new configurations. Rather than simply permuting a single candidate string, SA/GA maintains a whole collection of strings simultaneously. Without the mating, the algorithm is simply several annealing simulations running in parallel. The mating provides a kind of mixing between these separate simulations allowing for a more efficient exploration of configuration space.

On the other hand, SA/GA can be viewed as a genetic algorithm with an "environmental" pressure. As the temperature is lowered, fewer and fewer offspring which do not surpass their parents in maximum fitness are retained. This makes the genetic algorithm more "greedy" as the run progresses and causes it to search more aggressively for improvement than does an ordinary GA. However, even at very low temperatures, significant genetic diversity is maintained since only one of the offspring must be superior in order for both to be accepted.

In the next section, we discuss our implementation. In section 3, we discuss the theoretical background. In section 4 we discuss the application of the algorithm to the so-called GA-deceptive problem which is specially constructed to lead GAs to sub-optimal solutions. The paper concludes with a discussion of possibilities for future work.

47.2 The SA/GA Implementation

The general idea behind SA/GA can be implemented within the context of any of the many varieties of genetic algorithms and adds very little to the computational expense of the algorithm. We have embedded the SA/GA idea in the context of a very basic genetic algorithm (constant mutation rate, single point crossover, selection of both mates according to their fitness value). The temperature is initially set to a sufficiently large value that very few replacements are rejected. Several prescriptions for lowering the temperature, called annealing schedules, exist [van Laarhoven 87]: we choose to multiply the temperature by a factor κ, $0 < \kappa < 1$, every time a successful replacement is made (i.e., every time the offspring are accepted). This implementation is very simple : modern GAs often involve overlapping populations, two-point crossover, fitnesses based on a string's rank within the population and other refinements [Whitley 90]. The reason for choosing such a simple basis for the implementation of SA/GA is to allow us to test the basic idea of incorporating Metropolis sampling within a GA without introducing other sources of improvement in the GA.

47.3 Theoretical foundations

47.3.1 General schema theorem

Suppose that at generation t we have a population $I(t)$ consisting of $\mid I(t) \mid = N$ individuals. We divide the population into two subsets $I(t) = H(t) \bigcup H'(t)$, where the $\mid H(t) \mid$ individuals in $H(t)$ possess the schema H (a schema defines a subset of the search space, which can be represented as a string containing wild card symbols) and the $\mid H'(t) \mid$ in $H'(t)$ do not. $L(H)$ denotes the defining length of schema H (the number of potential breakpoints in a schema). L_{max} denotes the length of the strings. If both mates in the mating process are chosen by fitness, the evolution of the subpopulation $H(t) \subset I(t)$ is expressed in the following general schema theorem:

$$S(t+1) \geq r(H)S(t) - r(H)S(t)(1 - r(H)S(t))\mathrm{Prob}(H, H' \rightarrow H', H'),$$

with

$r(H) =$ the average fitness ratio of schema H,

$S(t) = \mid H(t) \mid / \mid I(t) \mid$

$\mathrm{Prob}(H, H' \rightarrow H', H') =$ probability of getting two children in H', if one parent is in H'.

1. For a standard genetic algorithm, $\mathrm{Prob}(H, H' \rightarrow H', H') = P_c P_D$ (the crossover and disruption probabilities respectively), where $P_D = L(H)/(L_{max} - 1)$. [Schaffer 87]

2. For SA/GA, $\mathrm{Prob}(H, H' \rightarrow H', H') = < A(H, H' \rightarrow H', H') > P_D$,where $< A(H, H' \rightarrow H', H') >$ = the average of the T-dependent Metropolis factor, when the children are in H', if one parent is in H'.

The schema theorem for SA/GA makes it seem that this peculiar average of the Metropolis factor is playing the role of a T-dependent cross-over probability. This is true in the schema theorem but a detailed examination of the derivation shows that it makes more sense to think of $< A(H, H' \rightarrow \mathrm{anything}) >$ as the effective crossover probability and the effective disruption probability as $(< A(H, H' \rightarrow H', H') > P_D)/ < A(H, H' \rightarrow \mathrm{anything}) >$.

The Schema Theorem for SA/GA shows that SA/GA is indeed a GA which takes the environment (temperature) into account. Most interpretations of the schema theorem for standard GA also apply to SA/GA. The essential difference with a standard GA is that SA/GA uses a dynamic crossover probability instead of a constant one. Moreover this dynamic crossover probability is competing with the disruption probability in an interesting way: SA/GA is more likely to do simple reproduction instead of crossover if the children are below a certain level of energy in comparison with their parents. Thus, as the temperature is lowered, the bias against long schemata, which is present in traditional GAs due to the dependence of P_D on $L(H)$, competes with SA/GA's greediness for fit schemata (i.e., if a long schema is very fit, crossover with its attendant disruption will not be allowed).

47.3.2 Asymptotic Convergence

The SA/GA algorithm is mathematically best described by means of a Markov chain [van Laarhoven 87]: a sequence of trials, where the outcome of each trial depends only on the outcome of the previous one. In the case of SA/GA each trial corresponds to the subsequent generation of a new population. A Markov chain is described by means of a set of conditional transition probabilities $P_{ab}(k-1,k)$ for each pair of possible configurations (a,b). $P_{ab}(k-1,k)$ is the probability that the outcome of the k-th trial is b, given that the outcome of the $(k-1)$-th trial is a. In the case of the SA/GA algorithm the transition probabilities depend on the value of the control parameter T, the analagon of the temperature in the physical annealing process. We can prove asymptotic convergence on the basis of a trivial argument, which can be easily formalized. Because we employ a non zero-mutation rate, any configuration can always be reached from any other configuration with a non-zero probability. So, somewhere along the line we will encounter the optimal configuration (the population containing N copies of the optimum we seek). With the temperature going to zero no worse configurations are accepted. So SA/GA will converge to the optimal configuration. In fact , the following theorem is valid.

Theorem 1 *An optimization algorithm that induces a Markov chain so that there is a non-zero probability to reach any configuration b from configuration a at any place in the chain (in other words : $P_{ab}(k-1,k) > c > 0$ for all a, b, k, and T), and so that the transition probability of the optimal configuration to the optimal configuration is 1 for $T = 0$, converges asymptotically to the optimal configuration.*

The schema theorem tells us that, as the temperature is lowered, the population will uniformly possess the fittest schema. Theorem 1 provides the additional information that as the temperature tends to zero, the fittest schema will be the optimal schema. The fact that the schema theorem by itself does not prove asymptotic convergence to optimality is the reason that deceptive problems can be constructed for GA's and not for SA's.

47.4 Empirical Tests

47.4.1 Deceptive problems

In order to investigate the effect of including Metropolis sampling in the GA, we have used the so-called GA-deceptive problems as tests [Goldberg 89b]. The 3-bit deceptive problem is, as its name implies, framed in terms of strings consisting of 3 bits. To every combination of bits is assigned a value which defines the evaluation function. The assignments are as follows: $f(111) = 30, f(000) = 28, f(001) = 26, f(010) = 22, f(100) = 14, f(110) = f(101) = f(011) = 0$.

The reason for the assignments is the following. The string "111" is the optimal string. However, any string derived from this one by changing a single bit has a fitness lower than any obtained by changing two bits while the string obtained by changing all three bits, "000", has the highest fitness of all non-optimal strings. Thus, the optimal value is surrounded by a very deep well with the local gradient always leading away from it. This in itself is not a very difficult problem since there are only 8 possible strings. To make the GA-deceptive problem, we use strings of $n3$ bits with the fitness of the string being the sum of the fitnesses derived from the n individual 3-bit problems. To make it even more difficult, and truly deceptive for a GA, the individual 3-bit problems are distributed across the string so that the first, $n+1-st.$ and $2n+1st.$ bits make up the first three bit problem; the second, $n+2-nd.$ and $2n+2-nd.$ bits the second 3-bit problem, etc. We have also tested SA/GA on a $n4$-bit problem defined in the same way in terms of a 4-bit deceptive problem with the necessary assignments given as: $f(1111) = 30, f(0000) = 28, f(0001) = 26, f(0010) = 24, f(0100) = 22, f(1000) = 20, f(0011) = 18, f(0101) = 16, f(0110) = 14, f(1001) = 12, f(1010) = 10, f(1100) =$

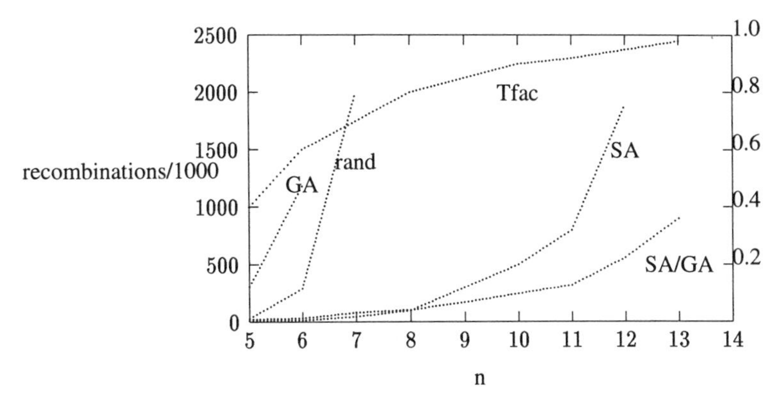

FIGURE 1: Comparison of SA/GA, SA, GA and a random walk on the n3-bit deceptive problems. Each point is an average over 10 runs and with "optimized" parameters. The optimal T_{fac} parameter is plotted against the right hand Y-axis.

$8, f(1110) = 6, f(1101) = 4, f(1011) = 2, f(0111) = 0.$

47.4.2 Parameter settings

In the runs that follow, the mutation rate was held fixed at a value $p_{mutation} = 0.1$, and only the population-size, N, and the annealing rate, κ, were varied. In choosing the initial temperature, a shortcut was used. We scaled all fitnesses by a factor of $30n$ so that they were always in the range $[0, 1]$. Thus, the maximum dE possible was 1.0 and we always started with a temperature $T = 2.0$. Among the quantities monitored was the best-of-run; when a best-of-run with fitness 1.0 was found, i.e. the optimal string, the run was terminated. A typical annealing schedule is to lower the temperature by multiplying it with $T_{fac} = 0.9$ after $10n$ reconfigurations have been accepted where n is the number of variables (in our case the number of bits). To smoothen the decrease in temperature we took $\kappa = {T_{fac}}^{1/c}$ and multiplied the temperature by this factor, every time a successful replacement is made (i.e., every time the offspring are accepted). This coincides more or less with multiplying the temperature with T_{fac} after c successful replacements. We also kept c constant ($c = 800$) for the nine instances of the 3-bit Problem (5×3-bit, 6×3-bit,...,13×3-bit). Under the above assumptions we get a large convex area in the plane (T_{fac}, N) where SA/GA finds the global optimum in more or less the same number of recombinations. This means that SA/GA appears to be robust in the settings of the parameters T_{fac} and N. SA/GA appeared especially insensitive to the setting of the population-size (for all instances of the 3-bit problem SA/GA performed equally well with a population of 100 as with a population of 1000). Therefore, we kept the population constant at 200 for the results that follow. For each instance of the 3-bit problem there is an interval of T_{fac} where SA/GA performs the best. This interval becomes narrower for the more difficult instances. The centres of these intervals are plotted on Figure 1. As one can see, SA/GA digests extreme greediness of the annealing schedule for the simpler problems (5×3-bit, 6×3-bit).

47.4.3 Performance of SA/GA

Figure 1 shows the performances of SA/GA (with ideal parameter settings) and of standard simulated annealing (with ideal parameter settings). With a population-size set to 1, SA/GA degenerates in a standard simulated annealing algorithm. SA/GA out performs standard simulated annealing on the deceptive problem. There appeared to be a strong relationship between the ideal temperature decrease (T_{fac}), the temperature, and the distribution of the energies of the members of the population. With an ideal T_{fac} the standard deviation of the population remains more or less constant (there is a gradual decrease just before the optimum is found) throughout a run. So the population maintains its diversity in consecutive generations. However, the difference between the maximum and the average in a population decreases with the temperature and at the same time remains comparable with the temperature (about two times the temperature). This enables the algorithm to "remember" good partial solutions (local optima in the neighbourhood of the global optimum). Empirical results subscribe the above interpretation. In the case of the $n3$-bit problem, we see that SA/GA is actually solving each 3-bit problem one after the other while trying to keep already solved 3-bit problems. This conservative tendency increases when the temperature decreases. Building further on the above observations we were capable of setting the ideal T_{fac} much quicker (without trying out a lot of different T_{fac}'s). If T_{fac} is set too low the difference between the maximum and the average in a population doesn't stabilise around the temperature and the standard deviation decreases. This makes SA/GA too conservative and it gets stuck in local optima. If T_{fac} is set too high it takes much longer for the difference between the maximum and the average to stabilise around the temperature.

With a high enough starting temperature and no temperature decrease ($T_{fac} = 1$), SA/GA degenerates in a standard genetic algorithm. Figure 1 also shows the number of runs needed by a standard genetic algorithm (population of 100) to successfully find the complete optimization. A genetic algorithm with a population size of 1 could be seen as doing a random walk. So it appears that a standard genetic algorithm with a population larger than one performs worse than a random walk. Actually the standard genetic algorithm only finds the global optimum by accident due to the non-zero mutation rate (in fact the GA is converging to the wrong 000 solution [Goldberg 89b]). The $n3$-bit problems allowed us to analyse SA/GA comfortably, because it requires relatively little effort of SA/GA. To further test the performance capabilities of SA/GA we submitted it to a much harder deceptive problem: the 10×4-bit problem (see section 4.1). For the 10×4-bit problem SA/GA consistently finds the complete optimization in an average (over ten runs) of 1.7 million recombinations. The ideal T_{fac} equaled 0.99, with population size 200. Similar results have been obtained using the GENITOR II algorithm. Compared to the performance of the GENITOR II algorithm [Whitley 90] SA/GA does somewhat better, since the GENITOR II algorithm does not always find the complete optimization.

Concluding the above observations, and interpretations we can safely say that SA/GA is not deceived by the deceptive problems. SA/GA can maintain the $0*..*0*..*0$ as well as the $1*..*1*..*1$ schemata (see section 3.1). Therefore the competition between 000 and 111 does not have to fall back on lower order schemata which favour 000.

47.5 Further research

A number of applications in machine learning lend themselves to a natural encoding in a GA representation. Obvious examples are the training of neural nets, and the construction of classifier systems [Goldberg 89a]. However, the performance of GA's on these type of problems has so far proven to be very poor [Whitley 90], [Bornholdt 92], [McCallum 90]. Because of its promising performance, we intend to use SA/GA for exactly these challenging problems. First we will test SA/GA on the "simpler" problem of the training of neural nets. After that we will construct a classifier system based on SA/GA.

47.6 REFERENCES

[Aarts 90] Aarts, E. H. L. and Korst, J. (1990) *Simulated Annealing and Boltzmann Machines. A Stochastic Approach to Combinatorial Optimization and Neural Computing*. John Wiley & Sons, Chichester.

[Metropolis 53] Metropolis, N., Rosenbluth, A., Rosenbluth, M., Teller, A., and Teller, E. (1953) "Equation of state calculations by fast computing machines", *J. Chem. Phys.*, **21**, 1087-1092.

[van Laarhoven 87] van Laarhoven, P. J. M. and Aarts, E. H. L. (1987) *Simulated Annealing: Theory and Applications*. D. Reidel Publishing Company, Dordrecht.

[Goldberg 89a] Goldberg, D. (1989) *Genetic Algorithms in Search, Optimization and Machine Learning*. Addison-Wesley, Reading, MA.

[Goldberg 89b] Goldberg, D., Korb, B. and Deb, K. (1989) "Messy genetic algorithms: motivation, analysis, and first results", *TCGA Report*, No. 89003.

[Lutsko 91] Lutsko, J. F. and Kuijpers, B. (1991) "On the induction of near-optimal decision trees", unpublished, and in this volume.

[Whitley 90] Whitley, D. and Starkweather, T. (1990) "Genitor II: a distributed genetic algorithm", *J. Expt. Artif. Intell.*, JETAI, **2**, 189-214.

[Schaffer 87] Schaffer, D. "Some effects of selection procedures on hyperplane sampling by genetic algorithms", *Genetic Algorithms and Simulated Annealing*, Davies, L., ed., Morgan Kaufmann, 89-103.

[Bornholdt 92] Bornholdt, S. and Graudenz D.. (1992) "General Asymmetric Neural Networks and Structure Design by Genetic Algorithms", *Neural Networks*, **5**, 327-334.

[McCallum 90] McCallum, R. and Spackman, K. A. (1990) "Using Genetic Algorithms to learn Disjunctive Rules from Examples", *Machine Learning, Proceedings of the seventh international conference on machine learning* , Porter, B. and Mooney, R., eds., 149-152.

48

A Tool for Model Generation and Knowledge Acquisition

Sally Jo Cunningham and Paul Denize

Department of Computer Science
University of Waikato
Private Bag 3105
Hamilton, New Zealand
sallyjo@waikato.ac.nz
pdenize@waikato.ac.nz

ABSTRACT Tools to automatically induce domain descriptions from examples are valuable aids for the knowledge acquisition stage of expert system construction. This paper presents a description of an algorithm that induces two domain descriptions: a conceptual model, which gives a broad understanding of variable interactions and their effect on the system, and a predictive model, which determines the system value associated with variable values input to the model. This induction algorithm is based on Entropy Data Analysis (EDA), which builds linear equations to approximate the system described by the training data.

Keywords: knowledge acquisition, model generation, machine learning.

48.1 Introduction

Systems that learn from sets of examples are valuable tools in the construction of expert systems. Knowledge acquisition has long been realized to be the bottleneck in expert system development. By automating the induction of general concepts or rules from available data, this bottleneck can be eased.

This paper describes the application of Entropy Data Analysis (EDA) to the problem of inducing rules from raw, empirical data. EDA, also known as k-systems analysis, is a relatively new technique for modeling multivariate numerical systems. Unlike more traditional statistical methods, which often must assume a model, EDA builds a model to fit the data. Moreover, nonlinearity does not adversely affect the construction of the model. EDA analysis of raw data produces a system model in the form of a set of equations that approximate the behaviour of the system function implicit in the data. The "true" system function may be arbitrarily complex, but can still be effectively approximated by EDA models.

In turn, this model may be analyzed to provide both a high-level conceptual description of the domain as well as a predictive set of production rules:

> The *conceptual domain description* reveals the combinations of factors (variable value combinations) in the data that influence the dependent variable, and the effect that these factors have on the dependent variable; i.e., "Low levels of rain and low temperatures adversely affect shrimp catches".

[1] *Selecting Models from Data: AI and Statistics IV.* Edited by P. Cheeseman and R.W. Oldford.

The *predictive production rules* are IF-THEN rules that predict the effect that specific independent variable values will have on the dependent variable; i.e., "IF the level of rain is between 0 and 0.7 inches and the temperature is less than 40 degrees Fahrenheit, then the shrimp catch is predicted to be about 1200 pounds".

48.1.1 Previous research

Entropy data analysis was developed as a modelling technique for general systems. It has been used as a technique for multivariate analysis (supplementing traditional statistical analysis) in several domains (see, for example, [Kn86], [Sh87]). The use of EDA as a method for obtaining both a semantic model and a "grammar" of rules describing a system was first suggested by Gary Shaffer in [Sh86]; this paper examines the applicability of EDA to machine learning, in particular to function induction.

Quinlan's ID3 algorithm is perhaps the best known ([Qu79], [Qu83]) induction algorithm. ID3 produces a decision tree, which is commonly transformed into a set of rules for incorporation into an expert system. Cendrowska [Ce87] noted that this transformation generally introduces redundancy into the rule set, and developed an algorithm that produces rules directly from data. Her PRISM algorithm is similar to EDA in that both use an entropic measure to determine the "important" attribute-value pairs in a system. Where PRISM uses this ranking of importance to create a minimal rule set, EDA instead calculates a set of linear equations that approximates the system behaviour. These equations may be used to create a rule set. Experimentation indicates that if the example set is complete (i.e., all possible combinations of all variable values are represented in the data instances), then PRISM and EDA will produce the same rule set. If the data is incomplete, then PRISM may produce a smaller rule set. PRISM is appropriate only for discrete classification tasks, which EDA-based induction is also appropriate for function induction. In addition, EDA also provides a conceptual model of the domain.

This paper is organized as follows: Section 2 describes the process of entropy data analysis; Section 3 presents a case study, and describes the production rules and conceptual description derived for the case; Section 4 discusses the effect of noise and sparse data on the quality of the model induced; and Section 5 presents conclusions about the use of EDA in induction. The "Anima Mundi" program (a trademark of the Seer Corporation, [Jo85] is used to perform the computations of EDA.

48.2 Entropy Data Analysis

Complete discussions of the mathematical foundations of EDA, the details of the EDA algorithm, and issues in efficiently implementing EDA are found in [Jo85b,Jo85c,Jo85d,Jo85e]. The discussion below summarizes the EDA algorithm. Consider a system with two or more independent variables and one system variable. Each combination of variable values is called a *factor*. EDA attacks the problem of discerning the information contributed by each variables or variable combination (factor) in determining the system variable value by the following algorithm:

1. Begin with a "flat" system–a system in complete entropy, in which all variable or variable combinations are considered equally likely. This system is represented by a set of linear equations based on the marginal probabilities for the system.

2. Measure the information content of the factors (by the classic information theoretic entropy measure), and choose the factor with the greatest information content.

3. Add the factor chosen by (2) to the system equations. Adding this factor will bring the flat system closer to a description of the "true system" (the system that produced the original data).

4. Test for convergence between the equations in (3) and the marginal probabilities found in the data. If convergence has occurred, then the equations in (3) approximate the "true system"; otherwise, go to step (2).

Usually only a small subset of the set of all possible factors are necessary to approximate the true system. Knowing these factors permits the storage of system information in a highly compressed form for reconstructing the original system to a desired degree of approximation.

48.3 A Sample Application – rye yield prediction

The following example, taken from the domain of agriculture, will illustrate the procedures involved in applying the general EDA algorithm described in Section 2 to the induction of a conceptual and a predictive model of a data set.

Rye crop yields, like many other crop yields, are known to be sensitive to rainfall and temperature. The following data was taken from [Du74]:

The independent variables rainfall and temperature are both continuous , with rainfall having the range [25,50] and temperature having the range [38.0,64.0]. The completeness of the data (the extent to which all variable value combinations are present) affects the effectiveness of the analysis. Since the two variables are continuous, the data therefore has a very low level of completeness. The solution to this problem is straightforward: to increase completeness, data values that are "close" to one another may be lumped together and represented by a single value. This "clustering" problem is well-studied in statistical analysis (see, for example, the discussion in [St80]).

We choose to group rainfall and temperature into clusterings representing low, medium, and high ranges for both rainfall and temperature:

Temperature:	[38.00,41.92]	[41.92,51.03]	[51.04,64.00]
Rainfall:	[25.00, 36.31]	[36.32, 45.43]	[45.44, 50.00]

We now wish to produce two types of information about rye crop yields:

1. a *conceptual description* of the weather conditions that tend to either depress or increase crop yield, and

2. a set of *predictive rules* that will let us determine, for a given temperature and rainfall, what our rye crop is likely to be.

Rainfall (inches)	*Temperature (Fahrenheit)*	*Rye crop yield (bushels/acre)*
45.0	54.1	21.0
47.0	61.6	20.0
33.0	50.8	21.0
39.0	52.1	24.0
30.0	50.2	20.0
28.0	57.1	12.5
41.0	55.7	19.0
44.0	57.6	23.0
31.0	50.1	23.0
29.0	38.0	19.0
34.0	56.2	21.0
27.0	51.5	12.0
42.0	54.1	21.0
35.0	46.7	27.0
43.0	60.8	17.5
39.0	56.9	26.0
31.0	60.3	11.0
42.0	54.6	24.0
43.0	53.5	26.0
47.0	64.0	18.5
25.0	45.7	15.5
50.0	61.5	16.5
45.0	59.7	18.0
34.0	53.2	20.5
29.0	45.1	22.0

TABLE 48.1: Rye crop yield data

48.3.1 Conceptual description

As described in Section 2 above, EDA proceeds by first determining the *factors* (variable and value combinations) with the greatest information content. These factors are added to the set of "flat" system equations (equations which assume that all factors are equally likely), and the effect of the factor on the system value may then be observed (whether the factor will increase or decrease the system value). When the above data was analyzed, the following factors were found to possess the greatest amount of system information:

Factor	*Effect on system value*
rainfall [25.00, 36.31] and temperature [51.04, 64.00]	negative, -19.9%
rainfall [36.32, 45.43] and temperature [51.04, 64.00]	positive, 14.2%
rainfall [25.00, 36.31] and temperature [41.93, 51.03]	positive, 11.4%

These factors capture a semantic model of the domain, and may be translated as:

If the rainfall is low and the temperature is high, then the crop yield will be lowered.
If the rainfall is moderate and the temperature is high, then the crop yield will be increased.

If the rainfall is low and the temperature is moderate, then the crop yield will be increased.

The degree to which the factor affects the system value can also be incorporated into the conceptual model by assigning a semantic interpretation to the percentage of increase/decrease of the system value. Based on personal experience, we assign the following semantic labels: a factor influencing the system value by more than 30% "strongly" increases or decreases the value, while an influence of less than 10% "slightly" affects the system value. The percentage levels and the labels are arbitrary, and may be modified to incorporate domain-specific information.

48.3.2 Predictive rules

As described in Section 2 above, factors are added to the "flat" system equations until an arbitrary degree of closeness to the original system is reached. We then possess a set of linear equations that describe the system to the desired accuracy, for those variable range values present in the original data. To predict the crop yield for a given pair of rainfall and temperature values, we determine the predicted crop yield for that pair (i.e., to determine the yield for low rainfall and low temperature, we solve the equations for the clusters "rainfall [25.00, 36.31]" and temperature [38.00, 41.92], and calculate the predicted crop yield of 19.00 bushels/acre).

By solving these equations for all possible combinations of the rainfall and temperature clustered values, the following table of predictive rules are derived:

Rainfall	*Temperature*	*Rye crop yield*
Low	Low	19.00
	Moderate	21.42
	High	15.20
Moderate	Low	19.22
	Moderate	19.22
	Low	21.95
High	Low	19.22
	Moderate	19.22
	High	18.33

But these predictive rules are of a coarse granularity. To obtain a finer grain for predictive power, we may re-cluster the variables rainfall and temperature into a larger number of clusters, each containing a smaller range of values.

We can then perform EDA analysis with this larger number of clusters, calculate new system equations, and use these new equations to calculate finer predictive rules. How fine a granularity is possible? A general rule is that the product of the number of clusters over all variables should not be more than 25% greater than the number of instances in the training set. If this threshold is exceeded, then the data will be too sparse to support meaningful induction.

48.4 Issues in Inducing Rules from Data

The accuracy and predictive power of the rules derived from data depends on the quality of the data. This data quality is determined by the completeness of the data and the amount of noise present in the data.

48.4.1 Completeness

A data set is *complete* if each possible combination of variable values is present in the training set. With 100% completeness (and an absence of noise), the system function can be perfectly reconstructed by EDA. In "real world" data, however, completeness is rarely possible. The effect of low levels of completeness on the accuracy of the conceptual model and predictive rules varies. If the "important" areas of the system function are present in the data (i.e., the local maxima, local minima, etc.), then meaningful analysis and rule induction is possible. Generally, we do not have this kind of knowledge about raw data. It is therefore ususally necessary to increase completeness by clustering the data. However, there is a trade-off between completeness and predictive power; the greater the completeness achieved by coarse clustering, the less predictive the rule set derived from analysis. In addition, clustering tends to add to the noise present in the data, since a cluster will often be associated with several system function values. As discussed below, however, this problem of function value contradiction is easily resolved.

Additionally, the EDA algorithm builds a system model that is predictive and descriptive only in the ranges of values actually found in the data (i.e., for the data in Table 1, we cannot construct predictive or descriptive models to determine the effect of a temperature of 75 degrees on rye crop yield). If the data is incomplete in the sense that extrapolation beyond the given data ranges is required, then EDA is not appropriate for system modelling. The system model constructed by EDA is based only on the information actually contained in the training data, and does not assume the additional information that would be necessary to permit extrapolation.

48.4.2 Noise

Noise in a data set arises from two sources: from mis-measurement, and from measuring the wrong variables. Noise is common in "real world" data, and some forms of noise are successfully managed by EDA induction.

Mis-measurement occurs when a value of a variable is incorrectly measured or missing. This problem may occur because the value is incorrectly perceived or transcribed, the measuring instrument is faulty, or for many other reasons. If the noise is randomly distributed throughout the data and the data is relatively complete, then the noise may be "smoothed out" by clustering. Noise in the variable values is reduced as the variable values are clustered. Noise in the system function value may lead to contradictory states in the data, where a single set of variable values is associated with multiple system function values. As noted above, the number of contradictory states tends to increase with clustering. Fortunately, a number of classic methods are available to resolve this contradiction: averaging the function values, choosing the mean value, choosing the mode value, etc. A more pernicious problem occurs, however, if the noise is not random; if, for example, a measuring instrument is biased to give consistently high readings. In this case, the system model derived can only be as good as the data. EDA, like other induction methods, is not capable of detecting a systematic bias, and will incorporate this bias into the derived system model.

Noise may also occur in the form of "residual variation", when factors additional to those recorded affect the values of the system function. In complex domains, this form of noise is highly likely; however, a simplified (and therefore less accurate) model of a domain may be necessary, as some factors may be unknown or not measurable.

A final form of noise is that of extraneous information–the presence of variables that do not affect the system function. EDA-based induction is particularly successful at dealing with this problem. Since the system equations derived by EDA are formed by choosing only those factors that contribute the greatest degree of information to the system (by the information theory entropy measure), any extraneous variables are not incorporated into the system model. These extraneous variables are effectively ignored.

48.5 Conclusions

EDA-based induction is appropriate for domains in which the values of the dependent variable are derived from an unknown function. This type of domain is not the norm for machine learning algorithms; most research has concentrated on discrete classification rather than function induction (see, for example, Quinlan's ID3 and its descendants discussed in [Qu83], [Qu79], [Qu93], [Ce87]). If the classification values for a domain can be linearly ordered, then EDA-based induction is possible for that domain. For example, if the classifications are "patient will die" and "patient will live", then we may represent the former possibility as a 0 and the latter as a 1. A cut-off in the range [0,1] determined which category a value generated by a predictive rule falls in-say, that any value greater than 0.5 indicates that the rule predicts that a patient will live.

Many classification values cannot be linearly ordered. Consider the classic Iris classification domain first referenced in [Fi36]. In this domain, sepal and petal lengths and widths are used to determine whether a given flower is an *Iris setosa, Iris versicolor*, or *Iris virginica*. There is no semantically meaningful ordering that can be applied so that, for example $setosa < versicolor < virginica$. If an arbitrary ordering is imposed on the discrete system values, EDA-based induction is appropriate for this type of classification problem only if the training data set is 100% complete, contains no noise (either from mis-measurement or residual variation), and contains mutually exclusive classifications. Since these requirements are rarely met by "real world" data, EDA-based induction is not considered appropriate for unorderable classification domains. If the domain does not fulfil these requirements, then the descriptive domain model tends to contain a very large number of factors that influence the system values only slightly (and hence the model is not very descriptive!), and the classifications produced by the predictive model tend to be artifacts of the arbitrary ordering (and hence there are a large number of mis-classifications).

Of course, the conceptual and predictive models induced can only be as good as the data provided. However, the EDA-based induction model is relatively robust in the presence of noise and incomplete data.

48.6 References

[Ce87] Cendrowska, Jadzia. PRISM: An Algorithm for Inducing Modular Rules. International Journal of Man-Machine Studies, vol. 27, 1987, pp. 349-370.

[Du74] Dunn, Olive Jean, and Virginia A. Clark. Applied Statistics: Analysis of Variance and Regression. Sydney: John Wiley and Sons, 1974.

[Fi36] Fisher,R.A. The use of multiple measurements in taxonomic problems. Annual Eugenics, 7, Part II, 179-188 (1936); also in Contributions to Mathematical Statistics, John Wiley 1950.

[Jo85a] Jones, Bush, and Jim Brannon. Anima Mundi User's Manual. Seer Corporation, 1985.

[Jo85b] Jones, Bush. Determination of Unbiased Reconstructions. International Journal of General Systems, vol. 10, 1985, pp. 169-176.

[Jo85c] Jones, Bush. A Greedy Algorithm for a Generalization of the Reconstruction Problem. International Journal of General Systems, vol. 11, 1985, pp. 63-68.

[Jo85d] Jones, Bush. Reconstructability Analysis for General Functions. International Journal of General Systems, vol. 11, 1985, pp. 133-142.

[Jo85e] Jones, Bush. Reconstructability Considerations with Arbitrary Data. International Journal of General Systems, vol. 11, 1985, pp. 143-151.

[Kn86] Knudsen, P.A. Emigration Patterns and Size Characteristics of Brown Shrimp Leaving a Louisiana Coastal Marsh. M.S. Thesis, Louisiana State University, Baton Rouge, 1986.

[Qu79] Quinlan, J. R. Discovering Rules from Large Collections of Examples: a Case Study. in Michie, D., Ed., Expert Systems in the Micro-Electronic Age. Edinburgh: Edinburgh University Press, pp. 168-201.

[Qu83] Quinlan, J.R. Learning Efficient Classification Procedures and Their Application to Chess Endgames. In Michalski, R.S., J.G. Carbonell, and T.M. Mitchell, Eds., Machine Learning: An Artificial Intelligence Approach. Palo Alto: Tioga, 1983, pp. 463-482.

[Qu93] Quinlam, J.R. C4,5 : Programs for machine learning. San Mateo CA.

[Sh86] Shaffer, Gary. Benthic Microfloral Production on the West and Gulf Coasts of the United States: Techniques for Analyzing Dynamic Data. Ph.D. thesis, Louisiana State University, USA, 1986.

[Sh87] Shaffer, Gary, and Peter Cahoon. Extracting Information from Ecological Data Containing High Spatial and Temporal Variability: Benthic Microfloral Production. International Journal of General Systems, vol. 13, 1987, pp. 107-123.

[St80] Steel, Robert G.D., and James H. Torrie. Principles and Procedures of Statistics: A Biometrical Approach, 2nd edition. McGraw-Hill Book Company, 1980.

49
Using knowledge-assisted discriminant analysis to generate new comparative terms

Bing Leng and Bruce G. Buchanan

Intelligent Systems Laboratory
Department of Computer Science
University of Pittsburgh
Pittsburgh, PA 15260

ABSTRACT In this paper we present a method - knowledge-assisted discriminant analysis - to generate new comparative terms on symbolic and numeric attributes. The method has been implemented and tested on three real-world databases: mushroom classification, letter recognition, and liver disorder diagnosis. The experiment results show that combining AI and statistical techniques is an effective and efficient way to enhance a machine learning system's concept description language in order to learn simple and comprehensible rules.

49.1 Introduction

It is known that vocabulary choice is a principal determinant of success in learning algorithms; and, when a concept description language is inadequate, learning can be hard or even impossible.[CHCB88] Constructive induction is a process by which a learning algorithm generates new terms through a set of operators, refines the language and thereby improves its performance.

Most previous work in constructive induction has focused on logical operators, *(and, or, not)*; for example, systems such as CITRE,[MR89] FRINGE,[Pag89] STAGGER,[Sch87b] DUCE,[Mug87] PLS0.[Ren85] However, little attention has been paid to relational operators $(>, =, <)$. Although BACON[LSBZ87] uses the equality operator to introduce new numerical terms in a noise-free context, - e.g., when a/b is a constant - no system can apply the relational operators to both symbolic and numeric features when noise is present. Relational operators are useful in cases where logical operators can cause unnecessary complexity in a concept description. More importantly, many real-world applications can meaningfully use comparisons among values of symbolic and numeric attributes. For example, a medical diagnosis system may use a condition that left-eye vision is more blurred than right-eye vision. For a learning system to find these kinds of relationships, when they are not specified in advance, it needs the ability to construct comparative terms.

Constructing new terms has been identified as a difficult problem because of its exponential nature.[DM83,BFOS84,LM88,Ren88] The real challenge is to develop an effective, efficient, and robust mechanism to generate human comprehensible comparative terms so that the concept description can be simplified and performance can be improved. One way to meet this challenge is to combine AI with statistical techniques, in particular, combining knowledge with discriminant analysis. Because of their complementary nature, combination is beneficial.

[1] *Selecting Models from Data: AI and Statistics IV.* Edited by P. Cheeseman and R.W. Oldford.

49.2 The problem

The problem we addressed here is to generate comparative terms from both symbolic and numeric attributes for a learning system, so that the concept description can be simplified and the performance can be improved. Two types of comparative terms are considered in this paper, and defined below:

Comparative Term: y_i **R** x_i: where y_i, x_i are the attributes, **R** is the relational operator, $(>,<,=)$.

Generalized Comparative Term: $\phi(a_1y_1,\ldots,a_my_m)$ **R** $\phi(b_1x_1,\ldots,b_nx_n)$ where $\phi(a_1y_1,\ldots,a_my_m)$, $\phi(b_1x_1,\ldots,b_nx_n)$ is a combination of the attributes.

Since understandability is an important concern, to make the comparative term meaningful, we have the following restrictions:

The symbolic attributes to be compared should have the same value range, otherwise, the comparison is metaphoric; the numeric attributes should have the same units; and the generalized comparative terms only compare combinations of numeric attributes and the combining function is sum.

To make a comparison, an ordering among the attributes' values is necessary. With numeric attributes, the comparison is relatively easy because values are ordered on the real number line; with symbolic attributes, there is no natural syntactic ordering. We have designed two methods for generating an ordering on symbolic values.[LB91,LB92a]

49.3 The approach

The approach is the combination of AI symbolic learning and statistical classification techniques, in particular, the use of knowledge and discriminant analysis.

Symbolic learning [BM78,Mit78,ML83,Qui86] is a heuristic search through the hypothesis space. Among many things, three characteristics make symbolic learning distinct from statistical classification technique: the use of prior knowledge, the comprehensibility, and a language of attribute-value pairs (orthogonal hyperplane) without assuming independence of attributes.

Discriminant analysis,[Fis36,Lac75,Han81,Kle80] on the other hand, is a mathematical method which derives a best possible linear combination of a set of numeric[3] attributes (arbitrary hyperplane) from a given data set in an efficient way, $O(n^2e)$, where n is the number of attributes, e is the number of examples. However, its applicability is limited on a non-homologous data set, i.e., when different relationships hold between attributes in different parts of an instance space.

Although neither technique is sufficient for all problems, the complementary nature of these two techniques makes it plausible to combine them together, namely, knowledge-assisted discriminant analysis (KADA). The essential idea is to use divide-and-conquer heuristics to guide a discriminant analysis algorithm to a proper subset of data, to apply it to derive a linear combination of a set of compatible attributes, and to use heuristics to convert the "best" attributes into comparative terms which are then passed to a symbolic learner, RL4.[CP90,Fu85,FB86,BM78]

[3] Although it can be applied to symbolic data by defining each attribute-value pair as an attribute with value 1 or 0, we lose the chance to find comparative terms.

49.3.1 KADA

KADA consists of two parts: prior knowledge and discriminant analysis algorithm itself. Prior knowledge includes information on plausible ranges of symbolic values; numeric value units; heuristics guiding discriminant analysis in a non-homologous data set; heuristics creating plausible new terms. Detailed descriptions of discriminant analysis can be found in the references.[Fis36,Lac75,Han81,Kle80] We now focus on two sets of heuristics.

Guiding heuristics

Two kinds of guiding heuristics have been used, both based on the idea of divide-and-conquer. However, one divides the data set based on clustering information, the other based on linear separability.

1. Clustering-based heuristics
   (a) Use a clustering algorithm to group the data together to form a clustering tree with each node containing the examples which are considered being close to each other;
   (b) Start from the root of the clustering tree, recursively apply the discriminant analysis algorithm on each node in the tree until one of the following is true: the line derived is "good enough" to separate the positive and negative examples; the node does not contain "enough examples". Where "good enough" and "enough examples" are two parameters defined by the prior knowledge of RL. When the algorithm stops with a good separation line, report that line.
2. Bi-splitting heuristics
   (a) Start from the whole data set, apply discriminant analysis algorithm on the set to get a linear line. Divide the set into two sub-sets based on that line. Then apply the discriminant analysis algorithm on these two sub-sets. This process is repeated until either a "good enough" line is found or there are not "enough examples" in the set. When the whole process stops, return all the lines derived.

Converting heuristics

The basic idea in converting a line to a comparative term can be stated as follows:

Sort the attributes based on their weights (coefficients) into a decreasing order. Select a sub-set of attributes from the first several ones such that they are compatible and the sum of their weights exceeds a pre-set threshold. By "compatible", we mean that they satisfy the restrictions defined above, that is, symbolic attributes have the same value range, while numeric attributes have the same units. When there is more than one attribute in the sub-set, we consider the following three cases:

1. if this sub-set contains symbolic attributes, we use the first two compatible ones, say, y_i and x_i, to make three comparative terms, namely, $y_i > x_i, y_i \geq x_i, y_i = x_i$.
2. if it contains only two numeric attributes, make the corresponding terms, y_i **R** x_i, where **R** is one of $>, \geq, =$.
3. if it contains more than two numeric attributes, make the corresponding generalized comparative terms, y_i **R** $b_0 + \sum b_i x_i$.

49.4 Experiments and results

The method has been implemented in LISP and tested using real data from three problem domains:[4] 1) predicting which mushrooms are edible, 2) classifying 26 letters, and 3) predicting categories of liver disorders. These three data sets are chosen to test symbolic comparative terms (mushroom domain), to test numeric comparative terms (letter recognition domain), and to test numeric generalized comparative terms (liver disorders domain).

In each data set, the examples were randomly divided into training and test sets. The rules were learned from the training set and tested on the later. This process was repeated ten times, the means and standard derivations were calculated, and a t-test was performed with 99% confidence limits.

Below, each data set is described briefly, followed by a presentation of the experimental results.

49.4.1 Mushroom data

The first experiment was conducted on the mushroom database. In this database there are 8124 examples with 4208 positive and 3916 negative with respect to edibility. Each example is described by 22 symbolic attribute. *The Audubon Society Field Guide to North American Mushrooms* states that there is no simple rule for determining the edibility of a mushroom.[Sch87a] The goal of this experiment was to test the utility of symbolic comparative terms - to see whether a rule with comparative terms can classify unseen mushrooms with reasonable predictive accuracy. Knowing that attribute *odor* is good discriminator, we have purposely deleted it, leaving more leverage for other attributes.

There are two sets of attributes having the same value ranges. One set has attributes *stalk-surface-above-ring* and *stalk-surface-below-ring*, with values from the range {*ibrous, scaly, silky*, and *smooth*}. The other has 5 attributes *cap-color, gill-color, stalk-color-above-ring, stalk-color-below-ring*, and *spore-print-color*, with values from a range of 13 color terms. We introduce these as prior domain knowledge.

The system started learning with the original attributes. When the performance was not satisfying, (thresholds: accuracy $\geq$ 90%, number of rules $\leq$ 10) KADA generated a set of plausible comparative terms which was then passed to the learning program RL. Using these additional comparative terms, RL was able to learn a better set of rules. One set contains three rules and gives a 96% predictive accuracy (true positive prediction = 92.8%, false positive rate = 0.3%).

r1: *(cap-color $\geq$ spore-print-color) & (gill-size = broad) $\rightarrow$ edible*

r2: *(cap-color < spore-print-color) &*
(spore-print-color < stalk-color-above-ring) $\rightarrow$ edible

r3: *(cap-color < spore-print-color) &*
(spore-print-color $\geq$ stalk-color-above-ring) &
(habitat = waste) $\rightarrow$ edible

The overall performance is given in table 49.1.

[4] All three data sets are from UCI machine learning database.

TABLE 49.1: Results on mushroom data set

System	Accuracy(%)	No.rules	No.conjuncts
RL with comp.term	96.0 ± 1.2	3 ± 1	7 ± 2
RL without comp.term	91.3 ± 0.5	9 ± 1	22 ± 2

TABLE 49.2: Results on letter recognition data set

System	Accuracy(%)	No.rules	No.conjuncts
RL with comp.term	83.7 ± 0.43	408 ± 4	3081 ± 33
RL without comp.term	84.0 ± 0.54	434 ± 5	3308 ± 40
GA	80.1	1040	na

In this table, the second column indicates the classification accuracy on the test data, which is stated as $(TP + TN)/(P + N)$, where TP is the number of positive examples covered, TN is the number of negative examples excluded, and $P + N$ is the total number of examples. The third column records the number of rules learned in the final theory, and the last the total number of conjuncts, where a conjunct is an attribute-value pair. The table shows the results of RL with and without the comparative terms, respectively. STAGGER's performance (with a full set of attributes) was 95% accuracy and formed 33 rules.[Sch87a] All the systems use 1,000 examples for training and the rest of examples for test. From the table, we can see that with the comparative terms, RL gets a simpler theory and better performance.

49.4.2 Letter recognition data

The second experiment was done on a letter recognition domain. The learning task in this domain is to identify 26 English letters from a set of black-and-white rectangular pixel displays. All together, 20,000 unique examples are available, each being described by 16 numerical attributes.

The objective of this experiment was to test the utility of numeric comparative terms and also to see how well this learning algorithm works compared with Holland-style adaptive classifier systems,[FS91] or genetic algorithm (GA). Before learning started, we put some prior knowledge of the compatibility among the attributes into the partial domain theory. Although the attributes are all numeric, some are more meaningful to be compared than others. For instance, attributes $height$ and $width$ were being defined as compatible rather than $height$ with $x - box$. Given the definitions of these 16 attributes, 7 compatible sets were defined, each with two attributes. A typical rule learned is shown below:

(x-ege > y-ege) & (x-ege ≤ 1) & (x2ybr > 7) & (xegvy > 6) → letter Y

The results are given in table 49.2, along with the GA's performance on this database.

Again, the results show that the comparative terms helped to simplify the final theory. Although the accuracies of both sets of RL's rules are almost the same, they are both better than that of GAs. RL used 10,400 examples for trainings and 9,600 examples for test in both cases; whereas GAs used 16,000 for training and 4,000 for test.

49.4.3 Liver disorders data

The last problem domain is predicting categories of liver disorders. The examples consisted of 345 male patient records. There are 6 attributes, 5 of them are blood tests which are thought to

TABLE 49.3: Results on liver disorders data set

System	Accuracy(%)
KADA	63.1 ± 5.7
C4.5	60.1 ± 5.8

be sensitive to liver disorders that might arise from excessive alcohol consumption.

The goal of this experiment is two-fold: 1) to test the generalized comparative terms; 2) to test KADA's ability as a stand alone learning system, when the predictive accuracy is the primary concern. The reason is that a recent comparison work in machine learning [SMT91] pointed out that symbolic learning systems are weak on numeric data. One explanation is that the orthogonal hyperplane - attribute-value pair representation often used in symbolic learning systems - may not be expressive enough to capture the functional relationships among a set of numeric attributes. This leads to an expectation that an arbitrary hyperplane might work better when this is necessary. Using KADA itself as a learning system is a way to verify this expectation. In addition, if KADA works well, this implies that the generalized comparative terms are useful.

Since there are no known learning systems results on this domain, we ran C4.5, a newer version of ID3,[Qui86] in parallel for the sake of reference. In the experiment, the parameters in C4.5 were set to their default values. Both systems used 75% of the data for training and the rest for testing. The results are shown in Table 49.3.

49.5 Discussion

The experimental results from three problem domains are encouraging: 1) comparative terms can make a simpler rule set or theory, 2) with comparative terms, RL's classification accuracy compares favorably to that of it without, 3) KADA is effective and robust in a diverse set of domains. In this section, we provide an analysis of KADA's performance.

More expressive language → simpler theory. The results from the first two experiments reflect the fact that RL with comparative terms can reduce the size of the final theory. A simple explanation is straightforward: a comparative term is made up by two primitive terms, when a single comparative term is used instead of two primitive ones, a conjunct is saved. Detailed explanation can be found in our technical report.[LB92b]

Knowledge (heuristics) → efficiency and comprehensibility. The effect of the prior domain knowledge is three-fold: 1) to tell KADA which two attributes can be compared so that the resulting comparative term is meaningful; 2) to constrain the search space so that generating new terms is feasible; 3) to direct discriminant analysis algorithm to a proper sub-set of data.

Different representations → diverse applications. We have shown that KADA can generate both symbolic and numeric comparative terms. However, a discriminant analysis algorithm can only deal with numeric data without artificial or arbitrary assignments of symbolic values to numbers. What brings the symbolic and numeric together is the idea of different representations. Researchers[FD89] in machine learning and AI have stressed the importance of maintaining different representations for different tasks. Our work provides further evidence of this.

49.6 Related work

Several systems are related to our work here in one way or another, we will have a closer look in this section, pointing out the similarities and the differences.

Considerable work has been done in AI on constructive induction. However, as pointed out in the introduction, most them use *and* and *or* as constructive operators. Our work in constructive induction is focussed on relational operators. While other work forms a compound term by disjunctions, we are trying to identify some underlying patterns which could be captured by the comparative terms.

Two other relevant systems are Utgoff's perceptron tree [Utg88] and CART's classification tree.[BFOS84] They also use statistical methods to find linear separators and use a strategy of divide-and-conquer. However, these two programs do not use prior knowledge nor do they do constructive induction.

49.7 Conclusion

In the paper we proposed a hybrid approach - knowledge-assisted discriminant analysis - which generates both symbolic and numeric comparative terms for a symbolic learner. The experiments presented above demonstrate that this method is effective and robust in a diverse set of real world domains. With the comparative terms, RL4's classification accuracy compares favorably to that of it without, and some other learning systems.

One strength of this approach is that we can take the complementary advantages from both sides - statistics offers an efficient discrimination technique while AI provides a way to incorporate knowledge into the system to make the former more useful.

Acknowledgments

Funding for this work has been received from the National Library of Medicine [grant No. R01-LM05104]. We are grateful to Prof. Ross Quinlan for a copy of the C4.5 program and to the people at UC Irvine who collect and make available the data.

49.8 References

[BFOS84] L. Breiman, J.H. Friedman, R.A. Olshen, and C.J. Stone. In *Classification and regression Trees*. Wadsworth, Belmont, CA, 1984.

[BM78] B.G. Buchanan and T.M. Mitchell. Model-directed learning of production rules. In D.A. Waterman and F. Hayes-Roth, editors, *Pattern-Directed Inference Systems*. Academic, New York, 1978.

[CHCB88] S. H. Clearwater, H. Hirsh, T. P. Cheng, and B. G. Buchanan. Overcoming inadequate concept description languages. Technical Report Tech. Rept. ISL-88-4, 1988.

[CP90] S. H. Clearwater and F. J. Provost. RL4: A tool for knowledge-based induction. Washington DC, 1990.

[DM83] T.G. Dietterich and R.S. Michalski. A comparative review of selected methods for learning from examples. In J.G. Carbonell R.S. Michalski and T.M. Mitchell, editors, *Machine Learning: An Artificial Intelligence Approach*, pages 41–82. Tioga, Palo Alto, CA, 1983.

[FB86] L.M. Fu and B.G. Buchanan. Learning intermediate concepts in constructing a hierarchical knowledge base. pages 659–666, Philadelphia, PA, 1986.

[FD89] Flann and T.G. Dietterich. Selecting appropriate representations for learning from examples. *Proceedings of the AAAI-89*, pages 460–466, 1989.

[Fis36] R.A. Fisher. The use of multiple measurement in taxonomic problems. *Ann. Eugenics.*, 7:179–188, 1936.

[FS91] P.W. Frey and D.J. Slate. Letter recognition using holland-style adaptive classifiers. *Machine Learning*, 6:161–182, 1991.

[Fu85] L.M. Fu. *Learning Object-Level Knowledge in Expert Systems*. Ph.D. Thesis, Stanford University, Stanford, CA, 1985.

[Han81] D.J. Hand. *Discrimination and Classification*. J. Weiley, New York, 1981.

[Kle80] W.R. Klecka. *Discriminant Analysis*. Sage Publications, Beverly Hill, CA, 1980.

[Lac75] P.A. Lachenbruch. *Discriminant Analysis*. Hafner Press, New York, 1975.

[LB91] B. Leng and B.G. Buchanan. Constructive induction on symbolic features: Introducing new comparative terms. pages 163–167. Morgan Kaufmann, June 1991.

[LB92a] B. Leng and B.G. Buchanan. Extending inductive methods to create practical tools for building expert systems. *4th International Conference on Tools for Artificial Intelligence*, November 1992.

[LB92b] Bing Leng and Bruch G. Buchanan. Using knowledge-assisted discriminant analysis to generate new comparative terms. Technical Report Tech. Rept. ISL-92-15, Univ. of Pittsburgh, Pittsburgh, PA., September 1992.

[LM88] K.F. Lee and S. Mahajan. A pattern classification approach to evaluation function learning. *Artificial Intelligence*, 36:1–25, 1988.

[LSBZ87] P. Langley, H.A. Simon, G.L. Bradshaw, and J.M. Zytkow. *Scientific Discovery: Computational Explorations of the Creative Process*. MIT Press, Cambridge, Massachusetts, 1987.

[Mit78] T. Mitchell. *Version Spaces: An Approach to Concept Learning*. Ph.D. Thesis, Stanford University, Stanford, CA, 1978.

[ML83] R.S. Michalski and J. Larson. Incremental generation of VL1 hypotheses: The underlying methodology and the description of the program AQ11. Technical Report Tech. Rept. ISG 83-5, Urbana, IL, 1983.

[MR89] C.J. Matheus and L.A. Rendell. Constructive induction on decision trees. pages 645–650, Detroit, MI, 1989.

[Mug87] S. Muggleton. DUCE, an oracle based approach to constructive induction. pages 287–292, Milan, Italy, 1987.

[Pag89] G. Pagallo. Learning DNF by decision trees. Detroit, MI, 1989.

[Qui86] J. R. Quinlan. Induction of decision trees. *Machine Learning*, 1:81–106, 1986.

[Ren85] L. Rendell. Substantial constructive induction using layered information compression: Tractable feature formation in search. pages 650–658, Los Angeles, CA, 1985.

[Ren88] L. Rendell. Learning hard concepts. *Proceedings of the Third European Working Session on Learning*, pages 177–200, 1988.

[Sch87a] J. Schlimmer. Concept acquisition through representational adjustment. Technical Report Tech. Rept. No. 87-19, 1987.

[Sch87b] J.C. Schlimmer. Incremental adjustment of representations in learning. pages 79–90, Irvine, CA, 1987.

[SMT91] J.W. Shavlik, R.J. Mooney, and G.G. Towell. Symbolic and neural learning algorithms: An experimental comparison. *Machine Learning*, 6:111–143, 1991.

[Utg88] P.E. Utgoff. Perceptron trees: A case study in hybrid concept representations. pages 601–606, St. Paul, MN, 1988.

Lecture Notes in Statistics

For information about Volumes 1 to 5
please contact Springer-Verlag

Vol. 6: S.S. Gupta and D.-Y. Huang, Multiple Statistical Decision Theory: Recent Developments. VIII, 104 pages, 1981.

Vol. 7: M. Akahira and K. Takeuchi, Asymptotic Efficiency of Statistical Estimators. VIII, 242 pages, 1981.

Vol. 8: The First Pannonian Symposium on Mathematical Statistics. Edited by P. Révész, L. Schmetterer, and V.M. Zolotarev. VI, 308 pages, 1981.

Vol. 9: B. Jørgensen, Statistical Properties of the Generalized Inverse Gaussian Distribution. VI, 188 pages, 1981.

Vol. 10: A.A. McIntosh, Fitting Linear Models: An Application of Conjugate Gradient Algorithms. VI, 200 pages, 1982.

Vol. 11: D.F. Nicholls and B.G. Quinn, Random Coefficient Autoregressive Models: An Introduction. V, 154 pages, 1982.

Vol. 12: M. Jacobsen, Statistical Analysis of Counting Processes. VII, 226 pages, 1982.

Vol. 13: J. Pfanzagl (with the assistance of W. Wefelmeyer), Contributions to a General Asymptotic Statistical Theory. VII, 315 pages, 1982.

Vol. 14: GLIM 82: Proceedings of the International Conference on Generalised Linear Models. Edited by R. Gilchrist. V, 188 pages, 1982.

Vol. 15: K.R.W. Brewer and M. Hanif, Sampling with Unequal Probabilities. IX, 164 pages, 1983.

Vol. 16: Specifying Statistical Models: From Parametric to Non-Parametric, Using Bayesian or Non-Bayesian Approaches. Edited by J.P. Florens, M. Mouchart, J.P. Raoult, L. Simar, and A.F.M. Smith, XI, 204 pages, 1983.

Vol. 17: I.V. Basawa and D.J. Scott, Asymptotic Optimal Inference for Non-Ergodic Models. IX, 170 pages, 1983.

Vol. 18: W. Britton, Conjugate Duality and the Exponential Fourier Spectrum. V, 226 pages, 1983.

Vol. 19: L. Fernholz, von Mises Calculus For Statistical Functionals. VIII, 124 pages, 1983.

Vol. 20: Mathematical Learning Models — Theory and Algorithms: Proceedings of a Conference. Edited by U. Herkenrath, D. Kalin, W. Vogel. XIV, 226 pages, 1983.

Vol. 21: H. Tong, Threshold Models in Non-linear Time Series Analysis. X, 323 pages, 1983.

Vol. 22: S. Johansen, Functional Relations, Random Coefficients and Nonlinear Regression with Application to Kinetic Data, VIII, 126 pages, 1984.

Vol. 23: D.G. Saphire, Estimation of Victimization Prevalence Using Data from the National Crime Survey. V, 165 pages, 1984.

Vol. 24: T.S. Rao, M.M. Gabr, An Introduction to Bispectral Analysis and Bilinear Time Series Models. VIII, 280 pages, 1984.

Vol. 25: Time Series Analysis of Irregularly Observed Data. Proceedings, 1983. Edited by E. Parzen. VII, 363 pages, 1984.

Vol. 26: Robust and Nonlinear Time Series Analysis. Proceedings, 1983. Edited by J. Franke, W. Härdle and D. Martin. IX, 286 pages, 1984.

Vol. 27: A. Janssen, H. Milbrodt, H. Strasser, Infinitely Divisible Statistical Experiments. VI, 163 pages, 1985.

Vol. 28: S. Amari, Differential-Geometrical Methods in Statistics. V, 290 pages, 1985.

Vol. 29: Statistics in Ornithology. Edited by B.J.T. Morgan and P.M. North. XXV, 418 pages, 1985.

Vol 30: J. Grandell, Stochastic Models of Air Pollutant Concentration. V, 110 pages, 1985.

Vol. 31: J. Pfanzagl, Asymptotic Expansions for General Statistical Models. VII, 505 pages, 1985.

Vol. 32: Generalized Linear Models. Proceedings, 1985. Edited by R. Gilchrist, B. Francis and J. Whittaker. VI, 178 pages, 1985.

Vol. 33: M. Csörgo, S. Csörgo, L. Horváth, An Asymptotic Theory for Empirical Reliability and Concentration Processes. V, 171 pages, 1986.

Vol. 34: D.E. Critchlow, Metric Methods for Analyzing Partially Ranked Data. X, 216 pages, 1985.

Vol. 35: Linear Statistical Inference. Proceedings, 1984. Edited by T. Calinski and W. Klonecki. VI, 318 pages, 1985.

Vol. 36: B. Matérn, Spatial Variation. Second Edition. 151 pages, 1986.

Vol. 37: Advances in Order Restricted Statistical Inference. Proceedings, 1985. Edited by R. Dykstra, T. Robertson and F.T. Wright. VIII, 295 pages, 1986.

Vol. 38: Survey Research Designs: Towards a Better Understanding of Their Costs and Benefits. Edited by R.W. Pearson and R.F. Boruch. V, 129 pages, 1986.

Vol. 39: J.D. Malley, Optimal Unbiased Estimation of Variance Components. IX, 146 pages, 1986.

Vol. 40: H.R. Lerche, Boundary Crossing of Brownian Motion. V, 142 pages, 1986.

Vol. 41: F. Baccelli, P. Brémaud, Palm Probabilities and Stationary Queues. VII, 106 pages, 1987.

Vol. 42: S. Kullback, J.C. Keegel, J.H. Kullback, Topics in Statistical Information Theory. IX, 158 pages, 1987.

Vol. 43: B.C. Arnold, Majorization and the Lorenz Order: A Brief Introduction. VI, 122 pages, 1987.

Vol. 44: D.L. McLeish, Christopher G. Small, The Theory and Applications of Statistical Inference Functions. VI, 124 pages, 1987.

Vol. 45: J.K. Ghosh (Ed.), Statistical Information and Likelihood. 384 pages, 1988.

Vol. 46: H.-G. Müller, Nonparametric Regression Analysis of Longitudinal Data. VI, 199 pages, 1988.

Vol. 47: A.J. Getson, F.C. Hsuan, {2}-Inverses and Their Statistical Application. VIII, 110 pages, 1988.

Vol. 48: G.L. Bretthorst, Bayesian Spectrum Analysis and Parameter Estimation. XII, 209 pages, 1988.

Vol. 49: S.L. Lauritzen, Extremal Families and Systems of Sufficient Statistics. XV, 268 pages, 1988.

Vol. 50: O.E. Barndorff-Nielsen, Parametric Statistical Models and Likelihood. VII, 276 pages, 1988.

Vol. 51: J. Hüsler, R.-D. Reiss (Eds.), Extreme Value Theory. Proceedings, 1987. X, 279 pages, 1989.

Vol. 52: P.K. Goel, T. Ramalingam, The Matching Methodology: Some Statistical Properties. VIII, 152 pages, 1989.

Vol. 53: B.C. Arnold, N. Balakrishnan, Relations, Bounds and Approximations for Order Statistics. IX, 173 pages, 1989.

Vol. 54: K.R. Shah, B.K. Sinha, Theory of Optimal Designs. VIII, 171 pages, 1989.

Vol. 55: L. McDonald, B. Manly, J. Lockwood, J. Logan (Eds.), Estimation and Analysis of Insect Populations. Proceedings, 1988. XIV, 492 pages, 1989.

Vol. 56: J.K. Lindsey, The Analysis of Categorical Data Using GLIM. V, 168 pages, 1989.

Vol. 57: A. Decarli, B.J. Francis, R. Gilchrist, G.U.H. Seeber (Eds.), Statistical Modelling. Proceedings, 1989. IX, 343 pages, 1989.

Vol. 58: O.E. Barndorff-Nielsen, P. Blæsild, P.S. Eriksen, Decomposition and Invariance of Measures, and Statistical Transformation Models. V, 147 pages, 1989.

Vol. 59: S. Gupta, R. Mukerjee, A Calculus for Factorial Arrangements. VI, 126 pages, 1989.

Vol. 60: L. Györfi, W. Härdle, P. Sarda, Ph. Vieu, Nonparametric Curve Estimation from Time Series. VIII, 153 pages, 1989.

Vol. 61: J. Breckling, The Analysis of Directional Time Series: Applications to Wind Speed and Direction. VIII, 238 pages, 1989.

Vol. 62: J.C. Akkerboom, Testing Problems with Linear or Angular Inequality Constraints. XII, 291 pages, 1990.

Vol. 63: J. Pfanzagl, Estimation in Semiparametric Models: Some Recent Developments. III, 112 pages, 1990.

Vol. 64: S. Gabler, Minimax Solutions in Sampling from Finite Populations. V, 132 pages, 1990.

Vol. 65: A. Janssen, D.M. Mason, Non-Standard Rank Tests. VI, 252 pages, 1990.

Vol. 66: T. Wright, Exact Confidence Bounds when Sampling from Small Finite Universes. XVI, 431 pages, 1991.

Vol. 67: M.A. Tanner, Tools for Statistical Inference: Observed Data and Data Augmentation Methods. VI, 110 pages, 1991.

Vol. 68: M. Taniguchi, Higher Order Asymptotic Theory for Time Series Analysis. VIII, 160 pages, 1991.

Vol. 69: N.J.D. Nagelkerke, Maximum Likelihood Estimation of Functional Relationships. V, 110 pages, 1992.

Vol. 70: K. Iida, Studies on the Optimal Search Plan. VIII, 130 pages, 1992.

Vol. 71: E.M.R.A. Engel, A Road to Randomness in Physical Systems. IX, 155 pages, 1992.

Vol. 72: J.K. Lindsey, The Analysis of Stochastic Processes using GLIM. VI, 294 pages, 1992.

Vol. 73: B.C. Arnold, E. Castillo, J.-M. Sarabia, Conditionally Specified Distributions. XIII, 151 pages, 1992.

Vol. 74: P. Barone, A. Frigessi, M. Piccioni, Stochastic Models, Statistical Methods, and Algorithms in Image Analysis. VI, 258 pages, 1992.

Vol. 75: P.K. Goel, N.S. Iyengar (Eds.), Bayesian Analysis in Statistics and Econometrics. XI, 410 pages, 1992.

Vol. 76: L. Bondesson, Generalized Gamma Convolutions and Related Classes of Distributions and Densities. VIII, 173 pages, 1992.

Vol. 77: E. Mammen, When Does Bootstrap Work? Asymptotic Results and Simulations. VI, 196 pages, 1992.

Vol. 78: L. Fahrmeir, B. Francis, R. Gilchrist, G. Tutz (Eds.), Advances in GLIM and Statistical Modelling: Proceedings of the GLIM92 Conference and the 7th International Workshop on Statistical Modelling, Munich, 13-17 July 1992. IX, 225 pages, 1992.

Vol. 79: N. Schmitz, Optimal Sequentially Planned Decision Procedures. XII, 209 pages, 1992.

Vol. 80: M. Fligner, J. Verducci (Eds.), Probability Models and Statistical Analyses for Ranking Data. XXII, 306 pages, 1992.

Vol. 81: P. Spirtes, C. Glymour, R. Scheines, Causation, Prediction, and Search. XXIII, 526 pages, 1993.

Vol. 82: A. Korostelev and A. Tsybakov, Minimax Theory of Image Reconstruction. XII, 268 pages, 1993.

Vol. 83: C. Gatsonis, J. Hodges, R. Kass, N. Singpurwalla (Eds.), Case Studies in Bayesian Statistics. XII, 437 pages, 1993.

Vol. 84: S. Yamada, Pivotal Measures in Statistical Experiments and Sufficiency. VII, 129 pages, 1994.

Vol. 85: P. Doukhan, Mixing: Properties and Examples. XI, 142 pages, 1994.

Vol. 86: W. Vach, Logistic Regression with Missing Values in the Covariates. XI, 139 pages, 1994.

Vol. 87: J. Møller, Lectures on Random Voronoi Tessellations. VII, 134 pages, 1994.

Vol. 88: J.E. Kolassa, Series Approximation Methods in Statistics. VIII, 150 pages, 1994.

Vol. 89: P. Cheeseman, R.W. Oldford (Eds.), Selecting Models From Data: Artificial Intelligence and Statistics IV. X, 487 pages, 1994.